Geoscience and Remote Sensing, New Achievements

Geoscience and Remote Sensing, New Achievements

Edited by

Pasquale Imperatore & Daniele Riccio

Geoscience and Remote Sensing, New Achievements,
Edited by Pasquale Imperatore & Daniele Riccio

Republished by InTech

Janeza Trdine 9, 51000 Rijeka, Croatia

Edition 2014

Additional hard copies can be obtained from orders@intechopen.com

Geoscience and Remote Sensing, New Achievements,
Edited by Pasquale Imperatore & Daniele Riccio
p. cm.
ISBN 978-953-7619-97-8

Preface

Our Planet is continuously monitored by powerful remote sensors nowadays operating in wide portions of the electromagnetic spectrum. Our capability of acquiring detailed information on the environment has been revolutionized by revealing its inner structure, morphology and dynamical changes. The way we now observe and study the evolution of the Earth's status has radically influenced even our perception and conception of the World we live in.

Within this framework, the past few years have been characterized by the rapid introduction of advanced remote sensors, the noticeable effort in modeling and simulation activities, the increasing development of data processing techniques, and the proliferation of applications in operative scenarios. These massive developments have opened new possibilities and proposed innovative products for applications relevant to atmosphere, ocean and land surfaces, also including pollution and disaster monitoring. Remote sensing applications now involve specialized and interdisciplinary approaches, thus requiring a remarkable background in geosciences, engineering and information technologies.

The aim of this book is to bring together contributions from experts in order to present new research results and prospects of the future developments in the area of geosciences and remote sensing; emerging research directions are discussed. The volume consists of twenty-six chapters, encompassing both theoretical aspects and application-oriented studies. An unfolding perspective on various current trends in this extremely rich area is offered.

The book chapters can be categorized along different perspectives: among others, use of active or passive sensors, employed technologies and configurations, considered scenario on the Earth, scientific research area involved in the studies.

About one-half of the contributions is specifically relevant to active sensors, whereas five chapters involve use of passive ones; the remaining contributions can be considered of general validity. In this book the most employed active sensor is the Synthetic Aperture Radar (SAR): indeed, the new generation of SAR sensors requires development of new techniques for reliable data processing and information extraction. In addition, SAR data are continuously available all over the Earth crust thus fully supporting applications to specific case studies. SAR techniques are presented with reference to both single- and multiple-pass configurations; three chapters are devoted to innovative interferometric SAR techniques. Use of scatterometers is taken into account in three chapters devoted to present recent developments and key applications for monitoring terrain and sea surfaces.

A remarkable degree of novelty is testified by the presence of ten chapters presenting theories or new approaches relevant to scattering and remote sensing. The remaining chapters discuss new applications making use of already existing theories.

Integration of data is an appealing issue whenever quite mature technologies are employed. Within this framework about one-half of the chapters combines data from more than one sensor for Earth observation. New ideas, applications or results for single sensors are presented in the remaining chapters.

Approximately one-third of the chapters proposes quite general approaches independent from the observed scenario; four contributions concentrate on observation of terrains, in some cases proposing application to flooding areas, two chapters deal with sea and sea winds, including pollution monitoring, two contributions take into account the sea ice scenario; finally, five chapters focus on sensing the Earth atmosphere: theoretical and applicative studies on modeling electromagnetic propagation through its constituents are presented.

With reference to the scientific area, one-half of the chapters deals with applicative issues testifying interdisciplinary approaches to remote sensing. The remaining chapters emphasize the role of popular specific disciplines usually applied to remote sensing: signal processing (seven chapters), informatics (five chapters), and electromagnetics (four chapters).

In conclusion, the volume collects contributions from researchers throughout the world, helping to broaden the views on modern application of remote sensing techniques. The collected material is up-to-dated and interests both engineers and academic researchers working in the field of geosciences and remote sensing.

The editors wish to express their thanks to the authors for their direct contribution, as well as the appreciation to Aleksandar Lazinica for his contribution to the project coordination.

Pasquale Imperatore
Daniele Riccio

Università di Napoli "Federico II"

Contents

1

Ground based SAR interferometry: a novel tool for Geoscience

Guido Luzi
University of Florence
Italy

1. Introduction

The word Radar is the acronym of Radio detection and ranging. Radar is an active instrument, which measures the echo of scattering objects, surfaces and volumes illuminated by an electromagnetic wave internally generated belonging to the microwave portion of the electromagnetic spectrum. It was born just before the second world war for detecting and ranging target for non-civilian scopes. In this case the requested spatial resolution was not so challenging for the technology available that time. The opening of new technological frontiers in the fifties, including the satellites and the space vehicles, demanded a better spatial resolution for application in geosciences remote sensing (RS). Synthetic aperture radar (SAR) technique was invented to overcome resolution restrictions encountered in radar observations from space and generally to improve the spatial resolution of radar images. Thanks to the development of this peculiar technique, the radar observations have been successfully refined, offering the opportunity of a microwave vision of several natural media. Nowadays SAR instruments can produce microwave images of the earth from space with resolution comparable to or better than optical systems and these images of natural media disclosed the potentials of microwave remote sensing in the study of the earth surfaces. The unique feature of this radar is that it uses the forward motion of the spacecraft to synthesize a much longer antenna, which in turn, provides a high ground resolution. The satellite SEASAT launched in 1978 was the first satellite with an imaging SAR system used as a scientific sensor and it opened the road to the following missions: ERS, Radarsat, ENVISAT, JERS till the recent TerraSARX and Cosmo-SkyMED. The measurement and interpretation of backscattered signal is used to extract physical information from its scattering properties. Since a SAR system is coherent, i.e. transmits and receive complex signals with high frequency and phase stability, it is possible to use SAR images in an interferometric mode. The top benefit from microwave observations is their independence from clouds and sunlight but this capability can weaken by using interferometric techniques. Among the several applications of SAR images aimed at the earth surface monitoring, in the last decades interferometry has been playing a main role. In particular, it allows the detection, with high precision, of the displacement component along the sensor–target line of sight.
The feasibility and the effectiveness of radar interferometry from satellite for monitoring ground displacements at a regional scale due to subsidence (Ferretti et al., 2001),

earthquakes and volcanoes (Zebker et al., 1994 , Sang-Ho, 2007 and Massonnet et al. 1993 (a)) and landslides (Lanari et al., 2004 ; Crosetto et al., 2005) or glacier motion (Goldenstein et al., 1993 ; Kenyi and Kaufmann, 2003) have been well demonstrated. The use of Differential Interferometry based on SAR images (DInSAR) was first developed for spaceborne application but the majority of the applications investigated from space can be extended to observations based on the use of a ground-based microwave interferometer to whom this chapter is dedicated. Despite Ground based differential interferometry (GBInSAR) was born later, in the last years it became more and more diffused, in particular for monitoring landslides and slopes.

After this introduction the first following sections of this chapter resume SAR and Interferometry techniques basics, taking largely profit from some educational sources from literature (Rosen 2000; Massonnet, 2003a; Askne, 2004, Ferretti, 2007). The following sections are devoted to the GBInSAR and to three case studies as examples of application of the technique.

2. General radar properties

2.1 The radar equation

Conventional radar is a device which transmits a pulsed radio wave and the measured time for the pulse to return from some scattering object, is used to determine the range. The fundamental relation between the characteristics of the radar, a target and the received signal, is called the radar equation, a relationship among radar parameters and target characteristics. Among the possible formulations we comment that indicated by the following expression:

$$P_R = P_T \frac{G_{tx} G_{rx} \lambda^2}{(4\pi)^3} \frac{\sigma}{R^4} \tag{1}$$

where P_t is the transmitted power, G_{tx} and G_{rx} are the transmitting and receiving gains of the two antennas, with respect to an isotropic radiator, σ is the radar cross section, R the distance from the target, λ is the pulse carrier wavelength. In (1) a factor which takes into account the reduction in power due to absorption of the signal during propagation to and from the target is neglected. This expression allows to estimate the power of the signal backscattered from a target at a known range, at a specific radar system configuration. The minimum detectable signal of a target, proportional to the received power P_R, can be estimated knowing the transmitted power, P_T, the antennas' characteristics and the system noise; of note that the range strongly influences the strength of the measuring signal. A radar image consists of the representation of the received signal in a two dimensional map, obtained through the combination of a spatial resolution along two directions, namely range and azimuth or cross-range, which correspond in a satellite geometry to cross-track and along the track directions. Normally the radar transmitting and receiving antennas are coincident or at the same location: in this case we speak about a monostatic radar and the measured signal is considered coming from the backward direction. In (1) we introduced the radar cross section, the parameter that describes the target behavior. The radar cross section of a point target is a hypothetical area intercepting that amount of power which, when scattered isotropically, produces an echo equal to P_R as received from the object. Consequently σ can be found by using the radar equation and measuring the ratio P_R/P_T and the distance R, supposing the system parameters λ , G_{tx}, G_{rx}, are known. In RS we are interested in the backscatter from extended targets then we normalize the radar cross section with respect to a horizontal unit area, and we define a backscattering coefficient, σ_0, usually expressed in dB. This fundamental information recorded by a radar is a complex number namely an amplitude and a phase value at a certain polarisation, electromagnetic frequency and incidence angle (Ulaby et al., 1984). The complex backscattering coefficient in SAR system is usually measured at four orthogonal polarisation states. Normally these polarization states are chosen to be HH (horizontal transmission and horizontal reception), HV (horizontal transmission and vertical reception) and analogously VH and VV. In this chapter we only consider the case of a single linear polarization, usually VV. Finally we remind that the Microwave portion of the electromagnetic spectrum is usually subdivided in bands, and Remote Sensing instrumentation mainly operates at L, S, C, X, Ku and Ka band, corresponding to the following intervals : L (1GHz-2GHz) S (2GHz-4GHz), C (4GHz-8GHz), X (8GHz-12 GHz), Ku (12-18 GHz) and Ka (26.5GHz-40 GHz) spanning in vacuum wavelengths from 30. cm to 8.mm. A radar signal is subject to a specific noise, due to the echoes coming from different parts of a reflecting body within a resolution cell which will have different phases and hence causing in the signal summation constructive or destructive interference between the different components. The resulting noise-like behaviour is called the *speckle noise*. To reduce the effect of speckle we may use filters. One way to reduce speckle is to use multi look processing which improves the S/N but worsening the spatial resolution (Curlander et McDonough, 1991). Temporal coherent averaging is possible in case of large number of images as in the Ground Bsed SAR Ground Based SAR - GBSAR case.

2.2 The range resolution

The range measurement is based on the fact that the signal echo is received after a delay of T=2R/c, where R is the distance to the scattering object and c is the speed of the electromagnetic pulse. In practice we use a pulse train where pulses are separated by a time T_{prf} ,corresponding to a pulse repetition frequency, PRF = $1/T_{prf}$. This means that we have an ambiguity problem: the measured radar echo can be caused by one pulse or the subsequent. This translates in the following expression: PRF < $c/2R_{max}$ which relates the maximum usable Range, R_{max}, to PRF. The range resolution is determined by the pulse width ΔT of the pulse where the factor 2 is caused by the radar pulse going back and forth. Figure 1 shows the working principle of range measurement through radar.

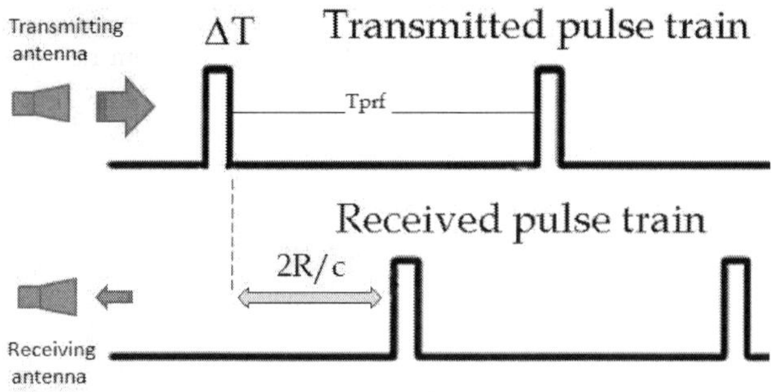

Fig. 1. The Radar functioning principle

The backscattered signal has an extension in time ΔT due to the pulse width and in order to obtain a good range resolution we need a short pulse. However, recalling Fourier transform properties, a short pulse width means a large frequency bandwidth. At the same time as dictated by the radar equation, at large distances, high amplitude is requested as the pulse energy determines the detection possibilities of the system i.e. its signal to noise ratio (S/N). This means that in designing a radar we are faced with the problem to want a long pulse with high energy and a wide bandwidth which implies a short pulse. To reduce these difficulties a signal processing technique, namely pulse compression, obtained by using a "chirp radar" (Ulaby et al., 1982) can be used. In this case the transmitted frequency is varying linearly with time and by correlating the return signal with a frequency modulated signal, a sharp peak is obtained for a distance related to the time offset. The resolution depends on the ability to sample sufficiently often the returned signals not to be aliased by the sampling rate.

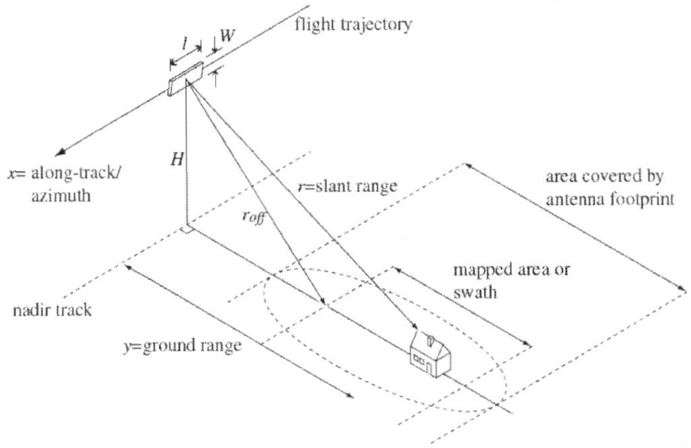

Fig. 2. SLAR geometry (after Mohr, 2005)

Active microwave RS observations usually employ a specific configuration: the side looking aperture radar (SLAR), whose line of sight (LOS) corresponds to a lateral view with respect to the track direction (see Figure 2). First it introduces a projection factor in the *range resolution* expression depending upon the incidence angle of the beam $\Delta r = \Delta T \cdot c/(2\sin\theta)$. Secondly a SLAR image suffers from some distortions due to *slant range* configuration resulting in errors related to the conversion of the measured *slant range* to the *ground range;* this contributes to make the radar image very different from the optical view (Rosen et al., 2000). When the surface is not flat, but we have topographic features, the terrain elevation distorts the distance to the radar sensor in such a way that slopes facing the radar appear shorter than they are when imaged in a normal map projection, while those that face away from the radar appear longer than in the map the latter are illuminated by the radar sensor very rarely: this is the *foreshortening effect.* Foreshortened areas appear brighter than their surroundings because the reflected radar energy from the slope is compressed to correspond to fewer pixels; when the slope of the terrain facing the radar is greater than the look-angle, the top of the slope is closer to the radar than the bottom we have a *layover*; finally *shadowing* can occur when terrain area cannot be illuminated and only system noise is imaged in the shadowed areas of radar images (Curlander and McDonough, 1991). These errors are of minor concern in observations where the slope area is imaged from below, that is to say in Ground Based cases.

2.3 The azimuth or cross-range resolution and SAR

The energy transmitted by a conventional radar is concentrated into a beam with an angular dimension, the field of view, θ_A, basically determined by the ratio between the operating wavelength and its mechanical size (Silver, 1986) and alike happens for the receiver which collects the energy coming from the antenna beam. In a radar image targets that differ from each other in their azimuth coordinates only, generate overlapping radar echoes and thus they cannot be distinguished. Conceptually azimuth location can be achieved by changing the viewing angle of a very directive antenna. In order to produce at a distance R a good *azimuth resolution*, R θ_A , in the along-track direction, we need short ranges and large antennas. At the same time to cover a wide swath, S, as requested e.g. in satellite geometry, we need a large θ_A meaning a small antenna. Viewing a target during the entire time it is within a beamwidth, determines a situation analogous to an artificially long antenna. If we acquire the amplitude and phase of the echoes an artificially narrow beamwidth in terms of resolution can be realized. The further a target is from the radar, the longer it is within the actual beamwidth, the longer the "antenna" and hence the narrower the resolution beamwidth. If the sensor is moving towards or away from the scattering object/surface, we can measure the velocity of the scattering object by measuring the Doppler effect which induces a frequency variation according to the apparent radial velocity of a certain scatterer on the ground. In order to make use of the forward motion, both the amplitude and phase of the return signal have to be recorded. The timing measurement is used to discriminate individual cells across the satellite track while the Doppler-induced variations in the frequency of the return signal are employed to provide the along track resolution. The SAR platform flies along a straight trajectory with a constant velocity illuminating a strip of terrain parallel to the flight track (see Figure 2). The data set can be stored in a two-dimensional array according to the SAR imaging geometry. The first step in SAR processing includes the pulse compression in range direction, usually denoted as *range compression*. The range compression is followed by the azimuth compression, which also yields the principle of the pulse compression technique. The azimuth chirp, which is approximately linear frequency modulated, is determined by the wavelength, the forward velocity and the slant range distance to the target. If all these parameters are known a priori, the reference function for a certain slant range distance is calculated to obtain a desired geometrical resolution after pulse compression in azimuth direction. A SAR image with a range independent azimuth resolution is obtained (Curlander and McDonough, 1991). Finally the azimuth compression is carried out. The final result of this acquisition and processing is a radar image with fine spatial resolution both in range and in azimuth directions: a few meter square cell from hundreds of kilometers.

3. SAR Interferometry from space

3.1 Introduction

Interferometry is a technique which use the phase information retrieved from the interaction of two different waves to retrieve temporal or spatial information on the waves propagation. First developed in optics, during the 20th century it has been later applied to radio waves and in the last decade to spaceborne SAR images. Since the SAR system is coherent, i.e. transmits and receive a complex signal with high stability, it is possible to use its interferometric signal, provided that propagation does not introduce decorrelation, namely a loss of information in irreversible way. This means that the scattered signal of the two images must be sufficiently correlated. We may combine images using different overpasses (multi-pass interferometry) where a baseline, a path difference due to satellite track separation, is present. In this case interferometric phase contains a contribution of topography which can be taken into account through the use of a digital elevation model (DEM). A simple scheme of how two images of the same area gathered from two slightly different across track positions, interfere and produce phase fringes that can be used to accurately determine the variation of the LOS distance is depicted in Figure 3. An interferogram is the map whose pixel values, s_i, are produced by conjugate multiplication of every pixel of two complex SAR images $I_{1,i}$, and $I_{2,i}$ in one image as shown in eq. 2a, where $I_{1,i}$ and $I_{2,i}$ are the complex pixel amplitudes, $R_{1,i}$ and $R_{2,i}$ are the two slant range coordinates, Bp,i is the baseline described by B_n and B_p, the baseline normal and parallel respectively to the line of sight, the last the only component affecting the phase, $\Phi noise,i$ is the phase noise that is due to speckle and thermal noise and usually including contribution from scattering too.

$$s_i = I_{1,i} I_{2,i}^{*} = \left| I_{1,i} I_{2,i}^{*} \right| e^{j\left[-\frac{4\pi}{\lambda}(R_{2i} - R_{1i}) - \Phi_{noise,i} \right]} = \left| s_i \right| e^{-j\varphi} \tag{2a}$$

$$s_i = \left| I_{1,i} I_{2,i}^{*} \right| e^{j\left[-\frac{4\pi}{\lambda}(B_{pi} - \Phi_{noise,i}) \right]} \tag{2b}$$

The amplitude of this product contains information on the noise of the phase observations and it is related to coherence, discussed in the next paragraph. Starting from the phase in equation (2b) and by assuming that the scene is stable, it is possible to derive a linear expression for the variations of the interferogram phase, between different pixels (Ferretti J., 2007; Askne J. et al., 2003):

$$\Delta\varphi = \frac{4\pi}{\lambda} B_n \Delta\vartheta = \frac{4\pi B_n}{\lambda R \tan\vartheta} \Delta R + \frac{4\pi B_n}{\lambda R \sin\vartheta} \Delta z + \Phi_{noise} + n \cdot 2\pi \tag{3}$$

Here Bn and R are defined above, $\Delta\theta$ is the difference in elevation angle, ΔR is the slant range difference and Δz is the altitude difference between pixels in the interferogram. The noise term is the phase noise, which determines how well the phase variations can be determined, also quantified by the coherence as described below.

The first term in (3) is purely a systematic effect that can easily be removed in the processing by applying "the flat earth compensation". In the second term there is a direct relation between the phase and the altitude in the image Δz. The last term represents the phase ambiguity induced by the modulo 2π phase registration. The ambiguity has to be removed in the processing by adding the correct integer number of 2π to each measured value. This is called phase unwrapping. If the 2π ambiguities are removed this phase difference can be used to calculate the off-nadir angle and the height variations i.e. a topographic map. As far as the problem of phase unwrapping is concerned, this topic is not tackled with in this chapter (see for instance Ghiglia & Romero, 1994). This factor can influence the choice of the operating frequency: long wavelengths can represent a good compromise between a moderate displacement sensitivity and a reduced occurrence of phase wrapping when the expected landslide velocity is high.

Baseline cannot increase over certain limit where the coherence is lost (baseline decorrelation effect). The use of the topographic effect which relates to the height of the portion of terrain corresponding to a pixel in the interferogram is one of the successful InSAR application, aiming at deriving a DEM of the imaged area (Zebker et al., 1986). It disappears for image pairs taken exactly from the same position (*zero baseline*). In this simpler case when further sources of phase variation are negligible the displacement of the ith point is recovered from the interferometric phase, φ_i by the following equation.

$$\Delta r_i = \frac{\varphi_i}{4\pi}\lambda \qquad (4)$$

In GBInSAR this is the ordinary configuration which provides "topography-free" interferogram and whose phase can be directly related to terrain movements.

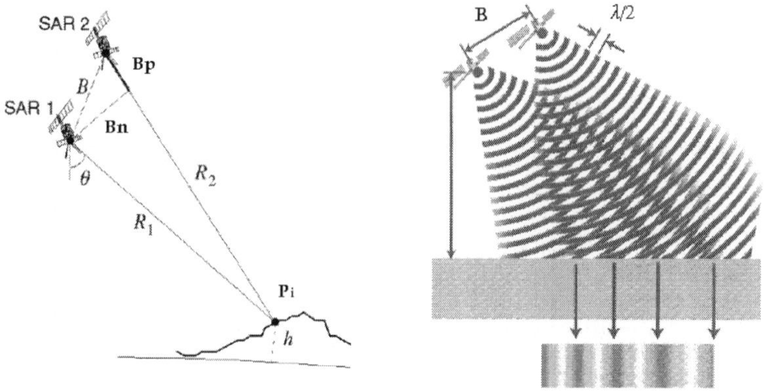

Fig. 3. (Left) InSAR geometry. The along the track direction is perpendicular to the graph plane. (Right) the rationale of the fringes formation due to baseline (Modified from Shang-Ho, 2008).

3.2 Coherence and phase

The statistical measurability of the interferometric phase from images collected at different times is related to its coherence (Bamler and Just, 1993). The spatial distribution of this parameter can be associated to the quality of the interferometric phase map. The interferometric coherence is the amplitude of the correlation coefficient between the two complex SAR images forming the interferogram. In a few words a common measure of the degree of statistical similarity of two images can be calculated through the following expression:

$$\gamma = \frac{\langle I_1 \cdot I_2^* \rangle}{\sqrt{\langle I_1 \cdot I_1 \rangle \langle I_2 \cdot I_2 \rangle}} = ce^{j\phi} \tag{5}$$

where c is coherence and the brackets < > mean the average value of the argument and ϕ is the corresponding interferometric phase, assuming the ensemble average can be determined by spatial averaging. The assumption that dielectric characteristics are similar for both acquisitions and have no impact on the interferometric phase cannot be assumed to have general validity and deserves a specific analysis taking into account the relevant conditions during each acquisition and in particular the time span between them (temporal baseline). E.g. vegetated area are usually rapidly decorrelating. On the other hand some features as buildings or artificial targets in coherence images may be stable over many years. Targets with such performances are called "permanent scatterers ©" (see Ferretti et al. 2001) and by using the phase of such reference points one may correct for the atmospheric screen effect with specific algorithm (Colesanti et al., 2003). In general the measured phase difference can be expressed as the summation of five different terms:

$$\Delta\varphi = \Delta\varphi_{base} + \Delta\varphi_{topo} + \Delta\varphi_{defor} + \Delta\varphi_{atm} + \Delta\varphi_{noise} \tag{6}$$

The first term $\Delta\phi_{base}$ is from baseline, $\Delta\phi_{topo}$ is due to topography, $\Delta\phi_{defor}$ is the ground deformation term, $\Delta\phi_{atm}$ is due to atmospheric propagation and $\Delta\phi_{noise}$ resumes random noise due different sources including the instrumental ones and variations occurring on the phase of the scattering surfaces. Limiting factors are due to delays in the ionosphere and atmosphere, satellite orbit stability variations occurred on the scattering surfaces during the time elapsed between the two acquisitions (Zebker et al., 1992). Although we normally say that microwaves are independent of clouds and atmospheric effects this is not entirely true and troposphere, and sometimes ionosphere, can affect the phase delay of waves and the accuracy of interferometric phase according to the water vapor and temperature fluctuations. Lastly it must be remembered that errors introduced by coregistration of the images can also affect coherence. The advantage of a ground based approach is mainly due to two factors: its zero baseline condition and its elevate temporal sampling both deeply reducing the decorrelation sources.

4. Ground Based SAR interferometry

4.1 The landing of a space technique

It is possible to acquire SAR images through a portable SAR to be installed in stable area. The motion for synthesizing the SAR image is obtained through a linear rail where a microwave transceiver moves regularly. Ground-based radar installations are usually at their best when monitoring small scale phenomena like buildings, small urban area or single hillsides, while imaging from satellite radar is able to monitor a very large area. As for satellite cases GBSAR radar images acquired at different dates can be fruitful for interferometry when the decorrelation among different images is maintained low. In ground based observations with respect to satellite sensors there is the necessity of finding a site with good visibility and from where the component of the displacement along the LOS is the major part. Recent papers have been issued about the feasibility of airborne (Reigber et al., 2003), or Ground Based radar interferometry based on portable instrumentation as a tool for monitoring buildings or structures (Tarchi et al. 1997), landslides (Tarchi et al., 2003b), (Leva et al. 2003), glaciers (Luzi et al. 2007). On the other hand satellite observations are sometimes not fully satisfactory because of a lengthy repeat pass time or of changes on observational geometry. Satellite, airborne and ground based radar interferometry are derived from the same physical principles but they are often characterized by specific problems mainly due to the difference of the geometry of the observation. A number of experimental results demonstrated the GBSAR effectiveness for remote monitoring of terrain slopes and as an early warning system to assess the risk of rapid landslides: here we briefly recall three examples taken from recent literature. The first is the monitoring of a slope where a large landslide is located. The second deals with an instable slope in a volcanic area where alerting procedures are a must. Finally an example of a research devoted to the interpretation of interferometric data collected through a GB SAR system to retrieve the characteristics of a snow cover is discussed.

4.2 The GB DInSAR instrumentation

Despite the use of the same physical principle, the satellite and ground based approaches differ in some aspects. In particular radar sensors of different kinds are usually employed mainly because of technical and operational reasons. While satellite SAR systems due to the need of a fast acquisition are based on standard pulse radar, continuous wave step frequency (CWSF) radar are usually preferred in ground based observations. The Joint Research Center (JRC) has been a pioneer of this technology and here the first prototype was born. The first paper about a GB SAR interferometry experiment dates back to 1999 (Tarchi et al., 1999), reporting a demonstration test on dam financed by the EC JRC in Ispra and the used equipment was composed of a radar sensor based on Vectorial Network Analyser (VNA), a coherent transmitting and receiving set-up, a mechanical guide, a PC based data acquisition and a control unit.

After some years a specific system, known as GBInSAR LiSA, reached an operative state and became available to the market by Ellegi-LiSALab company which on June 2003 obtained an exclusive licence to commercially exploit this technology from JRC. The use of VNA to realize a scatterometer, i.e. a coherent calibrated radar for RCS measurement, has been frequently used by researchers (e.g. Strozzi et al., 1998) as it easily makes a powerful tool for coherent radar measurements available. The basic and simplest schematic of the radiofrequency set-up used for radar measurements is shown in Figure 4 together with a simple scheme of the GBSAR acquisition. Advanced versions of this set-up have been realized in the next years to improve stability and frequency capabilities (Rudolf et al., 1999 and Noferini et al., 2005). This apparatus is able to generate microwave signals at definite increasing frequencies sweeping a radiofrequency band. This approach apparently different from that of the standard pulse radar owns the same physical meaning because a temporal pulse can be obtained after Fourier anti transforming the frequency data (the so called synthetic pulse approach).

The rapid grow of microwave technology occurred in the last years encouraged the development and realization of different instruments (Pipia et al., 2007 Bernardini et al., 2007); recently a ground based interferometer with a non-SAR approach has been designed with similar monitoring purposes (Werner et al., 2008). Data are processed in real time by means of a SAR processor. An algorithm combines the received amplitude and phase values stored for each position and frequency values, to return complex amplitudes (Fortuny J. and A.J. Sieber, 1994). The optimization of focusing algorithms has been recently updated by Reale et al, 2008; Fortuny, 2009. To reduce the effect of side lobes in range and azimuth synthesis (Mensa D.L. , 1991) , data are corrected by means of a window functions (Kaiser, Hamming etc), for range and azimuth synthesis. The attainable spatial resolutions and ambiguities are related to radar parameters through the relationships shown in Table 1. The accuracy of the measured phase is usually a fraction of the operated wavelength: by using centimetre wavelengths millimetre accuracy can be attained. As previously introduced, the phase from complex images can suffer from the ambiguity due to the impossibility of distinguishing between phases that differ by 2π. Single radar images are affected by noise and related interferometric maps must be obtained through an adequate phase stability between the pair of images: only pairs whose coherence loss can not affect the accuracy of the interferometric maps are usable. This task is of major difficulty when the considered time period is of the order of months.

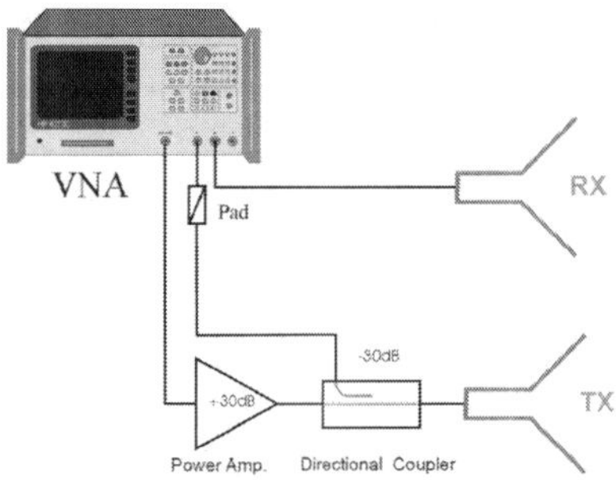

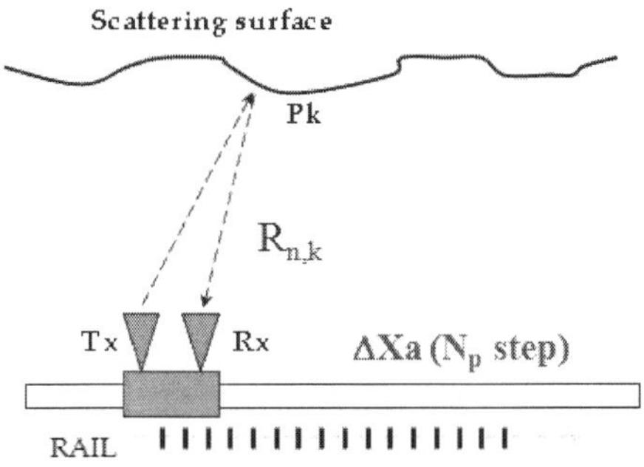

Fig. 4. A) Basic scheme of the RF section of the C band transceiver based on the Vectorial Network Analyser VNA. B) GB SAR acquisition through a linear motion.

A detailed analysis to the possible causes of decorrelation in the specific case of GBInSAR observations gathering many images per day for continuous measurements has been discussed by some researchers (Luzi et al., 2004 and Pipia et al. ,2007) while for campaigns carried out on landslides moving only few centimeters per year, when the sensor is periodically installed at repeated intervals several months apart over the observation period, a novel method has been proposed (Noferini et al. 2005).

Range resolution	$\Delta Rr = \frac{c}{2B}$
Azimuth resolution	$\Delta Raz = \frac{\lambda_c}{2L_x} \cdot R$
Non ambiguous range (m)	$R_{na} = \frac{c}{2\Delta f}$

Table 1. calculated resolution available from a CWSF radar observation; B radiofrequency bandwidth, λ_c in vacuum wavelength, Δf frequency step, Lx rail length, R range, c light velocity.

5. Examples of GB INSAR data collections

5.1. The monitoring of a landslide

This first example of how to benefit from the use of GBInSAR in Geoscience, is its employ as a monitoring tool for instable slopes, a well consolidated application largely reported in literature (Leva et al. 2003, Pieraccini et al., 2003, Tarchi et al., 2003a). The investigation and interpretation of the patterns of movement associated with landslides have been undertaken by using a wide range of techniques, including the use of survey markers: extensometers, inclinometers, analogue and digital photogrammetry, both terrestrial and aerial. In general, they suffer from serious shortcomings in terms of spatial resolution. GB SAR, thanks to its spatial and temporal sampling can overcome the restrictions of the conventional point-wise measurement. Here some results of an experimental campaign carried out through a portable GB radar to survey a large active landslide, the "Tessina landslide", near Belluno in north-eastern Italy are shown. In this site a exhaustive conventional networks of sensors fundamental to validate the proposed technique were at our disposal. For the same reason this site has been used by different research teams to test their instrumentation, starting since the first campaign carried out by JRC in 2000 (Tarchi et al., 2003a), following with University of Florence in Luzi et al. 2006 and later with Bernardini et al., 2007 and Werner et al., 2008. The GBInSAR monitoring executes analyzing maps of phase differences or equivalently displacements' map of the observed scenario, obtained from time sequences of SAR images.

5.2 The test site

The area affected by the landslide extends from an elevation of 1200 m a.s.l at the crown down to 610 m a.s.l. at the toe of the mudflow . Its total track length is approximately 3 Km, and its maximum width is about 500 m, in the rear scar area, with a maximum depth of about 50 m. Range measurements in different points were carried out through conventional instrumentation with benchmarks positioned in different locations as depicted in Figure 5, where a sight from the measurements facility is shown. Two of the optical control points correspond to high reflecting radar targets. In particular, point 1 refers to a passive corner reflector (PCR), an artificial target usually used as calibrator, which consists of a metal trihedral with a size of 50. cm. Point 2 is an active radar calibrator (ARC), specifically designed and built for this experimentation: an amplifier of the radar signal which allows acquisition of high reflection pixels on the radar image at far distances that are useful for amplitude calibration (radiometric calibration) and map geo-referencing. The GB radar instrumentation available for the experiments here reported consists of a microwave (C band) transceiver unit based on the HP8753D VNA, a linear horizontal rail where the antennas move while scanning the synthetic aperture, and a PC controlling the VNA, the antenna motion, the data recording, and all the other operations needed to carry out the measurement. Collected radar images are used for the calculation of the interferogram and converted into multi-temporal maps of the displacement component along the radar line of sight in geo-referenced raster format for GIS applications.

The measurement campaign on the Tessina landslide was continuously carried out between the 4th of June and the 9th of June 2004. The instrumentation was installed at an elevation of 997.3 m a.s.l., in a stable area on the opposite slope in front of the landslide, mainly visible at a minimum and maximum distance of 100. m and 500. m, respectively. The mechanical frame was fixed on a concrete wall. The radar image exhibits a fixed spatial resolution of 2 m along the range direction and a variable cross-range spatial resolution better than 6 m. The area selected for SAR imaging is a rectangle with size 400m per 1000m. The images obtained with the ground-based SAR system are usually projected as a two dimensional image of the scenario along two directions, range and azimuth, with a plane representation.

Fig. 5. View from the radar installation of the monitored area. Red figures indicate benchmarks for optical measuring (After Luzi et al., 2006).

The interpretation of bi-dimensional SAR images of a complex scenario, where terrain slope changes abruptly, is often unsatisfactory for comparison to an optical view. The availability of a DEM of the observed scene allows us to obtain SAR images on a three-dimensional space where radar and optical features are better detectable. Figure 6 shows an example of an intensity SAR image projected on the DEM: all three coordinates of the pixel are reconstructed. In this image the position of the radar is marked by a red dot; the signatures of the two high reflectivity targets, consisting of the passive corner reflector (PCR) and the active radar calibrator (ACR), used for referencing the map, are neat.

5.3 Data analysis

As previously discussed in GB SAR observations the main source of decorrelation is that one due to atmospheric propagation. At the C band radar frequencies the attenuation due to atmospheric path is low but the signal propagating through atmosphere suffers anyhow a time delay, mainly changing with air humidity and temperature fluctuations which ask for correction procedures of the acquired data. Briefly, the applied method consists of subtracting the phase value measured on a stable, highly reflecting reference point artificial or natural, from the measured phase of the selected pixel. In our case the characteristics of the observed scenario, mainly composed of sliding bare soil or by sparse vegetation, made it difficult to find stable natural scatterers. The passive corner reflector and the active radar calibrator were installed in two different positions along the upper contour of the landslide, and their positions were continuously checked by means of a theodolite to verify their effective stability. The PCR position, measured by theodolite, resulted stable along the entire duration of the campaign within +-1mm. The scarce vegetation on the main area under investigation allowed to get high coherence values.

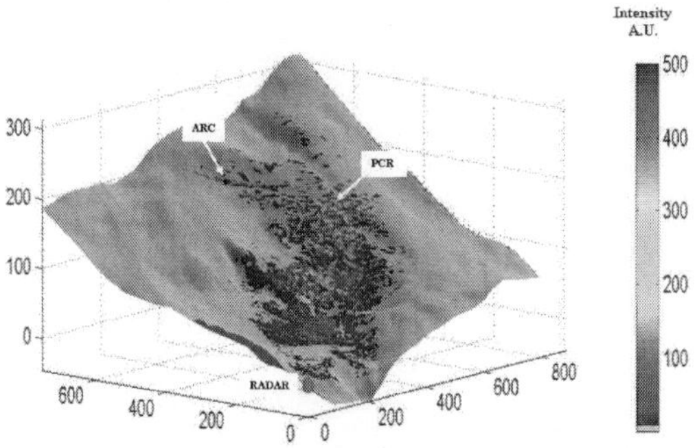

Fig. 6. Radar intensity image (arbitrary units) of the monitored slope obtained with data collected on 6 June 2004 and rendered on Digital Elevation Model of the slope. Two high reflectivity targets, the passive corn reflector (PCR) and the active calibrator (ARC) are indicated (After Luzi et al., 2006).

Displacements measured by the theodolite and corresponding values retrieved from radar data are plotted as a function of time in Figure 7. Some data gaps are due to interruptions during heavy rain events or small adjustments on the installation of radar targets. The measured phase of point 1 (PCR), whose position was confirmed to be stable within the millimetric accuracy of the optical instrumentation, is subtracted from the measured phases of the other points to take into account atmospheric induced error. Observing Figure 7, agreement appears viable and the displacements measured respectively through optical benchmarks and radar show similar trends. A noticeable discrepancy appears for the faster points (P10 and P17), whose corresponding pixels include inhomogeneous areas in terms of slope and surface characteristics. The uncertainty can be ascribed to the fact that the theodolite measures a single point, while radar data are obtained through a spatial averaging on an area of some meters. From these data a maximum 2.5mm/30′ displacement rate results. Regarding phase wrapping, this rate value ensures that the phase variation occurred between two subsequent measurements (< 30′) is small compared to the centimetre half-wavelength.
Moving from a point-wise analysis to the entire observed surface, the displacement of each pixel can be depicted in colour scale corresponding to different values in millimetres, making it possible to compare the radar data with an overlapped map of the scenario. In Figure 8 is shown the interferometric map obtained through a masking procedure which excludes areas with coherence lower than the 0.7 threshold. The geometry of observation

was never changed during the overall campaign, and approximately 300 images were collected, one every 16-18 minutes. The map in Figure 8 is obtained considering the data collected from 17h.48m GMT+1 to 22h.53m GMT+1 of the 6 June. As mentioned above, these data are very interesting because they refer to areas that are inaccessible for the placement of benchmarks. For example, we can monitor a minor central area where the movement rate is so high as to cause displacement of up to ten centimetres in 5 hours, while the rest of the landslide area shows a slower motion, about 1mm/hour. This map making available an estimate of the displacement along the LoS over the entire slope, can be the starting point to understand and analyze the behaviour of the landslide. Relationships between slope movement and other factors as rain rate, can be studied (Luzi et al., 2006) to understand landslide dynamic.

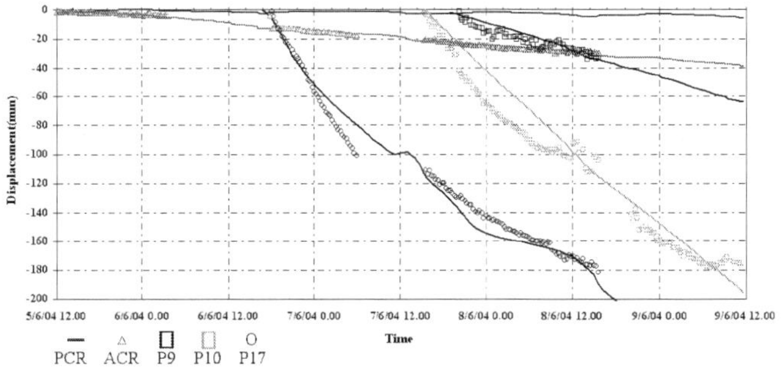

Fig. 7. Displacements measured by the theodolite (solid line) and corresponding values retrieved from radar data (symbols) for some reference points supplied with optical benchmarks, as a function of time. Figure points refers to Figure 5. (After Luzi et al., 2006)

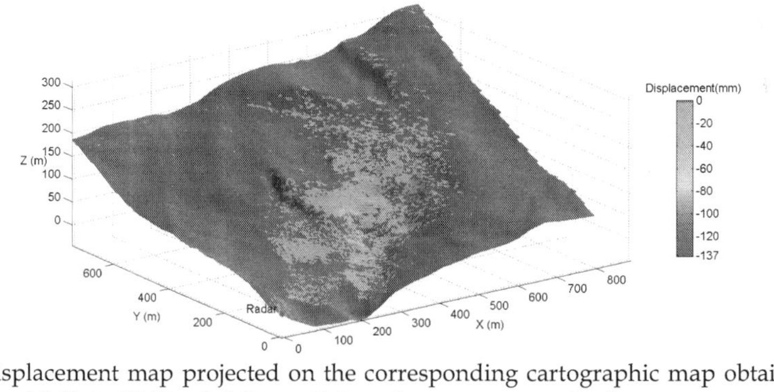

Fig. 8. Displacement map projected on the corresponding cartographic map obtained with data from 17:48 to 22:53 of 6 June. Colour bar represents displacement towards radar location (approaching), in mm (After Luzi et al., 2006).

5.4 Volcano deformations monitoring through GBInSAR monitoring

5.4.1 Introduction

Deformations monitoring through GB SAR has been applied in several different circumstances of slope instability. One of the most interesting case is the monitoring of a Volcanic area, presently in progress, and herein briefly described. When non-remote conventional approach can be inapplicable GB SAR can offer a good opportunity. To continuously monitor the behaviour of the morphological depression, known as Sciara del Fuoco, SdF, with alerting purposes, a GB-In SAR system, working at Ku band, was set up on the stable right flank of the Stromboli volcano in Italy. The monitoring started in March 2003 (Antonello et al., 2003) and ever since it is continuously acquiring. This lateral location was chosen due to the logistic impossibility to place the system in front of the unstable slope and permitted to follow the temporal and spatial evolution of the mass movement in the SdF and to obtain information about the crater area through interferometric maps acquired with ten minutes cadence. This monitoring was arranged as a consequence of the collapse of a large landslide which caused a tsunami on December 2002. More generally the presence of deformations in a volcanic area can be often related to volcanic activities. Stromboli volcano is characterized by a typical "Strombolian activity" which consists of very low energy explosions, every 10-15 minutes. The investigation and interpretation of the movement associated with deformations have been undertaken by using a wide range of techniques, including the use of survey markers, extensometers, inclinometers. However, they often incur serious shortcomings in terms of spatial or temporal resolutions. Although these techniques provide abundant datasets on movement styles, they are difficult to interpret in terms of the overall evolution of movement and cannot be installed in a risky area as the slope of an active volcano. GBDInSAR, can provide excellent spatial coverage and temporal resolution, and large movement events can be easily captured from remote.

5.4.2 The test site and the experimental data

The GB SAR installed in Stromboli Island, was designed by the Joint Research Centre of the European Commission (Rudolf & Tarchi, *1999*) and it is built and supplied by Ellegi/Lisalab company. Data are acquired from an elevation of 400 m a.s.l. and at an average distance from the target area of about 600 m. The instrument points up toward the NE Crater, with a 25° inclination angle of the radar antennas. It is continuously active since 20 February 2003 (Antonello et al., 2003; Antonello et al., 2007) and produces, on average, 120 images per day of the area under investigation (NE flank of crater and the upper part of the SdF). With an accuracy of the measurement of less than 1 mm it produces a synthesized radar image of the observed area every 12 minutes, with a pixel resolution of about 2 m in range, and 2 m on average in cross range. The interferometric analysis of sequences of consecutive images allows us to derive the entire displacement field of the observed portion of the SdF and of the crater along the LoS in the time interval. A negative displacement means a shortening of the LoS length. On the crater area this direction of movement corresponds to the inflation of the volcanic cone while, on the SdF, this is usually related to a local bulging or to the downslope sliding of the volcanoclastic material accumulated on the SdF slope.

Conversely, a positive value of displacement identifies a movement backward with respect to the sensor that on the crater area could be related to the deflation of the volcanic cone. As usual the radar image must be interpreted after a carefully understanding of the monitored area. In this case, as shown in Figure 9 different areas can be identified from the power image. In particular the SdF slope and the crater areas are well separated.

An example of an interesting and useful achievement from GB SAR data acquisition is here briefly recalled. Since 8 March 2007 the velocity recorded on the SdF increased again with movements toward the sensor. The interferogram highlighted a very high deformation rate (more than 300 mm/h), which exceeds the capability of the correct phase unwrapping. The arrangement of the interferometric fringes, clearly detectable in Figure 10, can be related to the bulging due to the opening of a new vent, actually occurred at 14.30 UT of 9 March. Following the method proposed by Voight (1988), Casagli et al. 2009 discuss how to predict in advance the opening of the vent.

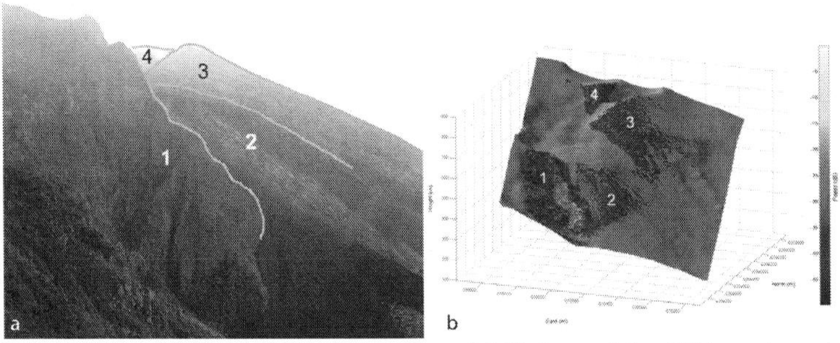

Fig. 9. Observed scenario from the radar system. (a) Picture of the SdF as viewed from the radar installation; (b) radar image projected on a DEM. Four main areas, as indicated by the numbering, can be identified: 1) the "Bastimento", the stable right flank of the SdF; 2) the upper part of the SdF; 3) the flank of the NE crater; 4) the outer part of the NE crater. The colour scale expresses the power of the backscattered signal.

Fig. 10. 3D model of the Stromboli Island superimposed a displacement map obtained from the GB-InSAR. Time interval: 11 minutes (from 11.17 UT and 11.28 UT 03.09.2007) showing a velocity greater than 300 mm/h enhanced through the fringes density (After Casagli et al., 2009).

5. 5 Interferometric phase and snow water equivalent

As a last example we report on a not yet consolidated but promising application: the use of GB SAR interferometry to retrieve of snow depth (SD) and snow water equivalent (SWE) of slopes. Information on the mass of snow through the knowledge of related parameters such as, SWE or SD, are important issues for climate studies, hydrology, and water resources managing. The spatial and temporal distribution of snow depth is one of the key parameters in the assessment of avalanche hazards, snow drift and avalanche modelling, and model verification. Most of the conventional methods including snow pits, probing or profiling, deliver point information and direct on site measurements are often risky in high mountains areas which are exposed to avalanche risk. Nevertheless the several RS available techniques for the measurement of SWE of dry snow is yet an open matter. The use of optical data is limited by adverse meteorological conditions and they are not well correlated to snow depth Microwave radiometry is very sensitive to the presence of snow on soil and is used for estimating SWE and melting/refreezing cycles at both basin scale (Macelloni et al., 2005). It does however have difficulty in distinguishing wet snow from wet soil and at lower frequencies usually suffers from a limited spatial resolution. As far as microwave active techniques are concerned, different algorithms have been developed and refined for use in multipolarization/multifrequency data sets (Shi et al., 2000 ; Nagler et al., 2000). The use of SAR images aimed at snow monitoring from satellite started since the 1990s (Bernier et al., 1998) but the use of differential SAR Interferometry, DInSAR, to monitor dry snow is a relatively recent application (Gunierussen et al., 2001; Oveishgram et al. 2007) and the use of ground based SAR sensors is also a novelty (Martinez et al., 2005). As far as the strongly related avalanche risk reduction and innovative study has been carried out by JRC summarized in the paper from Martinez et al., 2006.

5.5.1 The functioning principle

The use of SAR interferometry to evaluate snow mass characteristics, is based on relating the interferometric phase shift obtained from two or more SAR images to a change in the snow mass. Snow is a mixture of air, ice crystals and if melting, liquid water. In wet snow, a microwave signal suffers from attenuation due to the presence of liquid water and the interaction is complicated owing to the fact that even a very small amount of liquid water drastically influences the phase and amplitude of the backscattered field. When snow is dry, liquid water is absent and at longer wavelengths (L to C band) it can be considered almost transparent with a moderate volume scattering depending on observed frequency and the incidence angle. Higher frequencies showed a good sensitivity to dry snow properties but they have a limited penetration into snow cover. In the case of dry snow at low frequencies (lower than X band) sensitivity of the amplitude of backscattering to variations of the depth of a dry snow pack is weak (Strozzi et al., 1998). These considerations invited the start of some investigations about the retrieving of dry snow characteristics fom microwave interferometric data.

Dry snow is a mixture of air and ice crystals. The main processes of backscattering from a snow pack depicted in Figure 11 are: surface scattering at air-snow interface (1 in Figure 11), at the ground-snow interface (2 in Figure 11), and volume scattering at snow grains within the snow-pack (3 in Figure 11). Numerical backscatter simulations (Nagler et al., 2004) show that in the frequency range from L- to C-Band, surface scattering at the snow - ground interface is the dominating process. In this case the modifications of this signal due to scattering at the air-snow interface and within the snow volume are small compared to the phase shift resulting from the changes of the propagation path length through the snow pack due to refraction. The variation of the path length due to thickening of the snow pack can be measured in terms of the interferometric phase shift according to (Gunierussen et al., 2001):

$$\Delta\phi_{snow} = -\frac{4\pi}{\lambda}\cdot\delta z\cdot\left(\cos\theta - \sqrt{\varepsilon' - \sin^2\theta}\right) \quad (7)$$

where Δϕsnow is the interferometric phase (rad), λ the in vacuum wavelength(m), δz = z2 - z1 (m) corresponds to the change in the snow depth, z, between SAR data acquisition 1 and 2, θ is the local incidence angle, and ε is the snow permittivity, the physical parameter responsible for refraction and corresponding to the square of the refractive index used in optics. This parameter in the case of dry snow can be estimated through a third order polynomial function for ρs < 450 kg/m^3 (Matzler, 1996). For low local incidence angles (up to approximately 50°), the relationship between Δϕsnow and SWE, estimated as SWE= δz $<\rho_s>$, can be approximated to a linear relationship but due to the imaging geometry of GB SAR systems, slopes are often imaged at incidence angles above 50° and this linearization is not applicable. In this case (7) must be used in the not approximated form. Using in-situ point measurements or optionally an assumption on the snow density, δz can be derived but local variations of the snow density value will be reflected in the estimation of the snow height.

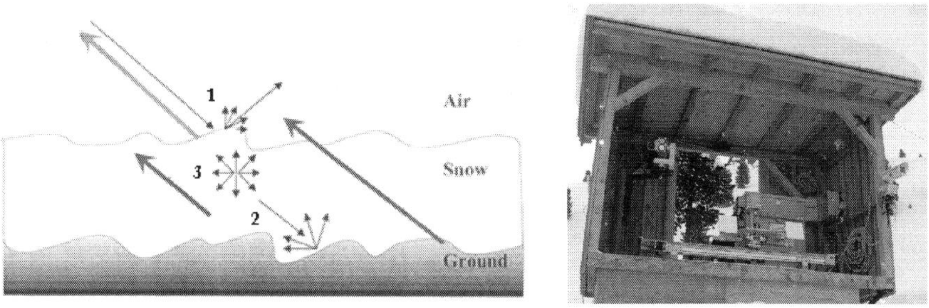

Fig. 11. To the left: simple scheme of the backscattering processes of a snow pack: surface scattering at air-snow interface 1); at the ground-snow interface 2); volume scattering at snow grains within the snow-pack 3). To the right a picture of the GB SAR apparatus and the TLS.

Also in this application we must care of decorrelation and as far as the coherence problem is concerned a snow pack can suffer from some decorrelation sources such as: melting, snow drift (wind erosion and deposition), snowfall, snow metamorphism and aging. Starting from satellite observations coherence over snow covered alpine terrain is lost in most cases after 1 and 3 days, and generally after 35 days (typical repeat pass of ERS) coherence is very low compromising any operational scopes. The measured phase difference of a pixel, consists not only of $\Delta\phi_{snow}$ estimated by (7), coming from the two-way propagation difference in the snow-pack, but also of other contributions such as the phase difference due to changes in propagation through atmosphere and a contribution coming from random noise. Neglecting the noise due to instrumental sources and taking advantage of some stable points as corner reflectors or stable natural targets, atmospheric effect can be evaluated. By estimating the phase as a difference and comparing them, atmospheric artefacts are drastically reduced. The most critical fact is that when snow is not fully dry, the suggested formula is inapplicable as the effect of melting and refreezing can drastically affect the phase as well. Finally phase wrapping can occur if the acquisition is not fast enough or in case of heavy snowfall but the high temporal sampling provided by GB SAR (approximately two images per hour) largely reduces this occurrence.

5.5.2 The test site

The test site was a high alpine area at approximately 2000 m elevation, which lies north of the main ridge of the Austrian Alps in Tyrol. The monitored area is an east-wards looking slope of the Trantaler Köpfe, which is located in the Wattener Lizum, Tuxer Alpen, Tyrol, Austria, about 20 km south-east of Innsbruck. The target region is a northeast oriented slope between the pinnacles of Tarntaler Köpfe (2767 m) and Lizumer Boeden (approx. 2020 m) at the bottom of the valley. The experiments were carried out within the FP6 EC project GALAHAD framework (Advanced Remote Monitoring Techniques for Glaciers, Avalanches and Landslides Hazard Mitigation) with the support of Department of Natural Hazards and Timberline, in Innsbruck, Austria (BFW) which in particular organized the ground truth data collections and several Laser scanner measurements (Schaffhauser A., 2009). The RS instruments, GB SAR and TLS, were installed on a concrete base at an altitude of 2041: a pictures is shown in Figure 11. Four automatic weather stations (AWS) were installed providing continuous measurements of the main meteorological parameters (temperature, wind, solar irradiation, snow height). The experimental campaign included two periods, namely winter 2006 and 2007. The first data collection lasted about three months, from the 9th of February 2006 to the 4th of April with only C band working. The second period was from the 1st of February 2007 to the end of April 2007, during which S band data acquisitions were also arranged.

5.5.3 Data analysis

In the first campaign with a C band operated GB SAR, already described in a previous paper (Noferini et al., 2005) was used. The same apparatus was upgraded over the winter of 2007 as to measure at S-Band as well. The illuminated area is about 1 km × 2 km wide, cross-range / range respectively. The synthesized image has a slant range resolution of about 7.5. m and 5m at C and S bands respectively and a cross-range resolution of about 30m (C band) and about 50m (S band) at 1500m distance from the radar. According to the previous considerations about coherence a deep analysis of its behaviour at the two different bands can be found in the cited paper by (Luzi et al., 2009). The final outcomes are that at C band coherence can be considered acceptable for a time interval between acquisitions approximately of 14 hours, while at S-Band this interval is definitely more than 2 days. This result confirms that temporal decorrelation affects C and S bands in different ways, that the latter is more suited for long temporal observations at low sampling rate. To validate the proposed interferometric technique, estimates of the snow depth retrieved by using the described model and the snow depth measurements obtained through the ultrasonic sensor at the AWS, were compared. A small plot inside the imaged area, located at 2160 m asl at about 1 km distance from the GB SAR is considered. The phase values for the selected points were obtained after focusing on an area 400.m x 1800.m wide, with a resulting pixel resolution of 2 m x 2 m. Figure 12 shows a data record from the 24th February (0:00h) 2006 to 1st March (0:00h) 2006: the snow depth measured by means of the ultrasonic sensor at the closest station is compared to the snow depth retrieved from interferometric data measured at the same time in some points. The points depicted in Figure 12 show the SD retrieved through the equations (1) and (2) for an incidence angle=60° and a snow density=100 kg/m^3. It is worth noting that the snow was dry with a low probability of melting. To retrieve snow depth from interferometric phase, and removing the atmospheric component, the measured values were subtracted from the phase measured on a passive corner reflector, which is a metal trihedral 0.5 m in size and located at a distance of 1766 m from the radar. Observing Figure 12, according to the assumed model, the snow fall induces a regular increase of the SD retrieved from interferometric phases and also taking into account the non-coincident location of the two measurements, we obtain a consistent agreement between the value retrieved from interferometric phase and those measured at the AWS. In the last part of the plot there is an inversion of the two curves: retrieved values are first lower than US values and then they get higher. A possible explanation is the settlement of the snow pack which can reduce the height of the snow measured by US without changing the SWE, while the values retrieved from interferometric phases stand, being sensitive to SWE. The agreement can be considered satisfactory if we take into account both the general variability of the snow depth and secondly the not-coincidental location of the ground truth with radar pixels.

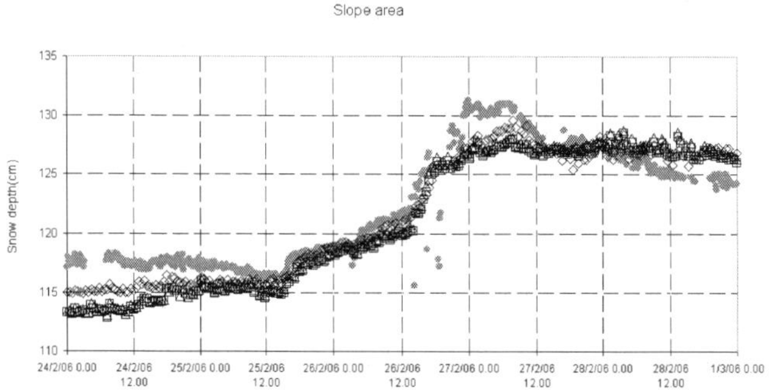

Fig. 12. Temporal record of the snow depth (grey filled points) measured through an US sensor and SD retrieved by means of GBInSAR at different points (Δ, ◊, □); elapsed time from 0h0m 24.02.2006 to 0h0m 01.03.2006 (After Luzi et al., 2009).

In winter the of 2006/2007 similar data were obtained confirming the effectiveness of the approach. The retrieval approach tested on the selected points has also been applied to the entire slope, the aiming at comparing TLS data and GB SAR observations. The local incidence angle for each pixel was calculated through the DEM of the observed area, provided to BFW by the Federal Office of Metrology and Surveying (10m resolution), assuming that the air to snow interface is parallel to terrain surface. Considering the same time interval elapsed between the two TLS scans (9 to 14 February 2007), a snow depth map was calculated both at C and S band. The results, corresponding to an area of 1000m x2000m in front of the GB SAR location, are shown in Figure 13A and 14B respectively. A circle locates in Figure 13 the area surrounding the automatic weather station where the data analysis is focused. The data are depicted on a section of the map together with a coherence map calculated at C band for the same area (Figure 13D). The difference in data coverage between Figure 13A and Figure 13B is due to the antenna pattern which at S band is coarser. The TLS map provided by BFW of the SD variation that occurred between the two dates is shown in Figure 13C: a SD increase of about 0.25 m is measured.

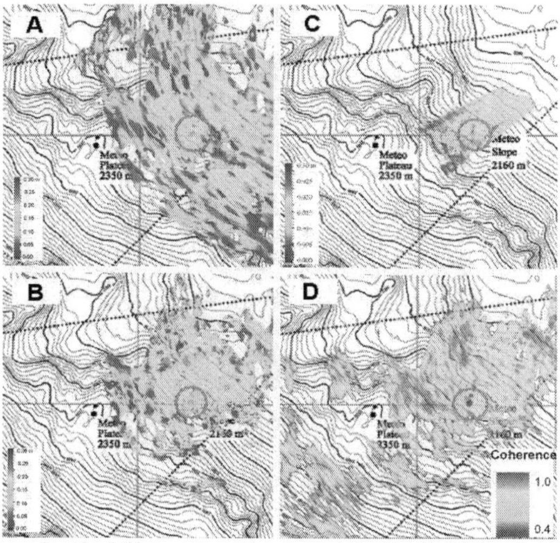

Fig. 13. A) Map of snow depth difference with respect to the initial value, obtained through cumulative interferogram starting 09.02.2007 and ending 14.02.2007: S band; snow density=100 kg/m^3 ; B) C band C) Snow depth difference compared to the initial value measured through TLS from 9 to14 February 2007; D) Coherence map calculated at C band corresponding to the time interval from 9 to14 February 2007. The green circle highlights the area where the US was placed (After Luzi et al., 2009).

Maps retrieved from microwave data (Figure 13A and Figure 13B) show a discontinuous texture compared to TLS; this is due to different factors: the coarser spatial resolution, a certain noise as testified by a barely homogeneous coherence behaviour (Figure 13D) and the presence of possible residuals of atmospheric effect after correction. At the same time and for the same area, the maps indicate similar SD values, with S band closer to TLS estimates and C band lower. It is worth noting that GB SAR and the TLS use a different time sampling; a TLS map is obtained by using two measurements (scans) only while the GB SAR differential phase is the result of the summation of an interferogram series acquired with an hourly sampling over the whole period, and secondly, their governing physical principle differs as well. TLS refers directly to the SD and it is affected by the first few millimetres of

the snow layer surface while through the microwave interaction (at large incidence angles), we are not able to separate depth and density effects.
Notwithstanding the difficulty of providing both lengthy data record in dry snow conditions and detailed knowledge of the observed snow characteristics, the obtained results confirmed the presence of a clearly measurable interferometric phase variation in relation to the growing height of the snow layer.

6. Conclusions

The brief introduction of the GBInSAR here presented is certainly incomplete but it was simply aimed at introducing the reader to this novel tool. In the discussed examples we focused on the slope monitoring because this is nowadays the most consolidated and operative use. At the same time we introduced the snow monitoring application as an opposite case where the technique is yet at a research stage. The spreading of new instrumentations, and the related issued papers, confirm that Remote Sensing community is more and more convicted that this technique can be very useful often providing a complementary information to the more popular spaceborne SAR interferometry. Some papers have been issued about the DEM retrieval and GBDInSAR but the application is still in progress.Finally a set of applications addressed to buildings and civil structures as bridges and dam, have not been tackled here but they represent further hopeful frontiers for GB SAR interferometry.

Acknowledgements

The author wishes to acknowledge the teams and the institutions who presently and formerly worked with him. The majority of them are authors and coauthors of the cited papers. Thanks are also due to the authors of the several sources used for the introductory part dealing with SAR and interferometry and apologies for missing.

7. References

Antonello, G., Casagli, N., Farina, P., Guerri, L., Leva, D., Nico, G., Tarchi, D. (2009). SAR interferometry monitoring of landslides on the Stromboli Volcano. *Proceedings of* FRINGE 2003 Workshop, 1-5 December 2003, ESA/ESRIN, Frascati, Italy.

Antonello G., Casagli N., Catani F., Farina P., Fortuny-Guasch J., Guerri L. , Leva D., Tarchi, D. (2007). Real-time monitoring of slope instability during the 2007 Stromboli eruption through SAR interferometry. *Proceedings of 1st NACL, Veil* (Colorado).

Askne J. (2003). Remote Sensing using microwaves. Available on web: www.chalmers.se/en/.

Bamler R. and Just. D. (1993). Phase statistics and decorrelation in SAR interferograms. *Geoscience and Remote Sensing Symposium*, 1993, IGARSS93 'Better Understanding of Earth Environment', 18-21 pp 980-984, August 1993.

Bernardini G., P. Ricci, F. Coppi (2007). A Ground Based Microwave Interferometer with imaging capabilities for remote sensing measurements of displacements, *7th Geomatic Week/3rd Int. Geotelematics Fair*, Barcelona, Spain, 2007 February 20-23

Bernier M and J.P. Fortin (1998). The potential of times series of C-band SAR data to monitor dry and shallow snow cover. *IEEE Trans. Geosci. Remote Sens.*, vol.36, no1, pp. 226-242, January 1998.

Casagli N., Tibaldi A., Merri A., Del Ventisette C., Apuani C., Guerri L., Fortuny-Guasch J., Tarchi D. (2003). Deformation of Stromboli Volcano (Italy) during the 2007 eruption revealed by radar interferometry, numerical modelling and structural geological field data. *Journal of Volcanology and Geothermal Research* 182 (2009) 182–200.

Colesanti C., Ferretti A., Novali F., Prati C., Rocca F. (2003) - SAR monitoring of progressive and seasonal ground deformation using the Permanent Scatterers Technique. *IEEE Trans. Geosci. and Remote Sens.*, 41 (7). pp. 1665-1701.

Crosetto M., Crippa B., Biescas E. (2005). Early detection and in-depth analysis of deformation phenomena by radar interferometry. *Eng Geol.*, 79 (1-2), pp. 81-91.

Curlander, J.C., McDonough, R.N., 1991. Synthetic Aperture Radar: Systems and Signal Processing. Wiley, New York, 672 pp.

Ferretti A., Monti-Guarneri A., Massonnet D., Prati C., Rocca F. (2007). InSAR Principles: guidelines for SAR Interferometry Processing and Interpretation, *ESA Publications ESTEC* Noordwijk NL; TM-19 February 2007, ed. K. Flecther ISBN 92-9092-233-8.

Ferretti, A., Prati, C., Rocca, F. (2001). Permanent scatterers in SAR interferometry. *IEEE Trans. Geosci. Remote Sens.*, 39 (1), 8 - 20.

Fortuny-Guash J. and A. J. Sieber (1994). Fast algorithm for near-field synthetic aperture radar processor. *IEEE Trans. Antennas Propagat.*, vol. 42, pp. 1458–1460, Oct. 1994

Fortuny-Guasch J. (2009). A Fast and Accurate Far-Field Pseudopolar Format Radar Imaging Algorithm, *IEEE Trans. Geosci. Remote Sens.* 47 (4), 1187 –1196 April 2009.

Goldstein R.M., Engelhardt H., Kamb B., Frolich. R.M. (1993) . Satellite radar interferometry for monitoring ice sheet motion: application to an Antarctic ice stream. *Science*, 262 (5139), 1525-1530.

Guneriussen T., K.A. Høgda, H. Johnson and I. Lauknes (2001). InSAR for estimating changes in snow water equivalent of dry snow, *IEEE Trans. Geosci. Remote Sens.*, vol. 39, no 10, 2101-2108, October 2001.

Ghiglia D.C. & Romero L.A. (1994). Robust two-dimensional weighted and unweighted phase unwrapping that uses fast transforms and iterative methods. *J. Opt. Soc. Amer. A*, Vol. 11, n. 1, pp. 107-117.

Herrera G., Fernandez-Merodo JA, Mulas, J., Pastor M , Luzi G, Monserrat O (2009). A landslide forecasting model using ground based SAR data: the Portalet case study. *Engineering Geology* Vol.: 105 Issue: 3-4 Pages: 220-230 MAY 11 2009

Kenyi L.W. and V. Kaufmann (2003). Estimation of rock glacier surface deformation using SAR intreferometry data. *IEEE Trans. Geosci. Rem. Sens.*, vol. 41, pp. 1512–1515, 2003.

Lanari R., Mora O., Manunta M., Mallorqui J.J., Berardino P., Sansosti E. (2004). A Small-Baseline approach for investigating deformations on full-resolution differential SAR interferograms. *IEEE Trans. on Geoscience and Remote Sensing*, 42 (7), 1377-1386.

Leva D., Nico G., Tarchi D., Fortuny-Guasch J., Sieber A.J.. Temporal analysis of a landslide by means of a ground-based SAR Interferometer. *IEEE Trans. Geosci. Remote Sens.*, vol. 41, no 4, Part 1, pp.745 - 752, April 2003.

Luzi G., Noferini L., Mecatti D. Macaluso G., Pieraccini M., Atzeni C., Schaffhauser A., Fromm R., Nagler T.. (2009). Using a Ground-Based SAR Interferometer and a Terrestrial Laser Scanner to Monitor a Snow-Covered Slope: Results From an Experimental Data Collection in Tyrol (Austria). *IEEE Transaction on Geoscience and Remote Sensing*, vol.47, no.2 , February 2009, Page(s): 382-393.

Luzi G., M. Pieraccini, D. Mecatti, L. Noferini, G. Macaluso, A. Galgaro, C. Atzeni, (2006), Advances in ground based microwave interferometry for landslide survey: a case study. *International Journal of Remote Sensing*, Vol. 27, No. 12 / 20 June 2006, pp. 2331 - 2350.

Luzi G., M. Pieraccini, D. Mecatti, L. Noferini, G. Macaluso, A. Tamburini, and C. Atzeni, (2007), Monitoring of an Alpine Glacier by Means of Ground-Based SAR Interferometry. *Geoscience and Remote Sensing Letters* , Vol. 4, No 3, July 2007 pp. 495-499.

Luzi, G., Pieraccini M., Mecatti D., Noferini L., Guidi G., Moia F., Atzeni C., (2004). Ground-Based Radar Interferometry for Landslides Monitoring: Atmospheric and Instrumental Decorrelation Sources on Experimental Data. *IEEE Trans. Geosci. Remote Sens.*, vol. 42, no 11, pp 2454 - 2466, November 2004.

Macelloni G., Paloscia S., Pampaloni P., Brogioni M., Ranzi R., Crepaz A, (2005).Monitoring of melting refreezing cycles of snow with microwave radiometers: the Microwave Alpine Snow Melting Experiment (MASMEx 2002-2003). *IEEE Trans. Geosci. Remote Sens.*, Vol.43, no 11, pp. 2431-2442, November 2005.

Massonnet D. and T. Rabaute (1993a). Radar interferometry: Limits and potential. *IEEE Trans. Geosci. Remote Sensing*, vol. 31, pp. 455–464, Mar. 1993.

Massonnet, D., Rossi, M., Carmona, C., Adragna, F., Peltzer, G., Feigl, K., Rabaute, T. (1993b). The displacement field of the Landers earthquake mapped by radar interferometry. Nature 364, 138– 142.

Martinez-Vazquez A., J. Fortuny-Guasch and U. Gruber, (2005). Monitoring of the snow cover with a ground-based synthetic aperture radar. *EARSeL Proceedings*, vol. 4, no. 2, pp.171-178, 2005.

Martinez-Vazquez A., J. Fortuny-Guasch (2006). Snow Cover Monitoring in the Swiss Alps with a GB-SAR. *IEEE Geoscience and Remote Sensing Society Newsletter*, pp.11-14, March 2006.

Mätzler C. (1996). Microwave permittivity of dry snow. *IEEE Trans. Geosci. Remote Sens.*, vol. 34, no 2, pp. 573 - 581, 1996.

Mensa D. (1991). High Resolution Radar Cross-Section Imaging. Artech House, Boston, 1991

Mohr J.J.(2005). SAR Light an introduction to Synthetic Aperture Radar. Version 2.0 August 9, 2005, NB 238 available on http://www.gfy.ku.dk/~cct/sat07/NB238.pdf

Nagler T. and H. Rott (2000). Retrieval of wet snow by means of multitemporal SAR data. IEEE *Trans. Geosci. Remote Sens.*, vol.38, no2 part 1, pp. 754-765 ,March 2000.

Nagler T. and H. Rott(2004). Feasibility Study on Snow Water Equivalent (SWE) retrieval with L-band SAR, Final report, ESA contract no. 16366/02/NL/MM, February 2004.

Noferini L., M. Pieraccini, D. Mecatti, G. Luzi, A. Tamburini, M. Broccolato, and C. Atzeni (2005). Permanent scatterers analysis for atmospheric correction in Ground Based SAR Interferometry. *IEEE Trans. Geosci. Rem. Sens.*, vol. 43, no 7, pp. 1459-1471, 2005.

Oveishgaram S and H. A. Zebker (2007). Estimating Snow accumulation from InSAR Correlation Observation. IEEE *Trans. Geosci. Remote Sens.*, vol. 45, no 1, pp. 10-20, 2007.

Pieraccini M., Casagli N., Luzi G., Tarchi D., Mecatti D., Noferini L. and C. Atzeni (2002). Landslide monitoring by ground-based radar interferometry: a field test in Valdarno (Italy). *International Journal of Remote Sensing,* 24 6, pp. 1385-1391.

Pipia L., Fabregas X., Aguasca A., Lopez-Martinez C., Mallorqui J., Mora O. (2007). A Subsidence Monitoring Project using a Polarimetric GB-SAR Sensor. *The 3rd Int. Workshop POLinSAR 2007* Frascati, Italy on 22-26 January 2007.

Reale D., Pascazio V., Schirinzi G., Serafino F., 3D Imaging of Ground based SAR Data. *Geoscience and Remote Sensing Symposium, 2008* IGARSS2008. IEEE International Volume 4, 7-11 July 2008.

Reigber A. and R. Scheiber. Airborne Differential SAR Interferometry: first results at L-Band. *IEEE Trans. on Geoscience and Remote Sensing,* 41, (6) pp. 1516-1520 June 2003.

Rosen P.A., Hensley S., Joughin I.R., Li F.K., Madsen S.N., Rodriguez E., Goldstein R.M. (2000). Synthetic aperture radar interferometry. *Proc. IEEE* 88 (3), 333–382.

Rudolf H., Leva, D. Tarchi, D. Sieber, A.J. (1999). A mobile and versatile SAR system. *Proc. IGARSS'99,* Hamburg, pp. 592–594.

Sang-Ho Yun (2008). *Volcano Deformation Modeling Using Radar Inteferomery.* ed. VDM Verlag Dr. Muller, 2008. ISBN: 97-3-4-9-3

Schaffhauser A., M. Adams, R. Fromm, P. Jörg, G. Luzi, L. Noferini, R. Sailer (2008). Remote Sensing based retrieval of snow cover properties, *Cold Regions Science and Technology* . 54 (2008), pp. 164-175.

Shi J. and J. Dozier (2000). Estimation of Snow Water Equivalence Using SIR-C/X-SAR, Part I: Inferring snow density and subsurface properties. *IEEE Trans. Geosci. Remote Sens.*, vol. 38, no 6, pp. 2465-2474, 2000.

Silver S. (1986). *Microwave Antenna Theory and Design.* Peter Peregrinus Ltd, London UK, 2nd edition 1986 ISBN 0 86341 017 0.

Strozzi T., U. Wegmüller and C. Mätzler (1999). Mapping Wet Snowcovers with SAR Interferometry. *Int. J. Remote Sens.*, Vol. 20, No. 12, pp. 2395-2403, 1999.

Strozzi T. , Matzler C. (1998). Backscattering Measurements of Alpine Snowcovers at 5.3 GHz and 35 GHz. *IEEE Trans. on Geoscience and Remote Sensing,* Vol. 36, No. 3, pp. 838-848 May 1998.

Tarchi, D., Ohlmer, E., Sieber, A.J. (1997). Monitoring of structural changes by radar interferometry. *Res. Nondestruct. Eval.* 9, 213– 225.

Tarchi, D., Rudolf, H., Luzi, G., Chiarantini, L., Coppo, P., Sieber, A.J. (1999). SAR interferometry for structural changes detection: a demonstration test on a dam. *Proc. IGARSS'99,* Hamburg, pp. 1522–1524.

Tarchi D., Casagli N., Fanti R., Leva D., Luzi, G. Pasuto A., Pieraccini M., Silvano S. (2003a). Landside Monitoring by Using Ground-Based SAR Interferometry: an example of application to the Tessina landslide in Italy, *Engineering Geology* 68, pp.15-30

Tarchi,D., Casagli, N., Moretti, S., Leva, D., Sieber, A.J. (2003b). Monitoring landslide displacements by using ground-based radar interferometry: Application to the Ruinon landslide in the Italian Alps, J. Geophys. Res., 108, 10.1-10.14.

Ulaby, F. T., R. K. Moore, and A.K. Fung, *Microwave Remote Sensing: Active and Passive,* Vol. II, Addison-Wesley, Advanced Book Program, Reading, Massachusetts, 1982,

Voight, B. (1988). Material science law applies to time forecast of slope failure. Landslide News, 3: 8-11.

Werner C., Strozzi T., Wesmann A., Wegmuller U. (2008), Gamma's portable radar interferometer. 13th FIG *Symposium on Deformation Measurements ad analysis*, LNEC, Lisbon May 12-16 2008.

Zebker H.A., Goldstein, R.M. (1986). Topographic mapping from interferometric synthetic aperture radar observations. J. Geophys. Res. 91, 4993- 4999.

Zebker H.A. and J. Villlasenor (1992). Decorrelation in interferometric radar echoes. IEEE *Trans. Geosci. Remote Sens.*, vol 30, no 10, pp. 950-959, 1992.

Zebker H.A. Rosen P.A. and S. Hensley (1997). Atmospheric effects in interferometric synthetic aperture Radar surface deformation and topographic maps. *J. Geophys. Res.* -Solid Earth, vol 102, N0 B4 , pp.7547-7563, April 10, 1997.

Zebker H.A., Rosen P.A., Goldstein R., Gabriel A., Werner C. (1994). On the derivation of coseismic displacement fields using differential radar interferometry: the Landers earthquake. *J. Geophys. Res.* 99, 19617- 19634.

2

Internet Surveillance Camera Measurements of Atmospheric Aerosols Concentration

C.J. Wong, M.Z. MatJafri, K. Abdullah and H.S. Lim
School of Physics, Universiti Sains Malaysia
11800 USM, Penang, Malaysia

1. Introduction

Nowadays, air pollution becomes a very serious problem with the rapid growth of industrialization and urbanization (Kim Oanh et al., 2006, Wu et al., 2006). This air pollution is not only continues to damage our environment, it also endanger our health (Pope et al., 2008, Pope et al., 2007, Banauch et al., 2006, Brunekreef et al., 2002). Evidence gathered to date indicates that the most harmful component of this pollution is the microscopic atmospheric aerosols with an aerodynamic diameter below 10 micrometers (PM_{10}) (Pope et al., 2008, Pope et al., 2007, Pope et al., 2004, Donaldson et al., 2000, Pope et al., 1995). Only particles less than 10 micrometers in diameter can be inhaled deep into the lungs, then embed themselves in the lungs to cause adverse health effects. These effects have been linked to respiratory disease, cancer and other potentially deadly illnesses. This is the reason for both the WHO and the United Nations have declared that atmospheric aerosols poses the greatest air pollution threat globally.

In order to monitor the levels of air pollution, so that early warning will be provided to prevent long exposure to this type of harmful air pollution. Many researchers attempt to develop more efficient techniques to monitor this atmospheric aerosols air pollution. This includes the techniques of Atmospheric Optical Thickness (AOT) and satellite images (Hadjimitsis, 2009, Hadjimitsis, 2008, Sifakis et al., 1992, Kaufman et al., 1983, Lim et al., 2009). Satellite images were normally used by researchers in their remote sensing air quality studies, but the main drawback of using satellite images is the difficulty in obtaining cloud-free scenes especially for the Equatorial region.

In order to overcome cloud-free scenes problem, aerial photographic imagery technique is used to obtain air pollution map. This technique utilizes fundamental optical theory like light absorption, light scattering and light reflection. This technique has long been used for visibility monitoring (Middleton, 1968, Noll et al., 1968, Horvath et al., 1969, Diederen et al., 1985). The continuous and rapid evolution of digital technologies in the last decade fostered an incredible improvement in digital photography technology, in information and communication technologies (ICT) and personal computer technology. This modern digital technology allows image data transfer over the internet protocol, which provides real time observation and image processing (Wong et al., 2009, Wong et al., 2007). This has made it possible to monitor real time PM_{10} air pollution at multi location. This is an attempt to fulfill the need for preventing long exposure to this harmful air pollution.

The object of this study is to develop a state-of-the-art technique to enhance the capability of the internet surveillance camera for temporal air quality monitoring. This technique is able to detect particulate matter with diameter less than 10 micrometers (PM_{10}). An empirical algorithm was developed and tested based on the atmospheric characteristic to determine PM_{10} concentrations using multispectral data obtained from the internet surveillance camera. A program is developed by using this algorithm to determine the real-time air quality information automatically. This development showed that the modern Information and Communications Technologies (ICT) and digital image processing technology could monitor temporal development of air quality at multi location simultaneously from a central monitoring station.

2. Description of the Algorithm

In this study, we developed an algorithm based on the fundamental optical theory, that is light absorption, light scattering and light reflection. This algorithm is used to perform image processing on the captured digital images to determine the concentration of atmospheric aerosols.

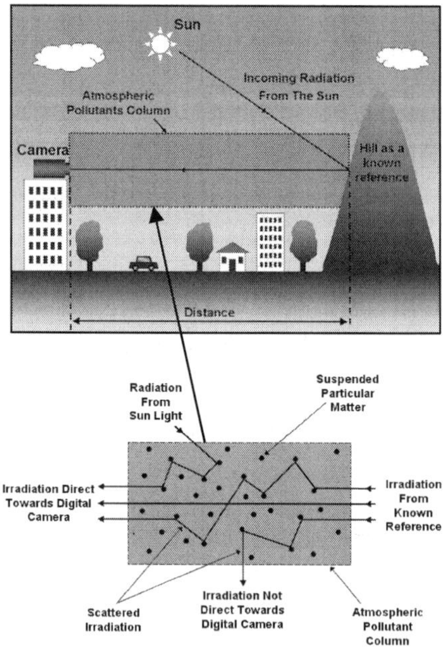

Fig. 1. The skylight parameter model to illustrate the electromagnetic radiation propagates from sunlight towards the known reference, and then reflected to propagate towards the internet surveillance camera penetrating through the interaction in atmospheric pollutant column.

Figure 1 shows the electromagnetic radiation path of ambient light propagating towards the internet surveillance camera, and then this electromagnetic radiation is reflected by a known reference target and penetrating through the ambient pollutant column. At the ambient pollutant column, this electromagnetic radiation encounters absorption and scatters. In a single scattering of visible electromagnetic radiation by aerosol in atmosphere, Liu et al. showed that the atmospheric reflectance due to molecules scattering, R_r is proportional to the optical thickness for molecules, τ_r (Liu et al., 1996). This atmospheric reflectance due to molecule scattering, R_r can be written as

$$R_r = \frac{\tau_r P_r(\Theta)}{4\mu_s\mu_v} \tag{1}$$

where $P_r(\Theta)$ is the scattering phase function for molecules, μ_v is the cosine of viewing angle, and μ_s is the cosine of solar zenith angle.
In the same paper, Liu et al. also showed that the atmospheric reflectance due to particles scattering, R_a is proportional the optical thickness for aerosols, τ_a (Liu et al., 1996). Later on, King et al. and Fukushima et al. have further confirmed this relationship (King et al., 1999, Fukushima et al., 2000). This particles scattering, R_a is

$$R_a = \frac{\tau_a P_a(\Theta)}{4\mu_s\mu_v} \tag{2}$$

where $P_a(\Theta)$ is scattering phase function for aerosols.

In year 1997, Vermote et al. showed that the atmospheric reflectance, R_{atm} is the sum of reflectance from particles, R_a and reflectance from molecules, R_r (Vermote et al., 1997). This atmospheric reflectance, R_{atm} can be written as

$$R_{atm} = R_a + R_r \tag{3}$$

By substituting equation (1) and equation (2) into equation (3), the atmospheric reflectance, R_{atm} also can be written as

$$R_{atm} = \frac{1}{4\mu_s\mu_v}\left[\tau_a P_a(\Theta) + \tau_r P_r(\Theta)\right] \tag{4}$$

Camagni et al. expressed the optical depth, τ in term of absorption, σ and finite path, s (Camagni et al., 1983). Equation (5) showed this optical depth, τ as

$$\tau = \sigma\rho s \tag{5}$$

where σ is absorption, ρ is density and s is finite path.

In the same paper, Camagni et al. also showed that this optical depth, τ is the sum of the optical depth for particle aerosols, τ_a and the optical depth for molecule aerosols, τ_r (Camagni et al., 1983). This optical depth, τ also can be written as

$$\tau = \tau_a + \tau_r \tag{6}$$

As the optical depths for particle aerosols, τ_a and for molecule aerosols, τ_r can be written in the form of equation (5). Thus the optical depths for particle aerosols, τ_a and for molecule aerosols, τ_r are written as

$$\tau_a = \sigma_a \rho_a s \tag{7}$$

$$\tau_r = \sigma_r \rho_r s \tag{8}$$

Equations (7) and (8) are substituted into equation (4). The atmospheric reflectance, R_{atm} become

$$R_{atm} = \frac{s}{4\mu_s\mu_v}\left[\sigma_a\rho_a P_a(\Theta) + \sigma_r\rho_r P_r(\Theta)\right] \tag{9}$$

R_{atm}, σ_a, σ_r, $P_a(\Theta)$ and $P_r(\Theta)$ are dependent on wavelength, λ, thus equation (9) can be expressed as

$$R_{atm}(\lambda) = \frac{s}{4\mu_s\mu_v}\left[\sigma_a(\lambda)\rho_a P_a(\Theta,\lambda) + \sigma_r(\lambda)\rho_r P_r(\Theta,\lambda)\right] \tag{10}$$

when ρ_a is particle aerosols concentration (PM_{10}), P and ρ_r is molecule aerosols concentration, G. Equation (10) can be written as

$$R_{atm}(\lambda) = \frac{s}{4\mu_s\mu_v}\left[\sigma_a(\lambda)\mathrm{P}P_a(\Theta,\lambda) + \sigma_r(\lambda)\mathrm{G}P_r(\Theta,\lambda)\right] \tag{11}$$

Equation (11) is extended into a two bands algorithm for wavelength, λ_1 and wavelength, λ_2. These two bands algorithm are as shown in equation (12) and equation (13).

$$R_{atm}(\lambda_1) = \frac{s}{4\mu_s\mu_v}\left[\sigma_a(\lambda_1)\mathrm{P}P_a(\Theta,\lambda_1) + \sigma_r(\lambda_1)\mathrm{G}P_r(\Theta,\lambda_1)\right] \tag{12}$$

$$R_{atm}(\lambda_2) = \frac{s}{4\mu_s\mu_v}\left[\sigma_a(\lambda_2)PP_a(\Theta,\lambda_2) + \sigma_r(\lambda_2)GP_r(\Theta,\lambda_2)\right] \quad (13)$$

where $R_{atm}(\lambda_i)$ is atmospheric reflectance, i = 1, 2 are the band numbers.

Solving equation (12) and (13) simultaneously and we obtain particle concentration of PM_{10}, *P* as

$$P = a_0 R_{atm}(\lambda_1) + a_1 R_{atm}(\lambda_2) \quad (14)$$

where a_j is algorithm coefficients, j = 0, 1 are then empirically determined.

From the equation (14); the PM_{10} concentration is linearly related to the atmosphere reflectance for band 1 and band 2. This algorithm was generated based on the linear relationship between τ and reflectance. Retalis et al. also found that the PM_{10} was linearly related to τ and the correlation coefficient for the linear model was better than exponential (Retalis et al., 2003). This means that reflectance was linear with the PM_{10}. In order to simplify the data processing, the air quality concentration was used in our analysis instead of using density, ρ, values.

3. Methodology

3.1 Equipment Set-Up

As shown in Figure 2, an internet surveillance camera was used as remote sensing sensor to monitor the concentrations of particles less than 10 micrometers in diameter. This internet surveillance camera is a Bosch's auto dome 300 series PTZ camera system. It is a 0.4 mega pixel (PAL) Charge-Couple-Device CCD camera, which allows image data transfer over the standard computer networks (Ethernet networks), internet. Therefore it can be used as a remote sensing sensor to monitor air quality.

Fig. 2. A 0.4 mega pixel (PAL) Charge-Couple-Device CCD, internet surveillance camera used in this study is a Bosch's auto dome 300 series PTZ camera system

This internet surveillance camera was calibrated by using a spectroradiometer with Pro Lamp light source and colour papers. This calibration enabled us to convert the digital numbers (DN) of the images captured by the internet surveillance camera to irradiance. The coefficients of calibrated internet surveillance camera are as listed below

$$L_R = 0.0003\, N_R + 0.0278 \quad (15)$$

$$L_G = 0.0004\, N_G + 0.0263 \quad (16)$$

$$L_B = 0.0004\, N_B + 0.0248 \quad (17)$$

where L_R is irradiance for red band ($Wm^{-2}\ nm^{-1}$), L_G is irradiance for green band ($Wm^{-2}\ nm^{-1}$) , L_B is irradiance for blue band ($Wm^{-2}\ nm^{-1}$), N_R is digital number for red band, N_G is digital number for green band and N_B is digital number for blue band.

The schematic set-up of the internet surveillance camera is shown in Figure 3. This set-up provides a continuous, on-line, real-time monitoring for air pollution at multiple locations. It is able to detect the present of particulates air pollution immediately, in the air and helps to ensure the continuing safety of environmental air for living creatures.

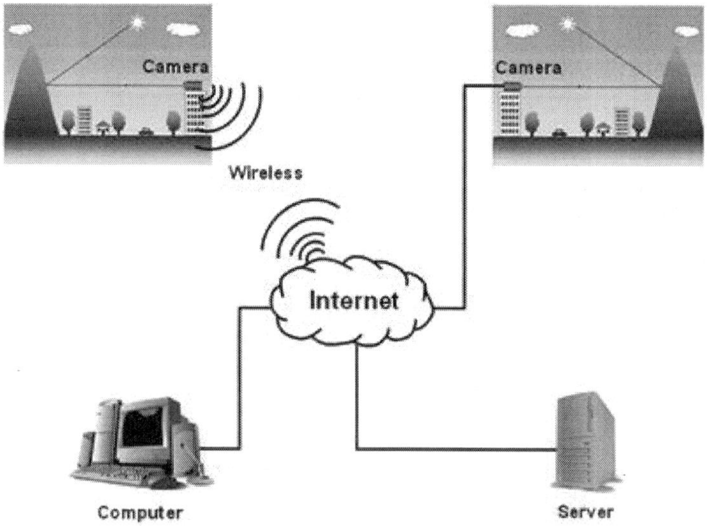

Fig. 3. The schematic set-up of internet surveillance camera as remote sensor to monitor air quality

3.2 Study Location

The internet surveillance camera was installed at the top floor of the Chancellery building, Universiti Sains Malaysia's campus. It is located at longitude of 100° 18′20.67″ and latitude of 5° 21′28.50″ as shown in Figure 4 and Figure 5. This internet surveillance camera is looking to the direction of the Penang bridge (Figure 5). As shown in Figure 5 and Figure 6, the reference target that we used in this study is green vegetation.

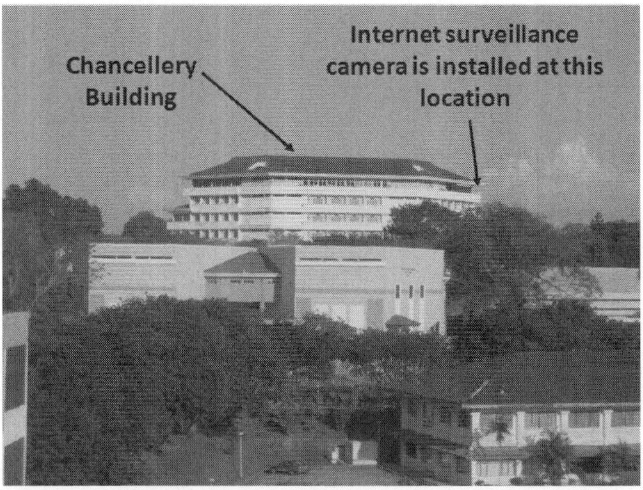

Fig. 4. The internet surveillance camera is installed at the top floor of Chancellery building in Universiti Sains Malaysia (USM).

Fig. 5. The satellite image showed the location of internet surveillance camera capture photograph and the location of the reference target

Fig. 6. The reference target of green vegetation captured by the internet surveillance camera

Figure 6 shows a sample from the digital images captured by the IP camera. The target of interest is the green vegetation grown on a distant hill. Digital images were separated into three bands (red, green and blue). Digital numbers (DN) of the target were determined from the digital images for each band. Equations 9, 10 and 11 were used to convert these DN values into irradiance.

4. Determine Algorithm Coefficients and Atmospheric Aerosol Concentration

A handheld spectroradiometer was used to measure the sun radiation at the ground surface. The reflectance values recorded by the sensor was calculate using equation (18) below.

$$R_S = \frac{L(\lambda)}{E(\lambda)} \tag{18}$$

where $L(\lambda)$ is irradiance of each visible bands recorded by the internet surveillance camera ($Wm^{-2}\ nm^{-1}$) [can be determined by equation (15), (16), (17)] and $E(\lambda)$ is sun radiation at the ground surface measured by a hand held spectroradiometer ($Wm^{-2}\ nm^{-1}$).
From the skylight model showed in Figure 1, the reflectance recorded by the internet surveillance camera (R_s) was subtracted by the reflectance of the known surface (R_{ref}) to obtain the reflectance caused by the atmospheric components (R_{atm}).

$$R_S - R_{ref} = R_{atm} \tag{19}$$

The DustTrak meter used to determine atmospheric aerosol concentration of PM_{10}. The relationship between the atmospheric reflectance and the corresponding atmospheric aerosol concentration data for the pollutant was established by using regression analysis as shown in Table 1. Thus, algorithm coefficients in equation (14) can be determined to calculate the atmospherics aerosol concentration of PM_{10}.

Algorithm	R^2	**RMS ($\mu g/m^3$)**
$PM_{10} = a_0 + a_1 R_1 + a_2 R_1^2$	0.5662	12
$PM_{10} = a_0 + a_1 R_2 + a_2 R_2^2$	0.2238	14
$PM_{10} = a_0 + a_1 R_3 + a_2 R_3^2$	0.4627	17
$PM_{10} = a_0 + a_1 \ln R_1 + a_2 (\ln R_1)^2$	0.4536	17
$PM_{10} = a_0 + a_1 \ln R_2 + a_2 (\ln R_2)^2$	0.1426	16
$PM_{10} = a_0 + a_1 \ln R_3 + a_2 (\ln R_3)^2$	0.5129	13
$PM_{10} = a_0 + a_1 (R_1 / R_3) + a_2 (R_1 / R_3)^2$	0.3196	15
$PM_{10} = a_0 + a_1 (R_1 / R_2) + a_2 (R_1 / R_2)^2$	0.3243	14
$PM_{10} = a_0 + a_1 (R_2 / R_3) + a_2 (R_2 / R_3)^2$	0.2983	15
$PM_{10} = a_0 + a_1 \ln(R_1 / R_3) + a_2 \ln(R_1 / R_3)^2$	0.5326	16
$PM_{10} = a_0 + a_1 \ln(R_1 / R_2) + a_2 \ln(R_1 / R_2)^2$	0.4283	12
$PM_{10} = a_0 + a_1 \ln(R_2 / R_3) + a_2 \ln(R_2 / R_3)^2$	0.2734	16
$PM_{10} = a_0 + a_1 (R_1 - R_2)/R_3 + a_2 ((R_1 - R_2)/R_3)^2$	0.3834	17
$PM_{10} = a_0 + a_1 (R_1 - R_3)/R_2 + a_2 ((R_1 - R_3)/R_2)^2$	0.4273	18
$PM_{10} = a_0 + a_1 (R_2 - R_3)/R_1 + a_2 ((R_2 - R_3)/R_1)^2$	0.3826	16
$PM_{10} = a_0 + a_1 (R_1 + R_2)/R_3 + a_2 ((R_1 + R_2)/R_3)^2$	0.4826	16
$PM_{10} = a_0 + a_1 (R_1 + R_3)/R_2 + a_2 ((R_1 + R_3)/R_2)^2$	0.5372	17
$PM_{10} = a_0 + a_1 (R_2 + R_3)/R_1 + a_2 ((R_2 + R_3)/R_1)^2$	0.6532	15
$PM_{10} = a_0 + a_1 (R_2 - R_1) + a_2 (R_2 - R_1)^2$	0.6215	17
$PM_{10} = a_0 + a_1 (R_2 - R_3) + a_2 (R_2 - R_3)^2$	0.3782	16
$PM_{10} = a_0 + a_1 (R_1 - R_3) + a_2 (R_1 - R_3)^2$	0.4725	13
$PM_{10} = a_0 + a_1 R_1 + a_2 R_2 + a_3 R_3$	0.7321	9
$PM_{10} = a_0 R_1 + a_1 R_3$ **(Proposed)**	**0.7852**	**6**

* R_1, R_2 and R_3 are the reflectance for red, green and blue band respectively for PM_{10}

Table 1. Regression results using different forms of algorithms to determine algorithm coefficients

Figure 7 shows three photographs of Penang Bridge at different atmospheric aerosol concentration level. These photographs were captured at around 10.30 am to 11.00 am but on different date. Photograph at Figure 7 (a) was captured during low atmospheric aerosol concentration. This atmospheric aerosol concentration level can be determined from the equation (14) after we determine the algorithm coefficients. The atmospheric aerosol concentration level for photograph at Figure 7 (a) is 34 ± 6 $\mu g/m^3$.

(a)

(b)

(c)

Fig. 7. Three photographs of Penang Bridge at different atmospheric aerosol concentration level

For Figure 7 (b) and Figure (c), the atmospheric aerosol concentration levels are 56 ± 6 μg/m³ and 93 ± 6 μg/m³ respectively.
The relationship between the atmospheric reflectance and the corresponding atmospheric aerosol concentration data for the pollutant was established by using regression analysis. The correlation coefficient (R^2) between the predicted and the measured PM_{10} values, and root-mean-square-error (RMS) value were determined. Figure 8 shows the correlation between the estimated measurement of atmospheric aerosol concentration by the internet surveillance camera and the measurement of atmospheric aerosol concentration by the DustTrak meter.

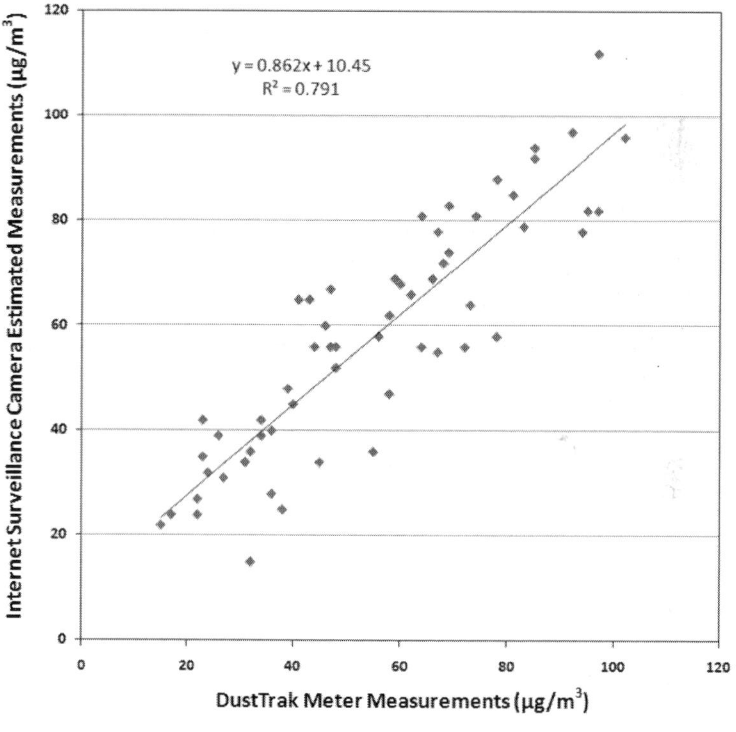

Fig. 8. Correlation coefficient and RMS error of the measured and estimated PM10 (μg/m3) values for the internet surveillance camera

The correlation coefficient (R^2) produced by the internet surveillance camera data set was 0.791. The RMS value for internet surveillance camera was ± 8 μg/m³.
Figure 9 shows the temporal development of real time air quality of PM_{10} in a day measured by the internet surveillance camera and DustTrak meter. The data were obtained on 21 Jul 2008 from 8.00am to 5.00pm.

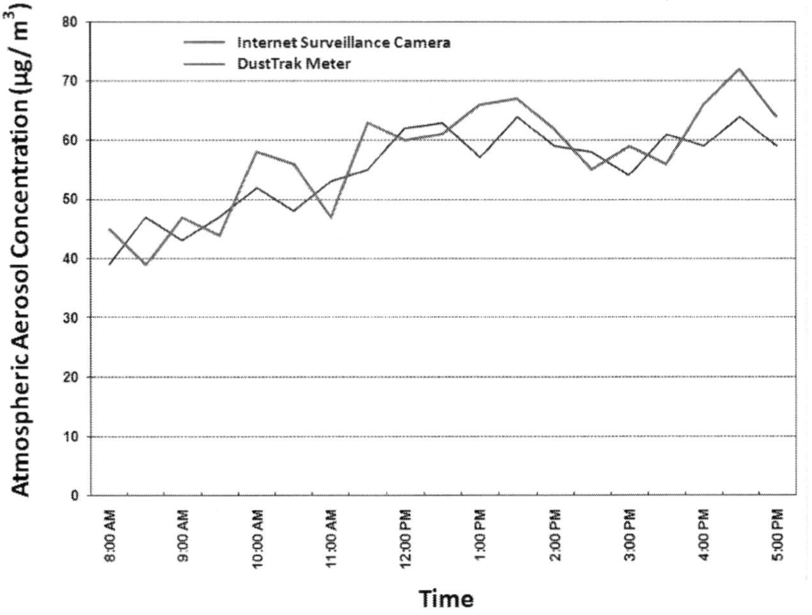

Fig. 9. Graph of atmospheric aerosol concentration concentration versus Time (21 Jul 2008)

5. Conclusion

This study has shown that by using image processing technique with new developed algorithm, internet surveillance camera can be used as temporal air quality remote monitoring sensor. It produced real time air quality information with high accuracies. This technique uses relatively inexpensive equipment and it is easy to operate compared to other air pollution monitoring instruments. This showed that the internet surveillance camera imagery gives an alternative way to overcome the difficulty of obtaining satellite image in the equatorial region and provides real time air quality information.

Acknowledgements

This project was supported by the Ministry of Science, Technology and Innovation of Malaysia under Grant 01-01-05-SF0139 "Development of Image Processing Technique via Wireless Internet for Continuous Air Quality Monitoring", and also supported by the Universiti Sains Malaysia under short term grant "Membangunkan Algorithma Untuk Pengesanan Pencemaran Udara Melalui Rangkaian Internet". We would like to thank the technical staff who participated in this project. Thanks are also extended to USM for support and encouragement.

6. References

Banauch, G.I.; Hall, C.; Weiden, M.; Cohen, H.W.; Aldrich, T.K.; Christodoulou, V.; Arcentales, N.; Kelly, K.J. & Prezant, D.J. (2006). Pulmonary function after exposure to the World Trade Center collapse in the New York Fire department. Am Respir Crit Care Med, Vol. 174, No. 3, 1 Aug 2006, Pages 312-19, PMID: 16864714

Brunekreef, B.; & Holgate, S.T. (2002). Air pollution and health. Lancet, Vol. 360, No. 9341, 19 Oct 2002, Pages 1233-42, PMID: 12401268.

Camagni, P. & Sandroni, S. (1983). Optical Remote sensing of air pollution, Joint Research Centre, Ispra, Italy, Elsevier Science Publishing Company Inc.

Charlson, R.J.; Horvath, H. & Pueschel, R.F. (1967). The direct measurement of atmospheric light scattering coefficient for studies of visibility and pollution. Atmospheric Environment (1967). Vol. 1, No. 4, July 1967, Pages 469-478, doi:10.1016/0004-6981(67)90062-5.

Diederen, H.S.M.A.; Guicherit, R. & HolLonder, J.C.T. (1985). Visibility reduction by air pollution in The Netherlands. Atmospheric Environment (1967). Vol. 19, No. 2, 1985, Pages 377-383, doi:10.1016/0004-6981(85)90105-2.

Donaldson, K.; Gilmour, M.I. & MacNee, W. (2000). Asthma and PM10. Respiratory Research. Vol. 1, No. 1, 3 July 2000, Pages 12–15, ISSN 1465-9921.

Fukushima, H.; Toratani, M.; Yamamiya, S. & Mitomi, Y. (2000). Atmospheric correction algorithm for ADEOS/OCTS acean color data: performance comparison based on ship and buoy measurements. Adv. Space Res, Vol. 25, No. 5, 1015-1024.

Hadjimitsis, D.G. (2009). Aerosol optical thickness (AOT) retrieval over land using satellite image-based algorithm. Air Qual., Atmos. Health. Vol. 2, No. 2, 25 March 2009., Pages 89-97, ISSN 1873-9318.

Hadjimitsis, D.G. (2008). Description of a new method for retrieving the aerosol optical thickness from satellite remotely sensed imagery using the maximum contrast value principle and the darkest pixel approach. Trans. GIS J. Vol. 12, No. 5, Oct 2008, Pages 633–644. doi:10.1111/j.1467-9671. 2008.01121.x.

Horvath, H. & Noll, K.E. (1969). The relationship between atmospheric light scattering coefficient and visibility. Atmospheric Environment (1967). Vol. 3, No. 5, Sept 1969, Pages 543-550, doi:10.1016/0004-6981(69)90044-4.

Kaufman, Y.J. & Fraser, R.S. (1983). Light extinction by aerosols during summer air pollution. J. of Climate & Appl. Meteorol. Vol. 22, No. 10, Oct 1983, Pages 1694–1706. doi:10.1175/1520-0450(1983)022<1694:LEBADS>2.0.CO;2

Kim Oanh, N.T.; Upadhyay, N.; Zhuang, Y.H.; Hao, Z.P.; Murthy, D.V.S.; Lestari, P.; Villarin, J.T.; Chengchua, K.; Co, H.X.; Dung, N.T. & Lindgren, E.S. (2006). Particulate air pollution in six Asian cities: Spatial and temporal distributions, and associated sources. Atmospheric Environment, Vol. 40, No. 18, June 2006, Pages 3367-3380, ISSN 1352-2310.

King, M. D.; Kaufman, Y. J.; Tanre, D. & Nakajima, T. (1999). Remote sensing of tropospheric aerosold form space: past, present and future, Bulletin of the American Meteorological society, 2229-2259.

Lim, H.S.; MatJafri, M.Z.; Abdullah, K; Wong, C.J. & Mohd. Saleh, N. (2009). Aerosol Optical Thickness Data Retrieval Over Penang Island, Malaysia, Proceeding of the 2009 IEEE Aerospace Conference, pp. 1-6, ISBN: 978-1-4244-2621-8, 7-14 March 2009, IEEE International, Big Sky, MT, USA.

Liu, C. H.; Chen, A. J. & Liu, G. R. (1996). An image-based retrieval algorithm of aerosol characteristics and surface reflectance for satellite images, International Journal Of Remote Sensing, 17 (17), 3477-3500.

Middleton, W.E.K. (1968). Vision Through the Atmosphere, University of Toronto Press, Toronto.

Noll, K.E.; Mueller, P.K. & Imada, M. (1968). Visibility and aerosol concentration in urban air. Atmospheric Environment (1967). Vol. 2, No. 5, Sept 1968, Pages 465-475, doi:10.1016/0004-6981(68)90040-1.

Pope, C.A. III; Renlund, D.G.; Kfoury, A.G.; May, H.T.; Horne, B.D. (2008). Relation of Heart Failure Hospitalization to Exposure to Fine Particulate Air Pollution The American Journal of Cardiology, Vol. 102, No. 9, 1 Nov 2008, Pages 1230-1234.

Pope, C.A. III. (2007). Mortality effects of longer term exposures to fine particulate air pollution: review of recent epidemiological evidence. Inhalation Toxicology, Vol. 19, No. 1, Suppl. 1, Pages 33–38, PMID: 17886048, ISSN 0895-8378.

Pope, C.A. III; Burnett, R.T.; Thurston, G.D.; Thun, M.J.; Calle, E.E.; Krewski, D. & Godleski, J.J. (2004). Cardiovascular mortality and long-term exposure to particulate air pollution: epidemiological evidence of general pathophysiological pathways of disease. Circulation Vol.109, No. 1, 6 Jan 2004, Pages 71–77, PMID: 14676145.

Pope, C.A. III; Dockery, D.W. & Schwartz, J. (1995). Review of epidemiological evidence of health effects of particulate air pollution. Inhalation Toxicol. Vol. 7, No. 1, Pages 1–18.

Sifakis, N. & Deschamps, P.Y. (1992). Mapping of air pollution using spot satellite data. Photogramm Eng Remote Sensing. Vol. 58, No. 10, Oct 1992, Pages 1433-1437, ISSN 0099-1112.

Sifakis, N.; Iossifidis, C. & Sarigiannis, D. (2006). High resolution 3D-mapping of urban air pollution using EO data. Proceeding of the 25th EARSeL Symposium, 6-11 June 2005, Porto Portugal, in: Global Developments in Environmental Earth Observation from Space (2006), André Marçal (ed.), pp 673-681.

Vermote, E. ; Tanre, D. ; Deuze, J. L. ; Herman, M. & Morcrette, J. J. (1997). 6S user guide Version 2, Second Simulation of the satellite signal in the solar spectrum (6S), [Talian laman web], [02/08/2005]. http://www.geog.tamu.edu/klein/geog661/handouts/6s/6smanv2.0_P1.pdf.

Wong, C.J.; MatJafri, M.Z.; Abdullah, K. & Lim, H.S. (2009a). Determination of Aerosol Concentration using an Internet Protocol Camera, Proceeding of the 2009 IEEE Aerospace Conference, pp. 1-7, ISBN: 978-1-4244-2621-8, 7-14 March 2009, IEEE International, Big Sky, MT, USA.

Wong, C.J.; MatJafri, M.Z.; Abdullah, K.; Lim, H.S. & Low, K.L. (2007). Temporal air quality monitoring using surveillance camera, Proceeding of the Geoscience and Remote Sensing Symposium, 2007. IGARSS 2007, pp. 2864 - 2868, ISBN: 978-1-4244-1211-2, Centre de Convencions Internacional de Barcelona, 23-27 July 2007, IEEE International, Barcelona, Spain.

3

Interferometric Imaging Technology for Microwave Radiometers

Ji Wu, Hao Liu, Jingye Yan, Cheng Zhang and Weiying Sun
Center for Space Science and Applied Research,Chinese Academy of Scienses
China

1. Introduction

Since 1957, when the first man made satellite opened the space age of human history, satellite earth observation has been taking advantage of high altitude to look at the globe and have applications in the areas of weather forecasting, oceanography, land survey and resource discovery, environmental and disaster monitoring, as well as applications in the area of defence. The observational frequency bands have coverage in microwave, through infrared and visible light, up to ultraviolet.

Among all frequency bands, microwave has special characteristics. Besides its all weather and all time ability, it also has the ability to penetrate the surface canopy and even the ground while the frequency is approaching the low end. It is also capable and easy to realize polarized measurement which increases the ability to identify the internal information of the observed targets. The spatial resolution for synthetic aperture radar has already reached the level in the order of meters or decimetres, comparable to optical sensors.

However, the active microwave sensors, such as synthetic aperture radar, have disadvantage in its high power consumption due to high transmission power and heavy mass which require many resources from spacecraft.

Another kind of microwave sensor is a passive sensor, also called a radiometer. It is basically a very high sensitivity noise receiver. By integrating the received noise from the observed scene over time, the characteristics of different scenes and targets, for example if the water content in clouds or soil changed, will be abstracted by the sensor through the variations of the integrated noise levels. The longer the integration is done, the more sensitivity the sensor would have. According to Planck's Law, all natural objects with temperature should have radio emissions independent of the Sun and any active radio illumination. This radio emission is called the brightness temperature and is related to both the physical temperature and the characteristics of the materials. The brightness temperature in the microwave bands are a very sensitive parameter to water, no matter whether it is in the air or in the soil. If different microwave band is applied, the brightness temperature is also sensitive enough to reflect the surface roughness that has a correlation length around the wavelength. Once the polarization information is acquired, more information can be abstracted from the emission such as ocean surface wind direction. Since the radiometer is a receiver only, it does not require high power. In fact, a radiometer takes much less power and mass resources from a spacecraft than that of active microwave sensors.

The passive microwave sensor also has disadvantages. Due to its nature as a passive sensor and the incoherent feature of the noise signal from microwave emission of the ground materials, the radiometer cannot apply the principle of synthetic aperture radar (SAR) to reach very high spatial resolution. Its spatial resolution is closely related to the physical size of the receiving antenna, i.e. the half power beam width of the antenna at the working frequency. For instance, if an L band (f_0 = 1.4 GHz) radiometer has a receiving antenna of 2 meters in diameter, and working in a low Earth orbit at an altitude of 800km, the spatial resolution on the surface of the ground is no better than 85km. For this reason, in spite of the many advantages of passive microwave remote sensing, microwave radiometers can only be used in the area of ocean, atmosphere and land soil moisture observations where no high spatial resolution is required.

The same difficulty had already been recognized in the area of astronomy where the location of stars with strong radio emission would likely be detected. From the 1950s to the 1960s, a technology called interferometric imaging had been developed (Thompson et al, 1986). The basic principle is to measure the scene indirectly, not in its spatial domain but in its spatial frequency domain, called the u-v plane. After the measurement, a Fourier transform is performed to get the original scene. The quantity which has been measured in the Fourier domain is called the Visibility Function by the astronomers. This name was carried on when this technology was introduced to the area of Earth observation in the 1980's (Ruf et al, 1988).

The measurements, or sampling, in the spatial frequency domain can be carried out with a basic tool of two coherent receiving channels with small element antennas and a complex correlator. The coherent receiving channels mean they use the same local oscillator. The complex correlator carries a multiplication function with the output signals from the two receivers and giving two outputs I and Q representing the real part and the imaginary part of the complex value of the multiplication output. The two small element antennas form a baseline in space. The length of it (the distance between the phase centres of the two antennas) represents the radius of a sampling point in the u-v plane, while the orientation of the baseline represents the polar angle of this sampling point in the u-v plane. With these two values, i.e. the module and polar angle, a sampling point is exclusively defined. In practice, one element can be used more than one time to take many samples while combining with other different elements and forming different baselines. In this way, the antenna array with small antenna elements can be much thinned. As astronomers did, a Y shaped two dimensional linear array can definitely represent a full 2-D array and there is no need to scan the antenna mechanically since field of view of this system is the element antenna pattern which already covers a wide area of interest.

Another attractive quality of the interferometric imaging technology is that the physical aperture of the thinned array of the interferometric imaging system can be reduced by half compared with traditional radiometers. A direct explanation of this is because the beam width of each of the grating-lobes of the interferometer is only 1/2 of a traditional two element array where the outputs of the two elements are added, not multiplied. Mathematical explanation will be given in the following section.

These are what interested Earth observation sensor engineers. A large thinned array can be folded during launch and redeployed in orbit and no mechanical scan during the observation is needed. Plus the characteristics of having only 1/2 of the traditional physical aperture, the interferometric imaging radiometer can reach a much higher spatial resolution than traditional radiometer technology can do. In order to avoid confusion with the already established phrase 'synthetic aperture' for active microwave radars, this technology is called interferometric imaging technology most of the time. Occasionally it is also called the interferometric synthetic aperture radiometer, or simply synthetic aperture radiometer.

In this chapter, we will start with the basic principles of this technology by talking about the spatial frequency property of an image. This approach is different from other documentation dealing with this topic and easy to understand by the readers with basic knowledge of Fourier transformation. In the third Section, we deal with the basic sampling technologies and its characteristics. In the fourth Section, the sampling technique is integrated into systems. Different sampling systems are introduced with emphases on time shared systems, particularly on the clock scan system which is our original contribution to this technology. In Section five, image reconstruction algorithms are introduced with the emphases on interpolation technology when the samples in the u-v plane are not on a rectangular grid. In Section six, some future applications of this new technology are introduced. All of them are associated with current real mission studies and some of them will certainly become real missions in the future.

2. Basic theorem

If we look down, what we see from space is the Earth. For middle or low resolution earth observation, the images that we get are mainly the scenes of the land or the oceans. In this Chapter, only gray scale image or fake colour image that represent brightness temperature levels are discussed.

The variations of the image's gray scale and texture represent the variations in spatial frequencies. Fast variations (with fine texture) correspond to high spatial frequencies, while gentle variations (with coarse texture) correspond to low spatial frequencies. The directionality of the texture variation represents the two-dimensional property of the spatial frequency. The contrast of the texture represents the intensity of the corresponding spatial frequency. The Fourier transform of a two-dimensional image is its spatial frequency spectrum, as illustrated in Fig. 1, in which the coordinates of the original spatial image are represented by (x-y), while the coordinate of the spatial frequency spectrum are represented by (u-v). The spatial frequency spectrum is also known as spatial frequency domain image, or u-v domain image.

The spatial frequency domain image is centred at the origin, and each point corresponds to a spatial frequency of the original image. For a vector from origin to a point in spatial frequency domain, the length represents the value of the spatial frequency, i.e. the spatial frequency becomes lower when the vector moves toward the origin, which can be defined as the low frequency component; and conversely, the spatial frequency becomes higher when the vector moves away from the origin, which can be defined as the high frequency component. As illustrated in Fig.2, the angle between the vector and v-axis in the spatial frequency domain corresponds to the angle between the image texture and the x-axis in the spatial domain, while the amplitude at the vector point in the spatial frequency domain represents the amplitude of the corresponded image texture in spatial domain. So, each point in spatial frequency domain contains its own practical physical meaning, and the information it takes covers all over the image in spatial domain, not only to one single point.

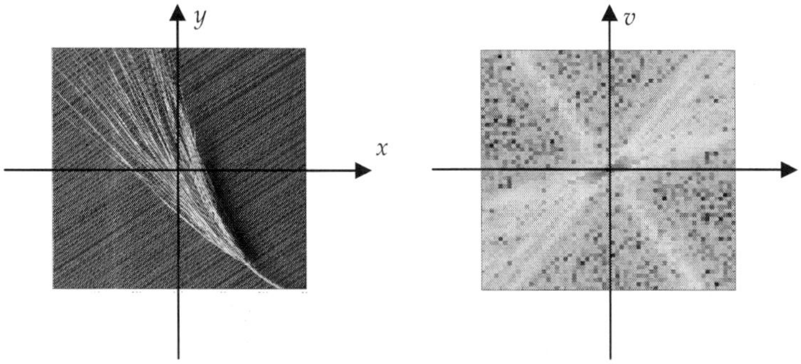

(a) (b)

Fig. 1. Original Spatial image (a) and its spatial frequency spectrum image (b)

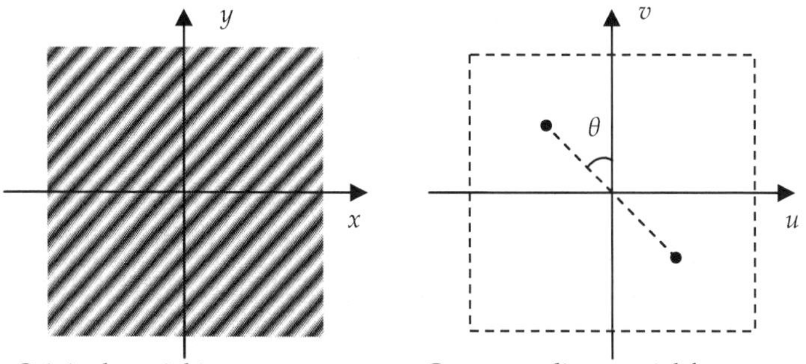

Original spatial image texture Corresponding spatial frequency

Fig. 2. Image texture and its spatial frequency

The basic principle of the interferometric passive imaging technology is to conduct measurement in spatial frequency domain, and then retrieve the original image in spatial domain by Fourier transform.

As same as the direct measurement on the original image in spatial domain, the spatial resolution of any measurement system is limited. An interferometric passive imaging system can get only limited samples in spatial frequency domain. The maximum u-v value sampled is the highest spatial frequency which corresponds to the highest spatial resolution of the retrieved image. According to the sampling principle, the least sampling spacing can also be determined.

According on the Fourier transform theory, the spatial frequency spectrum has the following properties:

1. Conjugate symmetry: $F(u,v) = F^*(-u,-v)$
2. Periodicity: $F(u,v) = F(u+M,v) = F(u,v+N) = F(u+M,v+N)$

The complex conjugate symmetry property implies that, when we conduct sampling in the spatial frequency domain, only half of the frequency spectrum in u-v plane is needed, and another half can be achieved mathematically by conjugate symmetry. In other words, once we sample a point in spatial frequency domain, we simultaneously get another sample at the conjugate symmetric point. By this method, the sampling efficiency can be significantly improved.

The conjugate symmetry property also explains why the interferometric imaging system has only 1/2 of the physical aperture than traditional ones but still get the same spatial resolution. This is because, due to the symmetry, the areas sampling points are doubled in the spatial frequency domain, the u-v plane. Once the Fourier transform is done, one gets the same number of points in the spatial domain, which means the retrieved image also gets double points. Put in simplified terms, this means measure one half, but get one. The size of the physical aperture in the interferometric sampling system is naturally reduced by half.

The periodicity property implies that, periodical extension, or namely aliasing, will appear after the Fourier transform in the retrieved image. Therefore, we must pay attention to the concerned image region, in which aliasing should not be included. In addition to selecting proper sample spacing, element antenna patterns can also be used to weigh the observation field of view, and then restrict the unwanted aliasing image.

3. Sampling measurement in spatial frequency domain

The purpose of sampling measurement in the spatial frequency domain is to get every point on a sampling grid in the spatial frequency domain, except the ones that can be obtained by complex conjugate symmetry. Since a single point in the spatial frequency domain corresponds to a specific image texture, as shown in Fig. 2, it can be measured by a two-element array which has a fringe-shaped beam. This array is also called a two-element interferometer, whose block diagram and fringe-shape pattern are illustrated in Fig. 3.

In the figure, the coherent receiving channels are simplified to two antennas only. The function of the complex correlator is to conduct complex multiplication of the signals received from the two antennas, with their original phases. The pattern of this interferometer is called the fringe function. If the length of the baseline is increased, the number of the grating-lobes of the Fringe Function increases too, and the width of each lobe decreases.

Since different spatial frequencies correspond to different image textures, the interferometers with different baseline lengths can measure the image texture characteristics corresponding to different Fringe Functions. When the baseline's length is increasing, the number of grating-lobes and the spatial frequency are both increasing. In other words, the long baseline interferometer can measure the high frequency component in the spatial frequency domain of the original image, while the short baseline interferometer can measure the low frequency component. So, by utilizing the combination of all the interferometers with different baselines' lengths and directions, the full sampling coverage on the spatial frequency domain of an original image can be achieved.

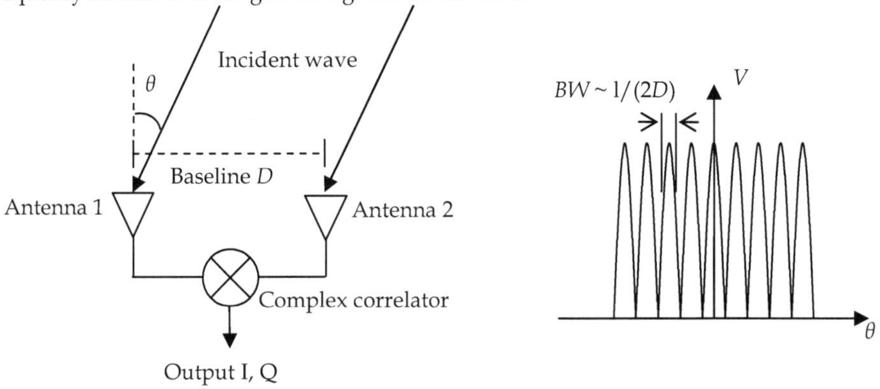

(a) Block diagram of an interferometer (b) Pattern of an interferometer

Fig. 3. Schematic diagram of two-element interferometer (a) and its pattern (b)

It should be mentioned here that the beam width of a lobe of the fringe function is 1/2D, which is half of the traditional two element array. This enables the interferometric imaging radiometer to use only 1/2 of the physical array aperture, i.e. the longest baseline in the system, to get the same spatial resolution that can be reached by the traditional radiometer.

Further theoretical analysis on the interferometric measurement can be applied. The ideal measurement output of a two-element interferometer is expressed as (Corbella et al, 2004),

$$V(u,v)=\iint \frac{T_B(\xi,\eta)}{\sqrt{1-\xi^2-\eta^2}}\exp(-j2\pi(u\xi+v\eta))dudv \tag{1}$$

in which $V(u, v)$ is the so-called visibility function, $(u, v) = (x_1-x_2, y_1-y_2)/\lambda$ is the baseline vector normalized by the wavelength, $(\xi, \eta)=(\sin\theta\cos\varphi, \sin\theta\sin\varphi)$ is the direction cosine of the incident wave, $T_B(\xi, \eta)$ is the brightness temperature distribution of the scene and the targets. The Fourier relationship between the visibility functions and the brightness temperature distribution is obvious. Each baseline corresponds to two spatial frequency sampling positions in the u-v plane, i.e. (u, v) and $(-u, -v)$. The brightness temperature distribution of the scene and the targets can be retrieved by Fourier transform after a full sampling measurement in the spatial frequency domain.

Only some discrete points in the spatial frequency domain can be sampled by interferometric measurement. According to Fourier transform principle, the maximum sample spacing $u_{\max}$ (longest baseline) determines the spatial resolution of the retrieved image, while the lowest sample spacing Δu (shortest baseline) determines the aliasing-free field of view (AF-FOV), as illustrated in Fig. 4.

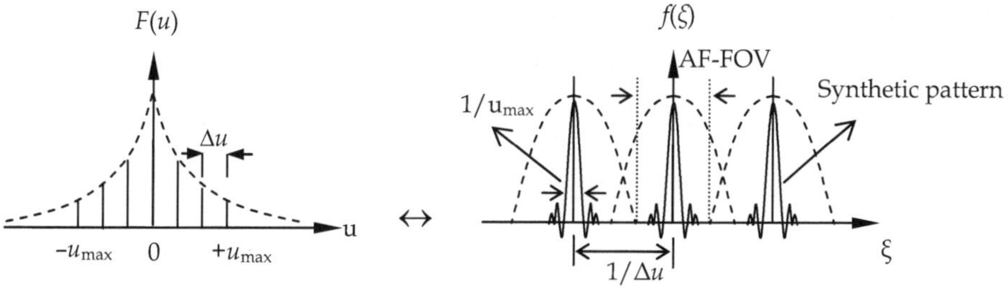

(a) Visibility function samples (b) Periodical extension of the retrieved image

Fig. 4. Visibility function sampling and image aliasing

In order to realize spatial resolution $\Delta\xi$, and also ensure no aliasing in the $2\xi_{max}$ region, the longest and shortest baselines of the interferometric measurement should satisfy the following requirements,

$$u_{\max} > \frac{1}{\Delta\xi}, \qquad \Delta u < \frac{1}{2\xi_{\max}} \tag{2}$$

Limited sampling coverage is equivalent to a sharp cut-off in the spatial frequency domain, which can cause the so-called Gibb's phenomenon. In general, the Gibb's phenomenon can be mitigated by a windowing operation on visibility functions. Some specific window functions can be applied on the measured visibility function samples, which can reduce the side-lobe level of the synthesized pattern, but also broaden the main beam and reduce the spatial resolution of the image. Some commonly used window functions include: Hanning, Hamming, Blackman, Kaiser and some other functions (Harris, 1978 and Anterrieu et al, 2002). Different window functions can achieve a different balance between the side-lobe level and main beam width. Another purpose of introducing window functions is to realize multi-resolution observation with the same observational dataset: to retrieve images with different spatial resolutions according to different application requirements (Ribo, 2003). Taking MIRAS of the SMOS mission as an example, the spatial resolution for land applications and ocean applications are different.

4. Passive interferometric imaging system

Array configuration plays a key role in an interferometric imaging system. One of the major goals of the interferometric imaging radiometer design is to take as much coverage of the u-v plane as possible while at the same time to use as few element antennas as possible. Since the observation target is located in the far field of the antenna array, the correlation output of any pair of antenna elements is equivalent if they compose the same baseline (both in length and direction) no matter where they are locally placed. This means that the correlation output is independent of the location from which the interferometer takes the visibility samples. To acquire the entire coverage of the harmonics in the spatial frequency domain, only one visibility is necessary for each harmonic. Therefore, the designer of an interferometric imaging radiometer system always tries to reduce the redundant baselines and use as few element antennas as possible. This is also one of the major jobs for the design of the system.

4.1 One dimensional push-broom imaging system

A one dimensional push-broom is a combined design of both traditional radiometers and interferometric imaging radiometers. In other words, the interferometric imaging technology is only applied in the cross track direction, the resolution in the along track direction is still obtained by traditional real aperture technology. Therefore, the element antennas of the array should be a fan beam with a broad beam in the cross track direction and narrow beam in the along track direction. For this kind of beam, the most common design is to have a slotted waveguide as the stick element antenna at the commonly used microwave bands. The waveguides are placed in the along track direction, as see in Fig. 5. In the cross track direction, the waveguides are arranged according to the design principle of interferometric technology which will be discussed below. This kind of imaging system is usually called 1 D push-broom imaging system. In this case, the spatial frequency domain reduced also from 2D to 1D, i.e. the u-v plane became u-axis only.

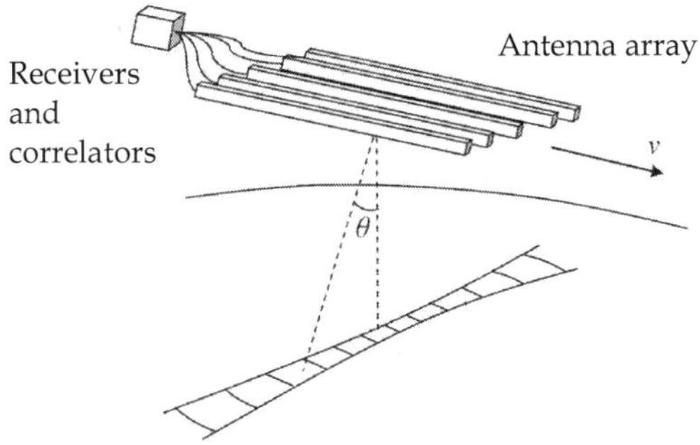

Fig. 5. One dimensional push-broom imaging system

As it is discussed in the previous sections, the ultimate objective of applying interferometric technology is to reduce the number of array elements. It is also known that the sampling of each special frequency component need not be duplicated, although there is always some redundancy in practice. An array configuration design example of a one dimensional push-broom imaging system is shown in Fig. 6. There one can see, with a 5 element antenna placed in a special way, all frequency components from 1 Δu up to 9 Δu are all sampled by taking different combinations of the element antennas. The dashed lines show the missing elements from a full array if a conventional array design was applied. You can also fund that there are redundant baselines, such as between elements 2 and 3, the baseline is also 3

Δu, the same that is shown between elements 3 and 4. However, the redundant baselines cannot be further eliminated, we can take advantage of having them by averaging the outputs given by the redundant baselines. In this way the sensitivity of the system will be increased.

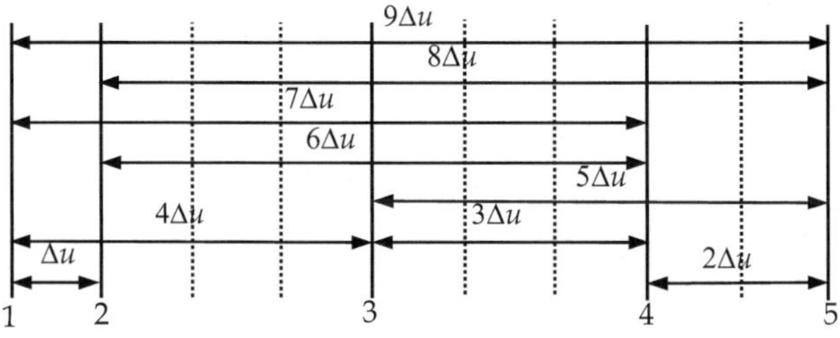

Fig. 6. 5 antenna elements comprise 9 baselines

In general, for a *N* elements antenna system, if we take the combination of any two of them, there are *N(N-1)/2* different ways. If those baselines are all independent from each other, i.e. if they cover from $1\,\Delta u$ to $N(N-1)/2\,\Delta u$ in the 1-D case, this system is called a zero redundancy array system. Unfortunately zero redundancy exists only if the number of antenna's elements is less than 4 for 1-D case, as shown in Fig.7.

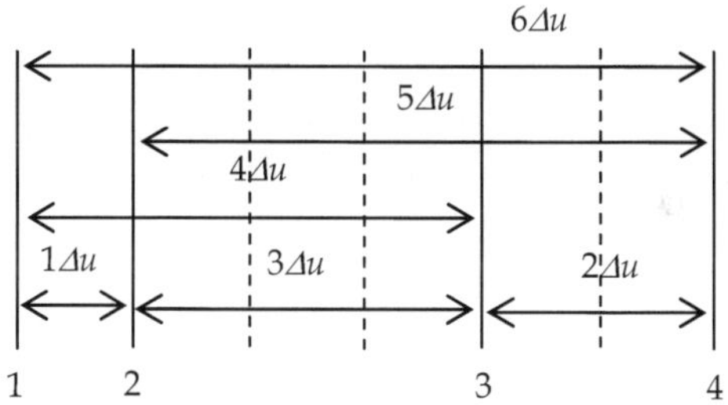

Fig. 7. A zero redundant thinned array with 4 element antennas

When there are more than 4 antenna elements, an optimization process is necessary in order to find the best design that uses as few elements as possible. The optimization process becomes complicated when the spatial resolution gets better thus the size of the array increases. A mathematic algorithm to compute the minimum redundancy array for any size has so far not been found (Ishiguro, 1980). An effective tool to search the minimum redundancy array is with the aid of a computer. Non-linear methods such as the simulated annealing algorithm and genetic algorithm have been developed to optimize the array configuration. Optimum configuration of element antenna positions with an array element $N \leqslant 30$ are given by (Ruf, 1993), which can be applied in different applications.

The first ever 1 D interferometric imaging system for Earth observation is the Electronically Scanned Thinned Array Radiometer - ESTAR. ESTAR was developed as part of the cooperative research between the NASA/Goddard Space Flight Center and the University of Massachusetts at Amherst (Le Vine, 1994, 2004). ESTAR is an L-band (1.4GHz) instrument. Five antenna elements comprise 7 baselines and result in an equivalent half power beam width (HPBW) of 7°. ESTAR flew on the NASA P-3B Orion aircraft in the 90's to demonstrate the imaging principles. Lots of measurement data were obtained for retrieval of soil moisture (Le Vine et al, 2001; Guha et al, 2003) and ocean salinity (Le Vine et al, 2000a). Results are consistent with values of soil moisture observed in-situ at the observed sites with previous measurements in the same area. ESTAR demonstrated that the passive interferometric technique is feasible. A follow-up project to ESTAR, called HydroSTAR, was proposed by NASA in 1998. It would have been a space borne 1-D interferometric imager for the Earth System Science Pathfinder (ESSP) mission (Le Vine, 1999). HydroSTAR was later cancelled due to technical risk and other none technical reasons.

One dimensional interferometric radiometers at higher frequency bands were built at the Center for Space Science and Applied Research (CSSAR), Chinese Academy of Sciences (CAS). Push-broom type C-band and X-band radiometers were completed in 2001 and 2004, respectively (Dong, 2000; Liu, 2004; Wu, 2005a), as shown in Fig.8. Flight tests were conducted and the brightness temperature images were acquired. The C-band interferometric radiometer works at 6.6GHz. Its 6 antenna elements form 10 baselines (zero baseline was not taken into account). The HPBW of the C-band radiometer is 4°. The X-band imager works at 9.4GHz with 8 antenna elements and 19 baselines resulting in a 2° HPBW.

(a) (b)

Fig. 8. One dimensional interferometric radiometers built by CSSAR

Another X-band interferometric radiometer, the X-Band Lightweight Rainfall Radiometer (LRR-X) is an airborne microwave sensor that is developed for the NASA Earth Science Technology Office by Goddard Space Flight Center and the University of Michigan (Ruf & Principe, 2002, 2003). LRR-X is intended to address several pressing issues related to the Global Precipitation Measurement (GPM) Mission. It is a science and technology testbed instrument. LRR-X is also a push-broom imager operating at 10.7 GHz with a ± 45 cross track field of view and a nominal 1.5° angular resolution at its nadir.

4.2 Two dimensional snapshot interferometric radiometer

One dimensional interferometric radiometer uses thinned array only in the cross track direction. Thus the dimension in along track direction is still large, as it can be seen from Fig.8. In order to further reduce the physical size of the antenna, two-dimensional interferometric radiometer has been developed. As it is discussed in sections 2 and 3, 2-D interferometric radiometer takes measurement in the full u-v plane. Therefore, it will have thinned antenna array and 1/2 apertures in both directions. If the sampling baseline can cover all angles and lengths within the upper limit that can be provided by the physical aperture, the image will be taken like a camera, or called snapshot imaging.

To provide full coverage with samplings distributed in grid in the spatial frequency domain, many array configurations are employable, such as T-shape, U-shape, Y-shape, Δ-shape and so on. Among those, T and U shaped arrays produce visibility function samples in a rectangular grid in the u-v plane, while the Y and Δ-shape arrays produce the visibility function samples in a triangular grid. Classical fast Fourier transform in rectangular grid is applicable for retrieving the brightness temperature image in the spatial domain. Hexagonal fast Fourier transform is also applicable in a triangular grid. Y-shape array has been adopted by many systems due to the unique advantage that the array thin factor is the best among all the 2-D configurations. However, none of those configurations can reach zero redundancy. Redundant baselines, particularly at the lower spatial frequency components, are used to average the noise, which is equivalent to an increase in the integration time.

2D-STAR was developed under NASA's Instrument Incubator Program (IIP) at the Goddard Space Flight Center. As the next generation instrument of ESTAR, 2D-STAR is an experimental instrument for evaluating techniques and applications of 2D aperture synthesis (Le Vine et al, 2000b, 2004). 2D-STAR works at L-band with two polarizations and does aperture synthesis in both directions (cross track and along track). Micro-strip patch antenna is selected as the element antenna, and the minimum element spacing Δu is 0.5λ. A square shaped full array of 11×11 elements provide the flexibility to compare between different array configurations, including Y, T, U, etc. The instrument made its maiden flight in 2002 and participated in a series of field tests in 2003 and 2004 flying over research sites in Alabama, Georgia and Oklahoma during SMEX-03 and SMEX-04 campaigns, respectively (Le Vine et al, 2007).

HUT-2D was developed by the Helsinki University of Technology (HUT), Finland (Rautiainen et al, 1999, 2003). Development began in the late 1990s and was completed in early 2006. It is an airborne L-band 2D Interferometric imaging system with dual polarizations. Its U-shape array consists of 36 elements and 575 baselines. The minimum element spacing Δu is 0.7λ. The angular resolution is between 5° and 7° . HUT-2D is recognized as one of the demonstrators of the SMOS (Soil Moisture and Ocean Salinity) mission for the ESA (European Space Agency) to validate the feasibility of retrieving soil moisture and ocean salinity using 2D interferometric radiometers. Successful ground and air based tests were conducted to study the interferometric theory and instrument calibration. The first test flight was carried out on May 29, 2006 (Kainulainen et al, 2007).

Soil Moisture and Ocean Salinity (SMOS) is the first space mission for an interferometric imaging radiometer. The Microwave Imaging Radiometer using Aperture Synthesis (MIRAS) was proposed to the European Space Agency (ESA) in 1994 and approved in 1998 as the only payload onboard the SMOS mission (Martin-Neira et al, 1994; Kerr et al, 2000). MIRAS adopts a Y-shape 2D array and consists of a central structure and three deployable arms, each of which has three segments, see Fig.9.

Fig. 9. Y-shaped antenna array of MIRAS and SMOS spacecraft of ESA

During launch, these arms are folded-up, but soon after separation from the launch vehicle they are gently deployed via a system of spring-operated motors and speed regulators. There are 69 antenna elements - the so-called LICEF receivers, which are equally distributed over the three arms and the central structure at a minimum distance of 0.875λ. The length of each arm is about 4 meters. The scheduled launch date of SMOS was postponed until November 2009. From an altitude of 755 km in Sun synchronous orbit, the element antenna will view an area of almost 3000 km in diameter. However, due to the interferometric principle and the Y-shaped antenna array, the field of view is limited to a hexagonal shape about 1000 km across called the 'alias-free zone'. This area corresponds to observations where there is no ambiguity in the phase-difference. The ground resolution is better than 50km for soil moisture and is better than 200km for ocean salinity.

Geostationary Synthetic Thinned Aperture Radiometer (GeoSTAR) is another Y-shaped snapshot interferometric imaging radiometer supported by NASA's IIP program and developed by the Jet Propulsion Laboratory (JPL) (Lambridtsen et al, 2004, 2006). This microwave atmospheric sounder will be onboard the geostationary GOES weather satellites operated by NOAA and will be in orbit by 2014. GeoSTAR will use the same channels as AMSU-A/B at 50GHz and 183GHz for oxygen and water vapour sounding. In total, 768 elements at 183GHz will be equally placed in a Y-shape structure. A more complicated array is selected for the 50GHz channel, 3 F-shaped arms with two kinds of aperture sizes are employed to avoid the structure interference with 183GHz and to improve the effective aperture ratio. Distance between adjacent elements is 3.5λ, which results in a field of view (FOV) of 17.5° from the geostationary orbit. In order to measure global temperature and humidity distribution, GeoSTAR is intending to provide continuous observations with a spatial resolution of 25~30km. It is expected to be the first microwave sounder in geostationary orbit for atmospheric measurement. A Y-shape demonstrator with 24 elements has been completed and a series of experiments are being conducted. The demonstrator has 4 channels within 50~55GHz (Tanner et al, 2006).

4.3 2-D time-shared scanning imaging system

The 2-D snapshot imaging systems with high time resolution is very suitable for the real-time imaging observations, particularly for the case of observing from a fast moving platform at low earth orbit. However, for non-real-time observations, for example, when the platform is relatively stationary compared to observational targets and the targets' natural radiation is slowly changing, it would be too expensive to apply the traditional 2-D snap short imaging systems. Furthermore, for space borne applications, the overall cost with many element and channels, the resources required in power supply and mass of the 2-D snap shot systems are all difficult obstacles to overcome.

Recently, the time-shared scanning scheme with low hardware complexity has attracted more and more attention. It has potential to overcome the problems caused by the bulky and complex hardware of snapshot systems. Time-shared scanning scheme basically refer to applying a few antenna elements to compose a simple array, and integrally moving the array or separately moving the elements to get more spatial frequency samples. After a scanning cycle period, a full u-v sampling coverage can be achieved, and the brightness temperature image can be reconstructed by inversing these saved u-v measurement data. Time shared imaging schemes can greatly reduce the number of antenna elements and receivers. Thus the system cost could be cut down and the imaging spatial resolution could also be further increased since the overall complexity of the system has been dramatically reduced.

Depending on the means of movement of the thinned array, the time shared scanning scheme can be divided into sliding scan and rotating scan. Sliding scan means the antenna elements slide along some guide rails back and forth during the sample measurement process. The guide rails of every antenna elements are usually straight lines. For example, in the T-shaped and Y-shaped sliding scans, see Fig. 10, each antenna element is sliding along its respective guide rail at a different pace. After carefully designing the movement strategy of each element, all the baselines that needed to cover the full u-v plane will be aligned during a scanning cycle period, which have the same sampling effect as a uniformly distributed array along the rails.

The sliding scan has some disadvantages. It requires a complicated mechanical control system. The heavy mechanical attrition in the sliding process will greatly shorten the lifetime of the system. More importantly, it is hard to keep the system's momentum balanced. All these disadvantages limit its space borne application. Therefore, it is only fit for the ground-based imaging systems.

The other time-shared sampling scan, the rotation scan, is implemented by rotating all the coplanar antenna elements around a fixed axis to obtain more observation baseline vectors and more u-v sampling points. It has obvious advantages when compared to sliding scan in that it adopts the simpler and more reliable rotation control components, generates less kinetic friction with high system stability, and is easier to achieve rotational balance. It is then more appropriate for space borne applications.

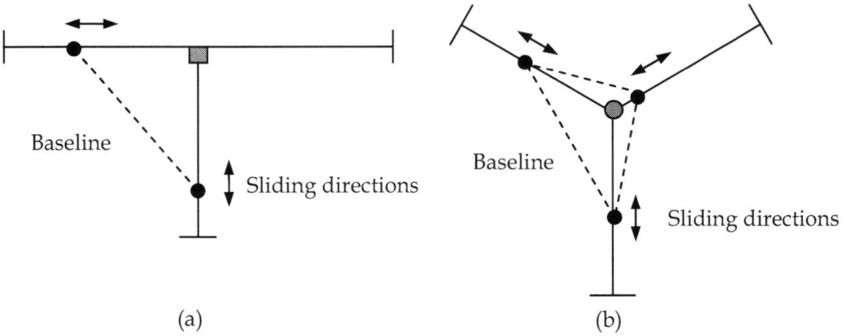

Fig. 10. Schematic diagrams of T-shaped sliding scan (a) and Y-shaped sliding scan (b)

There are generally two kinds of rotation scans, the synchronous rotation scan and the asynchronous rotation scan. The former one means the antenna array is integrally rotating around a fixed axis, and all the elements have the same angular rotation speed (Wu et al, 2005b), see Fig. 11. The latter one means the antenna elements are separately rotating around a same axis with different angular rotation speeds. In this type of rotational scan, the antenna elements are divided into two groups. The elements in one group have the same angular rotation speed and arm length, but different speeds to the elements in another group. This configuration is somewhat like a clock, especially when each group has only one antenna element, so it is also called a clock scan (Wu et al, 2007), see Fig. 12.

The sampling grid of synchronous rotation scan is composed by concentric circles, as shown in Fig. 11 (b). Each sampling circle corresponds to a physical baseline. An N elements array can form $N(N-1)/2$ physical baselines and scan out $N(N-1)/2$ sampling circles. The more uniform the concentric circles are, the better the image reconstruction is. Therefore, in order to get a rather uniform grid and reduce the system complexity at the same time, the antenna array with nonredundant linear distributed baselines is preferred. For linear arrays, there does not exist such an alignment with zero redundant baselines for N>4. For planar arrays, the array optimization is needed, and must resort to nonlinear iterative methods. Some good results of arbitrary planar array and circular array have been achieved by using simulated annealing method (Sun et al, 2005). Same as the traditional 2-D snapshot imaging system, the total number of antenna elements of the synchronous rotation scanning system is also determined by the required spatial resolution. Because its longest baseline, being about $N(N-1)/2 \cdot \Delta u$, is related to the number of antenna elements N, the system complexity of synchronous rotation scan can be reduced to about the square root of the 2-D snapshot system complexity, which is in the order of N^2.

The scanning tracks of the sampling point on the u-v plane of the clock scan scheme are some kind of spiralling curves, as shown in Fig. 12(b). The uniformity and the average gaps between the spirals are primarily determined by the speed ratio between the two antenna groups. As the speec ratio approaches 1, the more uniform the sampling grid and the longer the scanning cycle period would be. The largest and shortest baselines of a clock scan are respectively determined by the sum and difference of the lengths of the long arm and short arm. Any length of baselines between the shortest and largest baselines can

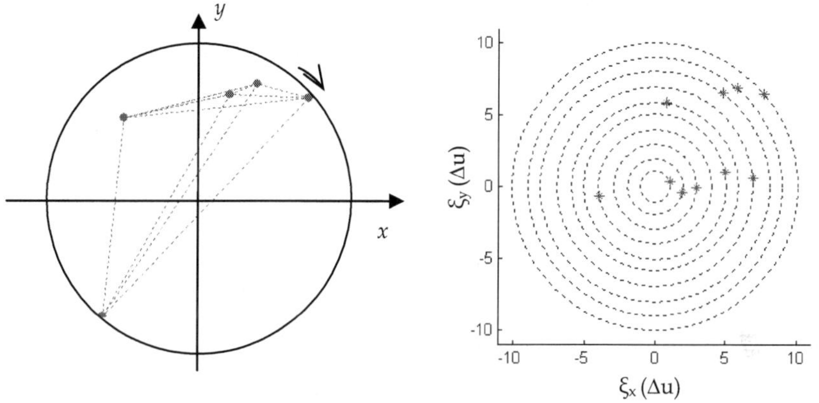

Fig. 11. Schematic diagrams of synchronous rotation scan scheme with 5-elements (a) and its sampling scanning trajectory (b)

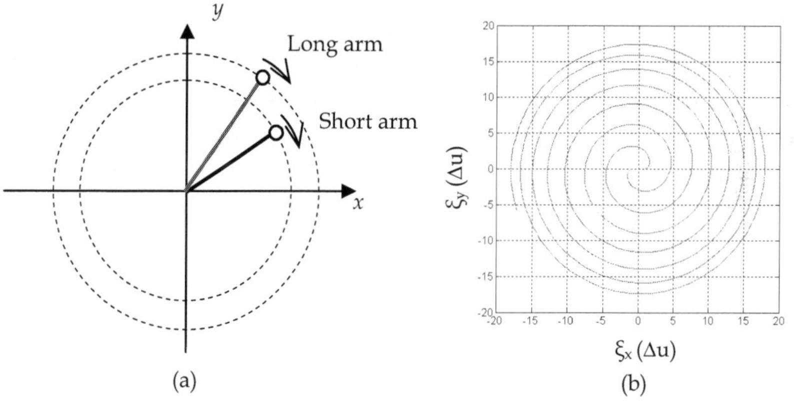

Fig. 12. Schematic diagrams of clock scan scheme with two elements (a) and its sampling scanning trajectory (b)

be created by the two antenna groups during their inconsistently rotating scan. Therefore, in some particular cases, the system complexity can be optimally reduced to the simplest limit of only two antenna elements and receivers. With the simplest two antenna elements configuration, the clock scan can still conveniently and efficiently achieve a full u-v sampling coverage. This is the most outstanding advantage of the clock scan. However, the scanning cycle period can be very long.

The number of arms of a clock scan system is determined by the image refresh time limit and system required brightness measurement sensitivity. Thus in order to increase the system time resolution and measurement sensitivity, the multi-arm configuration can be adopted, such as 2–2 arms, 4-4 arms and so on. A 4-4 arms system is shown in Fig. 13. The symmetric distributed multi-arm configurations are also helpful to keep the balance of the rotating system. According to the basic concepts of the clock scan, it can be extended to various specific transformations, such as step rotation scan, variable speed continuous rotation scan and swing rotation scan, etc.

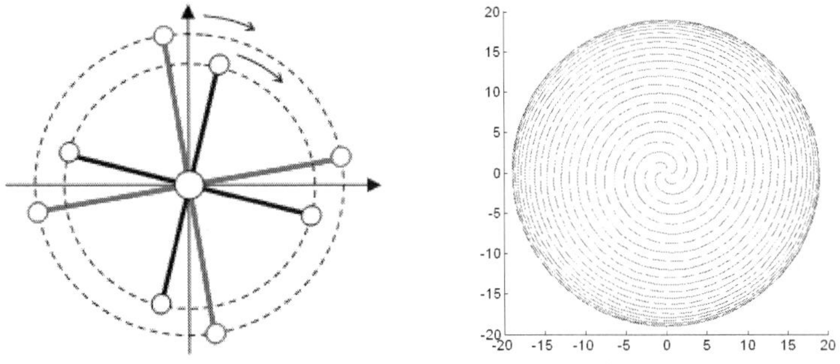

(a) (b)

Fig. 13. 4-4 arms clock scan system (a) and its sampling tracks on the u-v plane (b)

5. Brightness temperature retrieval methods

The visibilities or the samplings in the spatial frequency domain, measured by complex cross correlating between the signals received by two element antennas, are the primary measurements of synthetic aperture interferometric radiometers. They are related to the observed scene or target brightness temperature by a Fourier type transform. Accurate and fast inversion of the visibility data into the brightness temperature image is a critical step of the interferometric imaging technology. Due to different sampling schemes, there are still problems needing to be further explored.

In the ideal case, the visibilities and the brightness temperature are correlated by Fourier transform. However, in practice, the imperfections of the sampling process and various errors in receiving channels will influence the visibility measurement and damage the Fourier relationship. These system imperfections are mainly the channel imbalance and mutual coupling, antenna pattern distortion and mismatches, band width fringe washing effects and so on. All the distortions and amplitude/phase errors must be characterized and calibrated even though it is hard work. These calibration techniques include the distributed in-phase/out-phase noise injection technique applied in MIRAS system (Lemmetyinen et al, 2007) and I/Q vectors modulator technique applied in CAS-C/X system (Wu et al, 2004 and Liu et al, 2005b). After all the receiver channel errors have been corrected, the image reconstruction algorithms can be applied to the revised visibility data. In the following sub-sections, a few general brightness temperature retrieval approaches are introduced.

5.1 G-matrix inversion method

The G-matrix inversion method is a combination of system calibration and imaging reconstruction. It is very effective and convenient for small array systems, especially for 1-D imaging systems. It has been applied in the ESTAR system (Ruf et al, 1988 and Le Vine et al, 1994). The G-matrix inversion is based on numerical computation. Since the integral relation between the visibilities and brightness temperatures can be made discrete to a summation equation when considering that the brightness temperature distribution is represented by a sequence of discrete point sources. Then associate with the sampling theorem, the integral can be replaced by a vector product, and the set of visibility samples can be combined in matrix form,

$$V_{2N+1} = G_{(2N+1)\times M} \cdot T_M \tag{3}$$

where V_{2N+1} is the column vector of $2N+1$ visibility components including the zero-baseline and the real and imaginary part of other N baselines visibilities; T_M is the column vector of M discrete brightness temperature; G is the system modulation matrix including system errors. Each element of the G matrix corresponds to a measured spatial impulse response of each baseline correlation. The G matrix can also be recovered by deconvolving the measurements of known models of a brightness temperature scene. All the system imperfections and errors as well as the Fourier relations are embedded in the G matrix. Other calibration and revision works are not necessary any longer after the G matrix was measured.

The image reconstruction is a matter of inversing Equation (3). Generally the G matrix is not square, and the M brightness temperatures are as many as more than 3 times of the measured $2N+1$ visibility components. For this unconditioned problem, the inversion has to be computed by using a minimization algorithm in the least square sense. A direct way is to use the Moore-Penrose pseudo inverse method, which is expressed as

$$T = G^H (G \cdot G^H)^{-1} \cdot V \tag{4}$$

where H denotes the conjugate transpose. It is difficult to apply the G matrix method to a large 2-D array system, because the complexities of physically measuring the impulse response and mathematically inversing the G matrix are both increased with the square of the number of array elements. In addition, the drifts of receiver parameters require a periodic calibration. It is impractical to periodically refresh the large G matrix for space-borne applications.

5.2 Fourier Transform inversion method

When the receiver amplitude/phase errors can be calibrated and the system imperfection is small, the Fourier based inversion algorithms can be applied. The standard rectangular FFT can be directly applied to the rectangular sampling grid arrays, such as the U-shaped, T-shaped and X-shaped arrays. From signal theory it is known that the hexagonal sampling grid requires the minimum density of u-v samples to recover the image with a specified aliasing level. It has 13.4% less samples than rectangular sampling grid. Thus in order to process the visibilities sampled on hexagonal grids given by Y-shaped or triangular-shaped arrays, the hexagonal FFT algorithm has been developed (Camps et al, 1995 and 1997). Hexagonal FFT can use the standard rectangular FFT routines to quickly implement a reversible transform between the hexagonal grid in spatial domain and another hexagonal grid in spatial frequency domain without interpolation processes. It is noted that, the Y-shaped array is better than the triangular-shaped array because when having the same hardware complexity it has a larger sampling coverage and associated better spatial resolution. Therefore, as the optimal 2-D snapshot array configuration, the Y-shaped array was adopted by the MIRAS/SMOS system.

On the other hand, for the polar girds and spiral grids sampled by rotation scanning system, it seems no such fast direct algorithms exist to do Fourier transformations on such non rectangular grids. What one can do is to apply interpolation method to convert these irregular girds to a uniform grid that is proper for FFT routines and then use FFT to do the reconstruction. Such kind of methods are called interpolation based Fourier methods. There are mainly two interpolation based Fourier methods that have been proposed so far, the rectangular grid based Gridding method (Beatty et al, 2005) and the pseudo-polar grid based 1-D interpolation pseudo-polar FFT method (Zhang et al, 2007).

The Gridding method is a convolution based resampling technique which is widely used to convert the random non-uniform data to a rectangular grid. The procedure is that, firstly estimating the sampling density of the non-uniform data and compensating the data by dividing their density; then convolving the compensated data with a specific kernel function to recover the corresponding initial function and resample it on the required Cartesian grid, finally performing the standard IFFT to reconstruct the brightness temperature image. The density estimation and convolution kernel are the most important factors that are responsible for the reconstruction accuracy. Generally the Voronoi diagram method can be used to calculate the sampling density and the widely accepted Kaiser-Bessel kernel function can be used to do the convolution.

The interpolation pseudo-polar FFT method is specifically proposed for the polar grid of concentric circles sampled by synchronous rotation scanning system. Using two steps of 1-D interpolations, the angular interpolation and radial interpolation, the polar grid can be converted to a pseudo-polar grid, which is composed by concentric rectangles and equi-sloped rays, as shown in Fig. 14. Then applying 1-D FFT and fast FRFT (Fractional Fourier Transform), the spatial frequency data in pseudo-polar grid can be transformed to a Cartesian grid in the spatial domain. The interpolation pseudo-polar FFT method has a promising imaging performance by virtue of high accuracy of 1-D interpolations and fast computational operations.

5.3 Non linear iterative optimization method

For the systems without accurate calibration or missing some baselines, especially when large antenna pattern errors and large sampling non-uniformity exist, the inversion of the visibility equation dose not have an analytical solution. In this case, iterative optimization techniques, such as the Clean method and maximum entropy method (MEM), can be used to deal with this kind of problems.

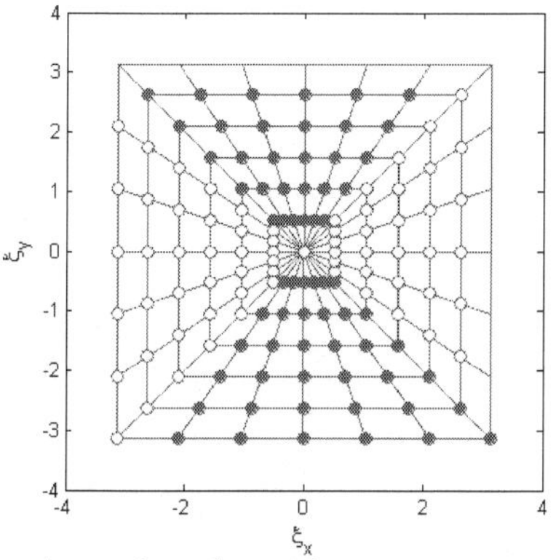

Fig. 14. Schematic diagram of pseudo-polar grid

The Clean method is the most popular method used in radio astronomy with interferometric imaging systems. Because of incomplete sampling in the u-v plane and atmosphere path distortion, the directly reconstructed image has large amount of noise. Consequently, the direct reconstruction is called dirty image and the equivalent array factor AF is called a dirty beam. The Clean method is to extract the useful information and remove the noise to obtain a high quality image. It first assumes the radio source is represented by a number of point sources, the stars, in an empty FOV, then it finds the position and strength of the most brilliant peak in the dirty image and subtracts the dirty beam centered at this point with a damping factor from the dirty image. This process is repeated until the residual image is no longer significantly above the noise-level. Finally the clean image is obtained by convolving the accumulated point sources model with an idealized clean beam, usually a Gaussian beam with the same half-power beam width as the dirty beam. From the principles it is clear that, the Clean method would have difficulties to directly applied for earth observation case because the earth targets appears as an extended thermal source filling the FOV and not a set of point sources. Although extensions of the Clean method for extended targets imaging are studied (Camps et al, 1998), regardless whether it is effective, it still requires much prior knowledge of the targets in the FOV.

The maximum entropy method (MEM) is another effective method applied to optimizing the synthetic aperture images in radio astronomy (Skilling & Bryan, 1984). It originated from maximum probability distribution in statistical physics, and particularly useful in solving the ill-posed problem with incomplete measurement data. The MEM tries to define a concept of entropy to characterize the solutions as well as prior knowledge and then obtains a solution which has the maximum entropy among all the solutions in a certain noise level. In order to successfully find the best solution being closest to the real one, enough constrains must be imposed to the infinite solution space, such as the positive constraint (all the elements are positive), the smoothness constraint (minimum variance of its second derivative) or other prior knowledge.

The Clean method and MEM method are both nonlinear, and consequently not easy to treat mathematically. It requires time consuming iterative procedures and large storage space.

Moreover, the prior knowledge of the observation scene is necessary even for extended targets. The practical applications of these nonlinear iterative methods for earth observations are not yet mature.

6. Application prospects

The interferometric passive microwave imaging system has a wide area of applications, including all the traditional microwave radiometer application areas and some other specific areas, such as high spatial resolution earth observation in geostationary Earth orbit or low frequency band, security detection of hidden weapons or other contrabands on the ground.

Geostationary earth orbit (GEO) microwave observation is an attractive potential application for interferometric imaging techniques. In order to implement accurate and timely monitoring of disastrous weathers such as typhoons, rainstorms or other severe convective weather events and increase weather forecasting precision, some prestigious government space agencies such as NASA and ESA are devoted in developing GEO meteorological satellite with microwave imaging system. China is also planning to launch a microwave imaging sounder on a GEO weather satellite. The main challenge of GEO applications is to realize a high spatial resolution at such a high orbit. One will meet great difficulty if they simply adopt the traditional real aperture microwave radiometer because the physical size of antenna will be too large even in millimeter band, not to mention how to scan it in order to get an image. Applying interferometric imaging technique is a promising way to overcome this problem. Especially if the scene becomes a slow moving/variable target, the time shared scan scheme, such as rotation scanning scheme, can be applied. This will further reduce the complexity of the system. Now the foresight study for rotation scanning interferometric millimeter atmospheric temperature imaging sounders for GEO satellite application have been carried out by NASA, ESA and also in China. The breakthroughs of this technology will help human beings understand the global weather system better and also giving more precise weather forecasts in the future.

Another important potential application of the interferometric imaging technique is the solar wind observation in solar polar orbit. The solar wind plays an essential role in shaping and stimulating planetary magnetospheres and ionic comet tails. It is a prime source of space weather. Coronal Mass Ejections (CMEs) produce the largest transient disturbances in the solar wind, which are closely related to solar activity, interplanetary shocks and geospace environment changes. With the development of Aerospace technology, precise forecasts of large disturbance events are being put on the agenda. Several solar-terrestrial space exploration projects are proposed in China, for example, the Solar Polar Orbit Radio Telescope (SPORT) project proposed by CSSAR/CAS (Center for Space Science and Applied Research, Chinese Academy of Sciences) (Wu et al, 2005c, 2006). In the SPORT mission, it is planned to launch a space-based observatory into solar polar orbit, and implement the first ever remote sensing measurements of the solar wind from the polar region above the sun outside of the ecliptic plane. The main objectives of SPORT are tracing the propagation of high density plasma clouds within about 1 square AU area near the ecliptic plane, imaging its intensity distribution and forecasting the propagation and intensity trend. Therefore, it is very helpful for geospace weather and environment forecast, and also for solar physics and plasma physics studies. The sensitive frequency of plasma clouds is very low. If using conventional real aperture radiometry technique, the required physical aperture of the antenna would be far exceed the mechanical production ability and rocket load capacity. Therefore, the interferometric imaging technique with rotational scanning scheme is adopted in SPORT project. It will apply the clock scan scheme with two sets of rotating arms. Each set of arms is composed of four long booms extended for 70-80 meters and 90 degree apart. The spacecraft is located in the rotational centre and will rotate together with one set of the arms. It is therefore a spin stabilized spacecraft. The rotation axis is pointing to the Sun all the time. More details can be read in (Wu et al, 2005c, 2006).

7. Conclusions

Interferometric imaging radiometer technology was introduced into the area of Earth observation from radio astronomy technology more than 20 years ago. Due to the differences between astronomy observation and that of the earth observation, the technology has gone through redevelopment over the past 20 years.

During redevelopment, there have been two main technical trends. The first one is to be more and more complicated by increasing the number of elements to reach higher spatial resolutions and at the same time to keep its radiometric sensitivity, such as the snapshot 2-D Y-shape designs. This technical trend has relied on the fast development of IC technology that enables us to integrate thousands of correlators into one chip. A representative of this technical approach is the MIRAS payload on SMOS mission by ESA. The other technical trend is trying to reduce the number of elements and using time shared scan scheme in order to cover the complete u-v plane. The initial driving force behind this trend is the necessity to keep the system design simple and manageable. Along this line, as described in this chapter, we have discovered the secrets inside the technology, i.e. the very basic configuration of the interferometric imaging system. It, in fact, can be represented by only two element antennas using the clock scan scheme. Any spatial resolution and radiometric sensitivity can be reached by using this two-element clock scan as basic building blocks in principle.

In November 2009, the first ever space mission SMOS using this new technology for earth observation will be launched. We expect a successful demonstration of this technology by SMOS mission. After this, in the near future, we expect to see more missions using this technology such as the geo-sounder on board of either a US weather satellite or a Chinese weather satellite since both are working on it. We would also expect the SPORT mission to be launched before 2020 using 4-4 element clock scan to take images of the interplanetary CMEs and also the universe background emission at 15MHz. It is certainly an exciting technology but perhaps not the dominating technology in the area of passive microwave remote sensing due to its complexity compared to tradition real aperture radiometers. Therefore, the traditional microwave radiometer will still exist. The users will select from both in order to keep the best performance over cost and technical feasibility.

8. References

Anterrieu, E., Waldteufel, P. & Lannes, A. (2002). Apodization functions for 2-D hexagonally sampled synthetic aperture imaging radiometers, *IEEE Transactions on Geoscience and Remote Sensing*, Vol.40, No.12, pp.2531-2542, ISSN: 0196-2892

Beatty, P.J., Nishimura, D.G. & Pauly, J.M. (2005). Rapid gridding reconstruction with a minimal oversampling ratio, *IEEE Transaction on Medical Imaging*, Vol.24, No.6, pp.799- 808, ISSN: 0278-0062

Camps, A., Bara, J., Corbella, I. & Torres, F. (1995). Visibility inversion algorithms over hexagonal sampling grids, *Proceedings of Soil Moisture and Ocean Salinity Measurements and Radiometer Techniques, ESA-ESTEC workshop,* pp.109–114, Noordwijk, The Netherlands, Apr. 1995, ESA document WPP87

Camps, A., Bara, J., Corbella, I. & Torres, F. (1997). The Processing of hexagonally sampled signals with standard rectangular techniques: application to aperture synthesis interferometer radiometer, *IEEE Transaction on Geoscience and Remote Sensing*, Vol.35, No.1, pp.183-190, ISSN: 0196-2892

Camps, A., Bara, J., Torres, F. & Corbella, I. (1998). Extension of the Clean technique to the microwave imaging of continuous thermal sources by means of aperture synthesis radiometers, *Journal of electromagnetic waves and applications,* Vol.12, No.3, pp.311-313, ISSN: 0920-5071

Corbella, I., Duffo, N., Vall-llossera, M., & Camps, A. (2004). The visibility function in interferometric aperture synthesis radiometry, *IEEE Transaction on Geoscience and Remote Sensing*, Vol.42, No.8, pp.1677-1682, ISSN: 0278-0062

Dong, X.L., Wu, J., Zhu, S.Y. et al. (2000). The design and implementation of CAS C-band interferometric synthetic aperture radiometer, *Proceedings of IGARSS'00,* Vol.2, pp.866-868, ISBN: 0-7803-8743-0, Honolulu, Hawaii, USA, July 2000, IEEE, New York

Guha, A., Jacobs, J.M., Jackson, T.J. et al. (2003). Soil moisture mapping using ESTAR under dry conditions from the southern great plains experiment (SGP99), *IEEE Transactions on Geoscience and Remote Sensing,* Vol.41, No.10, pp.2392-2397, ISSN: 0196-2892

Harris, F. J. (1978). On the use of windows for harmonic analysis with the discrete Fourier transform, *Proceedings of IEEE,* Vol.66, Issue.1, pp.51-83, ISSN: 0018-9219

Ishiguro, M. (1980). Minimum redundancy linear arrays for a large number of antennas, *Radio Science*, Vol.15, pp.1163-1170, ISSN: 0048-6604

Kainulainen, J., Rautiainen, K., Tauriainen, S., Auer, T., Kettunen, J. & Hallikainen, M. (2007). First 2-D interferometric radiometer imaging of the earth from an aircraft, *IEEE Geoscience and Remote Sensing Letters,* Vol. 4, No. 2, pp. 241-245, ISSN: 1545598X

Kerr, Y., Font, J., Waldteufel, P., Camps, A., Bara, J., Corbella, I., Torres, F., Duffo, N., Vallilossera, M. & Caudal, G. (2000). New radiometers: SMOS-a dual pol L-band 2D aperture synthesis radiometer, *IEEE Aerospace Conference Proceedings,* Vol.5, pp.119-128, ISBN: 0-7803-5846-5, Big Sky, MT, USA, Mar. 2000, IEEE New York

Lambrigtsen, B., Wilson, W., Tanner, A., Gaier, T., Ruf, C. & Piepmeier, J. (2004). GeoSTAR - a microwave sounder for geostationary satellites, *Proceedings IGARSS '04,* Vol. 2, pp. 777-780, ISBN: 0-7803-8742-2, Anchorage, AlasKa, USA, Sept. 2004, IEEE New York

Lambrigtsen, B. (2006). GeoSTAR: Developing a new payload for GOES satellites, *Proceedings of the 2006 IEEE/AIAA Aerospace Conference*, pp.1-9, ISBN: 0-7803-9545-X, Big Sky, Montana, Mar. 2006

Le Vine, D.M., Griffis, A.J., Swift, C.T. et al. (1994). ESTAR: a synthetic aperture microwave radiometer for remote sensing applications, *Proceedings of the IEEE*, Vol.82, No.12, pp.1787-1801, ISSN: 0018-9219

Le Vine, D.M. (1999). Synthetic aperture radiometer systems, *IEEE Transaction on Microwave Theory and Techniques*, Vol.47, No.12, pp.2228-2236, ISSN: 0018-9480

Le Vine, D.M., Koblinsky, C., Howden, S., et al. (2000a). Salinity measurements during the gulf stream experiment, *Proceedings of IGARSS'00*, Vol.6, pp.2537-2539, ISBN: 0-7803-6359-0, Honolulu, HI, USA, Jul. 2000, IEEE New York

Le Vine, D.M. et al. (2000b). Development of a two dimensional synthetic aperture radiometer at L-band, *Proceedings of IGARSS'00*, pp.994-2996, ISBN: 0-7803-6359-0, Honolulu, HI, USA, Jul. 2000, IEEE New York

Le Vine, D.M., Jackson, T.J., Swift, C.T. et al. (2001). ESTAR Measurements during the southern great plains experiment (SGP99), *IEEE Transaction on Geoscience and Remote Sensing*, Vol.39, No.8, pp.1680-1685, ISSN: 0196-2892

Le Vine, D.M., Haken, M. & Swift, C.T. (2004). Development of the synthetic aperture radiometer ESTAR and the next generation, *Proceedings of IGARSS'04*, Vol.2, pp.1260-1263, ISBN: 0-7803-8742-2, Anchorage, AlasKa, USA, Sept. 2004, IEEE New York

Le Vine, D. M., Jackson, T. J., Haken, M. (2007). Initial images of the synthetic aperture radiometer 2D-STAR, *IEEE Transaction on Geoscience and Remote Sensing*, Vol.45, No.11, pp.3623-3632, ISSN: 0196-2892

Lemmetyinen, J., Uusitalo, J., Kainulainen, J. et al. (2007). SMOS calibration subsystem, *IEEE Transactions on Geoscience and Remote Sensing*, Vol.45, No.11, pp.3691-3700, ISSN: 0196-2892

Liu, H., Wu, J. et al. (2004). The CAS airborne X-band synthetic aperture radiometer: system configuration and experimental Results, *Proceedings of IGARSS'04*, Vol.3, pp.2230-2233, ISBN: 0-7803-8742-2, Anchorage, AlasKa, USA, Sept. 2004, IEEE New York

Liu, H., Wu, J., Wu, Q. (2005). Analysis and calibration of the channels error of synthetic aperture radiometer, *ACTA Electronica Sinica*, Vol.33, No.3, pp.402-406, ISSN: 0372-2112

Martin-Neira, M., Menard, Y., Goutoule, J.M. & Kraft, U. (1994). MIRAS: A two-dimensional aperture synthesis radiometer, *Proceeding of IGARSS'94*, Vol.3, pp.1323–1325, ISBN: 0-7803-1497-2, Pasadena, CA, USA, Aug 1994, IEEE New York

Rautiainen, K., Valmu, H., Jukkala, P. et al. (1999). Four two-element prototype of the HUT interferometric radiometer, *Proceedings of IGARSS'99*, Vol.1, pp.234-236, ISBN: 0-7803-5207-6, Hamburg, Germany, Jun. 1999, IEEE New York

Rautiainen, K., Butora, R., Auer, T. et al. (2003). Development of airborne aperture synthetic radiometer (HUT-2D), *Proceedings of IGARSS'03*, Vol.2, pp.1232-1234, ISBN: 0-7803-7929-2, Toulouse, France, July 2003, IEEE New York

Ribo, S. (2003). Research on image validation and signal processing of aperture synthesis radiometry, *Internal ESTEC Working Paper*, No.2182, pp.81-92, European Space Research and Technology Centre, European Space Agency (ESTEC-ESA)

Ruf, C.S., Swift, C.T., Tanner, A.B. & Le Vine, D.M. (1988). Interferometric synthetic aperture microwave radiometry for the remote sensing of the earth, *IEEE Transaction on Geoscience and Remote Sensing*, Vol. 26, No.9, pp. 597–611, ISSN: 0196-2892

Ruf, C.S. (1993). Numerical annealing of low-redundancy linear arrays, *IEEE Transactions on Antennas and Propagation*, Vol.41, No.1, pp.85-90, ISSN: 0018-926X

Ruf, C., Principe, C. et al. (2002). Lightweight fainfall fadiometer STAR aircraft sensor, *Proceedings of IGARSS'02*, Vol.2, pp.850-852, ISBN: 0-7803-7536-X, Toronto, Canada, June 2002, IEEE New York

Ruf, C. & Principe, C. (2003). X-band Lightweight Rainfall Radiometer First light, *Proceedings of IGARSS'03*, Vol.3, pp.1701-1703, ISBN: 0-7803-7929-2, Toulouse, France, July 2003, IEEE, New York

Skilling, J. & Bryan, R.K. (1984). Maximum entropy image reconstruction: general algorithm, *Mon. Not. Roy. Astr. Soc.*, Vol.211, No.1, pp.111-124, ISSN: 0035-8711

Sun, W.Y., He, B.Y. & Wu, J. (2005). Optimization of fourier plane coverage of antenna arrays for SPORT, *PIERS Online*, Vol.1, No.5, pp.533-537, ISSN: 1931-7360

Tanner, A., Wilson, W., Lambrigsten, B., Dinardo, S., Brown, S., Kangaslahti, P., Gaier, T., Ruf, C., Gross, S., Lim, B., Musko, S. & Rogacki, S. (2006). Initial results of the Geosynchronous Synthetic Thinned Array Radiometer (GeoSTAR), *Proceedings of IGARSS'06*, pp.3968-3971, ISBN: 0-7803-9510-7, Denver, Colorado,USA, July 26, IEEE New York

Thompson A.R. & Schilizzi, R. T. (1986). *Interferometry and Synthesis Radio Astronomy*, John Wiley & Sons, ISBN-10: 0471806145, ISBN-13: 978-0471806141 , NewYork

Wu, J., Liu, H. et al. (2005a). Research Activity on Synthetic Aperture radiometry in CSSAR/CAS, *PIERS Online*, Vol.1, No.5, pp.538-542, ISSN: 1931-7360

Wu, J., Liu, H., Sun, W.Y. & Jiang, J.S. (2005b). Technological development and application prospect of synthetic aperture radiometer, *Remote Sensing Technology and Application*, Vol.20, No.2, pp.24-29, ISSN: 1004-0323

Wu, J., Liu, H., Sun, W.Y., Wang, C. & Shi, S. (2005c). Application of Synthetic Aperture Radiometer Technology in Solar Wind Remote Sensing, *Proceedings of PIERS 2005*, pp.23-26, ISSN 1559-9450, Hangzhou, China, August, 2005, The Electromagnetics Academy

Wu, J., Wang, C., Wang S. et al. (2006). Solar polar orbit radio telescope for space weather forecast, *Proceedings of the ILWS Workshop*, India, pp.215, ISBN: 81-87099-40-2, Goa, India, Feb. 2006, Quest Publications for ILWS

Wu, J., Zhang, C., Liu, H. et al. (2007). Clock scan of imaging interferometric radiometer and its applications, *proceedings of IGARSS'07*, pp. 244-5246, ISBN: 978-1-4244-1211-2, Barcelona, Spain, July 2007, IEEE New York

Wu. Q., Liu, H., & Wu, J. (2004). Image simulator for one-dimensional synthetic aperture microwave radiometer, *Proceedings of IGARSS'04*, Vol.6, pp. 3973- 3976

Zhang, C., Wu, J. & Sun, W.Y. (2007). Applications of pseudo-polar FFT in synthetic aperture radiometer imaging, *PIERS Online*, Vol.3, No.1, pp.25-30, ISSN1931-7360

4

Recent advances in the characterization of aerosol vertical distribution on a global scale

Francisco Molero
Atmospheric Pollution Unit, Department of environment CIEMAT
Madrid, Spain

1. Introduction

Understanding the climate of our planet and how human activities are changing it is one of the greatest scientific challenges facing humanity nowadays. The rise of global average surface temperature by 0.6 °K since the late 19th century, attributed to radiative forcing (i.e. perturbation of the Earth's energy balance) by anthropogenic greenhouse gases and aerosols, underline the current concern about consequences of climate system modification. A quantitative assessment of the human contribution to climate change requires not only to establish possible future emission scenarios, but also a synthesis of uncertainties along the cause-effect chain from emissions to temperatures. The up-to-date scientific understanding of the topic, summarized in the fourth assessment report of the Intergovernmental Panel on Climate Change, establish the very likely human contribution to the climate change, but it also points out the wide range of uncertainty inherent in current model prediction associated with the radiative effect of aerosols, still labelled in the report as low level of scientific understanding.

The three known effects of aerosols on climatic processes, namely the direct effect caused by scattering and absorption of solar radiation by aerosols, the indirect effect that produce the modification of cloud properties, such as size distribution of droplets, by aerosols acting as cloud condensation nuclei, and finally the recently investigated semi-direct effect due to evaporation of cloud droplets in aerosol-rich layers because of the rise in temperature caused by absorption of solar radiation by aerosols, still requires a better constraining estimates of their magnitude.

The characterization of these effects, taken into account the large variability in geographical and seasonal distributions of the different types of aerosols (marine, anthropogenic pollution, desert dust), requires the use of long-term, detailed global measurements from distributed networks of ground-based instruments and satellites, and also comprehensive regional experiments in clean and polluted environments.

The long-term monitoring of aerosol microphysical properties on a global scale will improve the current knowledge of the influence of aerosols on climate, allowing a more accurate understanding of how the climate system will respond to the anthropogenic forcing. This chapter will comment about recent advances in the characterization of aerosol vertical distribution by means of optical remote sensing systems coordinated in networks of continental scales and onboard satellites.

2. Effects of aerosols and clouds on climate change

The increasing concentration of greenhouse gases in the atmosphere has caused most of the warming observed worldwide over the last century. The fourth assessment report of the Intergovernmental Panel on Climate Change (Solomon et al., 2007) established the very likely human contribution to the climate change, but it also points out the wide range of uncertainty inherent in current model prediction associated with the radiative effect of aerosols and clouds, still labelled in the report as low level of scientific understanding. The complexity of aerosol processes and their interaction with clouds in our environment yields large uncertainties in quantitative understanding of their role in many of the major environmental issues. Not only an appropriate quantification of aerosol and clouds on a global scale, including vertical distribution, is required, but also, further studies are needed to tackle the difficulties to adequately representing the radiative properties of the aerosol-cloud interaction.

Suspended particulate matter in the atmosphere, commonly known as aerosol, have many sources ranging from sea spray and mineral dust, that are mechanically generated by wind at the Earth's surface, or biogenic aerosols such as pollen, mold spores, and airborne bacteria and viruses, to secondary aerosols like sulphates, nitrates and organics produced primarily by chemical reaction of gases in the atmosphere that condense to form particulate mass. Aerosols typically range in size from around 10 nanometres to around 100 micrometers and have limited lifetimes in the atmosphere. Aerosols in the lower part of the atmosphere normally last for only several days before being washed out by rain or settled by gravity. In the upper parts of the atmosphere, aerosols can persist much longer. Volcanic eruptions, that release tons of aerosols into the atmosphere, can propel these particles into the stratosphere, where they can persist for several years. These aerosols are created as a result of natural processes, and their sources and sinks have remained fairly stable in the last century, apart from the influence of human activities such as soil use modification. In addition to these sources, human activities can generate concentrations of aerosols far in excess of natural sources. In fact, there is compelling evidence that anthropogenic activity is increasing the concentration of tropospheric aerosols. Major cities produce large amounts of pollution aerosols as a result of industrial activity and automobile emissions. The burning of fossil fuels (primarily oil and coal) to produce energy emits large quantities of aerosols into the atmosphere. Fires set by humans around the world for agricultural purposes, such as clearing cropland, are also major sources of biomass burning aerosols. Unlike the long-lived greenhouse gases, which are distributed uniformly over the globe, aerosol distribution have substantial spatial and temporal variations, with largest concentration of pollution aerosols found near industrial regions in the northern hemisphere. The effects of aerosols tend to be localized and larger near their source regions, which makes it difficult to estimate the net global impact of aerosols on climate. Scientific evidence indicates that in regions with high anthropogenic aerosol concentrations, aerosol forcing is of the same magnitude, but opposite in sign, to the combined effect of all greenhouse gases. Also, this regionally-concentrated distribution can directly alter the general circulation patterns.

Estimating the effects of aerosols on climate is particularly challenging, because the radiative response to aerosol particles varies with size and chemical composition of the particles relative to the wavelength of the incident radiation (Dubovik, 2002). Experimental evidences has shown that all of these properties can and do change with time, such as when mineral-based desert dust moves over an urban area and black-carbon based aerosols attach to the mineral core. Aerosols influence the atmospheric energy budget through direct, semi-direct and indirect radiative effects. Direct effect is caused by the scattering and absorption of incoming solar radiation. Scattering of radiation reduces solar heating at the surface immediately below, causing regional "solar dimming" (Liepert, 2002), that somehow counteract some greenhouse-gas warming. The amount of cooling depends on the above mentioned aerosol parameters, but also on the type of underlying surface. Another likely consequence of the aerosol surface cooling is a reduction of evaporation and precipitation. To further complicate matters, the forcing can switch from negative values (cooling) in clear-sky conditions to positive values (warming) at a cloud cover of about 25%, because the energy distribution due to the presence of the cloud is different than a clear sky (Ramanathan et al. 2001).

The semi-direct effect is related with the absorption of solar radiation by aerosols, such as black-carbon and some mineral-based aerosols, that heats the surrounding atmosphere and can actually suppress the formation of clouds by elevating the atmospheric temperature, preventing the condensation of water vapor. This forcing can strengthen the low-level thermal inversion of the boundary layer, which can perturb low-level clouds, enhance aerosol lifetimes, and alter the boundary layer moisture. This has been observed in events related with the South Asian haze, where the warming in the aerosol layer can nearly totally desiccate stratocumulus cloud layers and alter the properties of the trade wind cumulus layer (Ackerman et al., 2000)

Finally, the indirect effect involves the influence of aerosols on the properties of clouds, such as the microstructure, dynamics, coverage and stability of cloud layers. An increase in aerosols, acting as cloud-condensation nuclei or ice nuclei, creates larger concentrations of clouds droplets, which leads to increased cloud lifetime and albedo. This is because normally the droplets grow, collide and coagulate until they grow large enough to fall as raindrops, but as the amount of aerosols in the cloud is relatively low, the cloud will consist of relatively fewer but larger droplets. If the amount of aerosols is increased, the cloud droplets that form tend to be smaller and more numerous; it takes longer for raindrops to form, and the clouds last longer. Clouds consisting of smaller droplets also reflect more sunlight back to space and contribute to increased cooling of the underlying surface (Twomey, 1977). As anthropogenic aerosols are most highly concentrated in the lower atmosphere, the indirect effect is expected to be most important in low-level clouds. The most obvious impact of clouds on the hydrological cycle is that of precipitation. In removing water from the atmosphere, precipitation modifies cloudiness and cloud structure. Moreover, the latent heating associated with precipitation is a driving force for atmospheric circulations. Precipitation frequencies can also be affected by aerosols, for instance by the suppression of raindrops in shallow and deep convective clouds, which are the major sources of thermodynamic forcing of the general atmospheric circulation (Levin, 2009).

The effect of clouds on the global radiative balance depends on the competition between the reflection of incoming solar radiation and the absorption of Earth's outgoing infrared radiation. The overall impact of high-altitude clouds is to warm the planet while the low-altitude clouds tend to cool it. Research to date suggests that, globally averaged, the overall cooling caused by clouds is more powerful than the warming they cause. Anyhow, these behaviours are strongly linked to microphysical processes in clouds because changes in those processes can modify the spatial extent, spatial distribution, and lifetimes of clouds, the water vapor distribution outside of clouds, and the fluxes of water and radiation through the atmosphere. Another reason why modelling clouds is difficult is that clouds

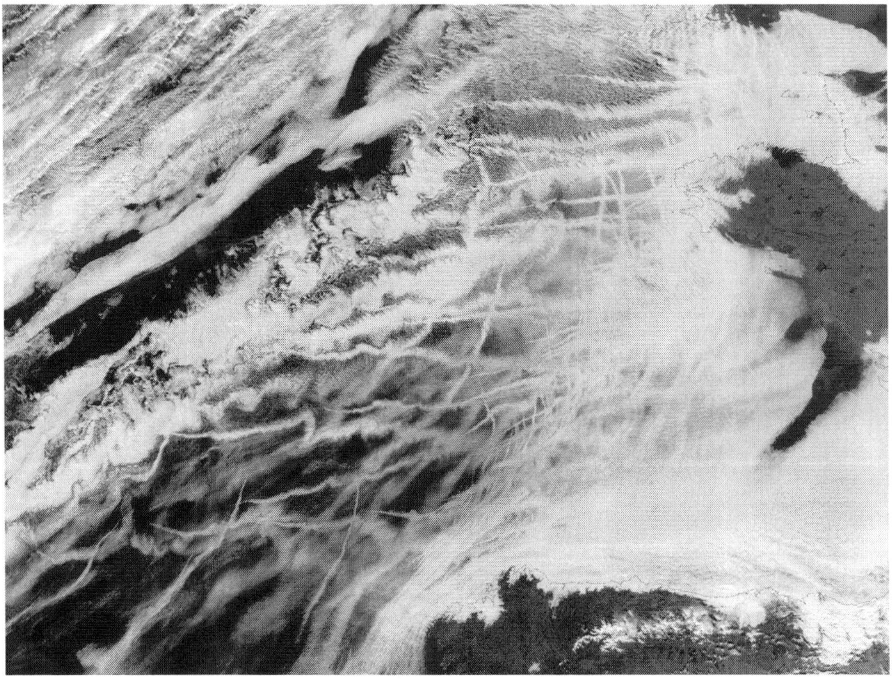

Fig. 1. Evidence of the indirect effect of aerosols: The image from the Moderate Resolution Imaging Spectroradiometer captured on January 27, 2003 and processed to create a true-color image using bands 1 (650 nm), 4 (555 nm), and 3 (469 nm) shows an unusually high number of ship tracks (thin white lines), visible in the clouds off of the coasts of France (Left) and Spain (Bottom). (Image courtesy of J. Descloitres, MODIS rapid response team, NASA/GSFC, Greenbelt, Maryland)

change almost instantaneously compared to the rest of the climate system and on very small scales. Most models are not capable of representing phenomena that change so rapidly or impact such a small portion of Earth at any given time (Baker, 1997).

Owing to the typically high altitude and often remote location, the nucleation of ice is the less well-understood process. Atmospheric ice formation at temperatures warmer than about 40 °K below the equilibrium freezing temperature, requires the presence of a special solid particle that acts as an ice nucleus, enhancing the stability of an ice embryo due to the presence of a surface. Materials that can act as ice nuclei include, but are not limited to, mineral dust, anthropogenic metal oxides, pollen and bacteria.

The clearest observational evidence for an indirect aerosol effect is provided by ship tracks, which are trails in ambient low-level clouds that result from the effluent from ships. Although ships are not significant sources of pollution themselves, they release enough aerosols from their smokestacks to modify overlying clouds. Those aerosols act as cloud condensation nuclei, which may either produce new cloud particles where none existed before, or may attract water from existing cloud particles, creating brighter clouds due to the enhancement of liquid-water content, possibly caused by suppression of drizzle, and smaller droplets. An example of this is shown in figure 1; in this satellite image of clouds over the Atlantic Ocean, the thin white lines are bright clouds consisting in small droplets that form due to aerosols emitted by ships. Another example is the pollution tracks as viewed by the Advanced Very High Resolution Radiometer satellite imagery (Rosenfeld, 2000). Perhaps the most significant aspect of the analysis of these pollution tracks is the conspicuous absence of them over the United States and Europe, implying that these regions are so heavily polluted that local sources can not be distinguished from the widespread pollution-induced narrow droplet spectra in those regions. A striking example is provided by anthropogenic lead-containing particles, one of the most efficient ice-forming substances commonly found in the atmosphere. Post-industrial emissions of particulate lead have been estimated to offset a proportion of the warming attributed to greenhouse gases, by "supercharging" pre-existing particles, making them highly efficient ice nuclei that allow clouds to form at lower altitudes (Cziczo, 2009). But after the regulation of tetraethyl lead, an additive to automotive petrol, in the mid-1980, total lead has dropped significantly, with a 20-fold decrease reported in the continental United States in the two decades since 1980. This might imply a drastic reduction in the offset, although a proper estimation of the climatic effect is still pending. Like the ship tracks and contrails produced by aircraft, the impact of those pollution tracks on regional and global climate is not yet known, but recent satellite observations of these phenomena are providing new information.

The main uncertainties in climate model simulations are due to the difficulties in adequately representing the interdependent microphysical, chemical, and dynamic processes that characterize the aerosol-cloud system in the atmosphere, that must be better understood in order to quantify the effects of anthropogenic aerosols on the albedos, emissivities, cloud-top temperatures and extent of clouds. Present models do not have sufficient sophistication in cloud microphysics to include those aerosol influences (Grabowski, 2009). As global climate models have not included all these complexities, the simulation of impact of aerosols on global climate is not possible yet.

First and foremost, a reliable global inventory of aerosol emission rates, lifetimes, global distribution and concentrations is urgently needed. The aerosol vertical distribution depends on the distributions of the emissions, on chemical production for secondary aerosols, on the distribution of clouds, precipitation and wet deposition processes, and finally, on the transport characteristics determined by the flow field. On the theoretical and experimental side, further investigations regarding the microphysical processes and radiative effects in clouds and aerosols are required. Another large uncertainty in climate studies is the effect of cloud multilayering. It has been observed that the largest variations in predictions of climate warming are due to the different ways the models specify how clouds are vertically distributed and overlap, which influence both the magnitude and vertical profile of heating in the atmosphere, and also the predicted precipitation. However, the interaction between aerosols and clouds is sufficiently complex that even cloud-resolving models have difficulty in accurately simulating their physics and dynamics. Therefore, detailed vertical profiles of aerosol and clouds will be relevant parameters to evaluate aerosol forcing, understand the effect of aerosol on cloud microphysics and precipitation, assess the degree of interaction of aerosols with the cloud layer and also the effect of clouds on aerosols, due to the major role of water in determining aerosol optical properties. The vertical profiles will also help to better understand aerosol lifetimes, their source regions through backtrajectory analysis and long-range transport events, which occur at elevated layers, decoupled from the ground.

3. Characterization of the aerosol vertical distribution on a global scale

As above-mentioned, altitude-resolved information on aerosols, including long-range transport in the free troposphere, and cloud observations including cirrus clouds, are essential for understanding the climate role of atmospheric aerosols. The most promising source of routine information on the vertical distribution of aerosols are lidar systems. Other possible options to provide altitude-resolved information are aerosol balloon sondes instruments and aircraft measurements, but they are not yet widely available and they are also less economical. The lidar technique, which stands for LIght Detection And Ranging, operates in a similar way as radar, but using light instead of microwaves. Lidars have been used for several years to determine the planetary boundary layer height because of the large gradient in aerosol concentration that occurs between the top of the boundary layer and the free troposphere. Therefore, ground-based lidars may fill the ongoing need for insight into the structure of the atmosphere and its variability with time. There are only a few lidar instruments planned for deployment on satellites, as it will be discussed below. In contrast, there are several research lidar measurement stations and networks that are well organized with high standards regarding quality control, ongoing development of new controlling measures, and data archival. The integration of aerosol lidar observations with other measurements by radiosonde, ozone sonde, sunphotometer and satellite is most useful allowing for a maximum synergy of information. Also, the informational content of lidar observations is greatly enhanced by air parcel trajectory analysis.

3.1 Light Detection And Ranging

Remote sensing by lidar has received wide application in investigation of atmospheric trace constituents, clouds, wind and temperature since its invention a few decades ago. The lidar technique provides information on several aerosol and clouds properties with high spatial and temporal resolution, working in a similar way as radar. The system typically consists of a laser transmitter and an optical receiver in parallel or collinear arrangement. Figure 2 shows a schematic depict of a lidar system with three emitting wavelengths, a typical configuration when Nd:YAG lasers, the most reliable and widely used type of laser, are used, that typically emits at 1064, 532 and 355 nm. The system transmits intense, short-duration light pulses of linear polarization at a high repetition rate into the atmosphere within the receiver field of view. The intensity of the light elastically backscattered by molecules and aerosols is measured versus time through the telescope receiver, collimating optics, a narrow bandpass filter for daylight suppression, and an appropriate detector. In the figure, several detection channels are shown, in a typical configuration of advanced lidar systems for measuring backscatter intensity, molecular or Raman signals and polarization components of the signal, as it will be discussed below. The signal profile will be stored by a fast analog-to-digital converter or by a photon counting device. Relative intensity data are accumulated separately from all altitude intervals for a selected averaging period, which may include thousands of individual laser shots.

The lidar of lowest complexity measures the aerosol backscatter signal at one wavelength. This lidar allows the retrieval of the aerosol backscatter coefficient, although critical assumptions have to be made in the inversion of the lidar signal in order to obtain aerosol optical properties. The procedure, with all its subsequent modifications and improvements, simply suffers from the fact that from only one observable (the energy returned as a function of time), two unknowns (the aerosol backscatter coefficient of the aerosol and the two-way transmission losses through the atmosphere due to light extinction by molecules and aerosols between transmitter, backscattering volume at a certain range and the receiver) must be determined, therefore, the system is underdetermined. Many techniques have been discussed in the literature to work around this difficulty (the slope method, Collis and

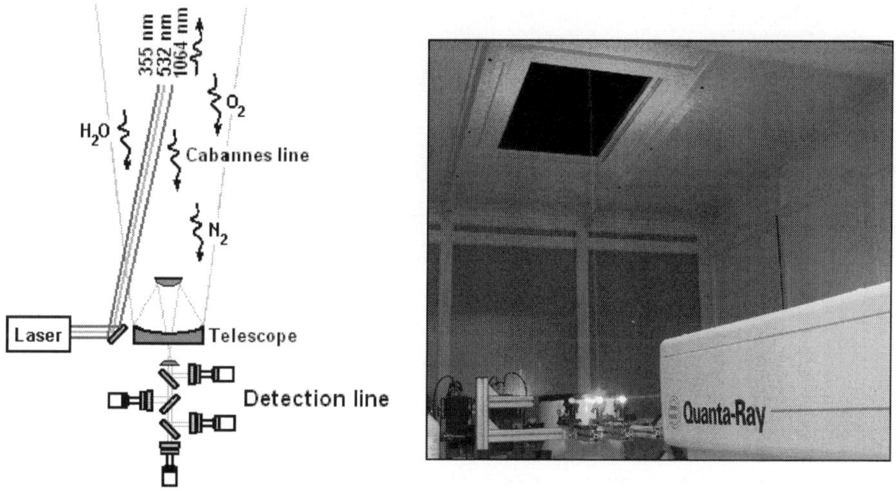

Fig. 2. Schematic of a lidar system showing the laser transmitter on the left, the optical receiver in biaxial configuration on the right. Several different returns are presented as photons. On the right, a photograph of the actual implementation of the Madrid station lidar, showing the laser on first term, three mirrors for the three different Nd:YAG wavelengths send to the atmosphere through the ceiling window, and the telescope and detection line behind.

Russell, 1976; the Bernoulli solution to the equation, Klett, 1981, Fernald, 1984; and column closure by the use of ancillary optical depth information, Welton et al., 2001). Nevertheless, the measurement remains only an estimate of either the backscatter or the extinction coefficient as long as elastic lidar-only data is available, because the relation between these magnitudes, that must be assumed constant in the above mentioned techniques, actually depends on the microphysical, chemical, and morphological properties of the aerosols, which, in turn, depend on relative humidity. Thus, even in the well-mixed layer, the relation might not be constant with height because the relative humidity usually increases with height. There is value, however, in the backscatter lidar, providing range to a target and structural information on the atmosphere once corrected for detector non-linearity, background and range. Basic lidar products are mixed-layer and cloud-base heights and qualitative information of aerosol layers in the free troposphere, visible and subvisible cirrus and stratospheric aerosol layers after major volcanic eruptions. Figure 3 shows the so-called quicklook or colour plot obtained by representing the range corrected signals taken along a certain period of time in a height-time display. It provides an overview of the atmospheric situation in terms of evolution of the boundary layer, lofted aerosol layer, like the Saharan dust intrusion at 5 Km, and cloud distribution, such as the cirrus clouds detected as a thick layer at the beginning of the measurements, 03:20 UTC, on the left part of the image, and disappears an hour later, right part.

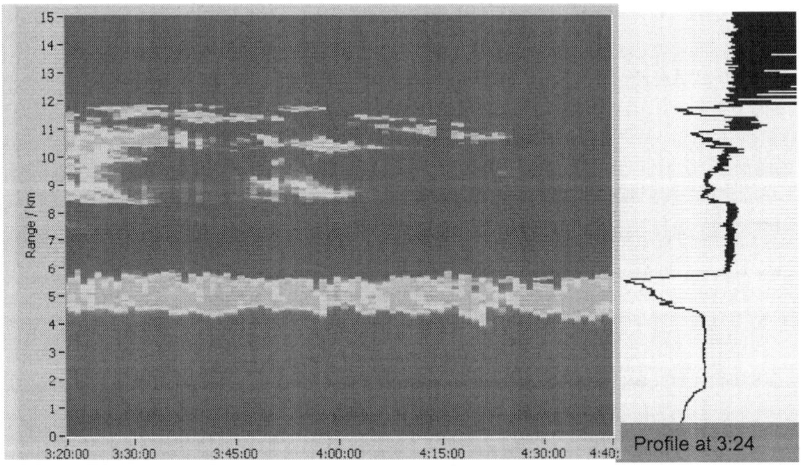

Fig. 3. Lidar profile and Quicklook at 532 nm, obtained over Madrid (40.45°N, 3.73°W) on the 21st of June, 2009, between 3:20 and 4:40 am. The background substracted and range-corrected signals shows relevant features at 5 Km, caused by a Saharan dust intrusion over the site, and between 8 and 12 Km, due to cirrus, that dissapear along the measuring time.

In the near-range, the signal is affected by losses due to the incomplete overlap of laser beam and receiver field of view. A good knowledge of this overlap is an important prerequisite for a proper retrieval of the backscatter coefficient in the lowest and usually most polluted part of the atmosphere. Due to this limitation, the reference range required in the inversion of the lidar signal is chosen at altitudes where the aerosol backscatter coefficient is negligible compared to the known molecular backscatter value, the so-called aerosol-free layer, and the equation is solved using the backward integration solution proposed by Klett (Klett, 1981).

This far-end calibration also leads to more stable solution than the near-end calibration, but clear air conditions are needed at some range to calibrate the signal. They are normally found in the free troposphere for laser wavelengths < 700 nm, but it might be critical for longer wavelengths (e.g., 1064 nm or the eyesafe wavelength of 1550 nm) because of weak Rayleigh scattering. The inversion procedure also includes the calculation of the molecular backscatter profile. Standard-atmosphere assumptions, nearby radiosonde data of temperature and pressure, or weather prediction model outputs for the lidar site are used to compute the entire Rayleigh scattering profile along the laser beam.

3.2 Advanced lidar systems

Accurate retrieval of extinction and backscatter profiles without making assumptions about the aerosol is only possible when measurements of two independent signals are performed. Recent developments of the technique, such as the use of the atmospheric nitrogen Raman return (Ansmann et al., 1990), or the broadening of the lidar return by molecules in a technique called High Spectral Resolution Lidar (HSRL) (Shipley et al., 1983), allow the independent determination of aerosol backscattering and extinction coefficient profiles. Both methods are based in the direct determination of the extinction coefficient profile by means of a channel that detects a pure molecular signal, and a second channel that provides the backscatter coefficient from a signal detecting aerosols and molecules. These advanced lidar systems determine the aerosol optical properties in a quantitative way and permit the estimation of main microphysical properties. Rapid progress in laser technology and data acquisition is supporting an increasing specialization in the design of lidar systems, leading to the point where systems can be built or certain purposes with high reliability and durability. Even several lidar systems for some specific applications are now becoming available commercially.

3.2.1. Raman lidar technique

The independent determination of backscatter and extinction coefficient profiles can be achieved by the measurement of pure molecular backscatter signal, because the molecular extinction profile and the backscatter coefficient can be calculated a priori with sufficient accuracy so that the aerosol extinction, the only remaining unknown, can be retrieved from the molecular signal. The use of the vibrational Raman scattering signal from nitrogen (or oxygen) offers the technologically easiest implementation, due to the large Raman shift (2331 cm^{-1}) that allows a reliable separation from the elastic aerosol signal with standard filters. Due to the low values of the nitrogen backscattering coefficient, that produce weak Raman signals, approximately by a factor of 500 compared to Rayleigh signals, this technique works best in the absence of the strong daylight sky background, therefore most Raman lidars operate only at nighttime in this mode. Anyhow, high power laser (>250 mJ/pulse) Raman lidars equipped with 0.3-nm interference filters to block sunlight, allow daytime operation at least throughout the convective boundary layer. For nighttime operation the filter bandwidth can be broad and the technical implementation is quite straightforward and it has been widely used, mainly at 355nm and 532nm, the second and third harmonics of the Nd:YAG laser. Recent applications of the technique provide the water vapor profile using its Raman signal (3652 cm^{-1}), which may be very interesting for studies of the aerosol-cloud interaction

Pure rotational Raman scattering by nitrogen and oxygen offers a scattering cross section that is about a factor of 30 higher than vibrational Raman scattering. The drawback is that the Raman shift is quite small, about 30 cm^{-1} only, so that separation from the elastic aerosol backscatter is more challenging, keeping in mind that out-of-band blocking has to be on the order of 10^{-8}. Both filter techniques and double grating polychromators have been demonstrated for this approach. In particular the combination with a Fabry-Perot comb filter can suppress daylight sufficiently to allow daytime operation. A more sophisticated setup also allows one to retrieve the temperature profile simultaneously.

3.2.2 High Spectral Resolution Lidar Technique

High Spectral Resolution Lidar (HSRL) is based on the Doppler broadening of the Rayleigh line, leading to an about 0.01cm^{-1} wide line surrounding the much narrower peak from aerosol scattering. In the HSRL technique, one channel detects the molecular signal by suppressing the centre part of the backscatter spectrum, containing the aerosol return, with an ultra-narrowband filter, generally an iodine vapour cell. A second channel records the total signal from aerosol plus molecular scattering. The combination of both signals allows determining extinction and backscatter profiles independently. The advantage of this technique is that it suitable for daytime operation because the Rayleigh scattering cross section of air is more than three orders of magnitude greater than that for vibrational Raman scattering. The drawback is high system complexity and high demands on system adjustment as well as on performance control.

Both techniques have been operated successfully for aerosol profiling. Data analysis schemes have been developed to retrieve vertical profiles of aerosol optical properties. The algorithms are well-tested and are nowadays almost routinely applied. Another important advantage of Raman lidars and HSRLs is that the profile of the backscatter coefficient is determined from a signal-ratio profile so that the overlap effect cancels out, regarding that the two channels are well adjusted and show the same overlap characteristics. As a consequence, the retrieval of the backscatter coefficient is possible down to heights rather close to the surface. In combination with Sun photometers, a comprehensive set of vertically and spectrally resolved optical properties can be determined. Anyhow, the requirements of expertise of the operating personnel with optical systems and inversion algorithms are still a limitation for the automation and network operation of these systems.

Once reliable extinction and backscattering coefficient profiles are obtained, the extension of the technique to multiple wavelengths offers the opportunity to determine vertically-resolved microphysical properties, such as size distribution parameters, volume concentrations and refractive index (Böckmann et al. 2005). During the past decade sophisticated inversion techniques have been developed and successfully tested that permit the retrieval of microphysical properties of aerosols from their optical properties provided by advanced multiwavelength lidar observations. For aerosol sizes in the typical range of the accumulation mode, measurements of the backscatter and extinction coefficients at the Nd:YAG wavelength (1064, 532 and 355nm) are necessary and sufficient to estimate the aerosol volume and surface density as well as the refractive index (Müller et al., 1999). The retrieval procedure is ill-posed and requires sophisticated regularization methods, so that presently the procedures are still experimental and applied for selected cases only. The low number of measured aerosol optical properties requires introducing physical and mathematical constraints in the inversion algorithms in order to come up with sensible microphysical aerosol parameters. These algorithms do not attempt to accurately derive the exact shape of aerosol size distributions, which might not be achievable even in the near future due to the low number of measured optical information and the lack of appropriate mathematical tools. However it is possible to derive mean parameters such as the effective radius of the aerosol size distribution with comparably high accuracy. At present it does not seem possible to fully retrieve aerosols in the so-called coarse mode of aerosol size distributions which is largely determined by aerosols from natural sources such as mineral dust. However, aerosols from anthropogenic activities are mainly present in the fine mode fraction which is accessible to the inversion algorithms (Müller et al., 2007).

The combination of advanced lidar products with Sun photometer data offers another approach, as the potential of Sun photometry to derive optical, microphysical, and radiative properties of aerosols is already well-documented (Eck et al., 1999). Also, depolarization observations can be used to improve the identification of aerosol types, such as desert dust. The emitted laser light is linearly polarized and the return signals are measured in two polarization channels which are parallel- and perpendicular-oriented to the laser polarization. From the linear total (aerosol + molecular) depolarization ratio of the scattering volume that is obtained from the ratio of the perpendicular- to the parallel-polarized signal component, the aerosol depolarization ratio can be calculated if the aerosol backscatter coefficient and the respective linear molecular depolarization ratio are known. Spherical aerosols as water droplets produce an aerosol depolarization ratio of almost zero in the case of 180° scattering. Dust aerosols cause a depolarization ratio of 25%-35%. Smoke, urban haze, and maritime aerosols show depolarization ratios of <10%. Ice aerosols (ice clouds) lead to depolarization ratios typically >40%-50% (at off-zenith laser beam angles). Present inversion algorithms assume spherical shape of the aerosols. Only recently efforts have been undertaken to introduce methods that allow for a characterization of aerosols of non-spherical shape, such as mineral dust. However, the underlying theoretical aspects of light-scattering by irregularly shaped aerosols still are in a rather exploratory status. Last but not least, profiles of microphysical aerosol properties can be derived with the available algorithms, however with an extreme consumption of computer and human operator time. Thus it is desirable to extend the available algorithms toward an efficient processing of profiles of optical data, which in turn delivers profiles of microphysical aerosol properties. Schemes with higher degree of automation are under development.

4. Networks of lidar stations

Following the example of AERONET, a global network of systematic column-integrated aerosol optical depth observations using surface-based sun-tracking photometers, different attempts exist to characterize the vertical distribution of aerosols on continental scales and further extent to global scale within the initiative of the Global Atmosphere Watch (GAW) aerosol programme (Bosenberg & Hoff, 2007). It is the goal of the GAW programme to coordinate and homogenize the different existing network in order to determine the spatio-temporal distribution of aerosol properties related to climate forcing and air quality on multi-decadal time scales and on regional, hemispheric and global spatial scales. Such initiative is being discussed nowadays under the frame of GAW Aerosol Lidar Observation Network (GALION).

Presently, it is not feasible to implement a global aerosol lidar network by installing a homogeneous set of systems at a number of stations selected for optimal coverage because advanced lidars are still relatively complex and delicate instruments requiring substantial efforts for operation. Instead, it is important to make use of existing systems at established stations, of the experienced operators of these systems, and of existing network structures.
Nowadays, several lidar networks perform regular measurements on continental scale to establish a comprehensive dataset of the aerosol vertical distribution and also assess volcanic, dust, fires or pollution events. Those networks include the Micro-pulse lidar network (MPLNet), the European Aerosol research lidar network (EARLINET), the Asian Dust Network (AD-Net), the Commonwealth of independent states lidar network (CIS-LiNet), Regional East Aerosol Lidar Mesonet (REALM) and the American LIdar Network (ALINE). A brief description of each of them follows:
The only tropospheric profiling network which can claim global coverage is the NASA MPLNET (http://mplnet.gsfc.nasa.gov/), designed for satellite validation and co-located with AERONET sites in order to produce quantitative aerosol and cloud products by synergy with sunphotometer measurements. The Micro-Pulse Lidar (MPL) was developed at NASA Goddard Space Flight Centre in the early 1990s and it consists of single-wavelength (523 nm), high repetition, low power, eye safe commercially available backscatter lidar capable of determining the range of aerosols and clouds in unattended operation mode. At present, MPLs are operated at 22 sites around the world. The combination of MPLNET consisting of low-cost, eyesafe, automated 532-nm backscatter lidars with AERONET (Holben et al., 1998), NASA's global network of more than 200 continuously running Sun photometers, is an example for a successful application of the lidar-photometer technique.
The European Aerosol Research Lidar Network, (EARLINET), (http://www.earlinet.org) is a voluntary association of institutions with an interest in aerosol science and a long-term commitment in vertical profiling of aerosol properties with advanced laser remote sensing. Presently EARLINET comprises 25 stations distributed over Europe. Instrumentation is rather inhomogeneous because most lidars existed before the network was established in 2000, but most systems are now equipped with at least one Raman channel for independent determination of extinction and backscatter. The main goal is to establish a climatology for the aerosol vertical distribution by building a quantitative comprehensive statistical database of the aerosol, therefore regular operation at three times per week has highest priority for all stations. Special studies of, for instance, Saharan dust outbreaks across the Mediterranean, distribution of smoke from wildfires, volcanic eruptions, air mass modification across Europe, diurnal cycle, or CALIPSO validation required numerous additional observations which were organized as necessary through corresponding alerting schemes. Quality assurance for hardware and software was performed through direct intercomparisons, tools for routine performance checks are under test.
The Asian Dust Network (AD-Net) (http://www-lidar.nies.go.jp/AD-Net/) is an international virtual community designed originally to track outbreaks of dust from China, Mongolia and Russia. Different instruments are involved, such as multi-wavelength Raman lidars in Tokyo and Gwanngju (Korea), HSRL system in Tsukuba (Japan) and automated two-wavelength polarization lidars in Japan (8), Korea (1), China (1), and Thailand (1), which are coordinated by the National Institute for environmental Studies (NIES) lidar

network, most of them co-located with skyradiometer from SKYNET. Those automated small and compact lidars are recently being upgraded with Raman channels.

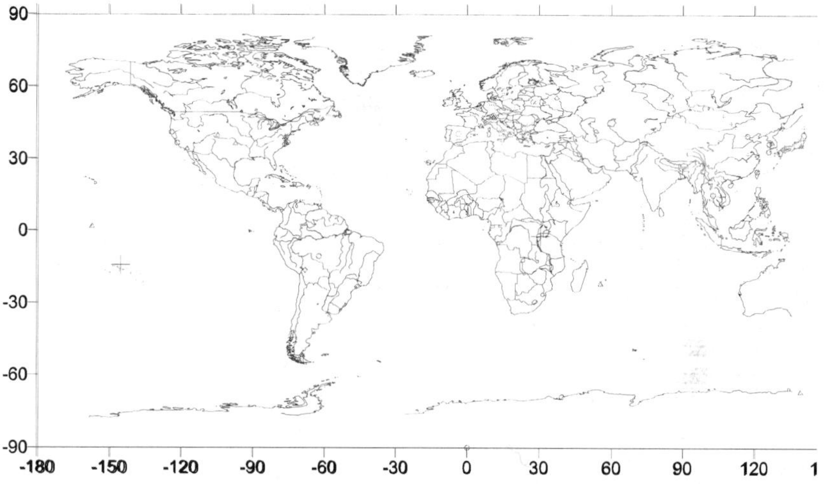

Fig. 4. Distribution of the existing networks stations on the globe. The different networks are indicated by the symbol and color: MPL-Net: Blue circles, EARLINET: green squares, AD-Net: red crosses, NDACC and REALM: red triangles, CisLiNet. Brown squares and ALINE: orange circles.

The Commonwealth of Independent States lidar network (CIS-LiNet) has been established by lidar teams from Belarus, Russia and Kyrgyz Republic. Its objective is carrying out lidar observation coordinated at the territory from Minsk to Vladivostok in cooperation with EARLINET and AD-Net. There is an aim to provide the lidar stations with sun radiometers, and include them in the global radiometric network AERONET.
The Regional East Aerosol Lidar Mesonet (REALM), is a network of lidar research groups on the east coast of the United States (http://alg.umbc.edu/REALM) operative since 2002, a collaboration of existing lidar facilities has attempted a network operation. But to date only two groups and three lidars have voluntarily contributed consistent data to the network with two other groups contributing campaign style activities.
Finally, the American LIdar Network (ALINE) is an informal agreement among the existing lidar groups in Latin America. It includes also research teams working to host lidar instruments in the near future.
Another network that employ lidars, but with a different research aim, is the Network for the Detection of Atmospheric Composition Change (NDACC, previously NDSC), has been monitoring the stratosphere and upper troposphere for at least 15 years. NDACC consists of more than 70 high-quality, remote-sensing research sites for observing and understanding the physical/chemical state of the stratosphere and upper troposphere and assessing the impact of stratospheric changes on the underlying troposphere and on global climate. Only

a subset of the stations actually contains lidars. In this subset, the lidars are designed primarily to profile O_3 in the stratosphere and stratospheric aerosols.
As it can be gathered, the operation of a global lidar network will initially depend completely on voluntary contributions from the various existing networks, most of which, themselves, are based on voluntary cooperation, plus contributions from individual stations. Therefore, the consistency of data across the network and its quality assurance might be compromised.

5. Satellite missions with lidar instruments onboard

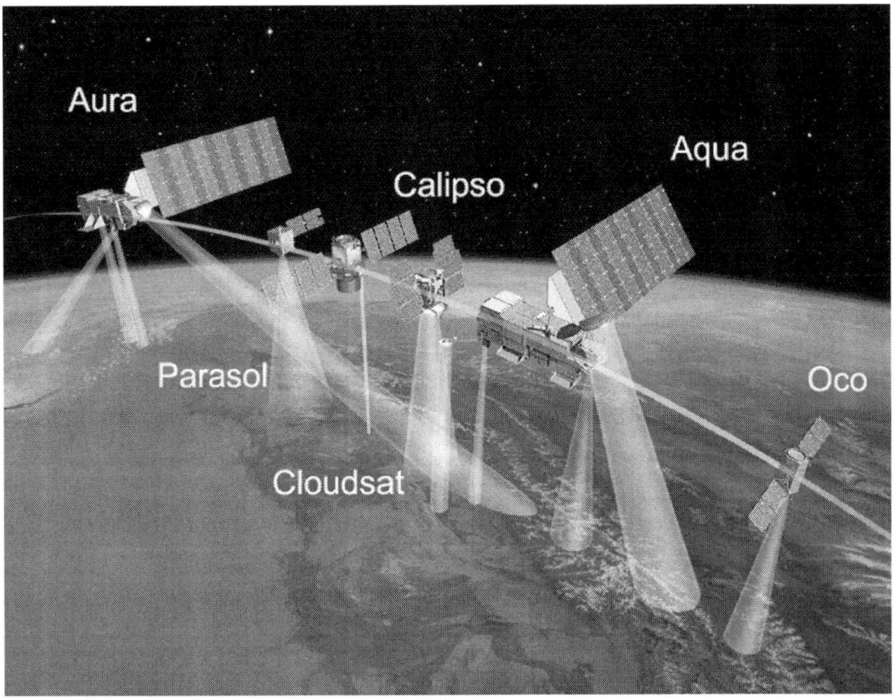

Fig. 5. The "A-train" constellation of satellites, with CALIPSO on the third place from the left. Sinergy with data provided by the radar onboard CloudSat, the polarization sensitive radiometer POLDER onboard Parasol and MODIS spectra onboard Aqua satellite will provide the most complete characterization of clouds and aerosols on global scale ever. Crédits : CNES/Ill. P. Carril

Lidars have been used as ground-based systems, on airborne platforms, and also from space, as a different strategy for aerosol vertical distribution characterization on a global scale. Unlike lidars that profile the atmosphere above at a single geographic location, spaceborne lidars allows the study of aerosols in regions that are difficult or impossible to explore by others means. The era of spaceborne lidar started with the Lidar In-space Technology Experiment, or LITE, (McCormick, 1997), a backscattering lidar system that flown in the cargo bay of the space shuttle "Discovery" in September 1994, to evaluate technological requirements of a spaceborne lidar and its scientific capability. The mission convincingly demonstrated the value of spaceborne lidar in retrieving the vertical structure of aerosol and clouds on a global scale and provided the backbone for more recent satellite-based lidars. In 2006, the French Centre National d'Etudes Spatiales and US National Aeronautics and Space Administration collaborate to launch the Cloud-Aerosol Lidar and Infrared Pathfinder Satellite Observations (CALIPSO) (Winker et al, 2007) spaceborne lidar, that collects profiles of the lidar attenuated backscattering coefficients of aerosols and clouds at 532 and 1064 nm from -2 to +40 kilometres above ground level using the CALIOP instrument. The CALIPSO satellite mission objective is to determine precisely the altitudes of clouds, aerosol layers and their overlap; identify the composition of clouds and the presence of "subvisible," or invisible, clouds; and estimate the abundance and sources of aerosols. It will provide the first global survey of cloud and aerosol profiles from space, with seasonal and geographical variations. CALIPSO will fly in formation with CloudSat and in concert with the other satellites of the "A-Train," a constellation of several Earth-observing satellites depicted in figure 5. The combination of data from the CloudSat radar, CALIPSO lidar, PARASOL radiometer and MODIS on the Aqua satellite observing the same spot on the ground within minutes, provides a rich source of information that can be used to assess the role of aerosols and clouds in climate, as well as reduce in some extent the ambiguity in deriving the aerosol profile from the lidar measurements (Kaufman et al. 2003).

CALIPSO will provide the first statistics on the global vertical distribution of aerosols and aerosol types and "subvisible" cirrus clouds (very thin clouds invisible to the naked eye). The synergy of CloudSat and CALIPSO measurements will provide information on the vertical structure of clouds and daily coverage of global cloud characteristics. It is expected to provide the first indirect but validated estimate of the contribution of clouds and aerosols to the vertical distribution of atmospheric warming, from which the first operationally based estimates of direct aerosol properties and uncertainties can be made.

Several examples of the global transport of aerosol were evident from spaceborne data. Aerosol from natural sources, such as Saharan dust, was measured on several orbits of LITE and CALIPSO. Although it has been known for quite sometime that large quantities of Saharan dust are transported across the Atlantic towards the Caribbean, the unique capabilities of spaceborne lidars proved ideal for tracking and quantifying the magnitude of these events, showing enormous plumes of hundreds of kilometres reaching altitudes above five kilometres or aerosol plumes generated by biomass burning in South America extending hundreds of kilometres from the source region, that make possible to assess their impact and quantify their contribution to long-range transport. Another example would be the measurement of anthropogenic aerosol leaving the Eastern United States and riding the "gulfstream highway" towards Europe (Hoff & Strawbridge, 1996).

The validation of the spaceborne instrument is required to assess the quality of the measurements, especially for aerosol types which have poorly or unknown lidar ratios which must be assumed in the CALIPSO aerosol retrieval since it is an elastic-type instrument. Network of ground-based lidar stations can support the aerosol observations from space by making targeted observations on CALIPSO overpass times. The better sensitivity of ground-based lidar systems can be used to confirm the sensitivity limits of the

satellite instruments characterize the atmospheric features missed by the satellite instruments and determine additional parameters.
In the near future, the European Space Agency will launch the Earth Explorer Atmospheric Dynamics Mission (ADM-Aeolus) (Reitebuch et al. 2004), that will provide global observations of wind profiles from space using an active Doppler Wind Lidar, and also The Earth Clouds Aerosols and Radiation Explorer (EarthCARE) Mission, with the ATLID instrument onboard, which will be used to map global distributions of aerosols. Both satellite platforms will be equipped with HSRLs, with very different characteristics than CALIPSO. That will not provide a homogeneous set of aerosol measurements. A global ground-based lidar network providing stable, long-term measurements will be necessary to provide a benchmark against which to reference multiple satellite instruments.

6. Conclusion

In summary, while there is considerable evidence supporting the hypothesis that human activity is modifying climate, further research is required to strengthen the physical understanding. The lack of precise knowledge about all the processes of importance to climatic change appears to be limiting progress in furthering our quantitative understanding of human impact of climate. The largest uncertainty in current model predictions is related with the radiative effects of aerosols. There is a need for long-term monitoring of radiative properties of the aerosols-clouds system with high temporal and vertical resolution, as it plays a crucial role in climate. Recent advances in lidar technology allow to foresee a system capable of monitor the aerosol-cloud interaction with high temporal and spatial resolution taking advance of the HSRL for daytime measurements and Raman signals to derive aerosol and water vapor properties. Employing this advanced technology to probe Earth's atmosphere is expected to reduce the uncertainties in the climate forecasts. Regarding networks of instruments, advanced aerosol lidar systems are still relatively complex and delicate instruments to operate, therefore substantial engineering effort is still required towards increased reliability and automated operation. As a future perspective, further developments in algorithms that simultaneously invert complementary data acquired with lidars, satellites and Sun photometers, allowing for a maximum synergy of information, will provide better retrievals of aerosol and cloud microphysical properties. Integration of innovative new satellite observations such as CALIPSO, ADM-Aeolous or EARTHCARE, field experiments, and laboratory studies with models will pave the way for breakthroughs in our understanding of how aerosols are modifying the environment.

7. References

Ackerman, A.S., Toon, O.B., Stevens, D.E., Heymsfield, A.J., Ramanathan, V. & Welton, E.J. (2000). Reduction of tropical cloudiness by soot. Science, 288, 1042-1047

Ansmann, A., Riebesell, M. & Weitkamp, C. (1990) Measurement of atmospheric aerosol extinction profiles with a Raman lidar, Opt. Lett., 15, 746-748.

Baker, M. B. (1997). Cloud Microphysics and Climate. Science, 276, 1072-1078.

Böckmann, C., Mironova, I., Müller, D., Scheidenbach, L., Nessler,R. (2005) Microphysical aerosol parameters from multiwavelength lidar. J. Opt. Soc. Amer. A, 22, 518-528

Bosenberg, J & Hoff, R. (2007) Plan for theimplementation of the GAW Aerosol Lidar Observation Network, GALION. WMO GAW Report No 178, Hamburg, March 2007

Collis, R. T. H. and Russell, P. B. (1976). Lidar measurement of particles and gases by elastic backscattering and differential absorption. Laser Monitoring of the Atmosphere, Hinkley, E. D., pp. 71–152, Springler-Verlag, New York.

Cziczo, D. J. et al. (2009). Inadvertent climate modification due to anthropogenic lead. Nature geoscience 2, 333-336

Dubovik, O., Holben, B. N., Eck, T. F., Smirnov, A., Kaufman, Y. J., King, M. D., Tanr'e, D., & Slutsker, I. (2002).Variability of absorption and optical properties of key aerosol types observed in worldwide locations, J. Atmos. Sci., 59, 590–608

Eck, T. F., Holben, B. N., Reid, J. S., Dubovik, O., Smirnov, A., O'Neill, N. T., Slutsker, I., and Kinne, S. (1999). Wavelength dependence of the optical depth of biomass burning, urban, and desert dust aerosols, J. Geophys. Res., 104, 31 333–31 349

Fernald, F.G. (1984). Analysis of atmospheric lidar observations: some comments, Appl. Opt. 23, 652-653

Grabowski, W. W. & Wang, L. (2009). Diffusional and accretional growth of water drops in a rising adiabatic parcel: effects of the turbulent collision kernel. Atmospheric Chemistry and Physics 9, 2335-2353

Hoff, R. M., & Strawbridge, K. B. (1996). LITE observations of anthropogenically produced aerosols, Advances in Atmospheric Remote Sensing with Lidar, pp 145-148. 18th International Laser Radar Conference (ILRC), Springer-Verlag, Berlin, Germany

Holben, B. N., Eck, T. F., Slutsker, I., Tanre, D., Buis, J P., Setzer,A, Vermote, E., Reagan, J. A., Kaufman, Y. J., Nakajima, T. Lavenu, F., Jankowiak, & Smirnov, A. (1998). AERONET - A federated instrument network and data archive for aerosol characterization, Remote. Sens. Environ., 66, 1-16.

Kaufman et al. (2003). Retrievals of profiles of fine and coarse aerosols using lidar and radiometric space measurements, IEEE Transactions on geoscience and remote science, 41, 1743-1754

Klett, J.D.(1981). Stable analytic inversion solution for processing lidar returns, Appl. Opt. 20, 211-220.

Levin, Z & Cotton, W. R. (2009). Aerosol Pollution impact on precipitation: A scientific review, Springer ISBN: 140208689X 9781402086892.

Liepert, B.G. (2002). Observed reduction of surface solar radiation at sites in the United States and worldwide from 1961 to 1990. Geophys. Res. Lett. 29, 61.1-61.4

McCormick, M.P. (1997). The flight of the lidar In-Space Technology Experiment (LITE). Advances in Atmospheric Remote Sensing with lidar. Selected papers of the 18th International Laser Radar Conference (ILRC), 141-144, Berlin, July 1996, Springer, Berlin.

Müller, D., Wandinger, U. and Ansmann, A. (1999). Microphysical particle parameters from extinction and backscatter lidar data by inversion with regularization: Simulation. Appl. Optics, 38, 2358-2368

Müller, D., Mattis, I., Ansmann, A., Wandinger, U., Ritter, C. & Kaiser, D. (2007). Multiwavelength Raman lidar observations of particle growth during long-range transport of forest-fire smoke in the free troposphere, Geophys. Res. Letts., 34, 2006GL027936

Ramanathan, V., Crutzen, P. J., Kiehl, J. T., Rosenfeld, D. (2001). Aerosols, Climate, and the Hydrological Cycle, Science 294, 2119-2124

Reitebuch, O., Chinal, E., Durand, Y., Endemann, M., Meynart, R., Morancais, D., Paffrath, U. (2004). Development of an airborne demonstrator for ADMAeolus and campaign activities. 22nd Int. Laser Radar Conference (ILRC), pp. 1007–1010. Matera, Italy, ESA SP-561,

Rosenfeld, D. (2000). Suppression of rain and snow by urban and industrial air pollution. Science 287,1793-1796

Shipley, S.T., Tracy, D.H., Eloranta, E.W., Trauger, J.T., Sroga, J.T., Roesler, F.L. & Weinman, J.A. (1983) A High Spectral Resolution Lidar to Measure Optical Scattering Properties of Atmospheric Aerosols, Part I: Instrumentation and Theory, Applied Optics, 23, 3716-3724

Solomon, S. (2007). Climate Change 2007: The physical Science Basis. Contribution of Working Group I to the Fourth assessment Report of the Intergovernmental Panel on Climate Change, Cambridge Univ. Press, Cambridge, UK

Twomey, S. (1977). Influence of pollution on the short-wave albedo of clouds. J. Atmos. Sci. 34, 1149-1152

Welton, E. J., Campbell, J. R. & Spinhirne, J. D. (2001) Global monitoring of clouds and aerosols using a network of micropulse LIDAR systems, Proc. LIDAR Remote Sensing for Industry and Environmental Monitoring, 4153, 151−158.

Winker, D. M., Hunt, W. H., and McGill, M. J. (2007). Initial performance assessment of CALIOP, Geophys. Res. Lett., 34, L19803, doi:10.1029/2007GL030135.

5

Blended Tools for Remote Sensing

Mu-Lin Wu
Diwan College of Management
Taiwan

1. Introduction

The mission of this chapter is to address that blended remote sensing tools can work more effectively than one single type of remote sensing software package, even several combined packages. In solving real world problems, remote sensing is usually not working alone. Geographic information systems (GIS), global positioning systems (GPS), and remote sensing are implemented simultaneously. Blended remote sensing tools can make GIS, GPS, and remote sensing more user-friendly. College students with blended remote sensing tools training can implement commercial remote sensing software packages more effectively. Productivity of college students can be promoted in critical thinking and problem solving by remote sensing technology.

National park management has to look into conservation, research, recreation, and environmental education simultaneously(Wu et al., 2001). One type of brand-name commercial software for national park management has its draw back. It was simply not good enough, even not user-friendly. Blended tools were then developed to solve many different types of problems relevant with environment and land. House management for land use enforcement at a watershed(Chang et al., 2001), water resource protection(Wu et al., 2001; Wu et al., 2002; Wu et al., 2003a), sewage management(Wu et al., 2003b), forest management for a county government(Wu et al., 2004), and remote sensing education at college(Wu et al., 2006; Wu et al., 2007) were real world problems have been solved by blended remote sensing tools since 2001.

The basic idea behind blended remote sensing tools is quite simple and straight forward. One software package can do a nice job and several software packages with a little bit of computer programming will do a much better job. Blended remote sensing tools can save several problems simultaneously with less budget and more efficiently. Blended tools can be customized to meet one particular technician's requirement in order to solve environmental problems in a period of time. Blended remote sensing tools were developed to provide different functions in a changing world. File format is the key component that blended remote sensing tools can be working smoothly among several different software packages and platforms. File compression is also a must when remote sensing is implemented to solve problems associated with high resolution images. Remote sensing application is usually a web type of job that file size has to be monitored all the time.

Computer programming is not a must when blended remote sensing tools were developed. But a little bit of computer programming would make blended remote sensing tools more user-friendly. Computer programming can leave to some specialists. In college remote sensing education, the instructors can ask some brilliant students to do computer programming. For the most of students, they would only modify several lines of computer codes. Trying to keep computer programming as simple as possible is critical when blended remote sensing tools would be working as expected. Easy to use and everyone likes to use are two important factors that blended remote sensing tools can be useful and operational. Working side by side with brand-name commercial software packages is usual a must for blended remote sensing tools. Software packages cited in this chapter are simple that they are available and useful.

2. How Blended Remote Sensing Tools Were Made

2.1 Problems Identified and Well Defined

Blended remote sensing tools for college education are a little bit different with those implemented in real world. The number of students enrolled in a remote sensing class is usually more than 30. Every student has one set of tools for homework assignment and computer jobs. Software packages providing full function that can be tried and tested for a period of time can only be implemented. Outside campus, budget is the major concern when a government agency would like to ask blended remote sensing tools be made. For research and teaching, software packages can be tried and tested are almost a must.

Problems can be solved by remote sensing technology alone are usually no needs for blended remote sensing tools to be made. In the first place, real world problems can be solved with GIS, GPS, and remote sensing simultaneously will find blended remote sensing tools useful. Remote sensing can not solve all kinds of problems but can solve some identified problems. Problems should be identified and well defined such that GIS, GPS, and remote sensing would work properly as expected.

Remote sensing and GIS software packages are available and powerful in the commercial market. GPS functions are usually provided with GIS software packages working in a mobile device for outdoor implementations. How many types of tools would be used is decided by what kind of problems to solve and how problems to be solved. Nine types of tools had been adopted for solving real world problems such as college remote sensing education (Wu et al., 2007a). Blended remote sensing tools were consisting of followings:

Remote sensing packages: ERMapper, PG-STEAMER(Pixoneer Geomatics, 2009); GIS packages: ArcMap, ArcPad; 3-D interactive graphics: Alice (Carnegie Mellon University, 2009); image enhancement: PG-STEAMER, SnagIt (Techsmith, 2009); image publishing for web mapping: MapViewSVG, Alice; advanced mathematics: MatLab; image file format editor: ECW header editor; image compression: ECW image compressor; development of application modules and database management: Visual Basic, Visual C++, Visual Basic.NET, Visual C#.NET (Wu et al., 2007a).

MapObject made by ESRI can be added using Visual Basic or Visual C# to make it as major components (Wu et al., 2007c; Wu et al., 2007d). The instructor has to do most of the computer jobs when students knew nothing about Visual C# or Visual Basic. The students only modified several lines of compute codes and jobs were done accordingly.

Teechart for Net can be implemented for statistical implementations (Steema Software, 2009). HyperCAM (Hyperionics Technology, 2009) can be used to capture the action from one's Windows screen and save it to an AVI (Audio-Video Interleaved) movie file.

Fig. 1. Alice web site. (http://www.alice.org/)

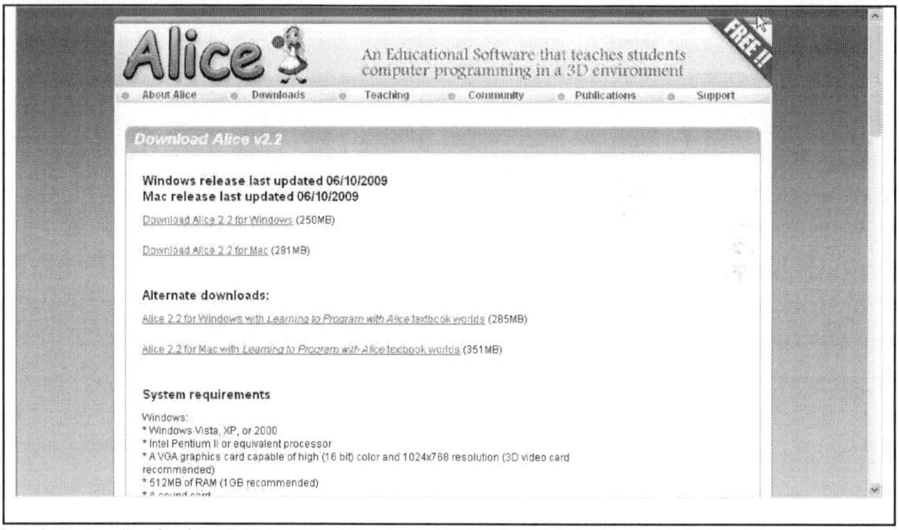

Fig. 2. Download Alice 2.2.

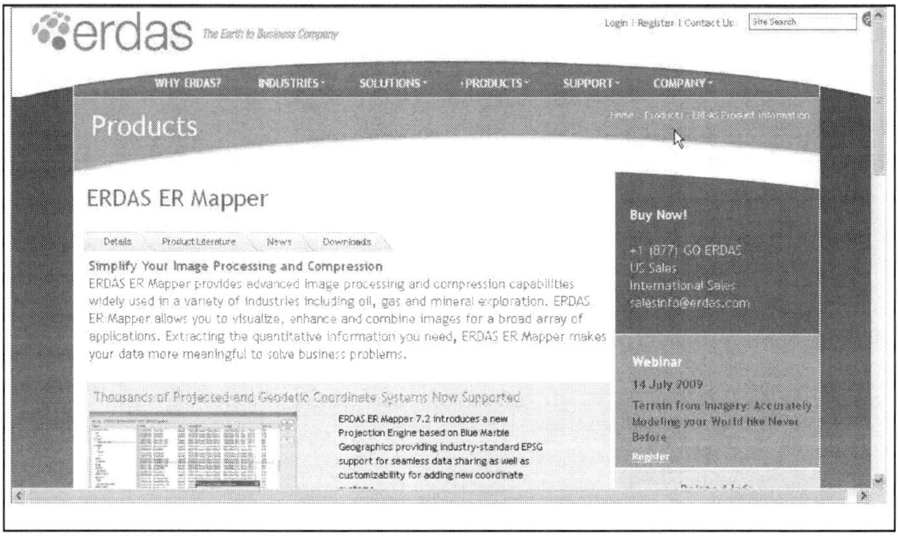

Fig. 3. ERDAS ER Mapper. (http://www.ermapper.com)

Fig. 4. PG-STEAMER downloads. (http://www.pixoneer.com/)

Fig. 5. SnagIt download.(http://www.techsmith.com/)

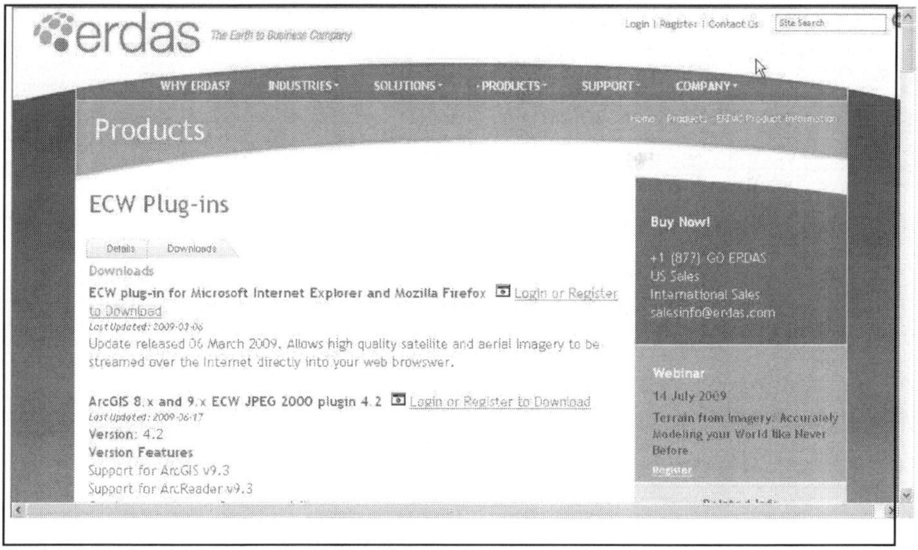

Fig. 6. ECW plug-in downloads. (http://www.ermapper.com)

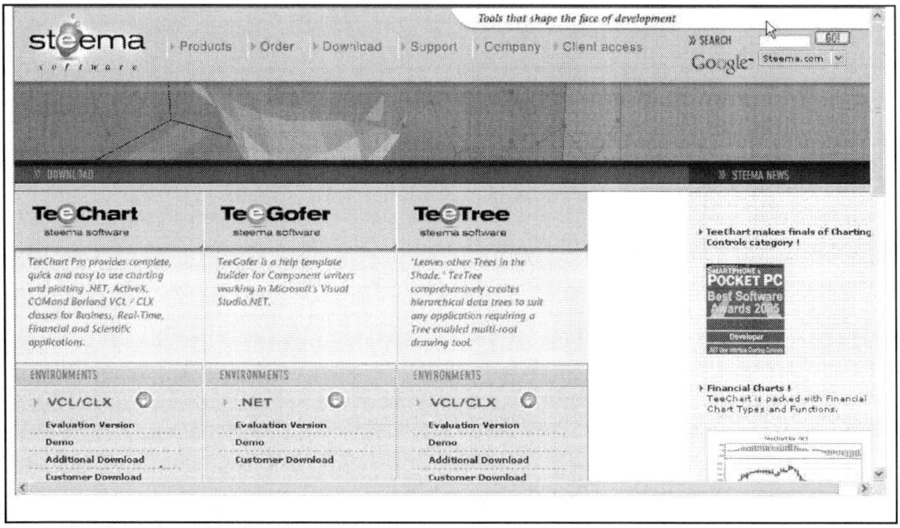

Fig. 7. Teechart downloads. (http://www.steema.com/)

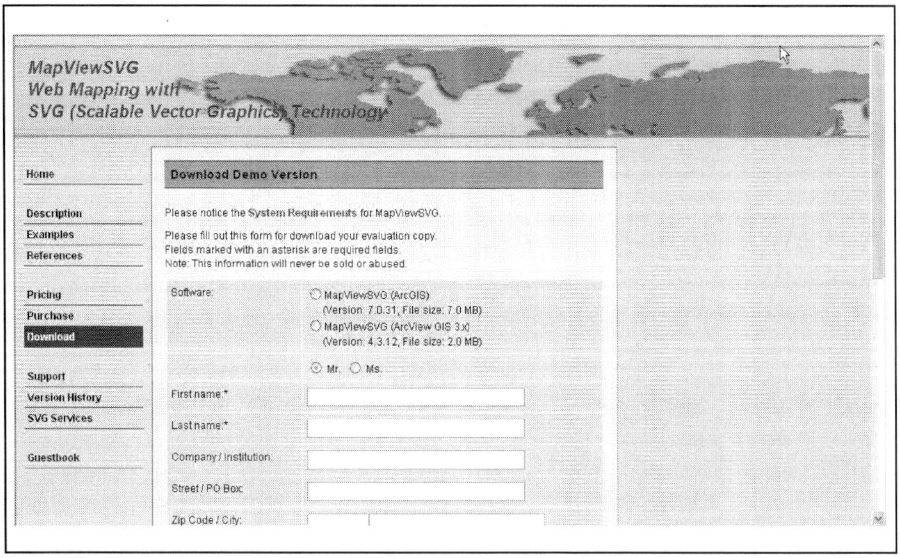

Fig. 8. MapViewSVG download. (http://www.mapviewsvg.com)

2.2 Working Procedure

Remote sensing software packages such as ERMapper or PG-STEAMER can be used for typical remote sensing implementations (Wu et al., 2007a). Digital images were manipulated and enhanced by PG-STEAMER or SnagIt to improve their information contributions. The coordinate information for a given pixel is very important for digital images to overlay with relevant maps. The ECW header editor program was used to retain and manipulate coordinate information of digital images. ArcGIS and ArcPAD made by ESRI were used to enhance integration of remote sensing and GIS. Database management in remote sensing was manipulated by computer programs using Visual Basic.NET and Visual C#.NET. Alice was used to make 3-D interactive graphics for web page implementations. Remote sensing images can be published for web mapping in SVG (scalable vector graphics) file format using MapViewSVG.

3. Functions Provided by Blended Remote Sensing Tools

Functions provided by blended remote sensing tools may not be confined to every single tool involved. Computer programming and system integration would make blended remote sensing tools more powerful. Image file format conversion is one of the basic functions which blended remote sensing tools can provide. Image compression is essential and critical when high quality of images was used, especial in the case of PDA (personal digital assistant). An ECW file can be small as 1.6 MB instead of its counter part, TIFF, 46 MB. Change detections can be done in a more easy way because of image enhancement and manipulations provided by blended remote sensing tools. Database management can be done in a simple and customized version when Visual Baisc.NET or Visual C#.NET was used. Web publishing is usually a must that blended remote sensing tools have to support. Management problems encountered in a government office or in the open fields would be solved by several different types of application modules. Blended remote sensing tools would serve as the major components in these application modules.

4. Application Modules for Solving Identified Management Problems

An image extraction module for quick review of images, a 3-D Application module, and a web application module were made by some college students (Wu et al., 2007a). The 3-D application module developed by one senior college student can be used for extractions of terrain animation done by PG-STEAMER, web pages showing interactive 3-D graphics done by Alice, and an image extraction program for remote sensing image files with ECW file format.

An image processing module developed by one college student and a statistical application module made by a college student using TeeChart were shown in reference (Wu et al., 2007c). A GIS application module made by a college student showing tracking and inquiry capability of one's home town indicated that blended remote sensing tools could provide GIS functions as well (Wu et al., 2007c).

A database management module made by a student using Visual Basic.NET which provided functions for easy and quick retrieval of digital images (Wu et al., 2007d). An application module for extraction of tourist attraction site information was also developed in the same publication.

House management in a government agency was usually working with traditional management information systems. Remote sensing can provide very detail information for a given house because of digital maps at a scale of 1:1000. One township was investigated house by house to tag house locations on digital orthophoto maps with its house address number. There were more than 6,500 houses and about 121,000 hectares had been investigated and created as house databases in 2006. The house databases were very useful in agricultural mapping and monitoring for a county government. Blended remote sensing tools were developed to make the house databases more useful in agricultural management (Wu et al., 2007b). Management on a house basis was easy with the help of a GPS device and PDA because it was not difficult to find the right path to get the right location of a given house. Change detection of house information in the period 2003 to 2006 could be done easily. This information was very important in urban planning and agricultural management. Field operations, agricultural mapping and monitoring, and web publishing were three more types of functions provided by blended remote sensing tools for a county government (Wu et al., 2007b).

5. Conclusions

Remote sensing education at college is usually done by textbooks teaching and computer job assignments. This chapter indicated blended remote sensing tools can be made for college teaching. Cases had been discussed in application modules made by college students using software packages in the blended remote sensing tools and computer programming with four types of computer languages. Digital image compression, file format conversion, image enhancement, database manipulation, interactive 3-D graphics, image retrieval and manipulations, terrain animation, GIS integration, and web publishing of remote sensing images were functions provided by the blended remote sensing tools. College students can have some experiences on system integration of these application modules to solve one given management problem with the help of computer programming. The blended remote sensing tools are good for college students to gain better computer programming and system integration capabilities in addition to traditional remote sensing capabilities.
House management for a county government using blended remote sensing tools had been discussed to indicate how a real world problem can be solved. Blended remote sensing tools can be modified based on request made by any customer. Blended remote sensing tools can make house management in agricultural monitoring and mapping more user-friendly and effectively. Agricultural mapping and monitoring can provide detailed information for a given house such as where it was, who lives there, and what can be done. Blended remote sensing tools can be used to make some application modules for agricultural mapping and monitoring to solve problems encountered in agricultural management. Blended remote sensing tools can solve real world problems which were well identified and defined. Blended remote sensing tools were designed to solve problems that are suitable for GIS, GPS, and remote sensing working together.

6. References

Carnegie Mellon University, (2009). Alice 2.2, http://www.alice.org/

Chang, Y.; Tai, S.; Song, D.; Liu, S. & Wu, M. (2001). House Management for Land Use Enforcement at a Watershed Using a Self-Developed Web-Based GIS, Proceedings of International Geoscience and Remote Sensing Symposium (IGARSS 2001), Sydney, Australia, July 2001, 3 pages.

Earth Resource Mapping Pty Ltd. (2002). ER Mapper ECW ActiveX SDK, Release 2.46, 304 pages.

Hyperionics Technology, (2009). HyperCAM, http://www.hyperionics.com/
Pixoneer Geomatics, (2009). PG-STEAMER, http://www.pixoneer.com/
Steema Software, (2009). Teechart for Net, http://www.steema.com/
Techsmith, (2009). SnagIt, http://www.techsmith.com/
Wu, M.; Lin, Y.; Yang, J. & Chung, C.(2001). National Park Management Using Self-Developed and Web-Based Geographic Information Systems, Proceedings of International Geoscience and Remote Sensing Symposium (IGARSS 2001), Sydney, Australia, July 2001, 3 pages.
Wu, M.; Chen, C. & Song, D. (2001). A Self-Developed and Web-Based Geographic Information System for Water Resource Protection, Proceedings of the 20th International Cartographic Conference (ICC 2001), Beijing, China, August 2001, 7 pages.
Wu, M.; Chen, C.; Tai, S.; Chou, W. & Huang, H. (2001). Remote Sensing Applications in Water Resource Protection. Proceedings of the 22nd Asian Conference on Remote Sensing (ACRS 2001), Vol. 1, pp. 706-711, Singapore.
Wu, M.; Chen, C.; Liou, S. & Wey, J. (2002). Mobile Geographic Information Systems for Water Resource Protection. Proceedings of Asian Conference on Remote Sensing (ACRS 2002), Kathmandu, Nepal, November 2002, 4 pages.
Wu, M.; Chen, C.; Liu, S. & Wey, J. (2003). Implementations of Remote Sensing, GIS, and GPS for Water Resources and Water Quality Monitoring, Proceedings of Asian Conference on Remote Sensing (ACRS 2003), BEXCO Busan, Korea, November 2003, 3 pages.
Wu, M.; Chen, C.; Chou, W. & Huang, H. (2003). Implementations of Geographic Information Systems on Sewage Management for Water Resources Protection, Proceedings of Asian Conference on Remote Sensing (ACRS 2003), BEXCO Busan, Korea, November 2003, 3 pages.
Wu, M.; Chen, S.; Fu, C.; Fan, J.& Fu, S.(2004). Development of a Government GIS for Forest Management at Hsinchu County in Taiwan. Proceedings of the International Symposium on Remote Sensing 2004 & 20th Anniversary of the Korean Society of Remote Sensing (2004 ISRS), October 2004, Jeju, Korea, 4 pages.
Wu, M.; Tai, S.; Chou, W.; Song, D.; Liu, S. & Yang, T. (2004). Task-Oriented GIS for Water Management at Taipei Water Resource District, Proceedings of the International Symposium on Remote Sensing 2004 & 20th Anniversary of the Korean Society of Remote Sensing (2004 ISRS), Jeju, Korea, October 2004, 3 pages.
Wu, M.; Liu, W.; Chang, Y. & Lu, J. (2004). GIS for Decision Support of Water Resource Management at Taipei Water Management Office in Taiwan, Proceedings of the International Symposium on Remote Sensing 2004 & 20th Anniversary of the Korean Society of Remote Sensing (2004 ISRS), Jeju, Korea, October 2004, 3 pages.
Wu, M.; Wang, Y.; Fang, C. & Huang, Y. (2006). An Image Information System for Management of Watersheds in Remote Sensing Education, Proceedings of IEEE Geoscience and Remote Sensing Society (IGARSS 2006) , pp.2017-2020, Denver, CO, U. S. A., July 31 to August 4, 2006.
Wu, M.; Chu, C.; Wang, Y.; Wong, D. & Hwang, M. (2007), Blended Tools for Remote Sensing Education, Proceedings of 2007 IEEE International Geoscience And Remote Sensing Symposium (IGARSS 2007), pp. 2217-2219, July 2007, Barcelona, Spain.

Wu, M.; Chu, C.; Wang, Y.; Wong, D. & Hwang, M. (2007), Blended Remote Sensing Tools for House Management, Proceedings of 2007 IEEE International Geoscience And Remote Sensing Symposium (IGARSS 2007), pp.1986-1989, Barcelona, Spain, July 2007.

Wu, M.; Chu, C.; Wong, D. ;Wang, Y. & Hwang, M. (2007), Remote sensing and GIS integration for college GIS education, Proceedings of The 28th Asian Conference on Remote Sensing (ACRS 2007), November 2007, Kuala Lumpur, Malaysia, 6 pages.

Wu, M.; Wong, D. ; Chu, C.; Wang, Y. & Hwang, M. (2007). Innovation problem solving methodologies for college remote sensing education, Proceedings of the 28th Asian Conference on Remote Sensing (ACRS 2007), Kuala Lumpur, Malaysia, November 2007, 5 pages.

6

Image Information Mining Systems

Inés María Gómez Muñoz and Mihai Datcu
German Aerospace Center (DLR)
Germany

1. Introduction

Image satellite sensors acquire huge volumes of imagery to be processed and stored in big archives. An example of such an archive is the German Remote Sensing Data Center (DFD) at Oberpfaffenhofen, Germany, that receives about hundreds of GigaBytes of data per day entailing 104 GigaBytes in the repository. To provide access to this data, web applications have been developed, e.g. the DLR EOWEB[1], to retrieve images according to meta information such as date, geographical location or sensor. Alexandria Digital Library[2] is another example of accessing remote sensed imagery through its meta information providing a distributed searching mechanism for retrieving geospatial referenced data collections. It is able to search different types of databases placed at different locations. The software enables to implement web clients as Globetrotter[3] or Gazetteer[4]. These systems based on meta information retrieval allow only constrained queries giving no information about the content, and consequently, no content based retrieval is offered.

At the conference on database techniques for pictorial applications that took place in 1979 in Florence, Italy, the pursued aim was the integration of databases with image processing. This idea evolved, in 1990, promoting a new field, called Content Based Image Retrieval (CBIR). In 1998, CBIR got married with Data Mining and Knowledge Database Discovery (KDD) emerging, in 2000, the Image Information Mining (IIM) field. This new domain requires expertise in image processing, database organization, pattern recognition, content-based retrieval and data mining: image processing indicates the understanding and extraction of patterns from a single image; content-based retrieval is characterized by retrieving images from the archive based on their semantic and visual contents; spatial data mining denotes the extraction of spatial relationships and patterns from remote sensed images not explicitly stored in an spatial database. An IIM system provides users the capability to deal with large collections of images by accessing into large image databases and also to extract and infer knowledge about patterns hidden in the images, so that the set

---

[1] http://eoweb.dlr.de:8080/servlets/template/welcome/entryPage.vm.

[2] http://www.alexandria.ucsb.edu/adl/.

[3] http://clients.alexandria.ucsb.edu/globetrotter/

[4] http://webclient.alexandria.ucsb.edu/client/gaz/adl/index.jsp.

of relevant images is dynamic, subjective and unknown. It enables the communication between heterogeneous source of information and users with diverse interests at high semantic abstraction.

In general, an IIM system presents two fundamental modules: a computationally expensive component where image processing and classification algorithms are executed, and an interactive part, where queries are introduced by the user and relevant images are retrieved. Fig. 1 represents the typical flow of a data in an IIM system: original data arrive at a feature extraction module, where main image characteristics are computed; then, these features are compressed and indexed in a database; in a second module, the archive is queried by the user for similar features computing similarity measurements for optimal image retrieval.

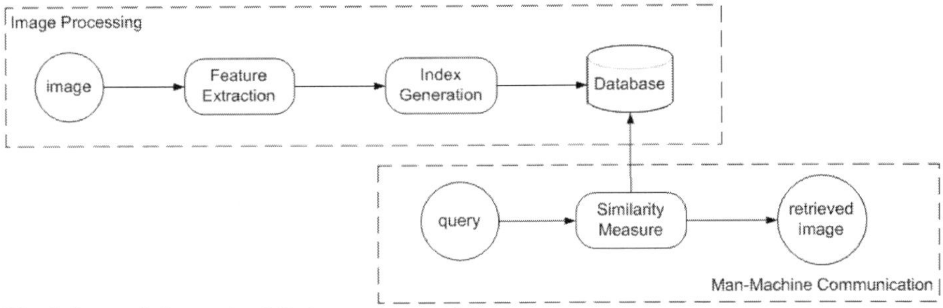

Fig. 1. Image Information Mining system architecture.

This chapter begins describing the generic concept and modules of an IIM system architecture, and Sec. 0 presents an overview of existing IIM systems.

2. Image Information Mining System Architecture

As depicted in Fig. 1, the generic concept of an IIM system requires several processing modules: extraction of properties from images, reduction and content indexation and communication between users and system. In this section, we present the state of the art of these modules giving an overview of existing techniques in these fields.

2.1 Feature Extraction

In general, by image we understand picture, thus relating it to the (human) visual perception and understanding. A picture is characterized by its primitive features such as colour, texture or shape at different scales. Thus, an image will be represented as a multidimensional feature vector acting as signature. Some classical techniques to characterize an image are the following:

- Colour: Colour information has been an important feature in image processing and computer vision. There exist different colour models or colour spaces, each one being useful for a specific application. A digital imaging system typically represents colour images in red, green, blue using the RGB space. Another one related with the perception of the colours by human beings is the HSV (Hue,

Saturation and Intensity) colour space. This one describes the property of the surface reflecting the light (hue), measures the colourfulness or whiteness (saturation) and the brightness (intensity) of colours. Often a full colour image providing the three colours (RGB) in each pixel is needed, being essential to interpolate missing colours with the information of neighbouring pixels. There are nonadaptive algorithms (Ray & Acharya, 2005) as nearest neighbour replication and bilinear interpolation, and adaptive algorithms (Ray & Acharya, 2005) based on pattern matching or edge sensing interpolation. On the other side, a common practise in image processing is the statistical analysis of colour histograms, due to the strong correlation between objects and colour in an image.

- Texture: Texture is a very interesting feature to characterize the spatial structure of an image. This is an active research field where parametric and non parametric methods are applied. Haralick's co-occurrence (Shanmugam et al., 1973) technique based on the computation of the gray-level co-occurrence matrix for several values of displacement, orientation and image quantization levels is an effective method in texture analysis. Other algorithms based on wavelet transformations as the computation of Gabor filter (Maillot et al., 2005) can also be applied.

- Shape: Shape of objects must be invariant to translation, rotation and scale of the image and is characterized in two senses: boundary-based, that considers the object outer contour, and region-based, where the whole shape region of the object is analyzed. In this sense, Fourier descriptors are suitable for transforming boundaries into shape features, and moment invariants for the extraction of geometric object region. A modified Fourier descriptor that preserves the invariance of geometric transformations and noise is proposed in (She et al., 1998). A common practise before applying shape techniques is to segment the image in small regions. Comaniciu (Comaniciu & Meer, 2002) presents this approach based on the mean shift method for density gradients estimation.

- Topology: topological properties of an image such as number of connected or disconnected components, do not change when an image is rotated, scaled, translated, stretched or deformed. One example of characterizing an image through its topological properties is the computation of the Euler number (Ray & Acharya, 2005). It is defined as the difference between number of connected components and number of holes in a binary image. An extension of the Euler number defined for binary images is the Euler vector (Ray & Acharya, 2005) that can be applied to gray-level images. Segmentation techniques may also help in the extraction of topological features.

2.2 Multidimensional Indexing

In the CBIR and IIM domain, the concept of multidimensional indexing differs from the one in a traditional database management system. In here, an index consists of the structure that provides access to the database in terms of record organization. In IIM, once an N-dimensional feature vector is obtained, images are assigned to a suitable content based description extracted from these features. These content descriptors are then organized into a data structure for retrieval.

In multidimensional indexing, the following items must be considered:

- Reduction of dimensionality: Due to the huge amount of images and extracted features, normally the dimensionality of the information at the indexing step is very high. This complicates the management of the feature vector rendering its computation very expensive. For this reason, mechanisms for reducing the dimension of the feature space must be considered. Among these methods, Karhunen-Loève transform (Ray & Acharya, 2005) and the Discrete Cosine Transform (DCT) (Khayam, 2003); (Watson, 1994) are often considered.
- Clustering: Extracted features with similar content must be grouped together through a classification algorithm. In this case, pixels containing similar features belong to the same class. Existing clustering techniques can be classified into two main groups: distance-based and model-based (Zhong & Ghosh, 2003) approaches. In the first group, we mention methods based on Euclidean and Mahalanobis distances, and to the second group belongs algorithms based on an a priori specified model, such as Gaussian mixture models or Markov chains.
- Data structure for content based retrieval: Once a clustering algorithm is performed, a data structure for indexing descriptors to semantic content must be selected. The common used methods are tree-based indexing techniques, as multidimensional binary search trees or R-trees, and hashing-based ones.

2.3 Content-Based Image Retrieval

Usually CBIR is limited by the semantic gap existing between signal classes and semantic labels. Li et al. (Li & Bretschneider, 2006) propose a context sensitive Bayesian network to infer the semantic concept of regions or classes. Semantic score functions based on region features (spectral and texture) are computed to link semantic concepts to regions. Tusk et al. (Tusk et al., 2002) suggest a Bayesian framework to cope with the semantic gap problem. They introduce a visual grammar that builds a hierarchical semantic model from pixel level to region and scene levels. Pixel-level characteristic provides classification by automatic fusion of primitive features; then, at region-level through a segmentation algorithm land cover labels are defined; and scene-level represents the spatial relationship among regions. Thus, the visual grammar consists of two learning steps, where naive Bayesian classifiers are applied: a probabilistic link between features and semantic labels, and a fuzzy modelling to link regions and scenes. Once the visual grammar is built, the image classification process aims at finding representative region groups that describe the scene. The procedure consists of modelling the labelled regions by a Dirichlet distribution based on the number of training examples containing a certain region group, and then, assigning the best matching class to image by using the maximum a-posteriori rule.

In order to provide the system the ability to search at query-time for images with similar features, a similarity metric for the comparison of objects or image properties must be defined. If we want a realistic measure, computer and human judgments of similarity should be generally correlated. If this condition is not met, images returned by the system will not be those desired by the user. These techniques are often based on distances or on a specific domain as histogram intersection, neural networks, shape measures or graph matching. Queries like "retrieve images containing an specific content" or "retrieve images that do not contain a particular object" can be asked to a CBIR system.

2.4 Semantic Learning for Content-based Image Retrieval

The main problem of using feature vectors for querying images with similar content is that often, the appearance of an image does not correspond to its semantic meaning, making the returned images only partially responds to the users query. Therefore, at object or region level, the highest level of abstraction, an image is represented by its objects, and a semantic label is assigned to each of them.

A common used technique to provide regions with semantic meaning is the manual annotation that, combined with a powerful segmentation method, can result in a good meaningful classification. Comaniciu (Comaniciu & Meer, 2002) proposes a colour image segmentation algorithm based on the mean shift that estimates density gradients, using a simple nonparametric procedure. Then, the users interactively identify the segmented regions by labelling the features. Because of hand-annotating images is tedious and human expensive, methods for learning image representations directly from data are investigated.

Fei-Fei and Perona (Fei-Fei & Perona) propose a Bayesian hierarchical model to learn and recognize natural scene categories through intermediate "themes". In there, the most complete scene category dataset found in the literature is used. An image is modelled as a collection of local patches (regions). Each patch is represented by a codeword from a large vocabulary of them obtained from all categories training examples. For each codewords in each category, a Bayesian hierarchical model is learnt, building a collection of Bayesian models. Then, to provide semantic meaning to an unknown image, first the image codewords are extracted, and then, they are compared with the predefined models, assigning the one which fits best. The main problem of the proposed algorithm is that, although it can learn intermediate themes of scenes with neither supervision nor human intervention, the categories are fixed, being not able to assign semantic meaning to other ones.

Another method that uses predefined lexicon of semantic concepts as trained data is the semantic pathfinder for multimedia indexing (Seinstra et al., 2006). In here, given a pattern x, part of a camera shot, the aim is to detect a semantic concept ω from shot i using probability $p(\omega \mid x_i)$. Each step in the semantic pathfinder analysis extracts x_i from data, and learns $p(\omega \mid x_i)$ for all ω in the semantic lexicon.

Maillot et al. (Maillot et al., 2005) propose a learning approach based on two steps: a feature selection step that chooses the most characterizing features for better visual concept detection, and a training phase using a Support Vector Machine (SVM), where positive and negative samples are required. Trying to solve the weaknesses of the learning approach like the lack of learning the spatial structure of semantic concepts, a further step is given, storing the visual knowledge that is the link between semantic concepts and sensor data in a symbol. This link is modelled as a fuzzy linguistic variable that enables the representation of imprecision, thus the image features are fuzzified a priori by a human expert, providing spatial relation representations and spatial reasoning.

In these articles, we find two facts that we try to avoid: On one hand, the lack of generalization by using a predefined lexicon when trying to link data with semantic classes. The use of a semantic lexicon is useful when we arrange an a priori and limited knowledge, and, on the other hand, the need of experts in the application domain to manually label the regions of interest.

An important issue to arrange while assigning semantic meaning to a combination of classes is the data fusion. Li and Bretschneider (Li & Bretschneider, 2006) propose a method where combination of feature vectors for the interactive learning phase is carried out. They propose an intermediate step between region pairs (clusters from k-means algorithm) and semantic concepts, called code pairs. To classify the low-level feature vectors into a set of codes that form a codebook, the Generalised Lloyd Algorithm is used. Each image is encoded by an individual subset of these codes, based on the low-level features of its regions.

Signal classes are objective and depend on feature data and not on semantics. Chang et al. (Chang et al., 2002) propose a semantic clustering. This is a parallel solution considering semantics in the clustering phase. In the article, a first level of semantics dividing an image in semantic high category clusters, as for instance, grass, water and agriculture is provided. Then, each cluster is divided in feature subclusters as texture, colour or shape. Finally, for each subcluster, a semantic meaning is assigned.

In terms of classification of multiple features in an interactive way, there exist few methods in the literature. Chang et al. (Chang et al., 2002) describe the design of a multilayer neural network model to merge the results of basic queries on individual features. The input to the neural network is the set of similarity measurements for different feature classes and the output is the overall similarity of the image. To train the neural network and find the weights, a set of similar images for the positive examples and a set of non similar ones for the negative examples must be provided. Once the network is trained, it can be used to merge heterogeneous features.

To finish this review in semantic learning, we have to mention the kind of semantic knowledge we can extract from EO data. The semantic knowledge depends on image scale, and the scale capacity to observe is limited by sensor resolution. It is important to understand the difference between scale and resolution. The term of sensor resolution is a property of the sensor, while the scale is a property of an object in the image. Fig. 2 depicts the correspondence between knowledge that can be extracted for a specific image scale, corresponding small objects with a scale of 10 meters and big ones with a scale of thousands of meters. The hierarchical representation of extracted knowledge enables answering questions like which sensor is more accurate to a particular domain or which are the features that better explain the data.

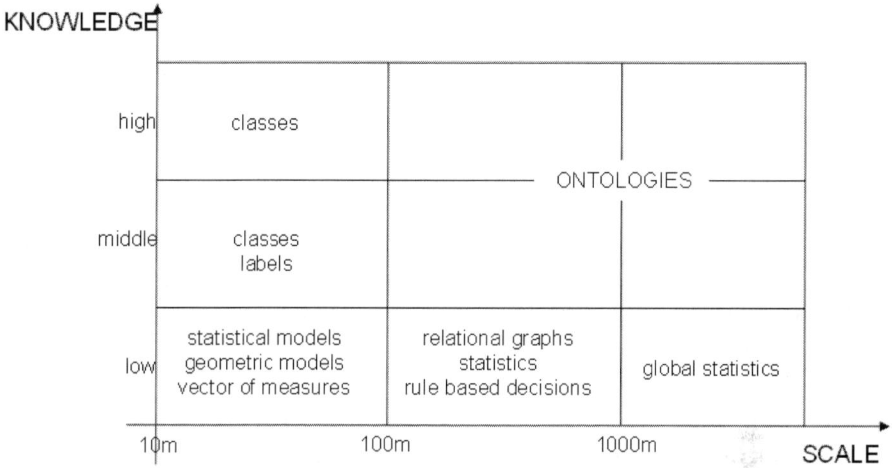

Fig. 2. Knowledge level in the hierarchy to be extracted depending on the image scale.

2.5 Relevance Feedback

Often an IIM system requires a communication between human and machine while performing interactive learning for CBIR. In the interaction loop, the user provides training examples showing his interest, and the system answers by highlighting some regions on retrieved data, with a collection of images that fits the query or with statistical similarity measures. These responses are labelled as relevance feedback, whose aim is to adapt the search to the user interest and to optimize the search criterion for a faster retrieval.

Li and Bretschneider (Li & Bretschneider, 2006) propose a composite relevance feedback approach which is computationally optimized. At a first step, a pseudo query image is formed combining all regions of the initial query with the positive examples provided by the user. In order to reduce the number of regions without loosing precision, a semantic score function is computed. On the other hand, to measure image-to-image similarities, they perform an integrated region matching.

In order to reduce the response time while searching in large image collections, Cox et al. (Cox et al., 2000) developed a system, called PicHunter, based on a Bayesian relevance feedback algorithm. This method models the user reaction to a certain target image and infers the probability of the target image on the basis of the history of performed actions. Thus, the average number of man-machine interactions to locate the target image is reduced, speeding up the search.

3. Existing Image Information Mining Systems

As IIM field is nowadays in its infancy, there are only a few systems that provide CBIR being under evaluation and further development. Aksoy (Aksoy, 2001) provides a survey of CBIR systems prior to 2001, and a more recent review is provided by Daschiel (Daschiel,

2004). In this section, we present several IIM systems for retrieval of remote sensed images, most of them being experimental ones.

Li (Li & Narayanan, 2004) proposes a system, able to retrieve integrated spectral and spatial information from remote sensing imagery. Spatial features are obtained by extracting textural characteristics using Gabor wavelet coefficients, and spectral information by Support Vector Machines (SVM) classification. Then, the feature space is clustered through an optimized version of k-means approach. The resulting classification is maintained in a two schemes database: an image database where images are stored and an Object-Oriented Database (OODB) where feature vectors and the pointers to the corresponding images are stored. The main advantage of an OODB is the mapping facility between an object oriented programming language as Java or C++, and the OODB structures through supported Application Programming Interfaces (API). The system has the ability of processing a new image in online mode, in such a way that an image which is not still in the archive is processed and clustered in an interactive form.

Feature extraction is an important part of IIM systems, however, it is computationally expensive, and usually generates a high volume of data. A possible solution would be to compute only those relevant features for describing a particular concept, but how to discriminate between relevant and irrelevant features? The Rapid Image Information Mining (RIIM) prototype (Shah et al., 2007) is a Java based framework that provides an interface for exploration of remotely sensed imagery based on its content. Particularly, it puts a focus on the management of coastal disaster. Its ingestion chain begins with the generation of tiles and an unsupervised segmentation algorithm. Once tiles are segmented, a feature extraction composed of two parts is performed: a first module consists of a genetic algorithm for the selection of a particular set of features that better identifies a specific semantic class. A second module generates feature models through genetic algorithms. Thus, if the user provides a query with a semantic class of interest, feature extraction will be only performed over the optimal features for the prediction, speeding up the ingestion of new images. The last step consists of applying a SVM approach for classification. While executing a semantic query, the system computes automatically the confidence value of a selected region and facilitates the retrieval of regions whose confidence is above a particular threshold.

The IKONA system[5] is a CBIR system based on client-server architecture. The system provides the ability of retrieving images by visual similarity in response to a query that satisfies the interest of the user. The system offers the possibility to perform region based queries in such a way that the search engine will look for images containing similar parts to the provided one. A main characteristic of the prototype is the hybrid text-image retrieval mode. Images can be manually annotated with indexed keywords, and while retrieving similar content images, the engine searches by keyword providing a faster computation.
IKONA can be applied not only for EO applications, but also for face detection or signature recognition. The server-side architecture is implemented in C++ and the client software in

[5] http://www-rocq.inria.fr/cgibin/imedia/cbir-gen.cgi

Java, making it independent from the platform where it runs. The only prerequisite on the client is to have installed a Java Virtual Machine.

The Query by Image Content (QBIC)[6] system is a commercial tool developed by IBM that explores content-based retrieval methods allowing queries on large image and video databases. These queries can be based on selected colour and texture patterns, on example images or on user-made drawings. QBIC is composed of two main components: database population and database query. The former deals with processes related to image processing and image-video database creation. The latter is responsible for offering an interface to compose a graphical query and for matching input query to database. Before storing images in the archive, they are tiled and annotated with text information. The manual identification of objects inside images can become a very tedious task, and trying to automatize this function, a full automatic unsupervised segmentation technique based on foreground/background models is introduced. Another method to automatically identify objects, also included in this system, is the flood-fill approach. This algorithm starts from a single pixel and continues adding neighbour pixels, whose values are under a certain threshold. This threshold is calculated automatically and updated dynamically by distinguishing between background an object.

Photobook (Picard et al., 1994) developed by MIT, is another content-based image and image sequences retrieval, whose principle is to compress images for a quick query-time performance, reserving essential image similarities. Reaching this aim, the interactive search will be efficient. Thus, for characterization of object classes preserving its geometrical properties, an approach derived from the Karhunen-Loève transform is applied. However, for texture features a method based on the Wold decomposition that separates structured and random texture components is used. In order to link data to classes, a method based on colour difference provides an efficient way to discriminate between foreground objects and image background. After that, shape, appearance, motion and texture of theses foreground objects can be analyzed and ingested in the database together with a description. To assign a semantic label or multiple ones to regions, several human-machine interactions are performed, and through a relevance feedback, the system learns the relations between image regions and semantic content.

VisiMine system (Aksoy et al., 2002); (Tusk et al., 2002) is an interactive mining system for analysis of remotely sensed data. VisiMine is able to distinguish between pixel, region and tile levels of features, providing several feature extraction algorithms for each level. Pixel level features describe spectral and textural information; regions are characterized by their boundary, shape and size; tile or scene level features describe the spectrum and textural information of the whole image scene. The applied techniques for extracting texture features are Gabor wavelets and Haralick's co-ocurrence, image moments are computed for geometrical properties extraction, and k-medoid and k-means methods are considered for clustering features. Both methods perform a partition of the set of objects into clusters, but with k-means, further detailed in chapter 6, each object belongs to the cluster with nearest mean, being the centroid of the cluster the mean of the objects belonging to it. However,

---

[6] http://wwwqbic.almaden.ibm.com/

with k-medoid the center of the cluster, called medoid, is the object, whose average distance to all the objects in the cluster is minimal. Thus, the center of each cluster in k-medoid method is a member of the data set, whereas the centroid of each cluster in k-means method could not belong to the set. Besides the clustering algorithms, general statistics measures as histograms, maximum, minimum, mean and standard deviation of pixel characteristics for regions and tiles are computed. In the training phase, naive Bayesian classifiers and decision trees are used. An important factor of VisiMine system is its connectivity to SPLUS, an interactive environment for graphics, data analysis, statistics and mathematical computing that contains over 3000 statistical functions for scientific data analysis. The functionality of VisiMine includes also generic image processing tools, such as histogram equalization, spectral balancing, false colours, masking or multiband spectral mixing, and data mining tools, such as data clustering, classification models or prediction of land cover types.

GeoIRIS (Scott et al., 2007) is another IIM system that includes automatic feature extraction at tile level, such as spectral, textural and shape characteristics, and object level as high dimensional database indexing and visual content mining. It offers the possibility to query the archive by image example, object, relationship between objects and semantics. The key point of the system is the ability to merge information from heterogeneous sources creating maps and imagery dynamically.

Finally, Knowledge-driven Information Mining (KIM) (Datcu & Seidel, 1999); (Pelizzari et al., 2003) and later versions of Knowledge Enabled Services (KES) and Knowledge-centred Earth Observation (KEO)[7] are perhaps the most enhanced systems in terms of technology, modularity and scalability. They are based on IIM concepts where several primitive and non-primitive feature extraction methods are implemented. In the last version, of KIM, called KEO, new feature extraction algorithms can easily plugged in, being incorporated to the data ingestion chain. In the clustering phase, a variant of k-means technique is executed generating a vocabulary of indexed classes. To solve the semantic gap problem, KIM computes a stochastic link through Bayesian networks, learning the posterior probabilities among classes and user defined semantic labels. Finally, thematic maps are automatically generated according with predefined cover types. Currently, a first version of KEO is available being under further development.

4. References

Aksoy, S. A probabilistic similarity framework for content-based image retrieval. *PhD thesis, University of Washington*, 2001.

Aksoy, S.; Kopersky, K.; Marchisio, G. & Tusk, C. Visimine: Interactive mining in image databases. *Proceedings of the Int. Geoscience and Remote Sensing Symposium (IGARSS)*, Toronto, Canada, 2002.

[7] http://earth.esa.int/rtd/events/esa-eusc 2004/;
http://earth.esa.int/rtd/events/esa-eusc 2005/;
http://earth.esa.int/rtd/events/esa-eusc 2006/;
http://earth.esa.int/rtd/events/esa-eusc 2008/

Chang, W.; Sheikholeslami, G. & Zhang, A. Semquery: Semantic clustering and querying on heterogeneous features for visual data. *IEEE Trans. on Knowledge and Data Engineering*, 14, No.5, Sept/Oct 2002.

Comaniciu, D. & Meer, P. Mean shift: A robust approach toward feature space analysis. *IEEE Trans. on Pattern Analysis and Machine Intelligence*, 24, No. 5, May 2002.

Cox, I. J.; Papathomas, T. V.; Miller, M. L.; Minka, T. P. & Yianilos, P. N. The Bayesian image retrieval system pichunter: Theory, implementation, and psychophysical experiments. *IEEE Trans. on Image Processing*, 9, No.1:20–37, 2000.

Daschiel, H. Advanced Methods for Image Information Mining System: Evaluation and Enhancement of User Relevance. *PhD thesis, Fakultät IV - Elektrotechnik und Informatik der Technischen Universität Berlin*, July 2004.

Datcu, M. & Seidel, K. New concepts for remote sensing information dissemination: query by image content and information mining. *Proceedings of IEEE Int. Geoscience and Remote Sensing Symposium (IGARSS)*, 3:1335–1337, 1999.

Fei-Fei, L. & Perona, P. A bayesian hierarchical model for learning natural scene categories. *Califorina Institute of Technology*, USA.

Khayam, S. A. The discrete cosine transform (dct): Theory and application. *Department of Electrical and Computer Engineering*, Michigan State University, 2003.

Li, J. & Narayanan, R. M. Integrated spectral and spatial information mining in remote sensing imagery. *IEEE Trans. on Geoscience and Remote Sensing*, 42, No. 3, March 2004.

Li, Y. & Bretschneider, T. Remote sensing image retrieval using a context-sensitive bayesian network with relevance feedback. *Proceedings of the Int. Geoscience and Remote Sensing Symposium (IGARSS)*, 5:2461-2464, 2006.

Maillot, N.; Hudelot, C. & Thonnat, M. Symbol grounding for semantic image interpretation: From image data to semantics. *Proceedings of the Tenth IEEE International Conference on Computer Vision (ICCV'05)*, 2005.

Manjunath, B. S. & Ma, W. Y. Texture features for browsing and retrieval of image data. *IEEE Trans. on Pattern Analysis and Machine Intelligence*, 18, No.8:837–842, 1996.

Pelizzari, A.; Quartulli, M.; Galoppo, A.; Colapicchioni, A.; Pastori, M.; Seidel, K.; Marchetti, P. G.; Datcu, M.; Daschiel, H. & D'Elia, S. Information mining in remote sensing images archives - part a: system concepts. *IEEE Trans. on Geoscience and Remote Sensing*, 41(12):2923–2936, 2003.

Picard, R. W.; Pentland, A. & Sclaroff, S. Photobook: Content-based manipulation of image databases. *SPIE Storage and Retrieval Image and Video Databases II*, No. 2185, February 1994.

Ray, A. K. & Acharya, T. *Image Processing, Principles and Applications*. Wiley, 2005.

Scott, G. J.; Barb, A. S.; Davis, C. H.; Shyu, C. R.; Klaric, M. & Palaniappan, K. Geoiris: Geospatial information retrieval and indexing system - content mining, semantics modeling and complex queries. *IEEE Trans. on Geoscience and Remote Sensing*, 45:839–852, April 2007.

Seinstra, F. J.; Snoek, C. G. M.; Geusebroek, J.M. & Smeulders, A. W. M. The semantic pathfinder: Using an authoring metaphor for generic multimedia indexing. *IEEE Trans. on Pattern Analysis and Machine Intelligence*, 28, No. 10, October 2006.

Shah, V. P.; Durbha, S. S.; King, R. L. & Younan, N. H. Image information mining for coastal disaster management. *IEEE International Geoscience and Remote Sensing Symposium*, Barcelona, Spain, July 2007.

Shanmugam, J.; Haralick, R. M. & Dinstein, I. Texture features for image classification. *IEEE Trans. on Systems, Man, and Cybernetics*, 3:610–621, 1973.

She, A. C.; Rui, Y & Huang, T. S. A modified fourier descriptor for shape matching in mars. *Image Databases and Multimedia Search, Series on Software Engineering and Knowledge Engineering*, Ed. S. K. Chang, 1998.

Tusk, C.; Kopersky, K.; Marchisio, G. & Aksoy, S. Interactive models for semantic labeling of satellite images. *Proceedings of Earth Observing Systems VII*, 4814:423–434, 2002.

Tusk, C.; Marchisio, G.; Aksoy, S.; Kopersky, K. & Tilton, J. C. Learning Bayesian classifiers for scene classification with a visual grammar. *IEEE Trans. on Geoscience and Remote Sensing*, 43, No. 3:581–589, march 2005.

Watson, A. B. Image compression using the discrete cosine transform. *Mathematica Journal*, 4, No.1:81–88, 1994.

Zhong, S. & Ghosh, J. A unified framework for model-based clustering. *Machine Learning Research*, 4:1001–1037, 2003.

7

Artificial Intelligence in Geoscience and Remote Sensing

David John Lary
Joint Center for Earth Systems Technology (JCET) UMBC, NASA/GSFC
United States

1. Introduction

Machine learning has recently found many applications in the geosciences and remote sensing. These applications range from bias correction to retrieval algorithms, from code acceleration to detection of disease in crops. As a broad subfield of artificial intelligence, machine learning is concerned with algorithms and techniques that allow computers to "learn". The major focus of machine learning is to extract information from data automatically by computational and statistical methods.

Over the last decade there has been considerable progress in developing a machine learning methodology for a variety of Earth Science applications involving trace gases, retrievals, aerosol products, land surface products, vegetation indices, and most recently, ocean products (*Yi and Prybutok*, 1996, *Atkinson and Tatnall*, 1997, *Carpenter et al.*, 1997, *Comrie*, 1997, *Chevallier et al.*, 1998, *Hyyppa et al.*, 1998, *Gardner and Dorling*, 1999, *Lary et al.*, 2004, *Lary et al.*, 2007, *Brown et al.*, 2008, *Lary and Aulov*, 2008, *Caselli et al.*, 2009, *Lary et al.*, 2009). Some of this work has even received special recognition as a NASA Aura Science highlight (*Lary et al.*, 2007) and commendation from the NASA MODIS instrument team (*Lary et al.*, 2009). The two types of machine learning algorithms typically used are neural networks and support vector machines. In this chapter, we will review some examples of how machine learning is useful for Geoscience and remote sensing, these examples come from the author's own research.

2. Typical Applications

One of the features that make machine-learning algorithms so useful is that they are "universal approximators". They can learn the behaviour of a system if they are given a comprehensive set of examples in a training dataset. These examples should span as much of the parameter space as possible. Effective learning of the system's behaviour can be achieved even if it is multivariate and non-linear. An additional useful feature is that we do not need to know a priori the functional form of the system as required by traditional least-squares fitting, in other words they are non-parametric, non-linear and multivariate learning algorithms.

The uses of machine learning to date have fallen into three basic categories which are widely applicable across all of the Geosciences and remote sensing, the first two categories use machine learning for its regression capabilities, the third category uses machine learning for its classification capabilities. We can characterize the three application themes are as follows: First, where we have a theoretical description of the system in the form of a deterministic model, but the model is computationally *expensive*. In this situation, a machine-learning "wrapper" can be applied to the deterministic model providing us with a "code accelerator". A good example of this is in the case of atmospheric photochemistry where we need to solve a large coupled system of ordinary differential equations (ODEs) at a large grid of locations. It was found that applying a neural network wrapper to the system was able to provide a speed up of between a factor of 2 and 200 depending on the conditions. Second, when we do not have a deterministic model but we have data available enabling us to empirically learn the behaviour of the system. Examples of this would include: Learning inter-instrument bias between sensors with a temporal overlap, and inferring physical parameters from remotely sensed proxies. Third, machine learning can be used for classification, for example, in providing land surface type classifications. Support Vector Machines perform particularly well for classification problems.

Now that we have an overview of the typical applications, the sections that follow will introduce two of the most powerful machine learning approaches, neural networks and support vector machines and then present a variety of examples.

3. Machine Learning

3.1 Neural Networks

Neural networks are multivariate, non-parametric, 'learning' algorithms (*Haykin*, 1994, *Bishop*, 1995, 1998, *Haykin*, 2001a, *Haykin*, 2001b, 2007) inspired by biological neural networks. Computational neural networks (NN) consist of an interconnected group of artificial neurons that processes information in parallel using a connectionist approach to computation. A NN is a non-linear statistical data-modelling tool that can be used to model complex relationships between inputs and outputs or to find patterns in data. The basic computational element of a NN is a model neuron or node. A node receives input from other nodes, or an external source (e.g. the input variables). A schematic of an example NN is shown in Figure 1. Each input has an associated weight, w, that can be modified to mimic synaptic learning. The unit computes some function, f, of the weighted sum of its inputs:

$$y_i = f\left(\sum_j w_{ij} y_j\right)$$

Its output, in turn, can serve as input to other units. w_{ij} refers to the weight from unit j to unit i. The function f is the node's activation or transfer function. The transfer function of a node defines the output of that node given an input or set of inputs. In the simplest case, f is the identity function, and the unit's output is y_i, this is called a linear node. However, non-linear sigmoid functions are often used, such as the hyperbolic tangent sigmoid transfer function and the log-sigmoid transfer function. Figure 1 shows an example feed-forward perceptron NN with five inputs, a single output, and twelve nodes in a hidden layer. A perceptron is a computer model devised to represent or simulate the ability of the brain to recognize and discriminate. In most cases, a NN is an adaptive system that changes its structure based on external or internal information that flows through the network during the learning phase.

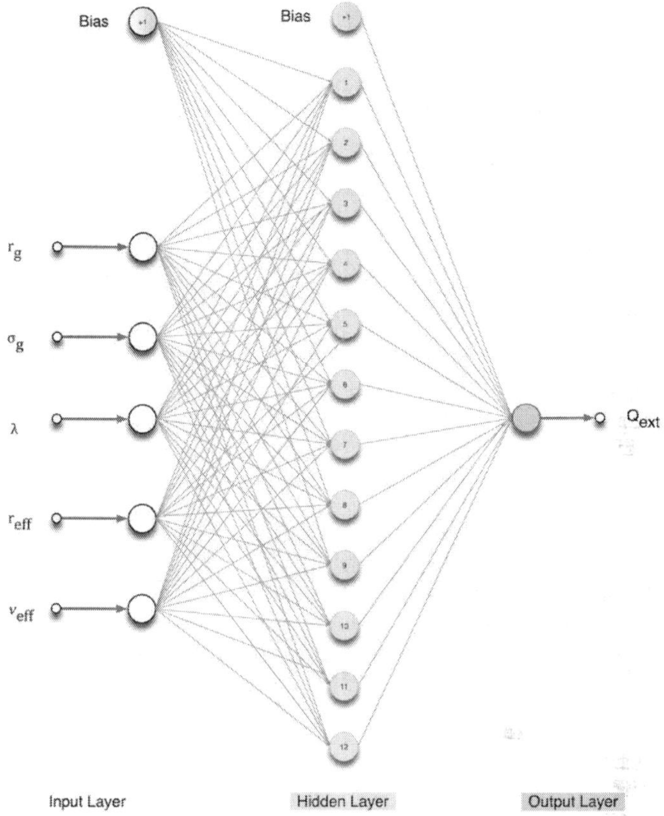

Fig. 1. Example neural network architecture showing a network with five inputs, one output, and twelve hidden nodes.

When we perform neural network training, we want to ensure we can independently assess the quality of the machine learning 'fit'. To insure this objective assessment we usually randomly split our training dataset into three portions, typically of 80%, 10% and 10%. The largest portion containing 80% of the dataset is used for training the neural network weights. This training is iterative, and on each training iteration we evaluate the current root mean square (RMS) error of the neural network output. The RMS error is calculated by using the second 10% portion of the data that was not used in the training. We use the RMS error and the way the RMS error changes with training iteration (epoch) to determine the convergence of our training. When the training is complete, we then use the final 10% portion of data as a totally independent validation dataset. This final 10% portion of the data is randomly chosen from the training dataset and is not used in either the training or RMS evaluation. We only use the neural network if the validation scatter diagram, which plots the actual data from validation portion against the neural network estimate, yields a straight-line graph with a slope very close to one and an intercept very close to zero. This is a stringent, independent and objective validation metric. The validation is global as the data is randomly selected over all data points available. For our studies, we typically used feed-forward back-propagation neural networks with a Levenberg-Marquardt back-propagation training algorithm (*Levenberg*, 1944, *Marquardt*, 1963, *Moré*, 1977, *Marquardt*, 1979).

3.2 Support Vector Machines

Support Vector Machines (SVM) are based on the concept of decision planes that define decision boundaries and were first introduced by Vapnik (*Vapnik*, 1995, 1998, 2000) and has subsequently been extended by others (*Scholkopf et al.*, 2000, *Smola and Scholkopf*, 2004). A decision plane is one that separates between a set of objects having different class memberships. The simplest example is a linear classifier, i.e. a classifier that separates a set of objects into their respective groups with a line. However, most classification tasks are not that simple, and often more complex structures are needed in order to make an optimal separation, i.e., correctly classify new objects (test cases) on the basis of the examples that are available (training cases). Classification tasks based on drawing separating lines to distinguish between objects of different class memberships are known as hyperplane classifiers.

SVMs are a set of related supervised learning methods used for classification and regression. Viewing input data as two sets of vectors in an n-dimensional space, an SVM will construct a separating hyperplane in that space, one that maximizes the margin between the two data sets. To calculate the margin, two parallel hyperplanes are constructed, one on each side of the separating hyperplane, which are "pushed up against" the two data sets. Intuitively, a good separation is achieved by the hyperplane that has the largest distance to the neighboring data points of both classes, since in general the larger the margin the better the generalization error of the classifier. We typically used the SVMs provided by LIBSVM (*Fan et al.*, 2005, *Chen et al.*, 2006).

4. Applications

Let us now consider some applications.

4.1 Bias Correction: Atmospheric Chlorine Loading for Ozone Hole Research

Critical in determining the speed at which the stratospheric ozone hole recovers is the total amount of atmospheric chlorine. Attributing changes in stratospheric ozone to changes in chlorine requires knowledge of the stratospheric chlorine abundance over time. Such attribution is central to international ozone assessments, such as those produced by the World Meteorological Organization (*Wmo*, 2006). However, we do not have continuous observations of all the key chlorine gases to provide such a continuous time series of stratospheric chlorine. To address this major limitation, we have devised a new technique that uses the long time series of available hydrochloric acid observations and neural networks to estimate the stratospheric chlorine (Cl_y) abundance (*Lary et al.*, 2007).

Knowledge of the distribution of inorganic chlorine Cl_y in the stratosphere is needed to attribute changes in stratospheric ozone to changes in halogens, and to assess the realism of chemistry-climate models (*Eyring et al.*, 2006, *Eyring et al.*, 2007, *Waugh and Eyring*, 2008). However, simultaneous measurements of the major inorganic chlorine species are rare (*Zander et al.*, 1992, *Gunson et al.*, 1994, *Webster et al.*, 1994, *Michelsen et al.*, 1996, *Rinsland et al.*, 1996,

Zander et al., 1996, *Sen et al.*, 1999, *Bonne et al.*, 2000, *Voss et al.*, 2001, *Dufour et al.*, 2006, *Nassar et al.*, 2006). In the upper stratosphere, the situation is a little easier as Cl_y can be inferred from HCl alone (e.g., (*Anderson et al.*, 2000, *Froidevaux et al.*, 2006b, *Santee et al.*, 2008)). Our new estimates of stratospheric chlorine using machine learning (*Lary et al.*, 2007) work throughout the stratosphere and provide a much-needed critical test for current global models. This critical evaluation is necessary as there are significant differences in both the stratospheric chlorine and the timing of ozone recovery in the available model predictions.

Hydrochloric acid is the major reactive chlorine gas throughout much of the atmosphere, and throughout much of the year. However, the observations of HCl that we do have (from UARS HALOE, ATMOS, SCISAT-1 ACE and Aura MLS) have significant biases relative to each other. We found that machine learning can also address the inter-instrument bias (*Lary et al.*, 2007, *Lary and Aulov*, 2008). We compared measurements of HCl from the different instruments listed in Table 1. The Halogen Occultation Experiment (HALOE) provides the longest record of space based HCl observations. Figure 2 compares HALOE HCl with HCl observations from (a) the Atmospheric Trace Molecule Spectroscopy Experiment (ATMOS), (b) the Atmospheric Chemistry Experiment (ACE) and (c) the Microwave Limb Sounder (MLS).

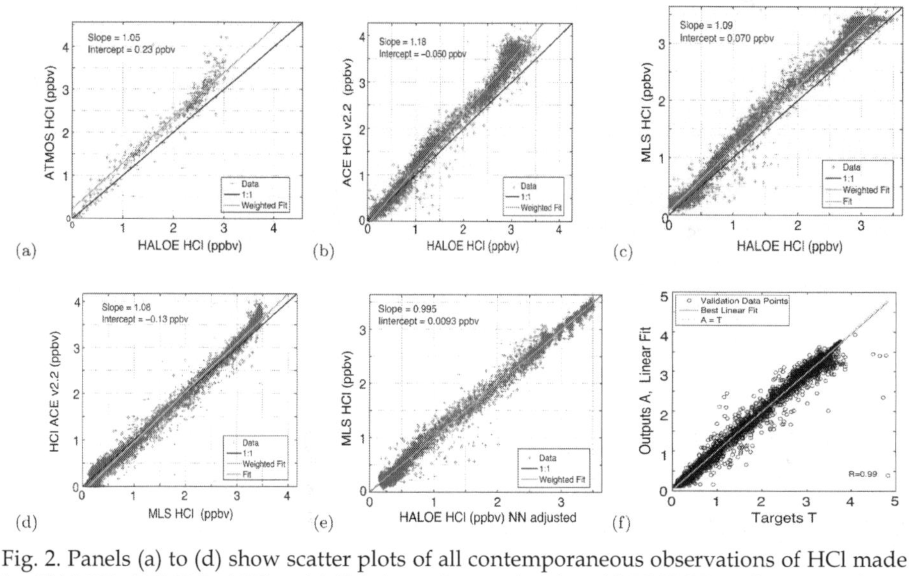

Fig. 2. Panels (a) to (d) show scatter plots of all contemporaneous observations of HCl made by HALOE, ATMOS, ACE and MLS Aura. In panels (a) to (c) HALOE is shown on the x-axis. Panel (e) correspond to panel (c) except that it uses the neural network 'adjusted' HALOE HCl values. Panel (f) shows the validation scatter diagram of the neural network estimate of $Cl_y \approx HCl + ClONO_2 + ClO + HOCl$ versus the actual Cl_y for a totally independent data sample *not* used in training the neural network.

A consistent picture is seen in these plots: HALOE HCl measurements are lower than those from the other instruments. The slopes of the linear fits (relative scaling) are 1.05 for the HALOE-ATMOS comparison, 1.09 for the HALOE-MLS, and 1.18 for the HALOE-ACE. The offsets are apparent at the 525 K isentropic surface and above. Previous comparisons among HCl datasets reveal a similar bias for HALOE (*Russell et al.*, 1996, *Mchugh et al.*, 2005, *Froidevaux et al.*, 2006a, *Froidevaux et al.*, 2008). ACE and MLS HCl measurements are in much better agreement (Figure 2d). Note, the measurements agree within the stated observational uncertainties summarized in Table 1.

Instrument	Temporal Coverage	Species	References	Median Observation Uncertainty
ACE v2.2	2004-2006	HCl, $ClONO_2$ ClO, and HOCl	*Bernath et al.* [2005]	8% (HCl), 30% ($ClONO_2$) >100% (ClO), >100% (HOCl)
ATMOS	1991, 1993, 1994	HCl, $ClONO_2$	*Zander et al.* [1992]	8% (HCl), 60% ($ClONO_2$)
Aura MLS v1	2004-2006	HCl, ClO and HOCl	*Froidevaux et al.* [2006a]	12% (HCl), 76% (ClO), >100% (HOCl)
CLAES v9	1991-1993	$ClONO_2$	*Roche et al.* [1993]	>100%
CRISTA	1994, 1997	$ClONO_2$	*Offermann et al.* [1999]	61%
HALOE v19	1991-2005	HCl	*Russell et al.* [1993]	4%
UARS MLS v5	1991-1999	ClO	*Waters et al.* [1996]	>100% (ClO)

Table 1. The instruments and constituents used in constructing the Cl_y record from 1991-2006. The uncertainties given are the median values calculated for each level 2 measurement profile and its uncertainty (both in mixing ratio) for all the observations made. The uncertainties are larger than usually quoted for MLS ClO because they reflect the single profile precision, which is improved by temporal and/or spatial averaging. The HALOE uncertainties are only estimates of random error and do not include any indications of overall accuracy.

To combine the above HCl measurements to form a continuous time series of HCl (and then Cl_y) from 1991 to 2006 it is necessary to account for the biases between data sets. A neural network is used to learn the mapping from one set of measurements onto another as a function of equivalent latitude and potential temperature. We consider two cases. In one case ACE HCl is taken as the reference and the HALOE and Aura HCl observations are adjusted to agree with ACE HCl. In the other case HALOE HCl is taken as the reference and the Aura and ACE HCl observations are adjusted to agree with HALOE HCl. In both cases we use equivalent latitude and potential temperature to produce average profiles. The purpose of the NN mapping is simply to learn the bias as a function of location, not to imply which instrument is correct.
The precision of the correction using the neural network mapping is of the order of ±0.3 ppbv, as seen in Figure 2 (e) that shows the results when HALOE HCl measurements have been mapped into ACE measurements. The mapping has removed the bias between the measurements and has straightened out the 'wiggles' in 2 (c), i.e., the neural network has learned the equivalent PV latitude and potential temperature dependence of the bias between HALOE and MLS. The inter-instrument offsets are not constant in space or time, and are not a simple function of Cl_y.
So employing neural networks allows us to: Form a seamless record of HCl using observations from several space-borne instruments using neural networks. Provide an estimated of the associated inter-instrument bias. Infer Cl_y from HCl, and thereby provide a seamless record of Cl_y, the parameter needed for examining the ozone hole recovery. A similar use of machine learning has been made for Aerosol Optical Depths, the subject of the next sub-section.

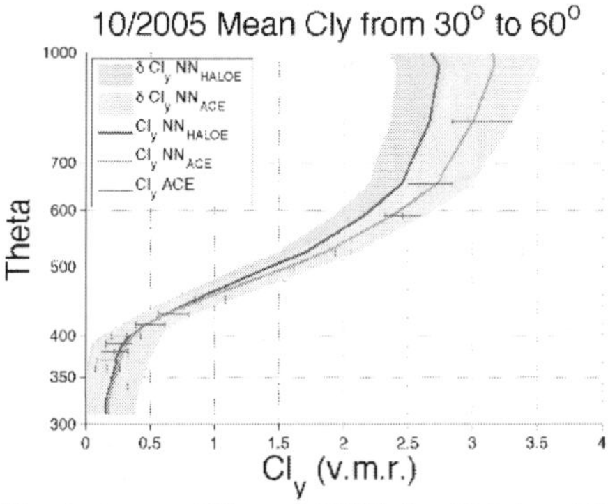

Fig. 3. Cl_y average profiles between 30° and 60°N for October 2005, estimated by neural network calibrated to HALOE HCl (blue curve), estimated by neural network calibrated to ACE HCl (green), or from ACE observations of HCl, $ClONO_2$, ClO, and HOCl (red crosses). In each case, the shaded range represents the total uncertainty; it includes the observational uncertainty, the representativeness uncertainty (the variability over the analysis grid cell), the neural network uncertainty. The vertical extent of this plot was limited to below 1000 K (≈35 km), as there is no ACE v2.2 ClO data for the upper altitudes. In addition, above ≈750 K (≈25 km), ClO constitutes a larger fraction of Cl_y (up to about 10%) and so the large uncertainties in ClO have greater effect.

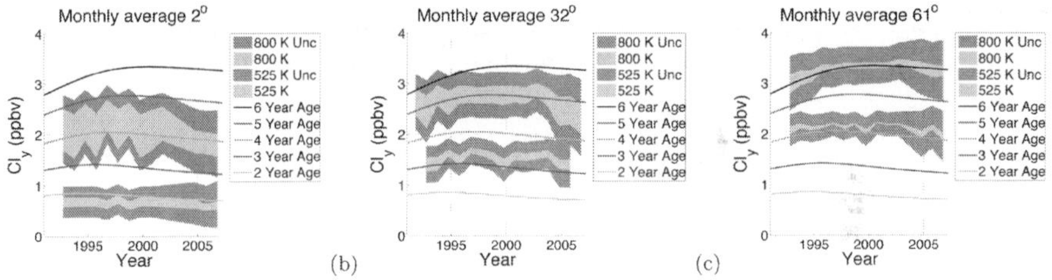

Fig. 4. Panels (a) to (c) show October Cl_y time-series for the 525 K isentropic surface (≈20 km) and the 800 K isentropic surface (≈30 km). In each case the dark shaded range represents the total uncertainty in our estimate of Cl_y. This total uncertainty includes the observational uncertainty, the representativeness uncertainty (the variability over the analysis grid cell), the inter-instrument bias in HCl, the uncertainty associated with the neural network inter-instrument correction, and the uncertainty associated with the neural network inference of Cl_y from HCl and CH_4. The inner light shading depicts the uncertainty on Cl_y due to the inter-instrument bias in HCl alone. The upper limit of the light shaded range corresponds to the estimate of Cl_y based on all the HCl observations calibrated by a neural network to agree with ACE v2.2 HCl. The lower limit of the light shaded range corresponds to the estimate of Cl_y based on all the HCl observations calibrated to agree with HALOE v19 HCl. Overlaid are lines showing the Cl_y based on age of air calculations (*Newman et al.*, 2006). To minimize variations due to differing data coverage months with less than 100 observations of HCl in the equivalent latitude bin were left out of the time-series.

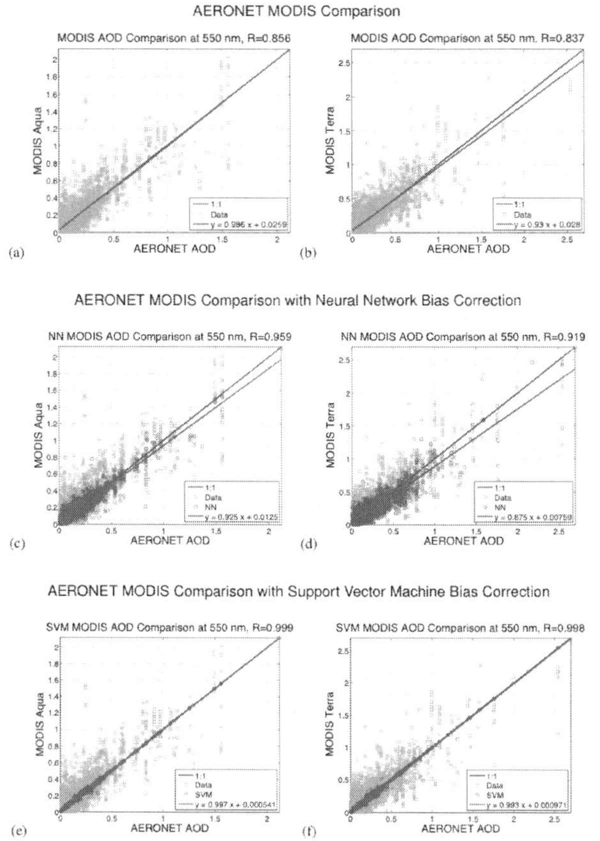

Fig. 5. Scatter diagram comparisons of Aerosol Optical Depth (AOD) from AERONET (x-axis) and MODIS (y-axis) as green circles overlaid with the ideal case of perfect agreement (blue line). The measurements shown in the comparison were made within half an hour of each other, with a great circle separation of less than 0.25° and with a solar zenith angle difference of less than 0.1°. The left hand column of plots is for MODIS Aqua and the right hand column of plots is for MODIS Terra. The first row shows the comparisons between AERONET and MODIS for the entire period of overlap between the MODIS and AERONET instruments from the launch of the MODIS instrument to the present. The second row shows the same comparison overlaid with the neural network correction as red circles. We note that the neural network bias correction makes a substantial improvement in the correlation coefficient with AERONET. An improvement from 0.86 to 0.96 for MODIS Aqua and an improvement from 0.84 to 0.92 for MODIS Terra. The third row shows the comparison overlaid with the support vector regression correction as red circles. We note that the support vector regression bias correction makes an even greater improvement in the correlation coefficient than the neural network correction. An improvement from 0.86 to 0.99 for MODIS Aqua and an improvement from 0.84 to 0.99 for MODIS Terra.

4.2 Bias Correction: Aerosol Optical Depth

As highlighted in the 2007 IPCC report on Climate Change, aerosol and cloud radiative effects remain the largest uncertainties in our understanding of climate change (*Solomon et al.*, 2007). Over the past decade observations and retrievals of aerosol characteristics have been conducted from space-based sensors, from airborne instruments and from ground-based samplers and radiometers. Much effort has been directed at these data sets to collocate observations and retrievals, and to compare results. Ideally, when two instruments measure the same aerosol characteristic at the same time, the results should agree within well-understood measurement uncertainties. When inter-instrument biases exist, we would like to explain them theoretically from first principles. One example of this is the comparison between the aerosol optical depth (AOD) retrieved by the Moderate Resolution Imaging Spectroradiometer (MODIS) and the AOD measured by the Aerosol Robotics Network (AERONET). While progress has been made in understanding the biases between these two data sets, we still have an imperfect understanding of the root causes. (*Lary et al.*, 2009) examined the efficacy of empirical machine learning algorithms for aerosol bias correction.

Machine learning approaches (Neural Networks and Support Vector Machines) were used by (*Lary et al.*, 2009) to explore the reasons for a persistent bias between aerosol optical depth (AOD) retrieved from the MODerate resolution Imaging Spectroradiometer (MODIS) and the accurate ground-based Aerosol Robotics Network (AERONET). While this bias falls within the expected uncertainty of the MODIS algorithms, there is still room for algorithm improvement. The results of the machine learning approaches suggest a link between the MODIS AOD biases and surface type. From figure 5 we can see that machine learning algorithms were able to effectively adjust the AOD bias seen between the MODIS instruments and AERONET. Support vector machines performed the best improving the correlation coefficient between the AERONET AOD and the MODIS AOD from 0.86 to 0.99 for MODIS Aqua, and from 0.84 to 0.99 for MODIS Terra.

Key in allowing the machine learning algorithms to 'correct' the MODIS bias was provision of the surface type and other ancillary variables that explain the variance between MODIS and AERONET AOD. The provision of the ancillary variables that can explain the variance in the dataset is the key ingredient for the effective use of machine learning for bias correction. A similar use of machine learning has been made for vegetation indices, the subject of the next sub-section.

4.3 Bias Correction: Vegetation Indices

Consistent, long term vegetation data records are critical for analysis of the impact of global change on terrestrial ecosystems. Continuous observations of terrestrial ecosystems through time are necessary to document changes in magnitude or variability in an ecosystem (*Tucker et al.*, 2001, *Eklundh and Olsson*, 2003, *Slayback et al.*, 2003). Satellite remote sensing has been the primary way that scientists have measured global trends in vegetation, as the measurements are both global and temporally frequent. In order to extend measurements through time, multiple sensors with different design and resolution must be used together in the same time series. This presents significant problems as sensor band placement, spectral response, processing, and atmospheric correction of the observations can vary significantly and impact the comparability of the measurements (*Brown et al.*, 2006). Even without differences in atmospheric correction, vegetation index values for the same target recorded under identical conditions will not be directly comparable because input reflectance values differ from sensor to sensor due to differences in sensor design (*Teillet et al.*, 1997, *Miura et al.*, 2006).

Several approaches have previously been taken to integrate data from multiple sensors. (*Steven et al.*, 2003), for example, simulated the spectral response from multiple instruments and with simple linear equations created conversion coefficients to transform NDVI data from one sensor to another. Their analysis is based on the observation that the vegetation index is critically dependent on the spectral response functions of the instrument used to calculate it. The conversion formulas the paper presents cannot be applied to maximum value NDVI datasets because the weighting coefficients are land cover and dataset dependent, reducing their efficacy in mixed pixel situations (*Steven et al.*, 2003). (*Trishchenko et al.*, 2002) created a series of quadratic functions to correct for differences in the reflectance and NDVI to NOAA-9 AVHRR-equivalents (*Trishchenko et al.*, 2002). Both the (*Steven et al.*, 2003) and the (*Trishchenko et al.*, 2002) approaches are land cover and dataset dependent and thus cannot be used on global datasets where multiple land covers are represented by one pixel. (*Miura et al.*, 2006) used hyper-spectral data to investigate the effect of different spectral response characteristics between MODIS and AVHRR instruments on both the reflectance and NDVI data, showing that the precise characteristics of the spectral response had a large effect on the resulting vegetation index. The complex patterns and dependencies on spectral band functions were both land cover dependent and strongly non-linear, thus we see that an exploration of a non-linear approach may be fruitful.

(*Brown et al.*, 2008) experimented with powerful, non-linear neural networks to identify and remove differences in sensor design and variable atmospheric contamination from the AVHRR NDVI record in order to match the range and variance of MODIS NDVI without removing the desired signal representing the underlying vegetation dynamics. Neural networks are 'data transformers' (*Atkinson and Tatnall*, 1997), where the objective is to associate the elements of one set of data to the elements in another. Relationships between the two datasets can be complex and the two datasets may have different statistical distributions. In addition, neural networks incorporate a priori knowledge and realistic physical constraints into the analysis, enabling a transformation from one dataset into another through a set of weighting functions (*Atkinson and Tatnall*, 1997). This transformation incorporates additional input data that may account for differences between the two datasets.

The objective of (*Brown et al.*, 2008) was to demonstrate the viability of neural networks as a tool to produce a long term dataset based on AVHRR NDVI that has the data range and statistical distribution of MODIS NDVI. Previous work has shown that the relationship between AVHRR and MODIS NDVI is complex and nonlinear (*Gallo et al.*, 2003, *Brown et al.*, 2006, *Miura et al.*, 2006), thus this problem is well suited to neural networks if appropriate inputs can be found. The influence of the variation of atmospheric contamination of the AVHRR data through time was explored by using observed atmospheric water vapor from the Total Ozone Mapping Spectrometer (TOMS) instrument during the overlap period 2000-2004 and back to 1985. Examination of the resulting MODIS fitted AVHRR dataset both during the overlap period and in the historical dataset will enable an evaluation of the efficacy of the neural net approach compared to other approaches to merge multiple-sensor NDVI datasets.

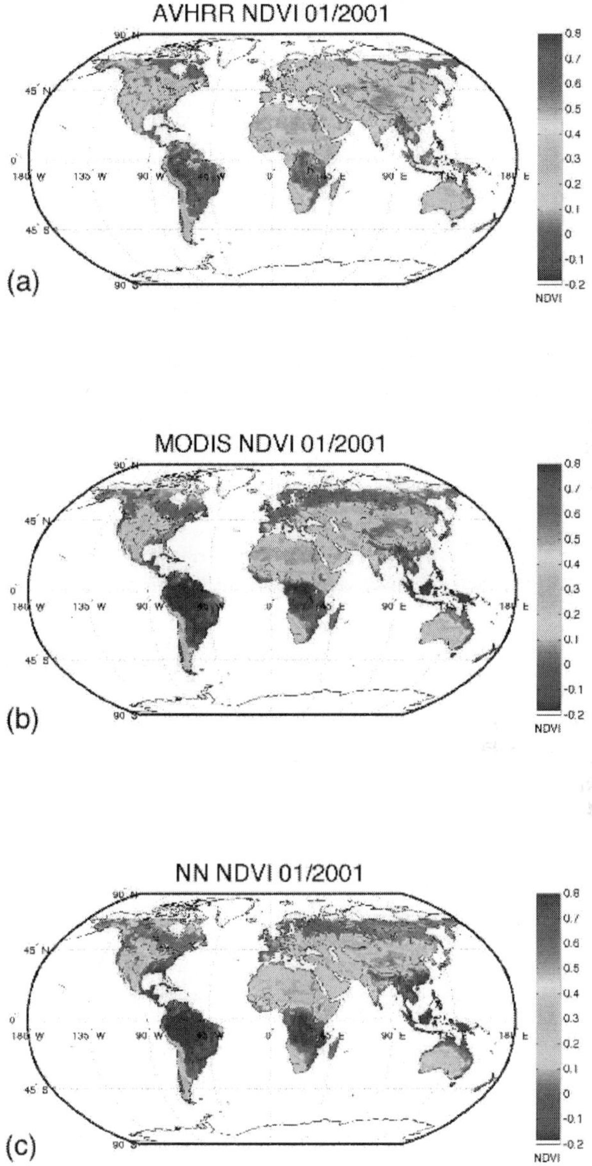

Fig. 6. A comparison of the NDVI from AVHR (panel a), MODIS (panel p), and then a reconstruction of MODIS using AVHRR and machine learning (panel c). We note that the machine learning can successfully account for the large differences that are found between AVHRR and MODIS.

Remote sensing datasets are the result of a complex interaction between the design of a sensor, the spectral response function, stability in orbit, the processing of the raw data, compositing schemes, and post-processing corrections for various atmospheric effects including clouds and aerosols. The interaction between these various elements is often non-linear and non-additive, where some elements increase the vegetation signal to noise ratio (compositing, for example) and others reduce it (clouds and volcanic aerosols) (*Los*, 1998). Thus, although other authors have used simulated data to explore the relationship between AVHRR and MODIS (*Trishchenko et al.*, 2002, *Van Leeuwen et al.*, 2006), these techniques are not directly useful in producing a sensor-independent vegetation dataset that can be used by data users in the near term.

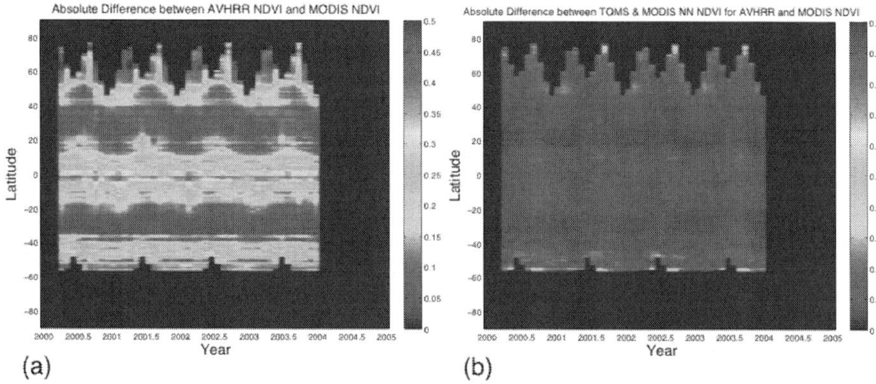

Fig. 7. Panel (a) shows a time-series from 2000 to 2003 of the zonal mean (averaged per latitude) difference between the AVHRR and MODIS NDVIs, this highlights that significant differences exist between the two data products. Panel (b) shows a time series over the same period after the machine learning has been used to "cross-calibrate" AVHRR as MODIS, showing that the machine learning has effectively learnt how to cross-calibrate the instruments.

There are substantial differences between the processed vegetation data from AVHRR and MODIS. (*Brown et al.*, 2008) showed that neural networks are an effective way to have a long data record that utilizes all available data back to 1981 by providing a practical way of incorporating the AVHRR data into a continuum of observations that include both MODIS and VIIRS. The results (*Brown et al.*, 2008) showed that the TOMS data record on clouds, ozone and aerosols can be used to identify and remove sensor-specific atmospheric contaminants that differentially affect the AVHRR over MODIS. Other sensor-related effects, particularly those of changing BRDF, viewing angle, illumination, and other effects that are not accounted for here, remain important sources of additional variability. Although this analysis has not produced a dataset with identical properties to MODIS, it has demonstrated that a neural net approach can remove most of the atmospheric-related aspects of the differences between the sensors, and match the mean, standard deviation and range of the two sensors. A similar technique can be used for the VIIRS sensor once the data is released.
Figure 6 shows a comparison of the NDVI from AVHR (panel a), MODIS (panel p), and then a reconstruction of MODIS using AVHRR and machine learning (panel c). Figure 7 (a) shows a time-series from 2000 to 2003 of the zonal mean difference between the AVHRR and MODIS

NDVIs, this highlights that significant differences exist between the two data products. Panel (b) shows a time series over the same period after the machine learning has been used to "cross-calibrate" AVHRR as MODIS, illustrating that the machine learning has effectively learnt how to cross-calibrate the instruments.

So far, we have seen three examples of using machine learning for bias correction (constituent biases, aerosol optical depth biases and vegetation index biases), and one example of using machine learning to infer a useful proxy from remotely sensed data (Cl_y from HCl). Let us look at one more example of inferring proxies from existing remotely sensed data before moving onto consider using machine learning for code acceleration.

4.4 Inferring Proxies: Tracer Correlations

The spatial distributions of atmospheric trace constituents are in general dependent on both chemistry and transport. Compact correlations between long-lived species are well-observed features in the middle atmosphere. The correlations exist for all long-lived tracers - not just those that are chemically related - due to their transport by the general circulation of the atmosphere. The tight relationships between different constituents have led to many analyses using measurements of one tracer to infer the abundance of another tracer. Using these correlations is also as a diagnostic of mixing and can distinguish between air-parcels of different origins. Of special interest are the so-called 'long-lived' tracers: constituents such as nitrous oxide (N_2O), methane (CH_4), and the chlorofluorocarbons (CFCs) that have long lifetimes (many years) in the troposphere and lower stratosphere, but are destroyed rapidly in the middle and upper stratosphere.

The correlations are spatially and temporally dependent. For example, there is a 'compact-relation' regime in the lower part of the stratosphere and an 'altitude-dependent' regime above this. In the compact-relation region, the abundance of one tracer is uniquely determined by the value of the other tracer, without regard to other variables such as latitude or altitude. In the altitude-dependent regime, the correlation generally shows significant variation with altitude.

A family of correlations usually achieves the description of such spatially and temporally dependent correlations. However, a single neural network is a natural and effective alternative. The motivation for this case study was preparation for a long-term chemical assimilation of Upper Atmosphere Research Satellite (UARS) data starting in 1991 and coming up to the present. For this period, we have continuous version 19 data from the Halogen Occultation Experiment (HALOE) but not observations of N_2O as both ISAMS and CLAES failed. In addition, we would like to constrain the total amount of reactive nitrogen, chlorine, and bromine in a self-consistent way (i.e. the correlations between the long-lived tracers is preserved). Tracer correlations provide a means to do this by using HALOE CH_4 observations.

Machine learning is ideally suited to describe the spatial and temporal dependence of tracer-tracer correlations. The neural network performs well even in regions where the correlations are less compact and normally a family of correlation curves would be required. For example, the methane CH_4-N_2O correlation can be well described using a neural network (*Lary et al.*, 2004) trained with the latitude, pressure, time of year, and CH_4 volume mixing ratio (v.m.r.). *Lary et al.* (2004) used a neural network to reproduce the CH_4-N_2O correlation with a correlation coefficient between simulated and training values of 0.9995. Such an accurate representation of tracer-tracer correlations allows more use to be made of long-term datasets to constrain chemical models. For example, the Halogen Occultation Experiment (HALOE) that continuously observed CH_4 (but not N_2O) from 1991 until 2005.

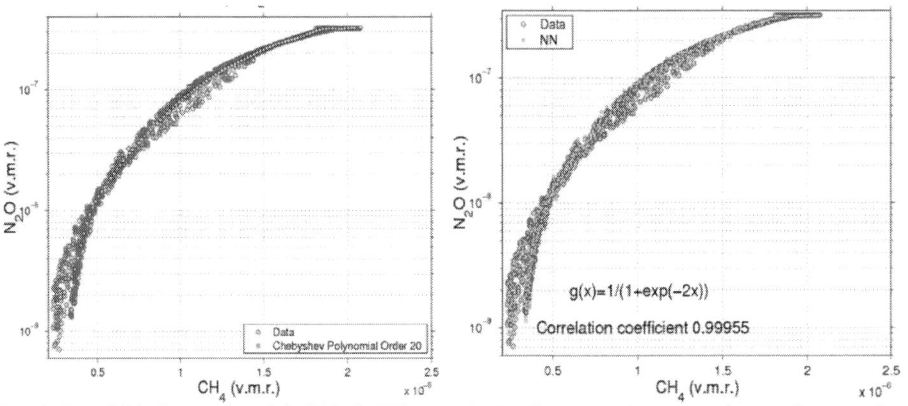

Fig. 8. Panel (a) shows the global N_2O-CH_4 correlation for an entire year, after evaluating the efficacy of 3,000 different functional forms for parametric fits, we overlaid the best, an order 20 Chebyshev Polynomial. However, this still does not account for the multi-variate nature of the problem exhibited by the 'cloud' of points rather than a compact 'curve' or 'line'. However, in panel (b) we can see that a neural network is able to account for the non-linear and multi-variate aspects, the training dataset exhibited a 'cloud' of points, the neural network fit reproduces a 'cloud' of points. The most important factor in producing a 'spread' in the correlations is the strong altitude dependence of the N_2O-CH_4 correlation.

Figure 8 (a) shows the global N_2O-CH_4 correlation for an entire year, after evaluating the efficacy of 3,000 different functional forms for parametric fits, we overlaid the best, an order 20 Chebyshev Polynomial. However, this still does not account for the multi-variate nature of the problem exhibited by the 'cloud' of points rather than a compact 'curve' or 'line'. However, in Figure 8 (b) we can see that a neural network is able to account for the non-linear and multi-variate aspects, the training dataset exhibited a 'cloud' of points, the neural network fit reproduces a 'cloud' of points. The most important factor in producing a 'spread' in the correlations is the strong altitude dependence of the N_2O-CH_4 correlation.

4.5 Code Acceleration: Example from Ordinary Differential Equation Solvers

There are many applications in the Geosciences and remote sensing which are computationally expensive. Machine learning can be very effective in accelerating components of these calculations. We can readily create training datasets for these applications using the very models we would like to accelerate.

The first example for which we found this effective was solving ordinary differential equations. An adequate photochemical mechanism to describe the evolution of ozone in the upper troposphere and lower stratosphere (UT/LS) in a computational model involves a comprehensive treatment of reactive nitrogen, hydrogen, halogens, hydrocarbons, and interactions with aerosols. Describing this complex interaction is computationally expensive, and applications are limited by the computational burden. Simulations are often made tractable by using a coarser horizontal resolution than would be desired or by reducing the interactions accounted for in the photochemical mechanism. These compromises also limit the scientific applications. Machine learning algorithms offer a means to obtain a fast and accurate solution to the stiff ordinary differential equations that comprise the photochemical calculations, thus making high-resolution simulations including the complete photochemical mechanism much more tractable.

For the sake of an example, a 3D model of atmospheric chemistry and transport, the GMI-COMBO model, can use 55 vertical levels and a 4° latitude x 5° longitude grid and 125 species. With 15-minute time steps the chemical ODE solver is called 119,750,400 times in simulating just one week. If the simulation is for a year then the ODE solver needs to be called 6,227,020,800 (or $6x10^9$) times. If the spatial and temporal resolution is doubled then the chemical ODE solver needs to be called a staggering $2.5x10^{10}$ times to simulate a year. This represents a major computational cost in simulating a constituent's spatial and temporal evolution. The ODEs solved at adjacent grid cells and time steps are very similar. Therefore, if the simulations from one grid cell and time step could be used to speed up the simulation for adjacent grid cells and subsequent time steps, we would have a strategy to dramatically decrease the computational cost of our simulations.

Neurological Acceleration of stiff ODE solver with error monitoring and dynamic training

Phase	Standard Stiff ODE Solver Action	Neural Network Action	Description
A	Always running	None	Collecting training data.
B	None	Initial training	Train *n* networks, one per constituent ODE.
C	Always running	Testing	Compare standard stiff ODE solver with neural network solution to train error networks, one per constituent ODE.
D	Always running	Update training	Update training of *n* ODE networks. Train *n* error networks giving neural network solution uncertainty.
E	Sometimes running	Operational with periodic training updates	**Operational.** When error networks indicate an error from the neural network solution that is less than the required threshold for each ODE use the neural network solution, otherwise use the regular stiff ODE solver. However, even when error network indicates an acceptable neural network solution periodically use the full stiff ODE solver solution to gather further training data to periodically update the training of all the networks.

Fig. 9. Strategy for applying a neural wrapper to accelerate the ODE solver.

Figure 9 shows the strategy that we used for applying a neural wrapper to accelerate the ODE solver. Figure 10 shows some example results for ozone after using a neural wrapper around an atmospheric chemistry ODE solver. The x-axis shows the actual ozone abundance as a volume mixing ratio (vmr) using the regular ODE solver without neural networks. The y-axis shows the ozone vmr inferred using the neural network solution. It can be seen that we have excellent agreement between the two solutions with a correlation coefficient of 1. The neural network has learned the behaviour of the ozone ODE very well. Without the adaptive error control the acceleration could be up to 200 times, with the full adaptive error control the acceleration was less, but usually at least a factor of two. Similarly, in Figure 11 the two panels below show the results for formaldehyde (HCHO) in the GMI model. The left panel shows the solution with SMVGear for level 1 at 01:00 UT and the right panel shows the corresponding solution using the neural network. As one would hope, the two results are almost indistinguishable.

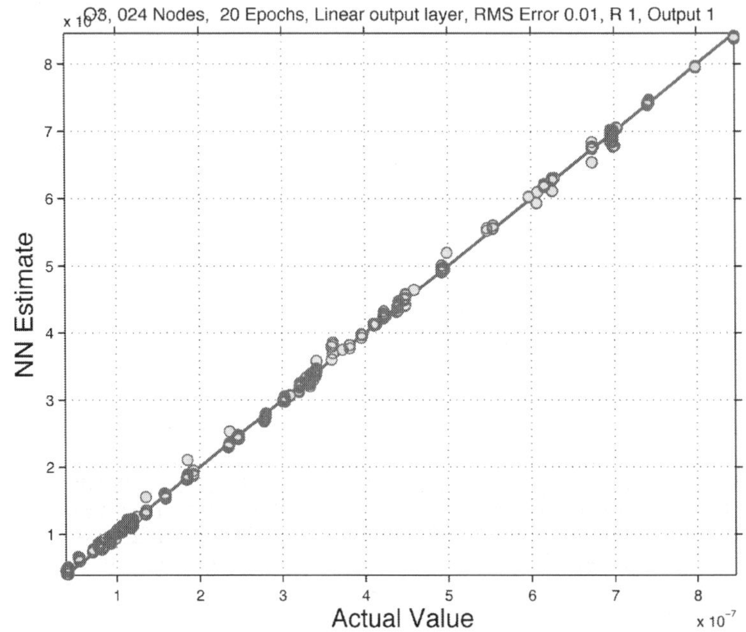

Fig. 10. Example results for using a neural wrapper around an atmospheric chemistry ODE solver. The x-axis shows the actual ozone v.m.r. using the regular ODE solver without neural networks. The y-axis shows the ozone v.m.r. inferred using the neural network solution. It can be seen that we have excellent agreement between the two solutions with a correlation coefficient of 1. The neural network has learned the behaviour of the ozone ODE very well.

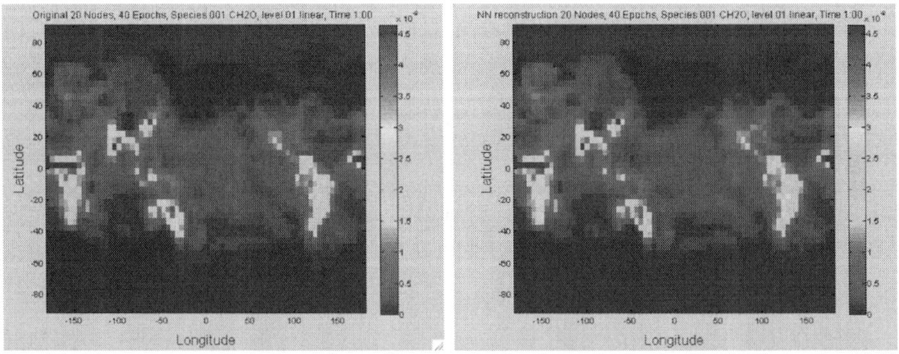

Fig. 11. The two panels below show the results for formaldehyde (HCHO) in the GMI model. The left panel shows the solution with SMVGear for level 1 at 01:00 UT and the right panel shows the corresponding solution using the neural network. As one would hope, the two results are almost indistinguishable.

4.6 Classification: Example from Detecting Drought Stress and Infection in Cacao

The source of chocolate, theobroma cacao (cacao), is an understory tropical tree (*Wood*, 2001). Cacao is intolerant to drought (*Belsky and Siebert*, 2003), and yields and production patterns are severely affected by periodic droughts and seasonal rainfall patterns. (*Bae et al.*, 2008) studied the molecular response of cacao to drought and have identified several genes responsive to drought stress (*Bailey et al.*, 2006). They have also been studying the response of cacao to colonization by an endophytic isolates of Trichoderma including Trichoderma hamatum, DIS 219b (*Bailey et al.*, 2006). One of the benefits to colonization Trichoderma hamatum isolate DIS 219b is tolerance to drought as mediated through plant growth promotion, specifically enhanced root growth (*Bae et al.*, 2008).

In characterizing the drought response of cacao considerable variation was observed in the response of individual seedlings depending upon the degree of drought stress applied (*Bae et al.*, 2008). In addition, although colonization by DIS 219b delayed the drought response, direct effects of DIS 219b on cacao gene expression in the absence of drought were difficult to identify (*Bae et al.*, 2008). The complexity of the DIS 219b/cacao plant microbe interaction overlaid on cacao's response to drought makes the system of looking at individual genes as a marker for either drought or endophyte inefficient.

There would be considerable utility in reliably predicting drought and endophyte stress from complex gene expression patterns, particularly as the endophyte lives within the plant without causing apparent phenotypic changes in the plant. Machine-learning models offer the possibility of highly accurate, automated predictions of plant stress from a variety of causes that may otherwise go undetected or be obscured by the complexity of plant responses to multiple environmental factors, to be considered status quo for plants in nature. We examined the ability of five different machine-learning approaches to predict drought stress and endophyte colonization in cacao: a naive Bayes classifier, decision trees (DTs), neural networks (NN), neuro-fuzzy inference (NFI), and support vector machine (SVM) classification. The results provided some support for the accuracy of machine-learning models in discerning endophyte colonization and drought stress. The best performance was by the neuro-fuzzy inference system and the support vector classifier that correctly identified 100% of the drought and endophyte stress samples. Of the two, the approaches the support vector classifier is likely to have the best generalization (wider applicability to data not previously seen in the training process).

Why did the SVM model outperform the four other machine learning approaches? We noted earlier that SVMs construct separating hyperplanes that maximize the margins between the different clusters in the training data set (the vectors that constrain the width of the margin are the support vectors). A good separation is achieved by those hyperplanes providing the largest distance between neighbouring classes, and in general, the larger the margin the better the generalization of the classifier.

When the points in neighbouring classes are separated by a nonlinear dividing line, rather than fitting nonlinear curves to the data, SVMs use a kernel function to map the data into a different space where a hyperplane can once more be used to do the separation. The kernel function may transform the data into a higher dimensional space to make it possible to perform the separation. The concept of a kernel mapping function is very powerful. It allows SVM models to perform separations even with very complex boundaries. Hence, we infer that, in the present application, the SVM model algorithmic process utilizes higher dimensional space to achieve superior predictive power.

For classification, the SVM algorithmic process offers an important advantage compared with neural network approaches. Specifically, neural networks can suffer from multiple local minima; in contrast, the solution to a support vector machine is global and unique. This characteristic may be partially attributed to the development process of these algorithms; SVMs were developed in the reverse order to the development of neural networks. SVMs evolved from the theory to implementation and experiments; neural networks followed a more heuristic path, from applications and extensive experimentation to theory.

In handling this data using traditional methods where individual gene responses are characterized as treatment effects, it was especially difficult to sort out direct effects of endophyte on gene expression over time or at specific time points. The differences between the responses of non-stressed plants with or without the endophyte were small and, after the zero time point, were highly variable. The general conclusion from this study was that colonization of cacao seedlings by the endophyte enhanced root growth resulting in increased drought tolerance but the direct effects of endophyte on cacao gene expression at the time points studied were minimal. Yet the neuro-fuzzy inference and support vector classification methods of analysis were able identify samples receiving these treatments correctly.

In this system, each gene in the plants genome is a potential sensor for the applied stress or treatment. It is not necessary that the genes response be significant in itself in determining the outcome of the plants response or that it be consistent in time or level of response. Since multiple genes are used in characterizing the response it is always the relative response in terms of the many other changes that are occurring at the same time as influenced by uncontrolled changes in the system that is important. With this study the treatments were controlled but variation in the genetic make up of each seedling (they were from segregating open pollinated seed) and minute differences in air currents within the chamber, soil composition, colonization levels, microbial populations within each pot and seedling, and even exact watering levels at each time point, all likely contributed to creating uncontrolled variation in the plants response to what is already a complex reaction to multiple factors (drought and endophyte). This type of variation makes accessing treatment responses using single gene approaches difficult and the prediction of cause due to effect in open systems almost impossible in complex systems.

5. Future Directions

We have seen the utility of machine learning for a suite of very diverse applications. These applications often help us make better use of existing data in a variety of ways. In parallel to the success of machine learning we also have the rapid development of publically available web services. So it is timely to combine both approached by providing online services that use machine learning for intelligent data fusion as part of a workflow that allows us to cross-calibrate multiple datasets. This obviously requires care to ensure the appropriate of datasets. However, if done carefully, this could greatly facilitate the production of seamless multi-year global records for a host of Earth science applications.

When it comes to dealing with inter-instrument biases in a consistent manner there is currently a gap in many space agencies' Earth science information systems. This could be addressed by providing an extensible and reusable open source infrastructure that gap that could be reused for multiple projects. A clear need for such an infrastructure would be for NASA's future Decadal Survey missions.

6. Summary

Machine learning has recently found many applications in the geosciences and remote sensing. These applications range from bias correction to retrieval algorithms, from code acceleration to detection of disease in crops. Machine-learning algorithms can act as "universal approximators", they can learn the behaviour of a system if they are given a comprehensive set of examples in a training dataset. Effective learning of the system's behaviour can be achieved even if it is multivariate and non-linear. An additional useful feature is that we do not need to know a priori the functional form of the system as required by traditional least-squares fitting, in other words they are non-parametric, non-linear and multivariate learning algorithms.

The uses of machine learning to date have fallen into three basic categories which are widely applicable across all of the Geosciences and remote sensing, the first two categories use machine learning for its regression capabilities, the third category uses machine learning for its classification capabilities. We can characterize the three application themes are as follows: First, where we have a theoretical description of the system in the form of a deterministic model, but the model is computationally *expensive*. In this situation, a machine-learning "wrapper" can be applied to the deterministic model providing us with a "code accelerator". Second, when we do not have a deterministic model but we have data available enabling us to empirically learn the behaviour of the system. Third, machine learning can be used for classification.

7. References

Anderson, J., Russell, J. M., Solomon, S. & Deaver, L. E. (2000) Halogen occultation experiment confirmation of stratospheric chlorine decreases in accordance with the montreal protocol. *Journal of Geophysical Research-Atmospheres,* 105, 4483-4490.

Atkinson, P. M. & Tatnall, A. R. L. (1997) Introduction: Neural networks in remote sensing. *International Journal of Remote Sensing,* 18, 699 - 709.

Bae, H., Kim, S. H., Kim, M. S., Sicher, R. C., Lary, D., Strem, M. D., Natarajan, S. & Bailey, B. A. (2008) The drought response of theobroma cacao (cacao) and the regulation of genes involved in polyamine biosynthesis by drought and other stresses. *Plant Physiology and Biochemistry,* 46, 174-188.

Bailey, B. A., Bae, H., Strem, M. D., Roberts, D. P., Thomas, S. E., Crozier, J., Samuels, G. J., Choi, I. Y. & Holmes, K. A. (2006) Fungal and plant gene expression during the colonization of cacao seedlings by endophytic isolates of four trichoderma species. *Planta,* 224, 1449-1464.

Belsky, J. M. & Siebert, S. F. (2003) Cultivating cacao: Implications of sun-grown cacao on local food security and environmental sustainability. *Agriculture and Human Values,* 20, 277-285.

Bishop, C. M. (1995) *Neural networks for pattern recognition,* Oxford, Oxford University Press.

Bishop, C. M. (1998) *Neural networks and machine learning,* Berlin; New York, Springer.

Bonne, G. P., Stimpfle, R. M., Cohen, R. C., Voss, P. B., Perkins, K. K., Anderson, J. G., Salawitch, R. J., Elkins, J. W., Dutton, G. S., Jucks, K. W. & Toon, G. C. (2000) An examination of the inorganic chlorine budget in the lower stratosphere. *Journal of Geophysical Research-Atmospheres,* 105, 1957-1971.

Brown, M. E., Lary, D. J., Vrieling, A., Stathakis, D. & Mussa, H. (2008) Neural networks as a tool for constructing continuous NDVI time series from AVHRR and MODIS. *International Journal of Remote Sensing,* 29, 7141-7158.

Brown, M. E., Pinzon, J. E., Didan, K., Morisette, J. T. & Tucker, C. J. (2006) Evaluation of the consistency of long-term NDVI time series derived from AVHRR, spot-vegetation, seawifs, MODIS and landsat etm+. *IEEE Transactions Geoscience and Remote Sensing,* 44, 1787-1793.

Carpenter, G. A., Gjaja, M. N., Gopal, S. & Woodcock, C. E. (1997) Art neural networks for remote sensing: Vegetation classification from landsat tm and terrain data. *IEEE Transactions on Geoscience and Remote Sensing,* 35, 308-325.

Caselli, M., Trizio, L., De Gennaro, G. & Ielpo, P. (2009) A simple feedforward neural network for the pm10 forecasting: Comparison with a radial basis function network and a multivariate linear regression model. *Water Air and Soil Pollution,* 201, 365-377.

Chen, P. H., Fan, R. E. & Lin, C. J. (2006) A study on smo-type decomposition methods for support vector machines. *Ieee Transactions on Neural Networks,* 17, 893-908.

Chevallier, F., Cheruy, F., Scott, N. A. & Chedin, A. (1998) A neural network approach for a fast and accurate computation of a longwave radiative budget. *Journal of Applied Meteorology,* 37, 1385-1397.

Comrie, A. C. (1997) Comparing neural networks and regression models for ozone forecasting. *Journal of the Air & Waste Management Association,* 47, 653-663.

Dufour, G., Nassar, R., Boone, C. D., Skelton, R., Walker, K. A., Bernath, P. F., Rinsland, C. P., Semeniuk, K., Jin, J. J., Mcconnell, J. C. & Manney, G. L. (2006) Partitioning between the inorganic chlorine reservoirs HCl and ClONO2 during the arctic winter 2005 from the ace-fts. *Atmospheric Chemistry and Physics,* 6, 2355-2366.

Eklundh, L. & Olsson, L. (2003) Vegetation index trends for the african sahel 1982-1999. *Geophysical Research Letters,* 30.

Eyring, V., Butchart, N., Waugh, D. W., Akiyoshi, H., Austin, J., Bekki, S., Bodeker, G. E., Boville, B. A., Bruhl, C., Chipperfield, M. P., Cordero, E., Dameris, M., Deushi, M., Fioletov, V. E., Frith, S. M., Garcia, R. R., Gettelman, A., Giorgetta, M. A., Grewe, V., Jourdain, L., Kinnison, D. E., Mancini, E., Manzini, E., Marchand, M., Marsh, D. R., Nagashima, T., Newman, P. A., Nielsen, J. E., Pawson, S., Pitari, G., Plummer, D. A., Rozanov, E., Schraner, M., Shepherd, T. G., Shibata, K., Stolarski, R. S., Struthers, H., Tian, W. & Yoshiki, M. (2006) Assessment of temperature, trace species, and ozone in chemistry-climate model simulations of the recent past. *Journal of Geophysical Research-Atmospheres,* 111.

Eyring, V., Waugh, D. W., Bodeker, G. E., Cordero, E., Akiyoshi, H., Austin, J., Beagley, S. R., Boville, B. A., Braesicke, P., Bruhl, C., Butchart, N., Chipperfield, M. P., Dameris, M., Deckert, R., Deushi, M., Frith, S. M., Garcia, R. R., Gettelman, A., Giorgetta, M. A., Kinnison, D. E., Mancini, E., Manzini, E., Marsh, D. R., Matthes, S., Nagashima, T., Newman, P. A., Nielsen, J. E., Pawson, S., Pitari, G., Plummer, D. A., Rozanov, E., Schraner, M., Scinocca, J. F., Semeniuk, K., Shepherd, T. G., Shibata, K., Steil, B., Stolarski, R. S., Tian, W. & Yoshiki, M. (2007) Multimodel projections of stratospheric ozone in the 21st century. *Journal of Geophysical Research-Atmospheres,* 112.

Fan, R. E., Chen, P. H. & Lin, C. J. (2005) Working set selection using second order information for training support vector machines. *Journal of Machine Learning Research,* 6, 1889-1918.

Froidevaux, L., Jiang, Y. B., Lambert, A., Livesey, N. J., Read, W. G., Waters, J. W., Fuller, R. A., Marcy, T. P., Popp, P. J., Gao, R. S., Fahey, D. W., Jucks, K. W., Stachnik, R. A., Toon, G. C., Christensen, L. E., Webster, C. R., Bernath, P. F., Boone, C. D., Walker, K. A., Pumphrey, H. C., Harwood, R. S., Manney, G. L., Schwartz, M. J., Daffer, W. H., Drouin, B. J., Cofield, R. E., Cuddy, D. T., Jarnot, R. F., Knosp, B. W., Perun, V. S., Snyder, W. V., Stek, P. C., Thurstans, R. P. & Wagner, P. A. (2008) Validation of Aura microwave limb sounder HCl measurements. *Journal of Geophysical Research-Atmospheres,* 113.

Froidevaux, L., Livesey, N. J., Read, W. G., Jiang, Y. B. B., Jimenez, C., Filipiak, M. J., Schwartz, M. J., Santee, M. L., Pumphrey, H. C., Jiang, J. H., Wu, D. L., Manney, G. L., Drouin, B. J., Waters, J. W., Fetzer, E. J., Bernath, P. F., Boone, C. D., Walker, K. A., Jucks, K. W., Toon, G. C., Margitan, J. J., Sen, B., Webster, C. R., Christensen, L. E., Elkins, J. W., Atlas, E., Lueb, R. A. & Hendershot, R. (2006a) Early validation analyses of atmospheric profiles from eos MLS on the Aura satellite. *IEEE Transactions on Geoscience and Remote Sensing,* 44, 1106-1121.

Froidevaux, L., Livesey, N. J., Read, W. G., Salawitch, R. J., Waters, J. W., Drouin, B., Mackenzie, I. A., Pumphrey, H. C., Bernath, P., Boone, C., Nassar, R., Montzka, S., Elkins, J., Cunnold, D. & Waugh, D. (2006b) Temporal decrease in upper atmospheric chlorine. *Geophysical Research Letters,* 33.

Gallo, K. P., Ji, L., Reed, B. C., Dwyer, J. & Eidenshink, J. C. (2003) Comparison of MODIS and AVHRR 16-day normalized difference vegetation index composite data. *Geophysical Research Letters,* 31, L07502-5.

Gardner, M. W. & Dorling, S. R. (1999) Neural network modelling and prediction of hourly NOx and NO2 concentrations in urban air in london. *Atmospheric Environment,* 33, 709-719.

Gunson, M. R., Abrams, M. C., Lowes, L. L., Mahieu, E., Zander, R., Rinsland, C. P., Ko, M. K. W., Sze, N. D. & Weisenstein, D. K. (1994) Increase in levels of stratospheric chlorine and fluorine loading between 1985 and 1992. *Geophysical Research Letters,* 21, 2223-2226.

Haykin, S. (2001a) *Kalman filtering and neural networks,* Wiley-Interscience.

Haykin, S. S. (1994) *Neural networks : A comprehensive foundation,* New York, Toronto, Macmillan.

Haykin, S. S. (2001b) *Kalman filtering and neural networks,* New York, Wiley.

Haykin, S. S. (2007) *New directions in statistical signal processing : From systems to brain,* Cambridge, Mass., MIT Press.

Hyyppa, J., Hyyppa, H., Inkinen, M., Engdahl, M., Linko, S. & Zhu, Y. H. (1998) Accuracy comparison of various remote sensing data sources in the retrieval of forest stand attributes. *Forest Ecology and Management.* Lake Buena Vista, Florida.

Lary, D. J. & Aulov, O. (2008) Space-based measurements of HCl: Intercomparison and historical context. *Journal of Geophysical Research-Atmospheres,* 113.

Lary, D. J., Muller, M. D. & Mussa, H. Y. (2004) Using neural networks to describe tracer correlations. *Atmospheric Chemistry and Physics,* 4, 143-146.

Lary, D. J., Remer, L., Paradise, S., Macneill, D. & Roscoe, B. (2009) Machine learning and bias correction of MODIS aerosol optical depth. *IEEE Trans. on Geoscience and Remote Sensing*

Lary, D. J., Waugh, D. W., Douglass, A. R., Stolarski, R. S., Newman, P. A. & Mussa, H. (2007) Variations in stratospheric inorganic chlorine between 1991 and 2006. *Geophysical Research Letters,* 34.

Levenberg, K. (1944) A method for the solution of certain problems in least squares. *Quart. Appl. Math.,* 2, 164-168.

Los, S. O. (1998) Estimation of the ratio of sensor degradation between noaa AVHRR channels 1 and 2 from monthly NDVI composites. *IEEE Transactions on Geoscience and Remote Sensing,* 36, 206-213.

Marquardt, D. W. (1963) An algorithm for least-squares estimation of nonlinear parameters. *Journal of the Society for Industrial and Applied Mathematics,* 11, 431-441.

Marquardt, D. W. (1979) Citation classic - algorithm for least-squares estimation of non-linear parameters. *Current Contents/Engineering Technology & Applied Sciences*, 14-14.

Mchugh, M., Magill, B., Walker, K. A., Boone, C. D., Bernath, P. F. & Russell, J. M. (2005) Comparison of atmospheric retrievals from ace and HALOE. *Geophysical Research Letters,* 32.

Michelsen, H. A., Salawitch, R. J., Gunson, M. R., Aellig, C., Kampfer, N., Abbas, M. M., Abrams, M. C., Brown, T. L., Chang, A. Y., Goldman, A., Irion, F. W., Newchurch, M. J., Rinsland, C. P., Stiller, G. P. & Zander, R. (1996) Stratospheric chlorine partitioning: Constraints from shuttle-borne measurements of [HCl], [clno3], and [ClO]. *Geophysical Research Letters,* 23, 2361-2364.

Miura, T., Huete, A. & Yoshioka, H. (2006) An empirical investigation of cross-sensor relationships of NDVI and red/near-infrared reflectance using eo-1 hyperion data. *Remote Sensing of Environment,* 100, 223-236.

Moré, J. J. (1977) The levenberg-marquardt algorithm: Implementation and theory. IN Watson, G. A. (Ed.) *Numerical analysis.* Springer Verlag.

Nassar, R., Bernath, P. F., Boone, C. D., Clerbaux, C., Coheur, P. F., Dufour, G., Froidevaux, L., Mahieu, E., Mcconnell, J. C., Mcleod, S. D., Murtagh, D. P., Rinsland, C. P., Semeniuk, K., Skelton, R., Walker, K. A. & Zander, R. (2006) A global inventory of stratospheric chlorine in 2004. *Journal of Geophysical Research-Atmospheres,* 111.

Newman, P. A., Nash, E. R., Kawa, S. R., Montzka, S. A. & Schauffler, S. M. (2006) When will the antarctic ozone hole recover? *Geophysical Research Letters,* 33.

Rinsland, C. P., Gunson, M. R., Salawitch, R. J., Michelsen, H. A., Zander, R., Newchurch, M. J., Abbas, M. M., Abrams, M. C., Manney, G. L., Chang, A. Y., Irion, F. W., Goldman, A. & Mahieu, E. (1996) ATMOS/atlas-3 measurements of stratospheric chlorine and reactive nitrogen partitioning inside and outside the november 1994 antarctic vortex. *Geophysical Research Letters,* 23, 2365-2368.

Russell, J. M., Deaver, L. E., Luo, M. Z., Park, J. H., Gordley, L. L., Tuck, A. F., Toon, G. C., Gunson, M. R., Traub, W. A., Johnson, D. G., Jucks, K. W., Murcray, D. G., Zander, R., Nolt, I. G. & Webster, C. R. (1996) Validation of hydrogen chloride measurements made by the halogen occultation experiment from the UARS platform. *Journal of Geophysical Research-Atmospheres,* 101, 10151-10162.

Santee, M. L., Mackenzie, I. A., Manney, G. L., Chipperfield, M. P., Bernath, P. F., Walker, K. A., Boone, C. D., Froidevaux, L., Livesey, N. J. & Waters, J. W. (2008) A study of stratospheric chlorine partitioning based on new satellite measurements and modeling. *Journal of Geophysical Research-Atmospheres,* 113.

Scholkopf, B., Smola, A. J., Williamson, R. C. & Bartlett, P. L. (2000) New support vector algorithms. *Neural Computation,* 12, 1207-1245.

Sen, B., Osterman, G. B., Salawitch, R. J., Toon, G. C., Margitan, J. J., Blavier, J. F., Chang, A. Y., May, R. D., Webster, C. R., Stimpfle, R. M., Bonne, G. P., Voss, P. B., Perkins, K. K., Anderson, J. G., Cohen, R. C., Elkins, J. W., Dutton, G. S., Hurst, D. F., Romashkin, P. A., Atlas, E. L., Schauffler, S. M. & Loewenstein, M. (1999) The budget and partitioning of stratospheric chlorine during the 1997 arctic summer. *Journal of Geophysical Research-Atmospheres,* 104, 26653-26665.

Slayback, D. A., Pinzon, J. E., Los, S. O. & Tucker, C. J. (2003) Northern hemisphere photosynthetic trends 1982-99. *Global Change Biology,* 9, 1-15.

Smola, A. J. & Scholkopf, B. (2004) A tutorial on support vector regression. *Statistics and Computing,* 14, 199-222.

Solomon, S., Intergovernmental Panel on Climate Change. & Intergovernmental Panel on Climate Change. Working Group I. (2007) *Climate change 2007 : The physical science basis : Contribution of working group i to the fourth assessment report of the intergovernmental panel on climate change,* Cambridge ; New York, Cambridge University Press.

Steven, M. D., Malthus, T. J., Baret, F., Xu, H. & Chopping, M. J. (2003) Intercalibration of vegetation indices from different sensor systems. *Remote Sensing of Environment,* 88, 412-422.

Teillet, M., Staenz, K. & Williams, D. J. (1997) Effects of spectral, spatial and radiometric characteristics on remote sensing vegetation indices of forested regions. *Remote Sensing of Environment,* 61, 139-149.

Trishchenko, A. P., Cihlar, J. & Li, Z. (2002) Effects of spectral response function on surface reflectance and NDVI measured with moderate resolution satellite sensors. *Remote Sensing of Environment,* 81, 1-18.

Tucker, C. J., Slayback, D. A., Pinzon, J. E., Los, S. O., Myneni, R. B. & Taylor, M. G. (2001) Higher northern latitude normalized difference vegetation index and growing season trends from 1982 to 1999. *International Journal of Biometeorology,* 45, 184-190.

Van Leeuwen, W., Orr, B. J., Marsh, S. E. & Herrmann, S. M. (2006) Multi-sensor NDVI data continuity: Uncertainties and implications for vegetation monitoring applications. *Remote Sensing of Environment,* 100, 67-81.

Vapnik, V. N. (1995) *The nature of statistical learning theory,* New York, Springer.

Vapnik, V. N. (1998) *Statistical learning theory,* New York, Wiley.

Vapnik, V. N. (2000) *The nature of statistical learning theory,* New York, Springer.

Voss, P. B., Stimpfle, R. M., Cohen, R. C., Hanisco, T. F., Bonne, G. P., Perkins, K. K., Lanzendorf, E. J., Anderson, J. G., Salawitch, R. J., Webster, C. R., Scott, D. C., May, R. D., Wennberg, P. O., Newman, P. A., Lait, L. R., Elkins, J. W. & Bui, T. P. (2001) Inorganic chlorine partitioning in the summer lower stratosphere: Modeled and measured [ClONO2]/[HCl] during polaris. *Journal of Geophysical Research-Atmospheres,* 106, 1713-1732.

Waugh, D. W. & Eyring, V. (2008) Quantitative performance metrics for stratospheric-resolving chemistry-climate models. *Atmospheric Chemistry and Physics,* 8, 5699-5713.

Webster, C. R., May, R. D., Jaegle, L., Hu, H., Sander, S. P., Gunson, M. R., Toon, G. C., Russell, J. M., Stimpfle, R. M., Koplow, J. P., Salawitch, R. J. & Michelsen, H. A. (1994) Hydrochloric-acid and the chlorine budget of the lower stratosphere. *Geophysical Research Letters,* 21, 2575-2578.

Wmo (2006) Scientific assessment of ozone depletion: 2006. WMO Global Ozone Res. and Monitor. Proj., Geneva.

Wood, G. A. R. (2001) *CACAO,* 620.

Yi, J. S. & Prybutok, V. R. (1996) A neural network model forecasting for prediction of daily maximum ozone concentration in an industrialized urban area. *Environmental Pollution,* 92, 349-357.

Zander, R., Gunson, M. R., Farmer, C. B., Rinsland, C. P., Irion, F. W. & Mahieu, E. (1992) The 1985 chlorine and fluorine inventories in the stratosphere based on ATMOS observations at 30-degrees north latitude. *Journal of Atmospheric Chemistry,* 15, 171-186.

Zander, R., Mahieu, E., Gunson, M. R., Abrams, M. C., Chang, A. Y., Abbas, M., Aellig, C., Engel, A., Goldman, A., Irion, F. W., Kampfer, N., Michelsen, H. A., Newchurch, M. J., Rinsland, C. P., Salawitch, R. J., Stiller, G. P. & Toon, G. C. (1996) The 1994 northern midlatitude budget of stratospheric chlorine derived from ATMOS/atlas-3 observations. *Geophysical Research Letters,* 23, 2357-2360.

8

C-Band Sea Ice SAR Classification Based on Segmentwise Edge Features

Juha Karvonen
Finnish Meteorological Institute (FMI)
Finland

1. Introduction

Sea ice is a very important issue in winter navigation in the Arctic waters and also in the Baltic Sea. The winter traffic in ice is assisted by ice breakers in ice covered areas with heavy ship traffic. In other ice covered areas, however, the vessels must have the capability to navigate in sea ice without aid, meaning that they have to be designed especially for ice conditions. Earlier e.g. the Finnish ice breakers used helicopters to find the best routes in sea ice. However, today the expensive helicopter use has been replaced by utilization of remote sensing data. And for the ships navigating in ice on their own, it is even more important to have valid and useful information of the current sea ice around the ship to find its route through ice. Especially Synthetic Aperture Radar (SAR) data are useful during the dark and often cloudy northern winters. Because C-band SAR, like Radarsat (1 and 2) and Envisat ASAR, is an active instrument it also works during the dark periods and has a suitable EM wavelength to propagate through the cloud cover. Our aim is to produce automatically generated sea ice products from SAR data, such that they are informative and easy to interpret. Sea ice types in SAR images are best described by the edges present in the images. The type and amount of edges give us information on the ice types in addition to commonly used areal backscattering values and statistics. Here we present some novel features for sea ice SAR data classification, mainly derived from detected edges.

The preprocessing of the SAR data is also an important step. First the data is rectified to a map projection, Pearson (1990). We use the Mercator projection, also used in nautical charts, in the Baltic Sea and the Polar Stereographic projection for the Arctic. Also an incidence angle grid over the SAR area in appropriate projection is generated from the data included in the SAR data. After applying a land masking and incidence angle correction, Karvonen et al. (2002), using the generated incidence angle grid, we perform an edge-preserving speckle filtering. Then we apply an intensity-based segmentation, and produce a multi-resolution presentation in three resolutions by combining adjacent segments with different criteria for different resolutions. We also detect the edges and corners from the full-resolution SAR data.

After the segmentation, the detected edges can be divided into two groups: segment edges (segment boundaries) and edges within segments. The segment edge features describe the contrast and shape of the segment, and the within-segment edge features give additional geophysically relevant information on the ice properties within the segment.

The segment shape gives information on the segment, e.g. ice floes typically can have polygonal or round shape, and ice ridges and cracks or leads typically are narrow but long segments,

often they are not straight lines, however. The segment shape can be described by shape features based on the segment edges. The segment shape features in our approach are computed from the ordered edge points sampled along the edge. Then the object shape can roughly be characterized by the polygon defined by the sampled set of the edge points, and multiple useful features based on this presentation can be computed for the ice type classification.
We have collected C-band SAR data, both Radarsat and Envisat ASAR data over the Baltic Sea and also over the Arctic Sea areas for our studies. The data are Radarsat ScanSAR wide mode data and Envisat ASAR wide swath mode data. These data have a resolution of about 100m and cover an area of 400-500 km wide, and thus are suitable for operational sea ice monitoring.

2. Preprocessing

The preprocessing chain consists of the georectification to Mercator (Baltic Sea) or Polar Stereographic projection (Arctic), incidence angle correction, Karvonen et al. (2002), land masking, speckle filtering, and multi-resolution segmentation.

2.1 Speckle Filtering

We have studied two iterative algorithms performing an edge-preserving filtering. One is based on an anisotropic mean and the other on anisotropic median computation. Both the algorithms have given very good results according to our visual judgment. The algorithms are shortly described in the following two subsections. In our experiments we have used 40 iterations for both the algorithms. We have not studied the defining of a sufficient number of iterations, and 40 iterations probably is a too high number and we could use less iterations to achieve the same performance in shorter time. One way to define the number of iterations automatically would be to have a threshold for the total change between iterations, and stop after the change has reduced to a smaller value than the given threshold.

2.2 Anisotropic Mean Filtering

The anisotropic mean algorithm iteratively computes weighted means within a small window S around each pixel. We have used a 3x3 window. The weights depend on the absolute difference between the pixel value $I_t(i,j)$, $(i,j) \in S$ and the mid-pixel value $I_t(r_0, c_0)$. The number of iterations (time) is denoted by the subindex t.

$$I_t(r_0.c_0) = \sum_{i=-1}^{1} \sum_{j=-1}^{1} I_{t-1}(r_0+i, c_0+j)/\Delta_{i,j}, \tag{1}$$

$$\Delta_{i,j} = |I_{t-1}(r_0, c_0) - I_{t-1}(r_0+i, c_0+j)|, \tag{2}$$

$$if\ \Delta_{i,j} = 0,\ then \Delta_{i,j} = 1, \tag{3}$$

$$if\ \Delta_{i,j} > T,\ then\ \Delta_{i,j} = \infty. \tag{4}$$

This is iterated from the time $T = 1$ until the desired amount of iterations, in our case 40, has been reached. At $t = 0$, then input is the original SAR image.

2.3 Anisotropic Median Filtering

The anisotropic median is also computed iteratively either using only non-edge points or edge points depending on whether the mid-point is non-edge or edge, respectively. This is also computed iteratively in a 3x3 window, S. So it is computed as

$$I_t = Median(I_{t-1}(i,j)),\ (i,j) \in S \backslash E \tag{5}$$

for the non-edge mid-points and

$$I_t = Median(I_{t-1}(i,j)),\ (i,j) \in S \cap E \tag{6}$$

for the edge points. E is the set of the edge points. This algorithm naturally requires an edge detection to be performed before running it. This is also iterated from the time $T = 1$ until the desired amount of iterations (40) has been reached. At $t = 0$, then input is again the original SAR image.

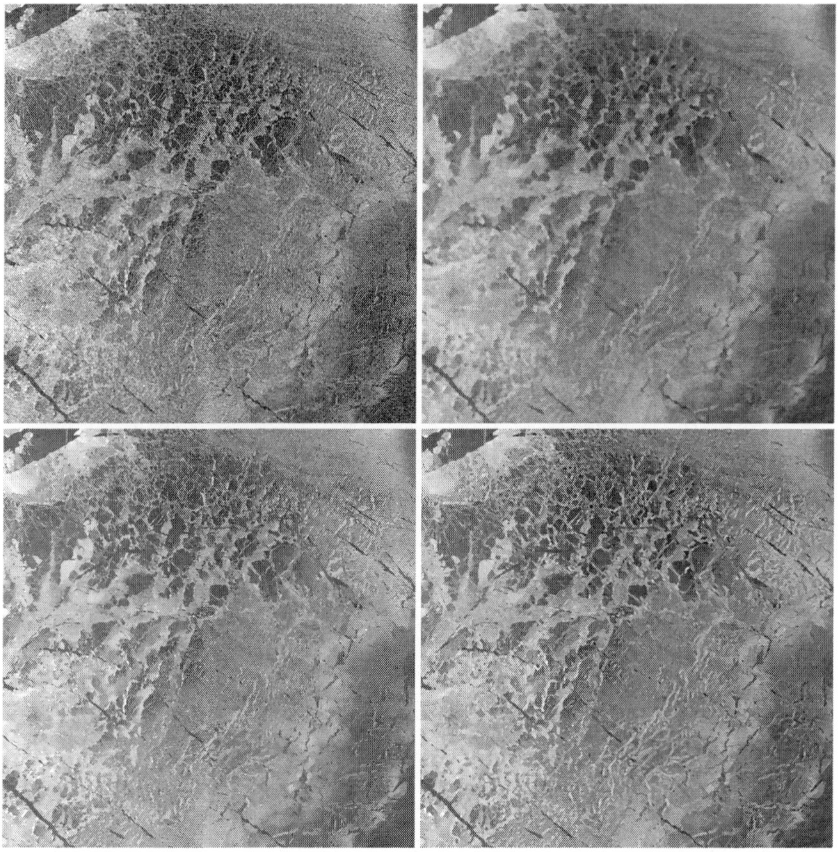

Fig. 1. An example of speckle filtering, the original Radarsat-1 image ($\approx$ 75x75 km, upper left), iterative 3x3 median (40 iterations, upper right), anisotropic mean ($T = 15$, 40 iterations, lower left), and anisotropic 3x3 median (40 iterations, lower right). The (isotropic) iterative median clearly blurs edges.

2.4 Segmentation

The segmentation algorithm we use is a K-means algorithm, Linde et al. (1980), applied to the pixel intensity values of the speckle-filtered SAR images. The values of K are typically in the range 4–8 for our SAR data. In the beginning the class means are initialized based on a cumulative data histogram computed from the image. Then the upper limits for K clusters are computed to produce K bins of equal amounts of samples and the initial class means are set to be in the middle of two adjacent limits, i.e. $m_k = 0.5(L_{k-1} + L_k)$, where L_i's are the limits between two adjacent data bins. After this initialization step, the K-means algorithm is iterated using only the image pixel values at the cluster (or segment) edges, in the sense of 8-neighborhood, from the previous iteration in the iterative computation. The iteration is repeated until no changes occur or a maximum number of iterations has been reached (to guarantee stopping).
A more sophisticated segmentation result could be achieved by adding more (texture) features. We are studying the inclusion of autocorrelation to the segmentation, but then we also need to exclude the values at the segment boundaries, because large changes at the segment edges cause high autocorrelation. Instead we should first perform an intensity-based segmentation and only after that divide the segments, if necessary, based on the texture feature.

2.5 Multi-resolution Approach

Because we here are using small scale-segments in the SAR images as features, it is necessary to have a multi-resolution presentation of the data. Then we can compute statistics of smaller-scale features over the larger scale segments, such that the results are statistically relevant.
The traditional multi-resolution approaches typically use some low pass filtering and builds a multi-resolution pyramid of the data. This naturally also reduces the accuracy of segment boundaries at the low resolutions. On the other hand, processing at the low resolutions is faster and less memory is required. However, we have here adopted a multi-resolution approach based on segmentation, segment sizes and contrasts between segments. We use three resolution layers generated by an algorithm which starts from the K-means segmentation results and then combines the adjacent segments up to a given size limit T_s (in pixels) to their neighbor segments which are larger than T_s (if they exist) if the edge contrast between the segments at the edge boundary is less than a contrast threshold T_c. The contrast threshold depends linearly on the segment area and varies between given values for the minimum and maximum segment sizes. At each iteration the smaller segments are joined to the segments larger than T_s, and after each iteration the values T_s and T_c are increased (T_s) and decreased (T_c) linearly starting from given parameter start values and ending up to given parameter end values. Finally we perform a joining of the small segments to larger segments such that all the segments smaller than a given threshold T_{tot} are joined to their neighbors. A sophisticated way of doing this is again to use an iterative method such that first the smaller segments are joined and finally the larger segments. The thresholds depend on the desired resolution level and on the image resolution. Higher size thresholds are used for the lower resolutions. A suitable value for the contrast start threshold is around 30–50 for our data, and the end value in the range 0–10.
The pseudocode of the joining algorithm looks this:

```
# Initialization of the thresholds, Tsz is segment size threshold
# and its initial value Tsz(0) is a smaller value than the final value Tsz(1).
# Tc is a inter-segment contrast threshold.
# Its initial value Tc(0) is a larger  value than its final value Tc(1).
Tsz = Tsz(0); Tc=Tc(0);
```

```
Sstep=(Tsz(1)-Tsz(0))/(Niterations);
Cstep=(Tc(0)-Tc(1))/(Niterations);

# This iteration joins the segments starting from the smaller segments
for (Niterations) do
  for (each segment)
    if ((segment_size < Tsz) AND (segment_contrast < Tc) AND
        (some_neighbor_segment_size >= Tsz)) then
      Join the segment to the closest larger segment (minimum edge contrast)
      Tsz = Tsz + Sstep; Tc = Tc - Cstep;
    endif
  end
end

# This iteration is just to guarantee that all the segments are joined
# It typically only has a very small affect (if it has).
while (no changes occur OR maximum count reached) do
  for (each segment)
    if ((segment_size < Tsz) AND (segment_contrast < Tc) AND
        (some_neighbor_segment_size >= Tsz)) then
      Join the segment to the closest larger segment (minimum edge contrast)
    endif
  end
end
```

One way to reduce resolution would also be to reduce the number of clusters (K) in the K-means clustering, i.e. to use less clusters for lower resolutions. We have made some studies of this approach also, but the work for finding optimal parametrization and integrating this with the current algorithm is still under construction.

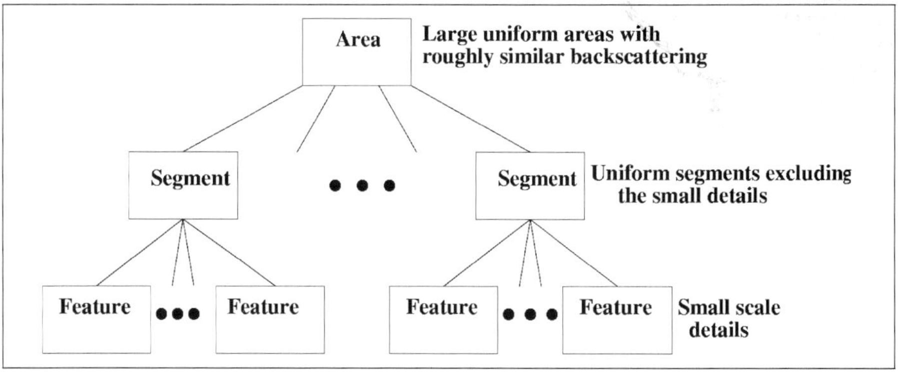

Fig. 2. The multi-resolution concept.

3. Edge Features

We have used the canny edge detection, Canny (1986), to detect edges in the SAR images. The Canny edge detector however only takes into account the local neighborhood in the thresholding. To get the connected edges better included we perform the Canny edge detection twice for one image, with two sets of thresholds, the high and low thresholds. If an edge resulting

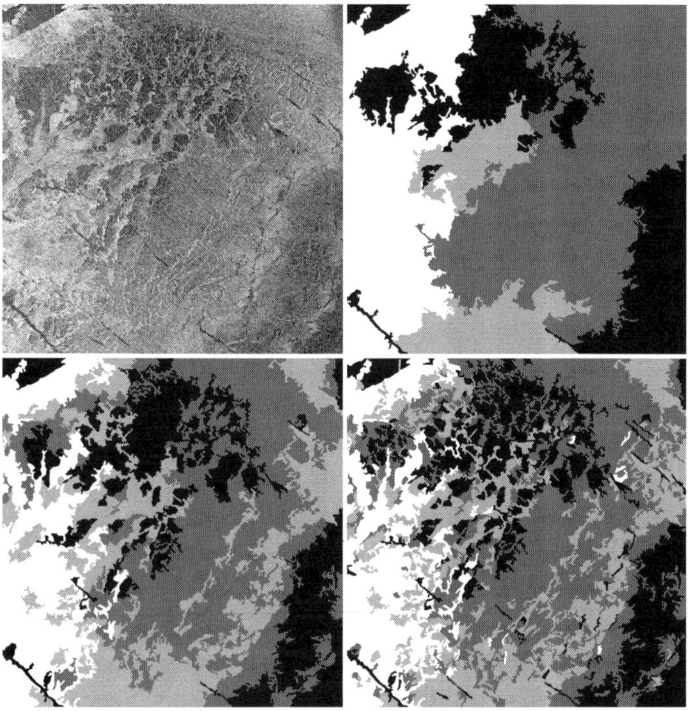

Fig. 3. A part of a Radarsat-1 SAR image (Baltic Sea, $\approx$ 75x75 km, upper left) and its segmentation in the three resolutions: low resolution (upper right), medium resolution (lower left) and high resolution (lower right).

from the Canny edge detection with the high parameter values is connected to an edge detected with the low parameters, then the edge from the detection with low parameter values is also included as an edge. We use the Canny algorithm with 5x5 pixel Gaussian smoothing and parameters $T_{lo} = 100$ and $T_{hi} = 120$ as the high Canny parameter values and $T_{lo} = 60$ and $T_{hi} = 100$ as the low Canny parameter values. The selection of these values is naturally dependent on the data scaling. These presented values seem to be a suitable selection for our SAR data. The edge detection is always computed for the SAR data before speckle filtering.
We divide the located edges into two categories, depending whether they are on a segment boundary area or inside the segment. The edge boundary area is defined as a the area of pixels which have other segments' pixels within its 8-neighborhood.

3.1 Segment Boundary Strength

The segment boundary strength can be defined in multiple ways. We can study the local gradients between the segments at the boundaries, or just simply check the amount of detected edge pixels at the segment boundary. The segment boundary strength can also give information on the segment. We utilize the segment edge contrast between adjacent segments in our

Fig. 4. A part of a SAR image ($\approx$ 25x25 km, left), detected edges (middle) and the corresponding structured edges (right), i.e. edges which are parts of larger edge segments than a given threshold, here 10.

segment joining algorithm. The segment boundary strength can also be used as a feature in segment classification, but here we mainly concentrate on the within-segment features.

3.2 Structure within Segments

The structure within segments is defined by the amount of different edge types within the segment. The edge is here said to be structured if the size of a uniform edge segment (i.e. connected edge pixels in the sense of 8-neighborhood) is larger than a given threshold T_e ($T_e > 1$), and unstructured (random edge) if the size is less or equal than T_e. If the segment size without segment boundaries is A, then we can compute three features related to the structuredness of the segment. The first is the degree of the segment random roughness or deformation

$$D_R = N_R/A, \tag{7}$$

the second is the degree of the segment structured randomness or deformation

$$D_S = N_S/A, \tag{8}$$

and the third is the relative randomness

$$D_{RS} = N_R/N_S. \tag{9}$$

N_R and N_S are the numbers of structured and random edges within the segment. The total relative number of segment edge pixels, which we here also call the segment deformation, is

$$D = (N_R + N_S)/A = N/A. \tag{10}$$

N is the total number of edge points in a segment.

3.3 Corners

We have also studied the occurrence of corner points at the segment boundaries and inside segments. To detect corners we have used a variant of the Harris (aka Harris-Stephens) edge detector, Harris & Stephens (1988). Instead of computing the Harris corner response function M_c, we have used the eigenvalues ($\lambda_1 > \lambda_2$) of the Harris matrix and thresholds T_{hi} and T_{lo} for the eigenvalues. If $\lambda_1 > T_{hi}$ at some image location (r,c), then (r,c) can be considered as an edge point, and if additionally $\lambda_2 > T_{lo}$, then it is a corner point. The feature we use is the relative amount of corners D_c computed as:

$$D_c = \frac{N_c}{N_S + N_R} = N_c/N. \tag{11}$$

The Harris algorithm could also be used for detecting edges instead of the Canny algorithm.

3.4 Segment Shape Features Based on Segment Edges

We have also studied some shape features of the segments. The segment shape is naturally described by the segment boundary. The segment edges or boundaries are estimated as polygons. For each segment we have used a constant (20 points) with equivalent distance between the points along the segment boundary to define the polygon. This approach is basically similar to the MPEG-7 shape descriptors, Bober (2001), but our features are different and better suitable for the random shapes of ice segment features. One simple feature is the segment length, l, which in our approach is estimated as the maximum length between two edge polygon corner points along the polygon edge. The shorter distance of the two alternatives of clockwise and counter-clockwise directions is the distance between a single pair of polygon corner points. The (average) segment width, w, can then be computed as

$$w = A/l, \tag{12}$$

where A is the segment area. The segment shape ratio R_s can then be computed as

$$R_s = l/w. \tag{13}$$

This feature is a scale-independent segment shape descriptor and is high for long and narrow segments and smaller for compact segments.
We also compute the segment edge contrast, C_e, i.e. the mean difference between the inside-segment edge points and outside-segment edge points

$$C_e = \sum_{k \in in} I_k/N_{in} - \sum_{k \in out} I_k/N_{out}. \tag{14}$$

The sums are computed along the segment edge, N_{in} and N_{out} are the numbers of the edge pixels inside and outside of the segment along the segment boundary, respectively. One more feature describing the curvature of a segment is computed as a count of those pairs of two adjacent polygon line segments for which the angle between the line segments exceeds a given angle α. If the coordinates of the three polygon corners defining the two adjacent polygon edge segments are (r_{k-1}, c_{k-1}), (r_k, c_k) and (r_{k+1}, c_{k+1}), the vectors to be compared are $p_1 = (\Delta r_1, \Delta c_1)$ and $p_2 = (\Delta r_2, \Delta c_2)$. The index k is computed in modulo N_p (circular) arithmetic such that no over or underflow occur. N_p is the number of polygon corners. The vector components are

$$\Delta r_1 = r_k - r_{k-1} \tag{15}$$

$$\Delta c_1 = c_k - c_{k-1} \tag{16}$$

$$\Delta r_2 = r_{k+1} - r_k \tag{17}$$

$$\Delta c_2 = c_{k+1} - c_k, \tag{18}$$

and the corresponding vector lengths l_1 and l_2

$$l_1 = \sqrt{\Delta r_1{}^2 + \Delta c_1{}^2} \tag{19}$$

$$l_2 = \sqrt{\Delta r_2{}^2 + \Delta c_2{}^2}. \tag{20}$$

The cosine of the angle between the polygon edge segments p_1 and p_2 is

$$\cos(\alpha) = \frac{< p_1, p_2 >}{l_1 l_2}. \tag{21}$$

We have set a threshold angle, T_α, for curvature i.e. the polygon is curved at the location (r_k, c_k) if $\alpha > T_\alpha$, and the total curvature R_c for a edge polygon is defined as the relation of the number of the curved polygon corner point locations N_c to the total number of the polygon corner points N_p:

$$R_c = N_c / N_p. \tag{22}$$

We have used the value $T_\alpha = \pi/3$ in our studies.
In figure 5 we show two artificial segments and their 20-point boundary polygons, and in table 1 the features based on the boundary polygons of these two segments are computed.

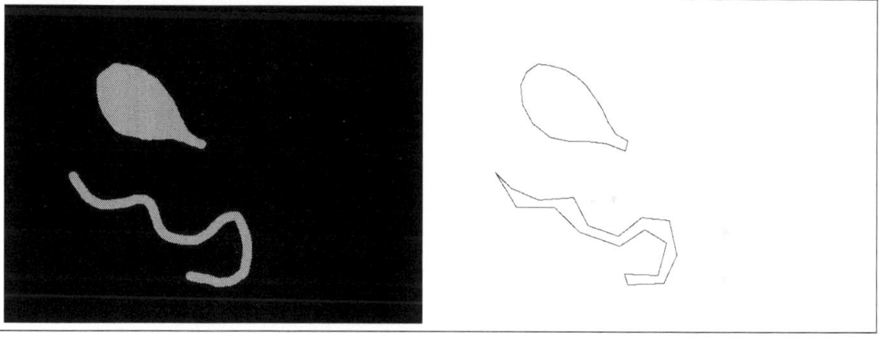

Fig. 5. An example of two artificial segments and their 20-point bounding polygons.

#	r_0	c_0	I_{in}	I_{out}	C_e	A	L	W	R_s	R_c
1	88	168	153	0	153	11783	229.27	51.39	4.46	2/20 = 0.1
2	251	104	153	0	153	7174	455.48	15.75	28.92	8/20 = 0.4

Table 1. Computed features for the artificial segments of Fig. 5.

3.5 Shape Features for the Small Segments

These features are not related to the edges, because the polygon estimation of the edge for small segments is not a very useful approach. We have used two measures of compactness instead. The first measure (C_{S1}) compares the overlapping of the actual segment and a sphere of the same size as the segment, with its center at the center of mass of the segment. The other measure (C_{S2}) finds the bounding sphere of the segment and the feature is the segment area divided by the bounding sphere area, A_{out}.

$$C_{S1} = A_{in} / A. \tag{23}$$

$$C_{S2} = A / A_{out}. \tag{24}$$

Both the features actually give similar information and we have used only the feature C_{S1} in our classification experiments. The interpretation is straightforward: If the feature values are close to one, the segment is compact and if they are close to zero, the segments shape is not compact. Thus we have used two thresholds, $T_{c1} < T_{c2}$. If $C_{S1} < T_{c1}$, the segment is classified to a long segment and if $C_{S1} > T_{c2}$ it is classified to a compact segment.

3.6 Other studied Edge Features

We also studied the directional edge strengths using the MPEG-7 edge filters, Manjunath et al. (2001), and the local direction distributions of the edges. The orientation of the SAR edges can not be used in the same way as for typical textures, i.e. by dividing the edges to vertically oriented, horizontally oriented and so on, because the SAR orientation depends on the imaging geometry and on the location, and similar ice fields can have edge direction distributions which are rotated with respect to each other. Because of this, we can not use an edge direction histogram as a SAR feature. But we can for example utilize a feature describing how oriented the edges in a SAR image are locally, i.e. whether there exist a locally dominant direction within a image window of a fixed size. Unfortunately they did not show very good classification performance for our SAR data. Only some features, like straight ship tracks or straight ice edges could be distinguished and these could also be located by other means, e.g. locating the structured edges and edge contrasts.

We have also computed edge segment size distributions withing segments and at the segment boundaries, but we have not studied their properties carefully yet. The division into structured and random edges, i.e. a two-valued distribution, is our current approach.

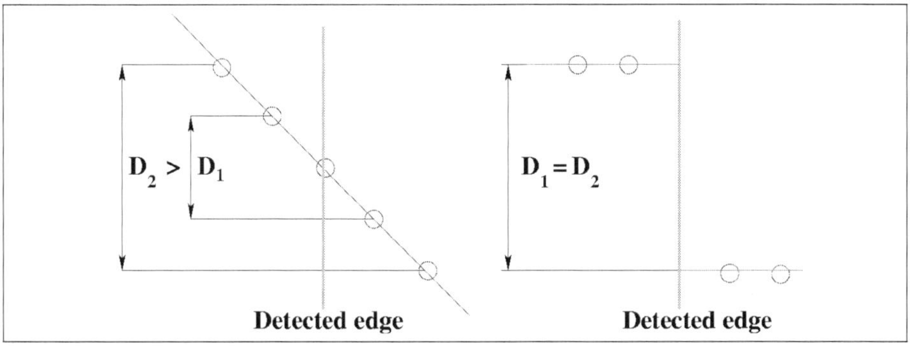

Fig. 6. A ramp edge and a sharp edge, the edge normal is horizontal in the image and the pixel value is in the vertical direction. For a sharp edge the intensity difference for both distances is about equal, and for an ideal ramp edge the intensity difference increases linearly as the distance increases.

We have also studied the division of segment and within-segment edges into sharp edges and ramp edges (smooth edges). The edge is considered as a sharp edge if at the edge $D_1 \approx D_2$,

$D_1 = I_1 - I_{-1}$, $D_2 = I_2 - I_{-2}$, i.e. the pixel values in the speckle filtered image at two distances, $l_1 < l_2$, along the edge normal on opposite sides of the edge are almost equal, and as a ramp edge if $aD_1 < D_2$, $a > 1.0$ is a given factor, see Fig. 6. The distribution of edge type to these two categories was also studied within the segments. The relation of the amounts of these two edge types can also be used to classify the segments, but the geophysical interpretation is still missing. At least it can be used to distinguish between smooth ice segments (like open water and fast ice) and deformed ice segments, as many other edge features, but its ability to provide complementary information is still vague. Intuitively it could be useful in distinguishing e.g. areas with (widely spaced) clear ridges from areas of rubble fields.

4. Some Classification Results

4.1 Open Water Detection

We have earlier used the segment-wise autocorrelation as an open water detector, see Karvonen et al. (2005). Our recent studies have shown that also edge information can be utilized in open water detection.

The relative amount of edges within segment D can be used to locate most of the open water area, but even better indicator for open water is the relative amount of structured edges D_S. In some cases open water can be mixed with level ice or fast ice areas. The classification can be further improved in some cases by using the relative amount of corners D_c as an additional feature. In general we can say that segment-wise D_S is a good open water detector, such that open water has very low values of D_S. Performed tests show that it works well for both the Baltic sea ice and for the Arctic Sea ice. We have two examples of this shown in Figs. 7 and 8. The ASAR mosaic of Fig. 8 has been composed by overlaying all the available ASAR data over the Kara sea area starting from November 2008. Multiple daily images were typically acquired, and this mosaic image describes the ice situation on January 23rd 2009.

4.2 Ice Classification Based on the Inside-Segment Edges

We have made studies with several different sets of edge features. The ratio of the total number of edges within segment and the segment area (D) represents the degree of deformation of the segment. However, this only feature can not always e.g. very well distinguish between open water and deformed ice areas. But including the relative amount of structured edges (D_S) and the relative number corners (D_C), the ice types can be rather well distinguished, see Fig. 9. This figure is a three channel image of the three features suitably scaled for visual inspection. In this figure over the the Gulf of Bothnia, Baltic Sea, the open water areas appear as brown areas and fast ice areas have more red color, indicating that these areas have relatively more corner points than the open water areas. The other ice areas mostly have different tones of green, the more deformed areas being brighter. This example shows the potential of using these three features together for sea ice SAR classification.

4.3 Ice Classification Based on the Segment Shape Features

More information from the data can be extracted by the segment shape classification. Here we only show one example of segment shape classification for one SAR window. The segments smaller than a given size threshold ($T_A = 3000$) have been located and classified to compact segments and non-compact ("long") segments and indicated with different colors in Fig. 10. The relative amounts, with respect to the segment area, of different types of these smaller segments withing medium-scale or large-scale segments (or areas) can then be computed, and we can then get information on the relative amount of cracks, ridges and other ice structures

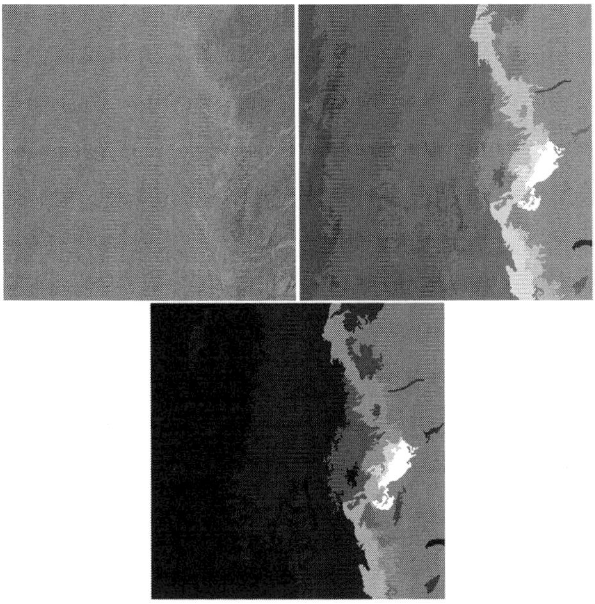

Fig. 7. A Radarsat-1 window over the ice edge, open water area is on the left side of the image (upper left), its D (upper right) and D_S images (lower middle). The open water areas appear as dark areas, especially in the D_S image, and the brash ice area at the ice edge appears bright in both edge images, indicating that it has relative much edge points.

(smooth or rough/ridged compact segments) within the larger areas. We have used an experimental set of parameters for the different segment classes as follows: for compact segments $R_s < 7$ and $R_c < 0.3$, for "long" segments $R_s > 11$ and $R_c < 0.4$. The edge contrast threshold applied was 5 for the dark segments and 15 for the bright segments, i.e. the contrast must exceed these values to be classified. These parameters are also experimental, and studying of ways to find better parameters is under construction.
Some examples of this classification are also given in Figs. 11 and 12. They show the relative amount of different features with different gray tones, the brighter values indicating higher occurrence of the specific feature type.
The relation of amount the edge types (sharp and ramp edges) can also be used as a feature, it is high in the areas of prominent features, e.g. ice floes, ridges with large enough spacing (depending on the SAR resolution) or cracks. This ratio can be used as an additional feature for refining the segment-wise classification. Here we show one example of this feature in Fig. 13 for the ASAR mosaic shown in Fig. 8

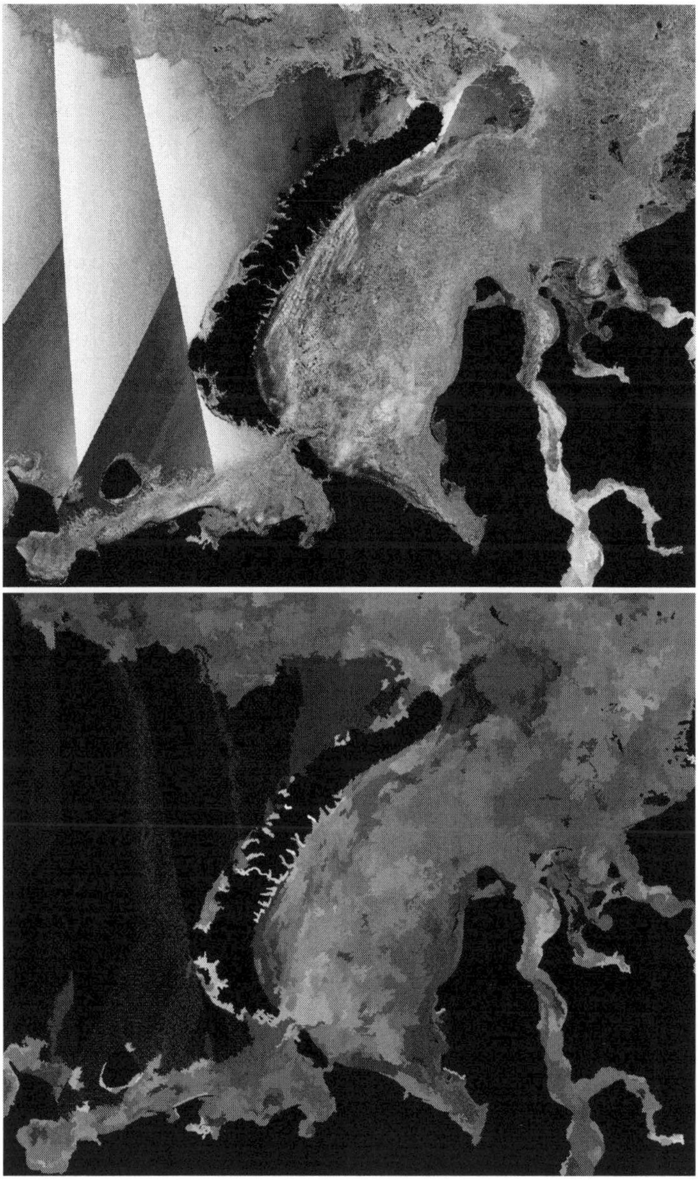

Fig. 8. A SAR image mosaic over the Kara Sea (Jan 23rd 2009, upper image) and the values of D_S over the area (lower), The areas of open water, mainly on the left side of the image have very low value of D_S.

Fig. 9. A Baltic Sea Radarsat-2 image (left) and a 3-feature classification result (RGB three channel presentation) in medium resolution (right), the used features are the relative number of corners (red), relative amount of edges (green), and relative amount of structured edges (blue), the total area covered by the SAR image is about 500x300 km.

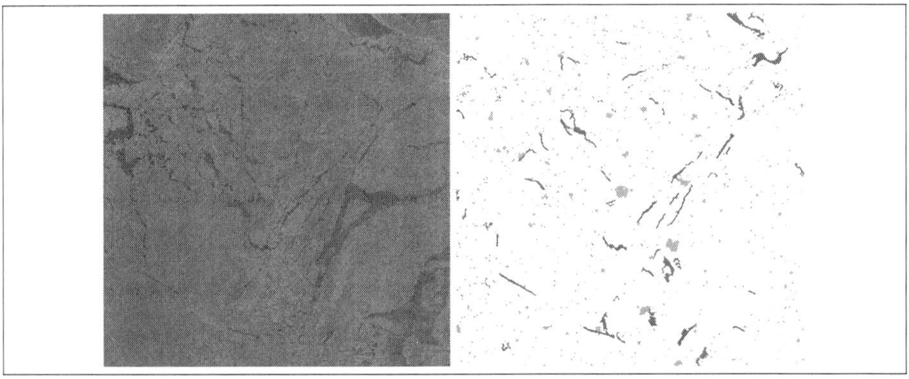

Fig. 10. A part of a Radarsat-1 SAR image (Baltic Sea, left), and the the classified features (for segments smaller than a threshold, i.e $A < T_A$, $T_A = 3000$ pixels in this example, right). The red segments have the edge contrast $C > T_{ctr2}$ and the blue segments $C < T_{ctr1}$, the segments drawn with lighter red and blue are classified based on the small segment algorithm. The total area covered by the image is about 75x75 km.

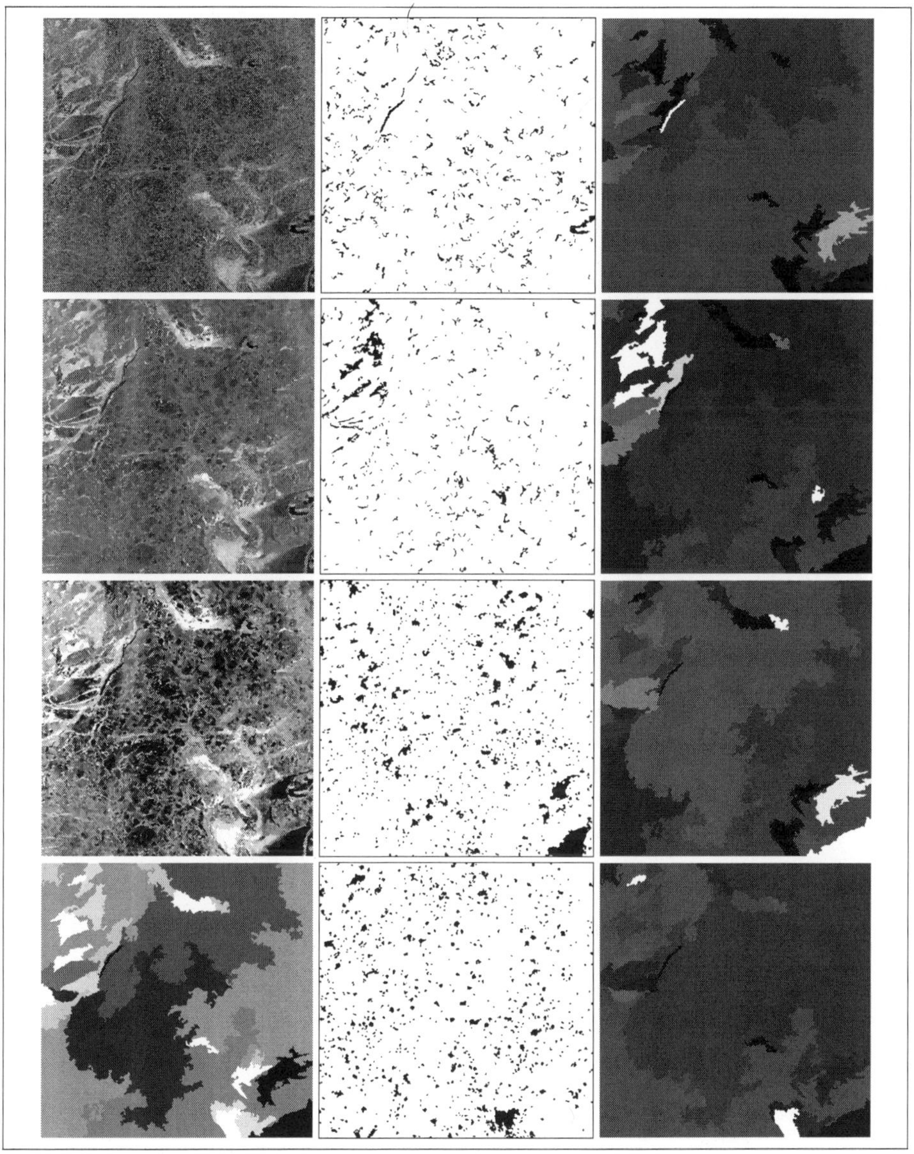

Fig. 11. Envisat ASAR image and detected class-wise features and their relative amounts in different image areas. In the first column from top towards bottom: the original SAR data, speckle-filtered (anisotropic median) data, segmentation. In the second column, the detected features from top towards bottom: dark long features, bright long features, dark compact features and bright compact features. In the third column the segment-wise (large-scale) amounts of different features corresponding to the second row features.

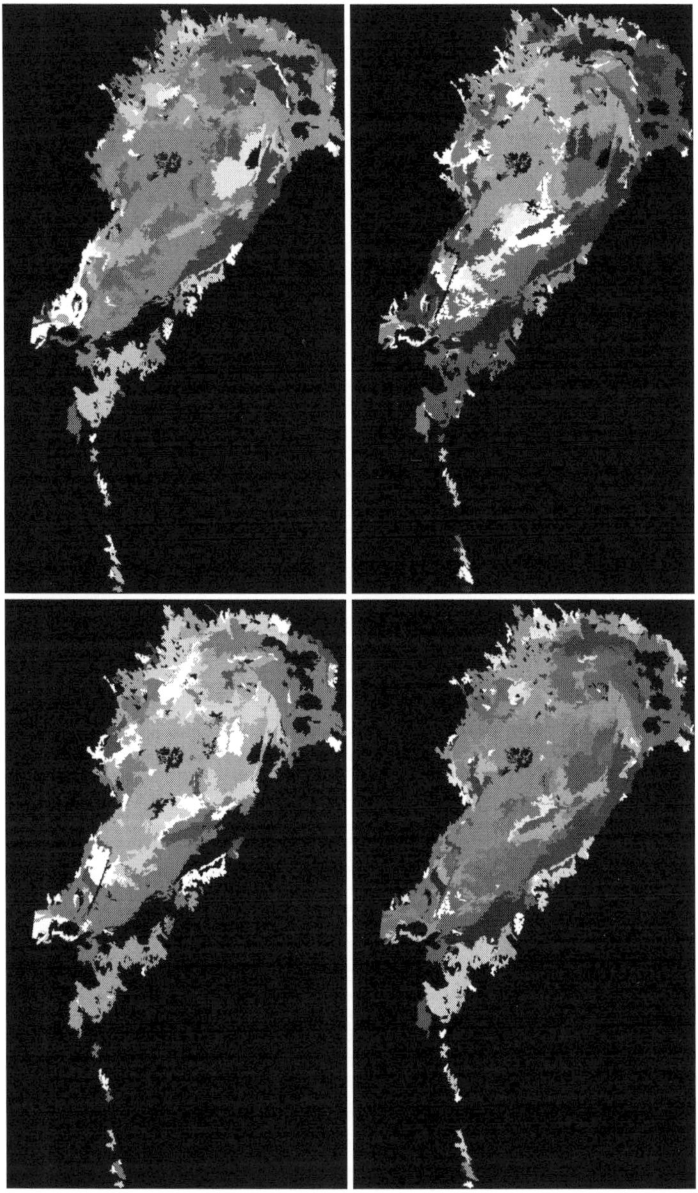

Fig. 12. Segment-wise (large scale) relative amounts of different feature types for the Radarsat-2 image (see Fig. 9): dark compact segments (upper left), bright narrow segments (upper right), dark narrow segments (lower left), bright compact segments (lower right).

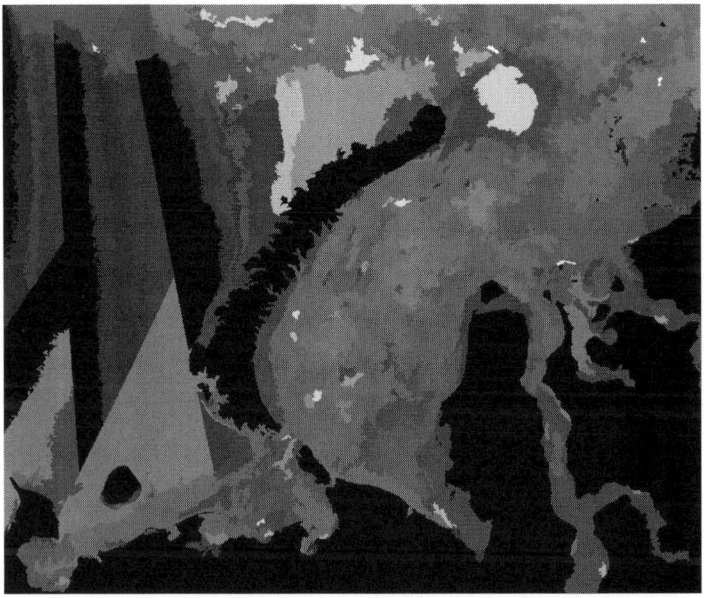

Fig. 13. The segment-wise ratio of structured to random edges for the ASAR mosaic of Fig. 8. The ice areas with many cracks, ice floes or other clearly distinguishing features have higher values. hight values and the other deformed fields, like rubble fields, have lower values. In the open water covered areas (on the left side of the image) the values can have large variations because there are only few edges in these areas, and a small change in the amount of edges of either type can cause large changes in the ratio.

5. Conclusion and Future Work

We have developed a whole sea ice SAR image processing and interpretation chain and demonstrated its usability. The basic idea is that most of the SAR information, in addition to the backscattering lies in the SAR edges. We have also found out that suitable combinations of our edge features can be used for sea ice SAR classification and they give us useful complementary information of the sea ice structure. We believe that we have not yet discovered the full potential of all the edge-related features and here only present some suitable features for SAR classification.

The speckle filtering using either anisotropic mean or median works well and the execution times are reasonable for operational SAR processing. We have not studied the optimal number of iterations, and probably still use too many iterations.

The multi-resolution approach also seems to work well, and gives reasonable segmentations and ice areas compared to visual interpretations. The selection of parameters naturally affects the results of the lower resolution images produced by the segment joining algorithm. The best results are achieved by using many iterations i.e. by increasing the joined segment size slowly, but this also increases the execution times.

The classification results have been promising. Many sea ice classes can be distinguished with very simple edge features, like the combination of amount of edges, and the relative amount of structured edges and the relative amount of corners. The methods can distinguish open water areas very well, and also different ice types and the areas with certain types of ice features (e.g. cracks or ridges) can be located. Not all the features are found, but when using large enough areas, the relative amounts of different features can be estimated.

The parametrization of the studied algorithms has been experimental and we must concentrate on better optimization of the parametrization. We are going to study an automated parameter extraction for given training data sets to reduce the work of experimental parameter definition. But even our experimental parameters have shown promising results and edge features are a very promising addition to SAR classification algorithms. These features will probably also be very useful for classification of other kinds of SAR data sets over land areas. We have studied these features only with a few images from three instruments Radarsat-1, Radarsat-2 and Envisat ASAR. In the next phase we are going to make tests for larger data sets, for example for a whole winter season in both Baltic Sea and Kara Sea, and also for other SAR instruments with different operating parameters (e.g. X- and L-band SAR).

The classification results have been evaluated against visual interpretation. Sea ice measurements are very difficult and expensive to carry out. Because the ice is typically moving, except in fast ice zones, multiple measurements should be made simultaneously (or temporally as close as possible) with the satellite passing time. Even making a few measurements is difficult and expensive, because typically a ship capable of operating in sea ice is required to get in the target area. And the ice properties can differ much in a relatively small area, less than a SAR pixel size. However, visual interpretation of the ice typing from SAR data by our sea ice experts has been very good compared to our occasional field campaign measurements and feedback from the Finnish ice breakers using this information, and we can consider it as good reference data.

6. References

Bober, M. (2001). Mpeg-7 visual shape descriptors, *IEEE Transactions on Circuits and Systems for Video Technology, Special Issue on MPEG-7* **11**(6): 716–719.

Canny, J. (1986). A computational approach to edge detection, *IEEE Trans. Pattern Analysis and Machine Intelligence* **8**(6): 679–698.

Harris, C. & Stephens, M. (1988). A combined corner and edge detector, *Proc. of Alvey Vision Conference*, Univ. of Manchester, pp. 147–151.

Karvonen, J., Simila, M. & Makynen, M. (2002). An iterative incidence angle normalization algorithm for sea ice sar images, *Proceedings of the IEEE International Geoscience and Remote Sensing Symposium (IGARSS'02)*, Vol. III, pp. 1524–1527.

Karvonen, J., Simila, M. & Makynen, M. (2005). Open water detection from baltic sea ice radarsat-1 sar imagery, *IEEE Geoscience and Remote Sensing Letters* **2**(3): 275–279.

Linde, Y., A. Buzo, A. & Gray, R. M. (1980). An algorithm for vector quantizer design, *IEEE Trans. Communication* **28**(1): 84–95.

Manjunath, B. S., Ohm, J.-R., Vasudevan, V. V. & Yamada, A. (2001). Color and texture descriptors, *IEEE Transactions on Circuits and Systems for Video Technology, Special Issue on MPEG-7* **11**(6): 703–715.

Pearson, F. (1990). *Map Projections: Theory and Applications*, CRC Press.

9

Early Warning And On-Line Mapping For Flood Events

D. Mioc [a, f], B. Nickerson [b], F. Anton [c], E. MacGillivray [d], A. Morton [d], D. Fraser [a], P. Tang [e] and A. Kam [a]

[a] *Department of Geodesy and Geomatics Engineering, University of New Brunswick, Canada*

[b] *Faculty of Computer Science, University of New Brunswick, Canada*

[c] *Department of Informatics and Mathematical Modeling, Technical University of Denmark, Denmark*

[d] *Emergency Measures Organization, Victoria Health Centre, 65 Brunswick Street, Fredericton, NB, Canada*

[e] *New Brunswick Department of Environment, Marysville Place, Fredericton, NB, Canada*

[f] *National Space Institute, Technical University of Denmark, Denmark*

1. Introduction

Floods are common natural disasters throughout the world. Each year they cause considerable damage to people's lives and properties. In the spring of 1973, the lower Saint John River in the Fredericton area (New Brunswick, Canada) experienced its worst ever

Fig. 1. The impact of flooding in Fredericton, New Brunswick in Spring, 2008

recorded flooding, resulting in economic losses of CAD 31,9 million, and the loss of one life (CIWD, 1974). At the peak of the flood, private houses and public churches were flooded, and roads and bridges were damaged.

Fig. 2. Flooding of St. John River in 2008

Since 1973 other floods have caused another three lives lost and more than CAD 68.9 million in damage.

Fig. 3. One house taken by the flood in 2008

In May 2008, heavy rains combined with melted snow have overwhelmed the St. John River, which is 673 kilometres long, and brought water levels to a height that many regions have not seen in more than three decades. Homes have floated off their foundation and travelled downstream, while 600 families and individuals have been evacuated (see Figures 1, 2 and 3). The determination of the financial cost of damages caused by this flooding is still not finalized.

Flood forecasting has been proven to reduce the property damage and the loss of lives (Sanders et al., 2005). The recent advances in forecasting for flood warning (Moore et al., 2005) have shown that is possible to integrate rainfall modeling and forecasting with flood forecasting and warning. The research paper on World Wide Web based hydrological model for flood prediction using GIS (Al-Sabhan et al., 2003) gives an excellent overview of current research advances and a new on-line available prototype that combined hydrological modeling with Internet technology.

However, in this research we didn't try to customize any of the existing flood forecasting models described in the literature as it is proven to be very difficult and very specific to the different modeling tools that are used (Al-Sabhan et al., 2003). Instead, we implemented the automatization of specific existing processes, workflows and modeling tools for flood forecasting and monitoring in the New Brunswick Department of Environment.

The Saint John River Forecast System operated by the Department of Environment Hydrology Centre is monitoring and predicting flood events along the Saint John River. The Hydrology Centre team uses hydrologic modeling software to predict water levels for the next 24 and 48 hours along the lower Saint John River Valley by incorporating climate data, weather forecast data, snow data and flow data.

However, the predicted water levels provided by this system cannot satisfy the requirements of the decision support system for flood events. The system neither directly displays the areas affected by flooding, nor shows the difference between two flood events. Based on the water levels, it is hard for users to directly determine which houses, roads, and structures will be affected by the predicted flooding. To deal with this problem, it is necessary to visualize the output from hydrological modeling in a Geographic Information System (GIS). GISs have powerful tools that allow the predicted flood elevations to be displayed as a map showing the extent of the inundation. After the interfaces for the visualization of the impact of flood events are designed, a computerized system is developed that predicts the extent of floods and dynamically displays near-real-time flood information for decision makers and the general public.

To improve flood prediction for Saint John River, we developed a Web GIS based decision support system for flood prediction and monitoring. In this paper, we present the methods for data integration, floodplain delineation, and online map interfaces. This paper is organized as follows: in Section 2, we briefly describe the Saint John River floodplain and in Section 3, we present hydrological modelling for flood forecasting. In section 4, we present the conceptual model of the flood prediction and monitoring system and in section 5, we explain the integration of hydrological modelling and GIS. Subsection 5.1 presents a Web-based interface for dynamic flood prediction monitoring and mapping that can dynamically display observed and predicted flood extents for decision makers and the general public. In section 6, we present our conclusions, and in section 7 our acknowledgments.

2. Saint John River Floodplain

The Saint John River lies in a broad arc across South-Eastern Quebec, northern Maine and western New Brunswick. Its Canadian portion extends from a point on the international boundary with the State of Maine in United States to the Bay of Fundy. It drains a total watershed area of 54.600 km^2. The river is about 700 km long, and the total fall from its headwaters to the city of Saint John is about 482 m. The slope of the river gradually decreases from about 1,5 metres per kilometre in the headwaters to 0,4 metres per kilometre in the reach above Fredericton (see Figure 4).

The study area of this research is the floodplain area along a 90 km long section of the river from Fredericton to Oak Point and the French, Grand, Indian and Maquapit lakes. Flooding has been a significant problem for this area for a long time. From the largest and best documented flood occurred between April and May 1973, the greatest flood damage areas are located within the proposed study area and include:

a) Fredericton south of the former CNR Bridge,
b) Nashwaaksis Subdivision,
c) East Bank downstream of the Princess Margaret Bridge, and
d) the Lincoln area (ENB-MAL, 1979).

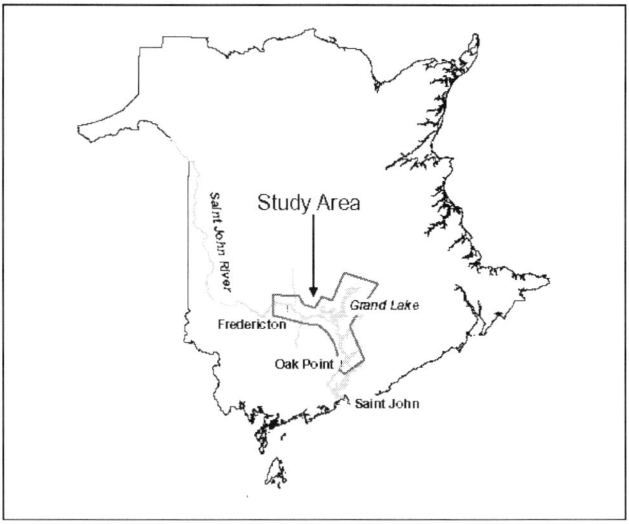

Fig. 4. Overview of Saint John River watershed

3. Hydrological Modelling for Flood Forecasting

Flood forecasting on the Saint John River is performed by the Hydrology Centre of the New Brunswick Department of Environment in co-operation from interprovincial and international agencies. Both hydrologic and hydraulic models are utilized in order to forecast water levels in the lower Saint John River. The basic component of the system is the U.S. Army Corps of Engineers' Streamflow Synthesis and Reservoir Regulation (SSARR) model. The Simulated Open Channel Hydraulics (SOCH) model of the Tennessee Valley Authority and the Dynamic Wave Operational (DWOPER) model (Fread, 1992; Fread, 1993; Fread and Lewis, 1998) of the National Weather Service are also used.

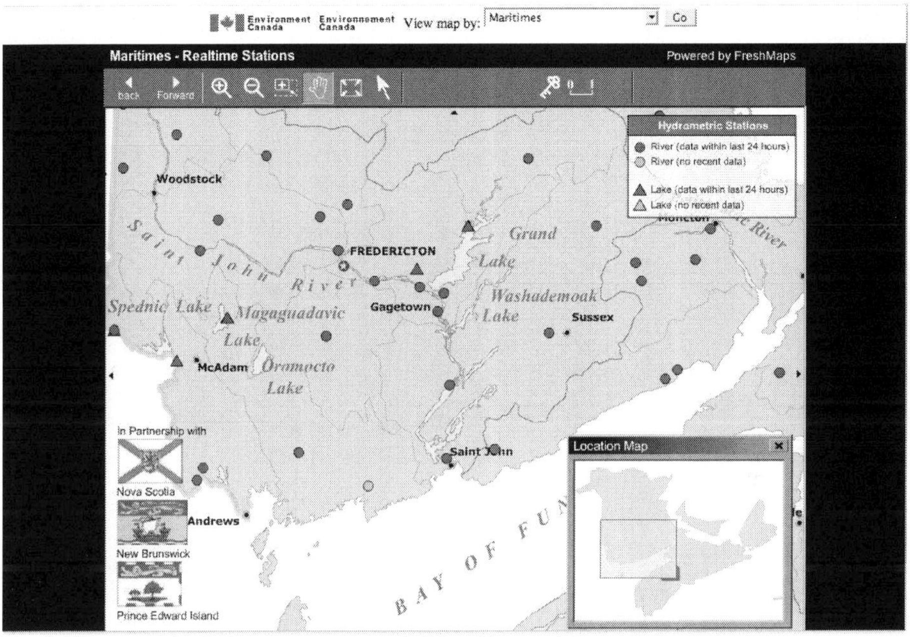

Fig. 5. The map of existing water gauges in New Brunswick

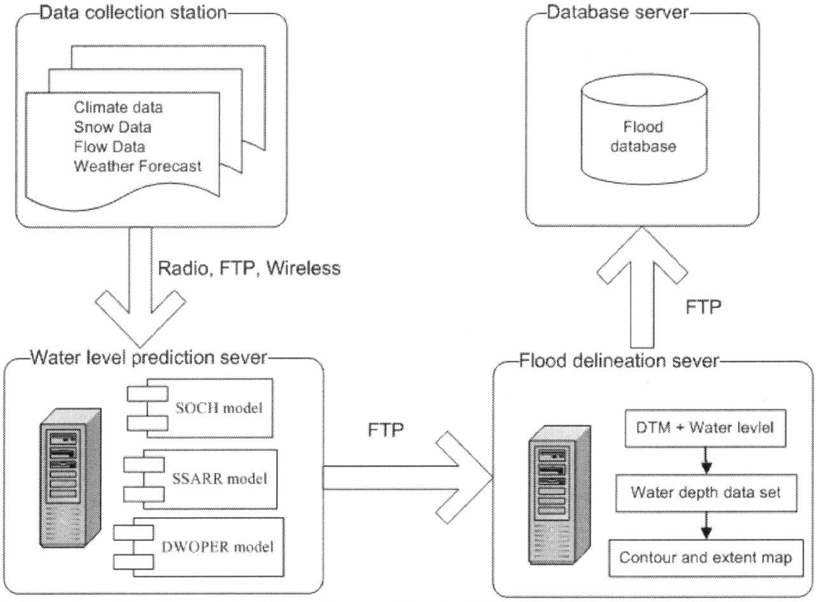

Fig. 6. The operational model of flood forecasting in New Brunswick River Watch

The Hydrology Centre monitors the water levels, stream flows and climate with partner agencies, and coordinates a co-operative snow survey with reports for the entire Saint John River Basin. There are networks of 25 stream-flow gauges, 16 water level gauges, and 43 climate stations throughout the Saint John River Basin (see Figure 5). The data are transmitted to the Hydrology Centre through a variety of telecommunication systems and protocols (see Figure 6). The data are processed and analyzed before being accepted as input data to the models.

Comparisons of predicted and actual water level observations over the last 10 years have shown that these forecasted river water levels have a 95% confidence level of 0,2 m. Thus, the hydrological modelling has very good flood prediction capabilities (Fread, 1993). However, the water levels predicted by the hydrological model cannot satisfy the requirements of the decision support system for flood events. Indeed, it is hard for users to directly determine which houses, roads, and structures will be affected by the predicted flooding, because the model neither directly displays the areas affected by flooding, nor shows the difference between two flood events. In order to overcome this problem, it is necessary to interface the hydrological modelling software with a Geographic Information System (GIS).

In the past decades, engineers have developed many methods for delineating floodplain boundaries. Most of these methods are manual, tedious, and labour-intensive. With the advent of robust computer tools (GIS) and high accuracy Digital Terrain Model (DTM), automated floodplain delineation is achievable. Recently, several management systems for floodplain delineation have been developed and applied in the flood risk areas. These include floodplain delineation using watershed Modeling System (WMS) (EMRL, 1998), Arc/Info MIKE11_GIS (DHI, 2004), and HEC-GeoRAS (Ackerman, 2005). In this project, we used all of the above systems with CARIS software in order to implement floodplain delineation. CARIS (Computer Aided Resource Information System) develops and supports rigorous, technologically advanced geomatics software for managing spatial and non-spatial data. CARIS software supports Triangulated Irregular Networks and offers advanced algorithms for Digital Terrain models, such as interpolating elevations for given coordinates. In the next sections, we will show how the integration of CARIS with hydrological modelling software allows us to generate floodplain maps.

4. Flood Prediction And Monitoring System

In order to improve the current flood prediction system for the Saint John River, a new research has been initiated. Several provincial organisations in New Brunswick (Emergency Measures Organisation, NB Department of Environment, River-Watch and the University of New Brunswick) have been actively involved in this new research project titled "Decision Support for Flood Event Prediction and Monitoring (FEPM)".
The main objective of this research project is to build up a decision support system to improve the prevention, mitigation, response, and recovery from flood events.

The New Brunswick Department of Environment Hydrology Center monitors a wide range of information on factors affecting flooding such as snow conditions, temperatures, precipitation patterns, water levels and stream flow conditions by using a wide variety of telecommunication systems ranging from satellites to the telephone.

The New Brunswick Department of Environment Hydrology Center team uses hydrologic modelling software (DWOPER[1]) to predict water levels for the next 48 hours along the lower Saint John River Valley by processing climate data, weather forecast data, snow data, and flow data from approximately 60 water level gauges in New Brunswick.
The design of the system allows near real-time imagery of actual flood conditions to be overlaid on the base mapping and existing imagery, as well as overlays indicating 100-year flood extents. Map layers of transportation networks, hydrographic features, property boundaries, municipal infrastructure (e.g. power lines, natural gas lines) and contour lines can also be visualized.

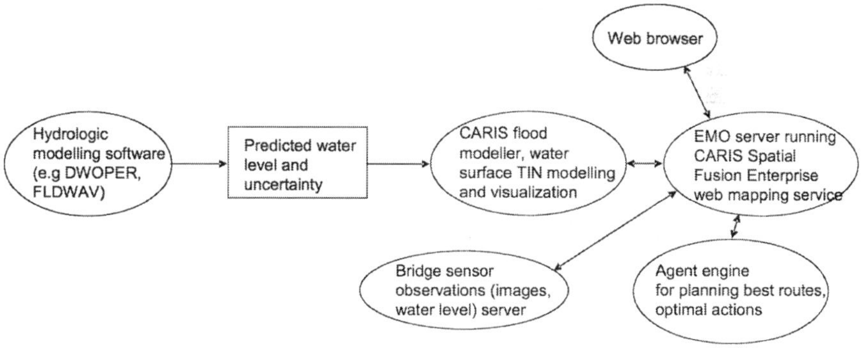

Fig. 7. Conceptual model of flood prediction and monitoring system

The final software products are integrated together within CARIS software as shown conceptually on Figure 7. Several provincial and research organisations in New Brunswick (University of New Brunswick, Emergency Measures Organization, NB Department of Environment, etc.) have been actively involved in the project. In this project, CARIS GIS software was used to implement floodplain delineation and online mapping.

5. Integration of Hydrological Modeling And GIS

The implementation that integrates hydrological modeling, Digital Terrain Modelling, and a GIS algorithm for floodplain delineation will be presented in the following section.

Floodplain delineation requires a high precision ground surface DTM. Analysis of available datasets shows that there are range and accuracy limitations among these datasets. It is therefore necessary to test and integrate these datasets in order to obtain a high accuracy Digital Elevation Model data. For this research, the accuracy of provincial elevation data

---

[1] DWOPER is a one-dimensional routing model developed by the Hydraulic Research Laboratory of the United States National Weather Service (Fread, 1992).

and the city of Fredericton data were analyzed. High accuracy control points can be used to evaluate the accuracy of DTM data. This procedure is implemented by using CARIS GIS tools. Firstly, we generated a TIN model from elevation data (see Figure 8). Then using the CARIS GIS comparative surface analysis tool, the differences between the elevations of the control points and the interpolated elevation of the corresponding points were calculated. Finally, the statistical accuracy was computed using the methodology developed during previous floods in New Brunswick (CIWD, 1974) and (ENB-MAL, 1979).

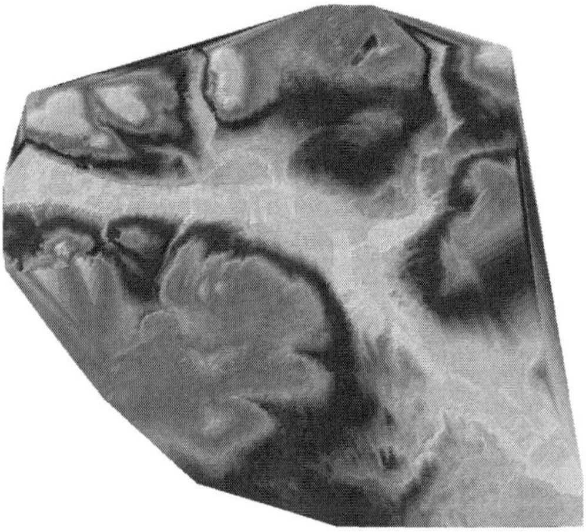

Fig. 8. DTM of the lower Saint John watershed

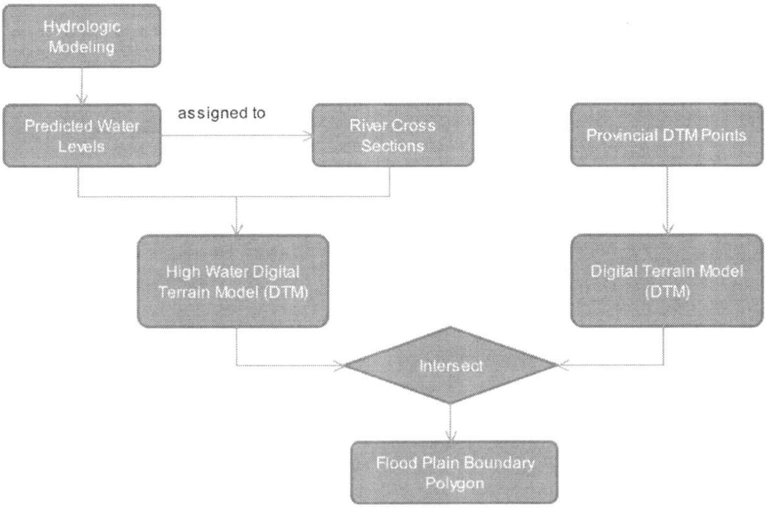

Fig. 9. The workflow of the calculation of predicted floodplain

To support near real-time flood modelling, we developed the procedures for transmitting real time water level data from the New Brunswick Department of Environment - River Watch to the end users (see Figure 13). The water level data from the output of flood modelling by the Hydrology Centre in the Department of Environment are transmitted via FTP. The timestamp of new data is checked every 30 minutes for upload in the database. Then, the water level data are transferred to the FEPM Web Page for generating and displaying gauge bar graphs.

At the same time, the water level data are accessed by the software module for flood plain computations (Mioc et al., 2008). With the advent of robust GIS tools and high accuracy Digital Terrain Model (DTM), automated floodplain delineation is achievable (Noman et al., 2003). As shown on Figure 9, the most significant inputs for automated floodplain delineation[2] are the DTM (see Figure 8) and the water levels on the sections shown on Figure 10. The process considers the DTM and water levels at different locations to determine the direction and extent of flow over a floodplain for a given hydrologic event.

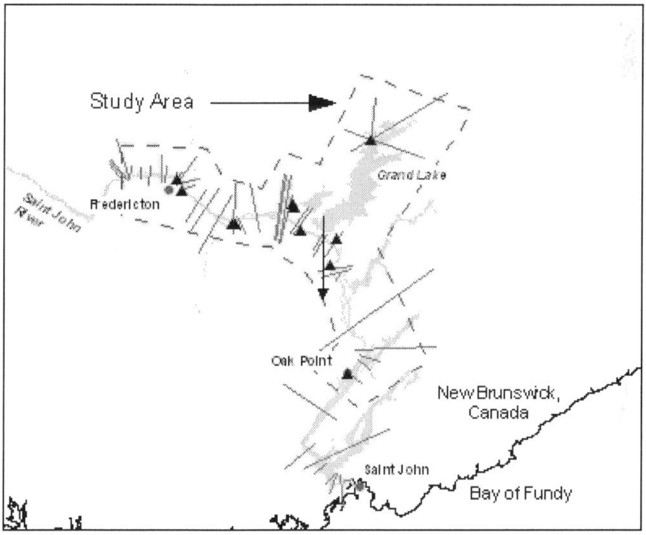

Fig. 10. Modeling water level surface using cross sections

The floodplain depth dataset is the primary output of this process. It indicates the high water mark and the depth of water inside the floodplain polygon, and is generated by computing the height difference between the water surface TIN with the ground surface DTM data. Based on the obtained flood depth data, the floodplain extent and depth maps can be generated. The intermediate parts of the process involve geo-referencing the water levels, extending the water levels to the probable floodplain area, and creating a TIN of the water surface. CARIS GIS allows users to create an irregular TIN or regular grid DTM, to calculate the accurately differences between two TINs or regular grid DTMs, to interpolate contours using a DTM, and

[2] Automated floodplain delineation is an excellent tool for producing floodplain extent maps (Noman et al., 2001; Noman et al., 2003).

to display the DTM using the CARIS 3D Viewer program. These software functionalities were used for the development of the algorithm for floodplain delineation.

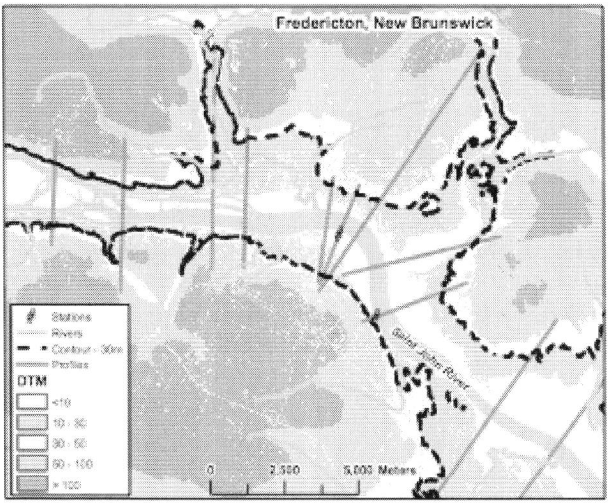

Fig. 11. Detailed zoom of the DTM and the floodplain of the area with the cross sections around Fredericton

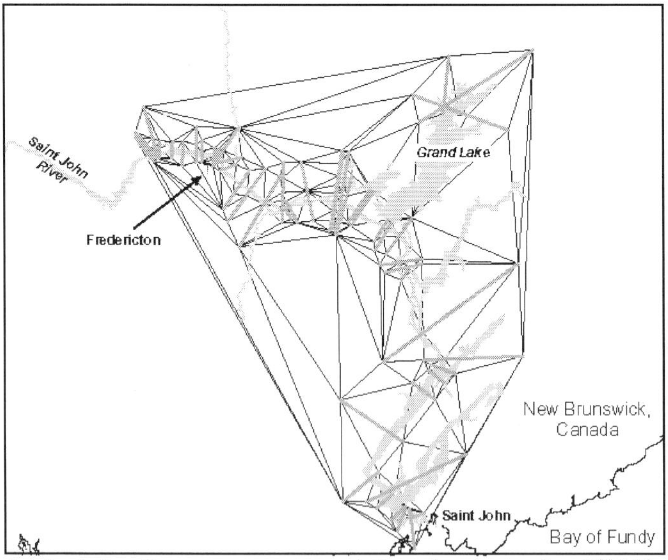

Fig. 12. Floodplain delineation process - computation of water surface TIN (the thick edges are constrained Delaunay edges corresponding to cross sections while the thin edges are Delaunay edges that do not correspond to cross sections)

CARIS software provides an effective spatial analysis tool that calculates floodplain delineation and facilitates the mapping of flood events. As an example of floodplain delineation, Figure 11 shows the cross-sections used for the flooding event that took place in the Spring of 1973. Figure 12 shows the constrained Delaunay triangulation (see (Okabe et al., 2000) for an introduction to constrained Delaunay triangulations) used to compute the water surface. The flood plain is computed by interpolating linearly elevations and flood depths in the triangles. The cross sections are guaranteed to be present as edges of the triangulation, because they are the constrained edges. The cross sections are spaced in order to better evaluate the influence of confluents and effluents and their spacing decreases with the curvature of the river.

5.1 Development of a Web-based interface for dynamic flood prediction monitoring and mapping

CARIS Spatial Fusion was used to develop software for integration of satellite imagery and dynamic flood maps. Web map Interfaces that dynamically display maps of current and predicted flood events were developed and implemented.

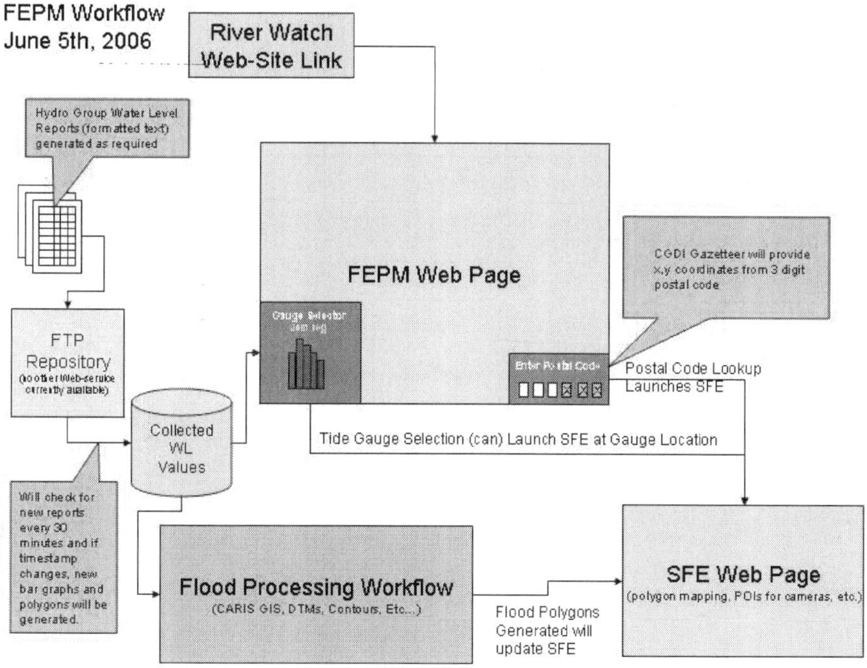

Fig. 13. Flood data processing diagram

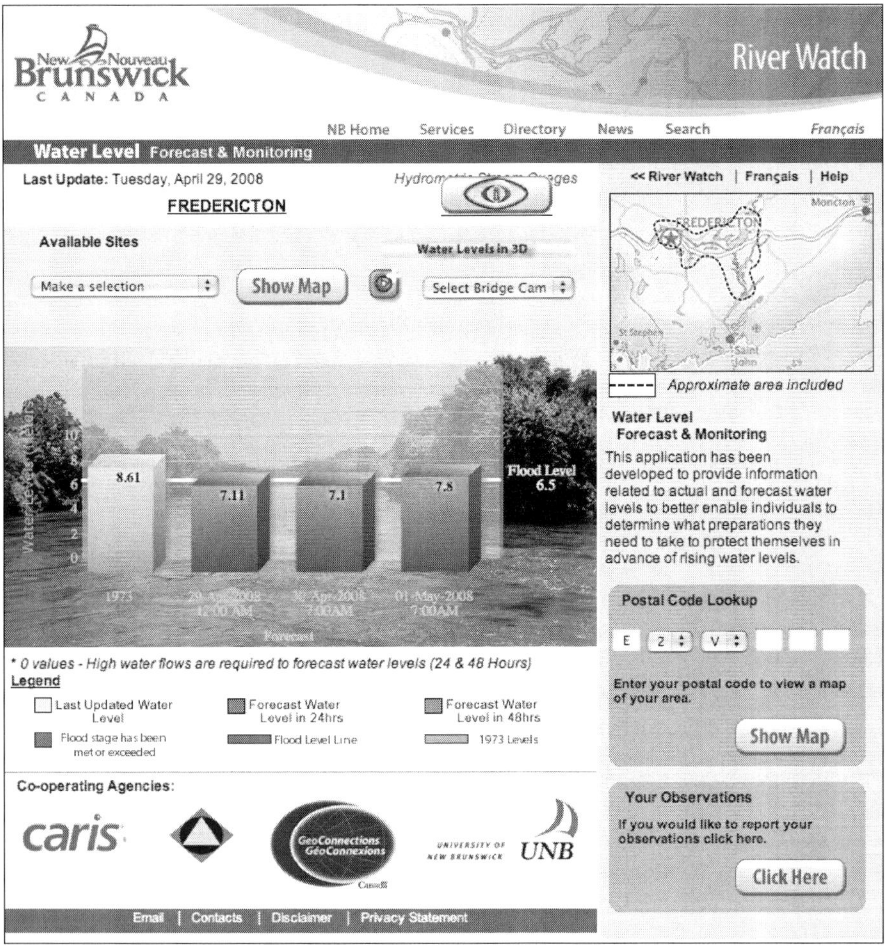

Fig. 14. New Brunswick River Watch Web site for flood warning in lower St. John River watershed

The architecture of the flood data processing is shown on Figure 13. The Web GIS software that we developed, allows for a spatial query based on 6-digit postal code (see Figure 14), so the users will be able to easily locate their area of interest. The web site allows one to display historical flood maps for twenty and hundred years average as well as for the catastrophic flood of 1973 (see Figure 15).

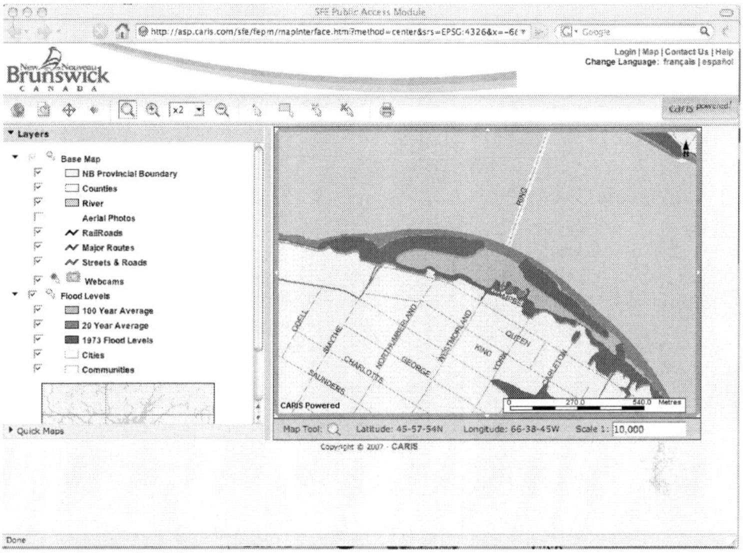

Fig. 15. Existing historical flood maps

The Web-GIS interface is also designed to calculate the flood polygon of current and predicted flood plains and display them as prediction maps (next 24 hours and 48 hours - see Figure 16). Each layer of the web map is separate, allowing the overlay and visualization of transportation networks, hydrographical features, property boundaries, municipal infrastructure and contour lines.

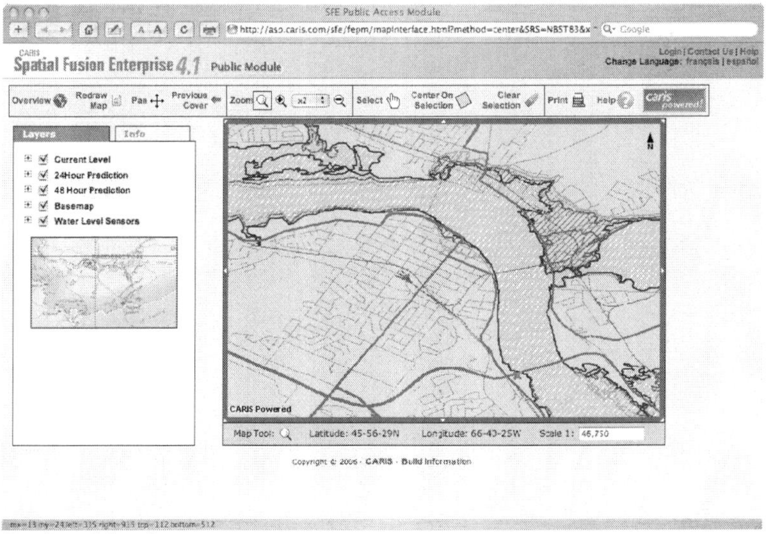

Fig. 16. Predicted flood maps

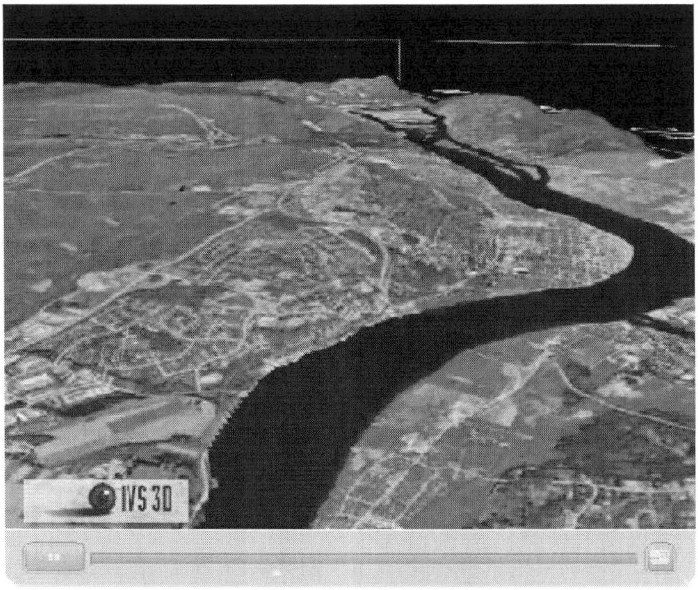

Fig. 17. Visualization of the flood in 1973 - Fredericton area

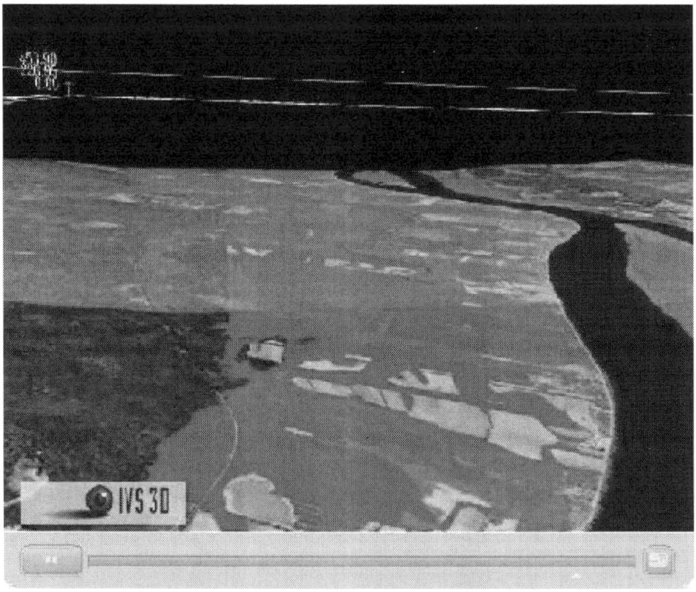

Fig. 18. Visualization of the flood in 1973 - rural area

To better understand the spread and the impact of floods, the three-dimensional visualization of the flood of the Spring of 1973 was implemented (see Figures 17 and 18) using IVS3D[3] software.
It allows users to visualize the major flood event that happened in the Spring of 1973 via "fly-through" animation. In this application the advanced software (from Interactive Visualization Systems) for dynamic visualization is used to interactively show the areas affected by the record high flooding in 1973.
The basic map layers are integrated with orthophotos and flood areas to create this realistic visualisation tool using IVS3D.

6. Conclusions

The Decision Support System for Flood Event Prediction and Monitoring implemented with web-mapping interfaces facilitates monitoring and prediction of flood events. It provides a basis for early warning and mapping of flood disasters. The general public can access the web site and browse the information in their area of interest. They can also visualize the impact of the flood events on the area where they live.
This research paper presents the integration of the DWOPER hydraulic model with the CARIS GIS system to dynamically compute and display near-real-time flood warning in the lower Saint John River valley. The main phases of development and implementation of a web-based GIS software for flood monitoring and prediction are presented as well.
With satellite imagery and a digital elevation model of the flood plain area, we can access a web-based prediction that models current flood events, and that can show how the water progresses based on the output from hydrological modelling for the next 24 and 48 hours along the lower Saint John River Valley.
This research provides the foundation for a revised decision support system that can result in improvements in the prevention, mitigation, response, and recovery from flood events along the lower Saint John River.
Further research is needed to improve the accuracy of digital terrain models by using LiDAR data, which will in turn improve the accuracy of hydrological modelling.

Acknowledgments

The authors would like to acknowledge the generous contribution of time, materials and resources to this project by the New Brunswick Department of Transportation.
This project was financially supported in part by the N.B. Emergency Measures Organization and the Canadian Department of Natural Resources Geoconnections program as well as the University of New Brunswick and the New Brunswick Innovation Foundation (NBIF).
The IT Division of the City of Fredericton provided datasets available for this project. The New Brunswick Department of Environment has provided data and expertise related to

[3] The software (IVS3D) has been developed to allow the users to explore, analyze, manipulate and gain knowledge from their data by representing very large complex information in the best possible way - in an intuitive fashion - in the way that we perceive the real world everyday. This virtual reality allows new insight to be rapidly gained and more information to be extracted from the underlying data. (Source: http://www.ivs3d.com/companyinfo/about_ivs.html)

hydrological modelling, and the NB Emergency Measures Organization helped with their expertise and additional funding for this project. CARIS provided the GIS software used in this project and contributed to the research project by providing the implementation of the web site and of all the related software components within CARIS Spatial Fusion. IVS provided the virtual reality software used in this project.

7. References

Ackerman, C.T. (2005). *HEC-GeoRAS; GIS Tools for support of HEC-RAS using ArcGIS*. Redlands: ESRI, 204 pages.

Al-Sabhan, W., Mulligan, M. & Blackburn G.A. (2003). A real-time hydrological model for flood prediction using GIS and the WWW. *Computers, Environment and Urban Systems*, Vol. 27, pp. 9-32.

Canada Inland Waters Directorate, Atlantic Region [CIWD]. (1974). *New Brunswick Flood*, April-May, 1973, Ottawa: Inland Waters Directorate, Atlantic Region, 114 pages.

Danish Hydraulic Institute [DHI]. (2004). *MIKE11 GIS reference and user manual.*

Environmental Modeling Research Laboratory [EMRL]. (1998). *Watershed modeling system (WMS) reference manual and tutorial.*

Fread, D.L. (1992). *Flow Routing, Chapter 10, Handbook of Hydrology*. (Ed. E.R. Maidment). New York: McGraw-Hill. pp. 10.1-10.36.

Fread, D.L. (1993). NWS FLDWAV Model: The Replacement of DAMBRK for Dam-Break Flood Prediction. Dam Safety'93, *Proceedings of the 10th Annual ASDSO Conference*, Kansas City, Missouri, pp. 177-184.

Fread, D.L., & Lewis, J.M. (1998). *NWS FLDWAV MODEL: Theoretical description and User documentation*. Hydrologic Research Laboratory, Office of Hydrology, National Weather Service (NWS), Silver Spring, Maryland, USA, 335 pages.

Environment New Brunswick & MacLaren Atlantic Limited, New Brunswick [ENB-MAL]. (1979). *Canada-New Brunswick Flood Damage Reduction Program: Hydrotechnical Studies of the Saint John River from McKinley Ferry to Lower Jemseg*. Fredericton, 116 pages.

Mioc, D., Nickerson B., McGillivray, E., Morton A., Anton, F., Fraser, D., Tang, P., & Liang, G. (2008). Early warning and mapping for flood disasters. *Proceedings of 21st ISPRS Conference, China, Bejing*, 2008, 6 pages.

Moore, R.J., Bell, V.A., & Jones, D.A. (2005). External Geophysics, Climate and Environment - Forecasting for flood warning. *C. R. Geoscience*, Vol. 337, pp. 203-217.

Noman, N.S., Nelson, E.J., &. Zundel, A.K. (2001). Review of automated floodplain delineation from digital terrain models. *Journal of Water Resources Planning and Management*, Vol. 127, No. 6, pp. 394-402.

Noman, N.S., Nelson, E.J., & Zundel, A.K. (2003). Improved Process for Floodplain Delineation from Digital Terrain Models. *Journal of Water Resources Planning and Management*, Vol. 129, No. 5, pp. 427-436.

Okabe, A., Boots, B., Sugihara, K., & Chiu, S.N. (2000). *Spatial Tessellations - Concepts and Applications of Voronoi Diagrams*. (2nd ed.). Chichester: John Wiley. 671 pages.

Sanders, R., Shaw, F., MacKay, H., Galy, H., & Foote, M. (2005). National flood modeling for insurance purposes: using IFSAR for flood risk estimation in Europe, *Hydrology and Earth System Sciences*. Vol. 9, No. 4, pp. 449-456.

10

Potential of C-Band Multi-polarized and Polarimetric SAR Data for Soil Drainage Classification and Mapping

Mohamed A. Niang[1], Michel C. Nolin[1] and Monique Bernier[2]
[1]*Pedology and Precision Agriculture Laboratories, Agriculture and Agri-Food Canada*
[2]*Institut National de la Recherche Scientifique, Eau-Terre-Environnement (INRS-ETE)*
Québec (Canada)

1. Introduction

Routine soil surveys determine many soil properties but soil drainage and soil moisture indicators (soil permeability, hydrologic soil group, etc.) are often the most relevant for agro-environmental purposes. They influence in-field crop yield and affect several environmental processes, i.e., soil erosion, nutrient and pesticide transport by runoff water and leaching, nitrification, and greenhouse gas production. Up-to-date information is needed to guide watershed and site-specific crop management but conventional soil survey procedures are time-consuming and expensive. New technologies, such as remote and proximal sensing, feature in many key soil science applications and are particularly effective for mapping soil drainage (Niang et al., 2007 and 2006; Bell et al., 1994 and 1992; Lee et al., 1988; Levine et al., 1994; Cialella et al., 1997; Campling et al., 2002; Liu et al., 2008). Since soil drainage is often related to other properties, such as soil water content and texture (Kravchenko et al., 2002), it can be mapped using both optical and radar remote sensing. When cloud-free optical imagery is not available, radar imaging is the best option. For most soil conditions, soil surface moisture and vegetative growth and development are considered to be indicators of soil drainage. However, these factors also affect radar backscatter. Knowledge of soil and crop types is needed to understand the relationship between radar backscatter and soil drainage (Smith et al., 2006). Soil drainage is a dynamic process that can by defined only by integrating several factors: water-holding capacity, hydraulic conductivity, and seasonal variation in water-table depth (Jennifer et al., 2001). Thus, radar remote sensing data must be acquired under appropriate environmental conditions to effectively determine soil drainage classes. Most studies on agricultural applications of radar remote sensing involve single frequency and single polarization data. Unless such data are combined in multi-temporal series, they are limited to a few values (McNairn and Brisco, 2004). Only a few studies suggest that multi-polarization radar could provide valuable and timely information by delineating homogenous soil zones (McNairn and Brisco, 2004; van der Sanden, 2004). One recent study showed that soil drainage could be mapped on an in-field scale by using high-resolution optical and C-band SAR data from

CV-580 (Liu et al., 2008). However, the relationship between polarimetric parameters and soil drainage indicators was not investigated.
This chapter evaluates the feasibility of using multi-polarization and polarimetric C-band SAR data from CV-580 to determine soil drainage classes. First, a Wishart supervised classification was applied to determine land use. Multi-polarized channels (HH, VH, VV, RR, RL LL, ψ45°, and ψ135°) were synthesized and backscattering coefficients were derived for the reference soil profile pits acquired from the soil survey. These data were analysed for designated land use and drainage classes. Polarimetric parameters from the Cloude and Pottier (1997) decomposition were also analysed. Linear discriminant analysis was applied on each land use using soil drainage classes from soil survey. The feasibility of using this approach with RADARSAT-2 data to map soil drainage is discussed.

2. Materials and methods

The study area was located at the Bras d'Henri watershed (167 km^2) near Quebec City (Figure 1). During the summers of 2004 and 2005, intensive soil surveys were conducted over the entire watershed at two prospecting scales (1:20,000 and 1:40,000). A total of 1612 soil profiles were collected for updating and upgrading the soil map and a recommended sampling method based on random transects (Nolin et al., 1994) was adopted. Five out of seven soil drainage classes, ranging from well-drained to very-poorly drained, were identified in the watershed according to the soil surveyor's expert knowledge. These data were used as a validation dataset. Polarimetric C-band SAR data from CV-580 were acquired in November 2005 over the study area. An area of 101 km^2, with 1045 soil profiles, was covered (Table 1). The SAR data were radiometrically calibrated by the Canadian Center of Remote Sensing (Murnaghan, 2005). Radiometric accuracy was less than 1dB for the four polarizations. Slant range resolution was 4 × 4 m (azimuth × range) for the SIR-C product and incidence angles were large (33° to 67°).

The polarizations (HH, HV, VV, RR, LL, ψ45°, and ψ135°) were synthesized from the SAR data using Polarimetric WorkStation software (Touzi and Charbonneau, 2004). These polarizations were used as ancillary data for soil drainage classification. The assumption was that, under given soil moisture conditions, variation in the radar backscatter would indicate soil drainage states. This relationship between polarizations and soil moisture has been documented (Baronti et al., 1995). Linear cross-polarizations (HH, VV, and HV) refer to the power from the transmitted linear wave (H or V) recorded by the sensor and repolarized into the orthogonal polarizations (V or H). With fully polarimetric data, nonlinear polarizations, such as circular and elliptical polarizations, can be synthesized.

For the left-handed circular waves (L), the electric field vector rotates counter-clockwise; for the right-handed circular waves (R), the vector rotates clockwise. The responses of circular co-polarizations (RR and LL) are associated with the volume or multiple scattering and the handedness of received wave, while the cross-polarization (RL) is associated with smooth surfaces (McNairn et al., 2002). Elliptical polarizations can also be derived from polarimetric data by defining the orientation (ψ) and ellipticity (χ) angles. The two polarizations (ψ45° and ψ135°) used in this study correspond at χ=0° and ψ=45°, and χ=0° and ψ135°, respectively, and were positively correlated to soil moisture (McNairn et al., 2002).

Another unique feature provided by the fully polarimetric data is the possibility of separating different scattering distributions by sub-matrices associated with the specific scattering properties of point, or distributed scatters and their ensembles. Several polarimetric target decomposition theorems have been proposed, including: coherent (Krogager, 1992; Cameron and Leung, 1990), non-coherent (Huynen, 1965; Holm and Barnes, 1988), model-based decomposition (Freeman and Durden, 1998), eigenvector-based decomposition (van Zyl, 1989; Cloude, 1992), and others as critically associated in the literature (Cloude, 1992; Cloude and Pottier, 1996 and 1997).

The $H/A/\underline{\alpha}$ (entropy H, anisotropy A, and mean alpha angle $\underline{\alpha}$) decomposition theorem proposed by Cloude and Pottier (1997) was used in this study to determine soil drainage classes. This decomposition is based on eigenvalues extracted from the coherency matrix. The average target scattering matrix estimated with these parameters allows the surface roughness to be separated from the soil dielectric constant (Hajnesk et al., 2001), which depends linearly on soil moisture.

For an accurate extraction of the radar backscatter, a 5 × 5 Lee filter window (Lee, 1986) was applied to reduce the speckle noise. The Wishart supervised classification ($H/A/\underline{\alpha}$) was also applied to the SAR data to assess the characterization of soil drainage classes for agricultural areas. The training areas were designated on a RGB composite color and the classification was performed with three scattering mechanisms (single bounce scattering, double bounce scattering, and volume diffusion) derived from the Freeman decomposition.

A principal component analysis (PCA) was used to reduce the effects of collinear data and matrix dimensions. The contribution of synthetized polarization and polarimetric parameters on soil drainage classification was assessed by forward stepwise discriminant analysis.

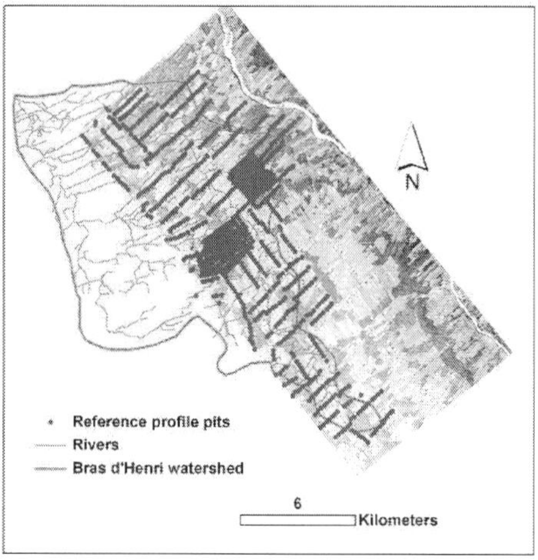

Fig. 1. Study area: The Bras d'Henri watershed (QC, Canada) superimposed with the CV-580 SAR; geo-referenced soil profile pits labelled as points (.) represent the training / validation dataset

Drainage class	Count	%
D3	53	5.1
D4	71	6.8
D5	303	29.0
D6	495	47.4
D7	123	11.8
Total	1045	100.0

Table 1. Distribution of soil profile pits by drainage classes

3. Results and discussion

Figure 2 shows the meteorological conditions during November 2005. Two days before the CV-580 overpass, cumulative precipitation for November was about 30mm and mean air temperature was about 5°C. These conditions can cause water saturation on certain soil types; soil water infiltration capability and drainage states could be shown after these dates.

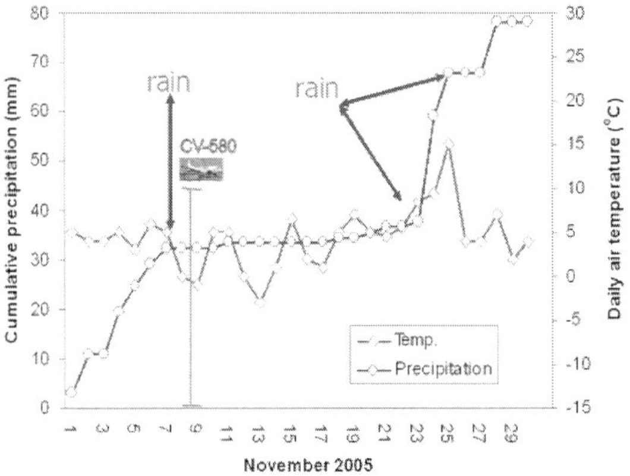

Fig. 2. Meteorological conditions during the acquisition of CV-580 SAR data

Figure 3 shows the Pauli color-coded image of part of the watershed which consists of agricultural areas, forest, wetland, and urban zones. It was obtained by assigning the red color R=|HH-VV|, green color G=|HV|, and blue color B=|HH+VV|, respectively. The Freeman decomposition using R=$|P_{DB}|$, G=$|P_v|$, and B=$|P_s|$ for red, green, and blue is presented in Figure 4. The P_{DB}, P_v, and P_s parameters are powers scattered by the double-bounce, volume, and surface components of their covariance matrices, respectively. The Freeman decomposition uses similar characteristics as the Pauli-based decomposition, but provides a more realistic representation because scattering models are used for dielectric surfaces.

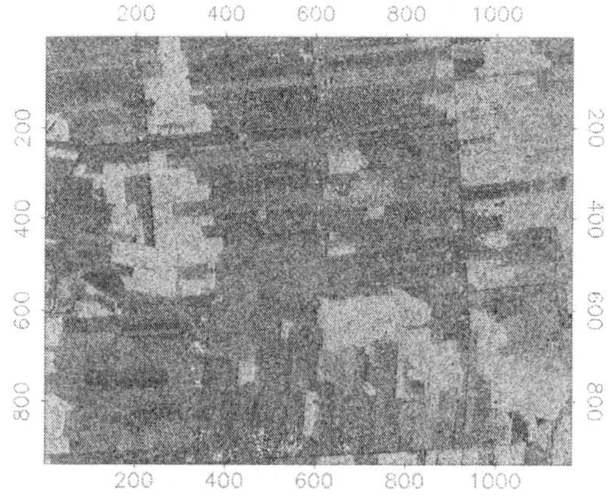

Fig. 3. Image obtained using the Pauli decomposition with R=|HH-VV|, G=|HV|, B=|HH+VV|

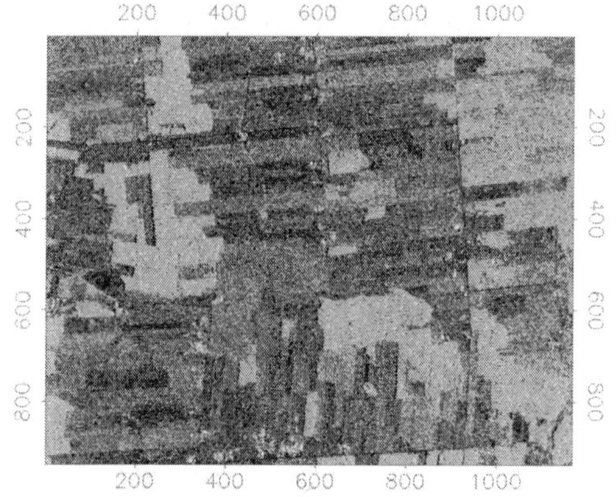

Fig. 4. Image obtained using the Freeman decomposition with R=| P_{DB} |, G=| P_v |, and B=| P_s |

Two types of agricultural areas, with P_s (blue) the predominant scattering mechanism and dark areas, appear to have the same contribution from the three scattering mechanisms in Figure 4. Some training areas were designated in these areas for the Wishart supervised classification (H/A/$\underline{\alpha}$). This study focused on these agricultural areas.

Classification results are given in Figure 5. The overall classification accuracy was 86%. Urban areas had a lower classification accuracy (63%) but the two designated agricultural areas had good classification accuracy: 86% for agricultural Area 1 and 93% for agricultural Area 2.

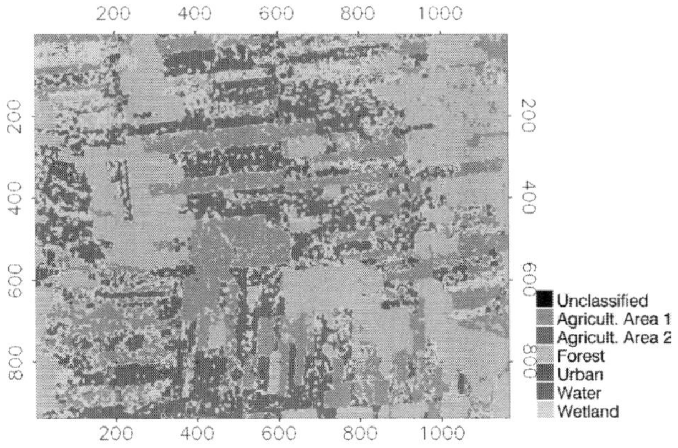

Fig. 5. Wishart supervised classification (H/A/α)

Land use	Forest	Agricultural Area 1	Agricultural Area 2	Water	Wetland	Urban
Forest	90	1	3	0	0	5
Agricultural Area 1	2	86	2	0	0	10
Agricultural Area 2	6	0	93	0	0	1
Water	0	0	5	91	0	2
Wetland	1	4	2	1	91	1
Urban	7	23	7	0	0	63

Table 2. Confusion matrix (%). Rows represent user-defined groups while columns represent segmented clusters

3.1 Multi-polarization analysis

The relationship between the mean backscattering with soil drainage classes and synthesized polarizations is given in Figures 6 and 7 for the two designated agricultural areas. An increase of the backscattering signal is observed for these areas, and for all polarizations, as a function of soil moisture as depicted by theoretical models IEM (Fung, 1994), SPM and GO (Ulaby et al., 1986), and soil drainage classes. For the agricultural Area 1, there are significant differences (about 4dB) between well-drained (D3+D4) and poorly-drained soil classes (D5+D6) (Figure 6). Unfortunately, there were no data for the very poorly-drained soil class (D7). For the agricultural Area 2, three sets of drainage classes can be distinguished: D3, D4+D5+D6, and D7. The VV backscatter is slightly greater than the HH polarization for the two areas. According to Dobson and Ulaby (1986), this result may be due to the high incidence angle (33° to 66°) of acquisition for the SAR image. Therefore, differences in the mean backscattering signal behaviour for the first three soil drainage classes in the two agricultural areas indicate that scattering mechanisms differ among these fields. Meteorological conditions prevailed until the date of harvest and, therefore, land cover was highly variable: ploughed fields (forage, cereal, soybean, or harvested corn), harvested but not ploughed fields with crop residue, and unharvested corn fields. In this context, polarimetric parameters may be useful for analysing the physical properties (dielectric and geometric) of the two agricultural fields.

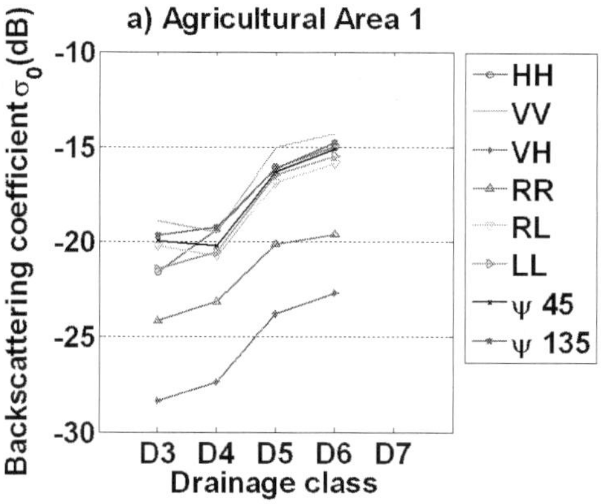

Fig. 6. Relationship of the backscattering coefficient with drainage classes in agricultural Area 1 for the polarizations

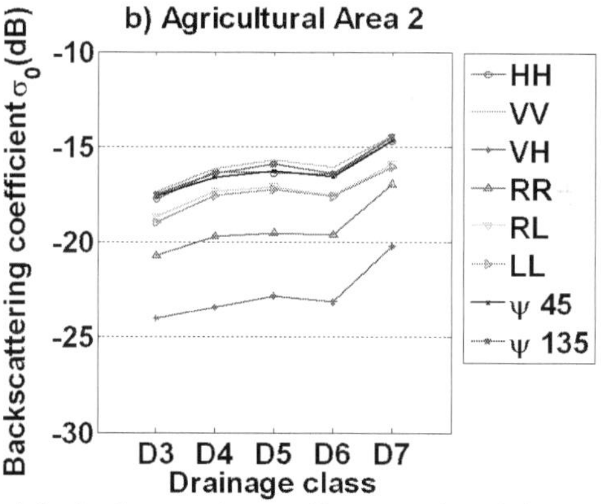

Fig. 7. Relationship of the backscattering coefficient with soil drainage classes in agricultural Area 2 for the polarizations

3.2 Polarimetric analysis

Figures 8, 9, and 10 give the $\underline{\alpha}$, H, and A parameters extracted from the 3 × 3 coherency matrix, respectively. The $\underline{\alpha}$ is the main parameter used to identify the dominant scattering mechanism. Entropy, H, is a statistical descriptor of the randomness of the scattering process which can be interpreted as the degree of statistical disorder. Anisotropy, A, measures the relative importance of the second and the third eigenvalues. It is a complementary parameter to entropy and can be used as a discriminating index when $H > 0.7$.

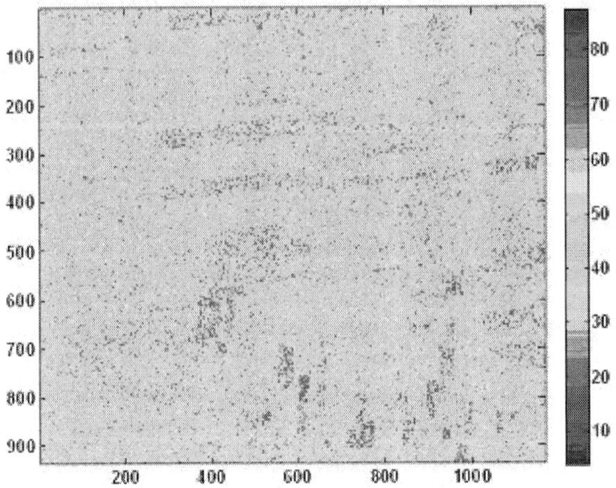

Fig. 8. Map showing the variation of the parameter, $\underline{\alpha}$ (°)

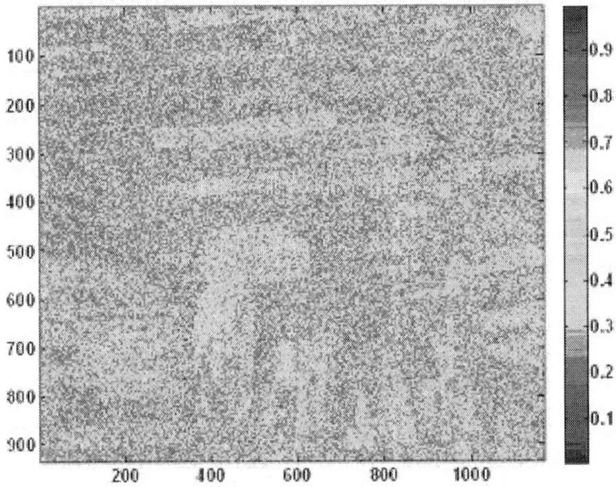

Fig. 9. Map showing the variation of the entropy parameter, H

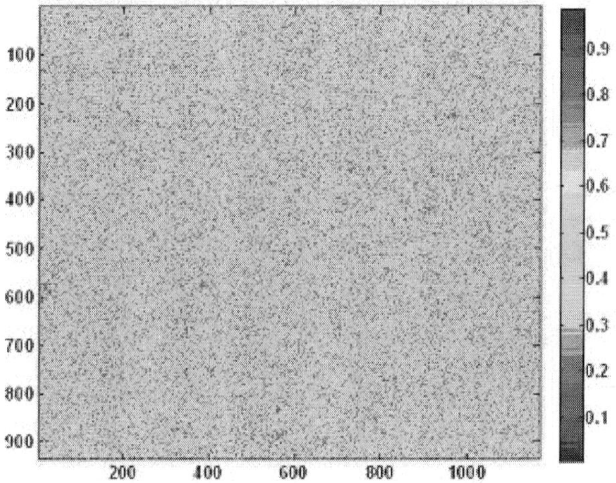

Fig. 10. Map showing the variation of the anisotropy parameter, A

The relationship of these parameters with the soil drainage classes for the two agricultural areas is given in Figures 11 and 12. The low values of the $\underline{\alpha}$ parameter ($15° \leq \underline{\alpha} \leq 31°$) for the two agricultural areas suggest that the scattering corresponds to the single-bounce scattering produced by the rough surface. Therefore, the low values of the entropy ($0.05 \leq H \leq 0.1$) for agricultural Area 1 indicate that the scattering process corresponds to a pure target. For agricultural Area 2, the values ($0.05 \leq H \leq 0.2$) indicate the presence of other scattering mechanisms weighted by the corresponding eigenvalues (λ_1, λ_2, λ_3). The values of the anisotropy were greater ($A \geq 0.6$) in the agricultural Area 2 for all drainage classes, whereas in the agricultural Area 1, the A parameter was in the medium range. This result confirms the presence of two scattering mechanisms with a dominant process and a secondary one with medium probability ($\lambda_3 \approx 0$). In agreement with the Freeman decomposition (Figure 4), we conclude that the agricultural Area 1 was composed mostly of bare soils while the agricultural Area 2 consisted of less dense harvested fields with short plants. Under conditions similar to those of agricultural Area 1, the combination of $\underline{\alpha}$ and H parameters appears to be sufficient for identifying soil drainage classes (Figure 11).

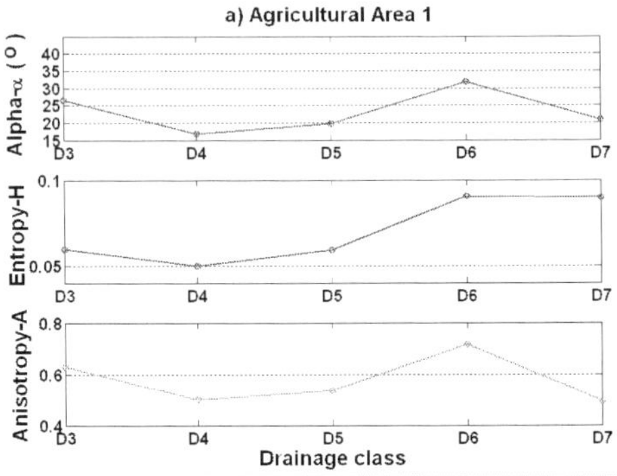

Fig. 11. Relationship of parameters α, H, and A with soil drainage classes in the agricultural Area 1

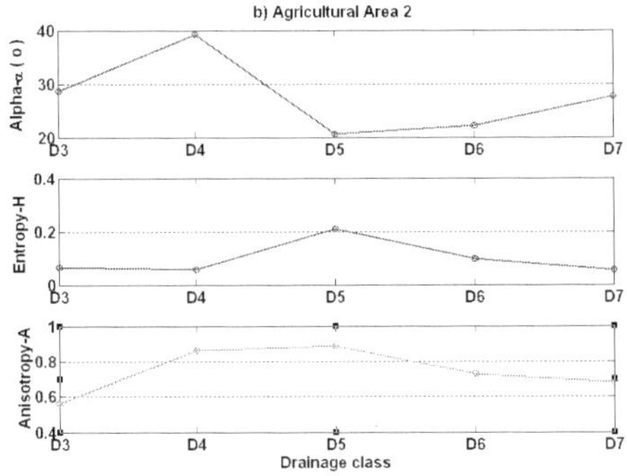

Fig. 12. Relationship of parameters α, H, and A with soil drainage classes in agricultural Area 2

3.3 Discriminant analysis

In an approach to reduce collinearity, a principal component analysis (PCA) was applied on the multi-polarization dataset. The first two components, consisting of 94% of total variance of the information from the multi-polarization dataset, were selected for further analyses. The contribution of the multi-polarization data and the polarimetric parameters was studied using the following discriminant analysis criteria: Wilks' lambda, F to remove, significant level associated to the partial Lambda, and the tolerance (Table 3).

The multivariate solution shows that the two PCA factors (PC1 for the agricultural Area 1, and PC2 for the agricultural Area 2) extracted from the multi-polarization dataset were the principal discriminators for soil drainage classes with a p-level near 0 and good tolerance. Anisotropy was the polarimetric parameter that had the most significant contribution to the discriminant function with good tolerance. The two other polarimetric parameters (α and H) contributed to the classification model but their tolerance was weak.

Variables	Wilks' λ	F to remove	p-level	Tolerance
$\underline{\alpha}$	0.91	1.81	0.13	0.25
A	0.91	2.32	0.06	0.89
H	0.91	1.86	0.12	0.24
PCA 1	0.95	14.19	0.00	0.99
PCA2	0.92	5.9	0.00	0.91

Table 3. Contribution of the selected parameters to the model classification

The results of the classification by discriminant analysis are presented in Table 4. For the two agricultural areas, poorly-drained soils (D6) were clearly identified (73–97%). Imperfectly-drained soils (D5) were detected in the agricultural Area 1 with an overall classification accuracy of 63%. The well-drained class (D3) was detected with an overall classification accuracy of only 33%. The meteorological conditions which prevailed during the acquisition of polarimetric SAR data, and the small proportional size of soil drainage classes in the agricultural areas, can partially explain the relatively low performance of the classification by discriminant analysis.

Drainage class	Agricultural Area 1		Agricultural Area 2	
	Proportion	Percent correct	Proportion	Percent correct
D3	0.13	33.3	0.06	0.0
D4	0.04	0.0	0.07	0.0
D5	0.35	62.5	0.30	2.4
D6	0.48	72.7	0.48	96.7
D7	0.00	0.0	0.09	19.2
Total	1.00	60.9	1.00	49.4

Table 4. Mean alpha, anisotropy, and entropy by drainage class for two agricultural areas

4. Conclusion

This research demonstrates the feasibility of using multi-polarization and polarimetric data for soil drainage classification and mapping. The new generation of SAR technology, with its high spatial resolution of fully polarimetric SAR data from RADARSAT-2, could help to clarify the complex relationship between soil drainage and radar backscatter and subsequently be a resource for developing operational models. Because of the dynamic changes of soil drainage, meteorological conditions and land use must be determined prior to analysis. Spring acquisition of polarimetric SAR data should produce better classification results for predicting soil drainage classes than late fall acquisition because weather conditions and land use would be less variable.

Acknowledgments

This project was funded by the Canadian Space Agency and Agriculture and Agri-Food Canada through a Government Related Initiatives Program (GRIP). We would like to acknowledge the contribution of partners to the project. We would also like to acknowledge the research assistants at Pedology and Agriculture Precision Laboratories for data collection.

5. References

Baronti S.; Del Frate F.; Ferrazzoli P.; Paloscia S.; Pampaloni P. & Schiavon, G. (1995). SAR polarimetric features of agricultural areas. *International Journal of Remote Sensing*, Vol. 14, pp. 2639-2656

Bell, J.C.; Cunningham, R.L. & Havens, M.W. (1992). Calibration and validation of a soil-landscape model for predicting soil drainage class. *Soil Sci. Soc. Am. J.*, Vol. 56, pp. 1860-1866

Bell, J.C.; Cunningham, R.L. & Havens, M.W. (1994). Soil drainage class probability mapping using a soil-landscape model. *Soil Sci. Soc. Am. J.*, Vol. 58, pp. 464-470

Cameron, W.L. & Leung, L.K. (1990). Feature-motivated scattering matrix decomposition, *Proc. IEEE Radar Conf.*, pp. 549-557, Arlington, VA, May 7-10, 1990

Campling, P.; Gobin, A. & Feyen. J. (2002). Logistic modeling to spatially predict the probability of soil drainage classes. *Soil Sci. Soc. Am. J.*, Vol. 66, pp. 1390-1401

Cialella, A.T.; Dubayah, R.; Lawrence, W. & Levin, E. (1997). Predicting soil drainage class using remotely sensed and digital elevation data. *Photogramm. Eng. Remote Sens.*, Vol. 63(2), pp. 171-178

Cloude, S.R. (1992). *Uniqueness of Target Decomposition Theorems in Radar Polarimetry. Direct and Inverse Methods in Radar Polarimetry, Part 1*, Boerner, W-M, (Ed.), pp. 267-296, Kluwer Academic Publishers, Dordrecht, Netherlands

Cloude, S.R. & Pottier, E. (1996). A review of target decomposition theorems in radar polarimetry. *IEEE Trans. GRS*, Vol. 34(2), pp. 498-518, Mar. 1996

Cloude, S.R. & Pottier, E. (1997). An entropy-based classification scheme for land applications of polarimetric SAR. *IEEE Trans. GRS*, Vol. 35(1), pp. 68-78

Dobson, M.C. & Ulaby, F.T. (1986). Active microwave soil moisture research. *IEEE Trans. GRS*, Vol. GE-24(1), pp. 23-36

Freeman, A. & Durden, S.L. (1998). A three-component scattering model for polarimetric SAR data, *IEEE Trans. GRS*, Vol. 36(3), pp. 963-973.

Fung, A.K. (1994). *Microwave Scattering and Emission Models and their Applications*, Artech House, Norwood, USA

Hajnsek, J. (2001). Inversion of surface parameters (soil moisture & roughness) using polarimetric SAR. Doctoral Thesis, *FSU, Jena, Germany, 2001* October 17 (ISSN 1434-8485 ISRN DLR-FB-2001-30)

Holm, W. & Barnes, R.M. (1988). On radar polarization mixed target state decomposition techniques, *IEEE Proceedings 1988 National Radar Conference*, pp. 249-254, Ann Arbor, MI, USA

Huynen, J.R. (1965). Measurement of the target scattering matrix, *Proc. IEEE, 53(8)*, pp. 936-946 (also see: ibid, (1960), Radar target sorting based on polarization signature analysis, Lockheed, LMSD-288216, Sunnyvale, CA [AFCRC-TN-60-588, May 1960])

Kravchenko, A.N.; Bollero, G.A.; Omonode, R.A. & Bullock, D.G. (2002). Quantitative mapping of soil drainage classes using topographical data and soil electrical conductivity. *Soil Sci. Soc. Am. J.*, Vol. 66, pp. 235-243

Krogager, E. (1992). Decomposition of the Sinclair Matrix into Fundamental Components with Application to High Resolution Radar Target Imaging. Direct and Inverse Methods in Radar Polarimetry, *Part 2*, Boerner, W-M., (Ed.), pp. 1459-1478, Kluwer Academic Publishers, Dordrecht, The Netherlands: (also see: ibid, 1993, "Aspects of Polarimetric Radar Target Imaging", *Doctoral (Dr. Sci. Tech.) Thesis*, Technical University of Denmark, Electromagnetics Institute, Lyngby, Denmark [check format]

Lee, J.S. (1986). Speckle suppression and analysis for synthetic aperture radar images. *Optical Engineering,* Vol. 25(81), pp. 636-643

Lee, K.-S.; Lee, G.B. & Tyler, E.J. (1988). Determination of soil characteristics from Thematic Mapper data of a cropped organic-inorganic soil landscape. *Soil Sci. Soc. Am. J.*, Vol. 52, pp. 1100-1104

Levine, E.R.; Knox, R.G. & Lawrence W.T. (1994). Relationships between soil properties and vegetation at the Northern Experimental Forest, Howland, Maine. *Remote Sens. Environ.*, Vol. 47, pp. 231-241

Liu, J.; Pattey, E.; Nolin, M.C.; Miller, J. & Ka, O. (2008). Mapping within-field soil drainage using high resolution remote and proximal sensing data. *Geoderma,* Vol. 143, pp. 261-272

McNairn, H. & Brisco, B. (2004). The application of C-band polarimetric SAR for agriculture: A review. *Can. J. Remote Sens.*, Vol. 30(3), pp. 525-542

McNairn, H.; Duguay, C.; Brisco, B. & Pultz, T.J. (2002). The effect of crop residue characteristics on polarimetric radar response. *Remote Sensing of Environment,* Vol. 80, pp. 309-320

Murnaghan, K. (2005). CV580 Quality report. Acquisition 04-03. Date: 9-Nov-2005. Line : 3002. Pass: 02. Location : Ste-Foy, Québec", 2p. 2005

Niang, M.; Bernier, M.; Nolin, M.C.; Ka, O. & Liu, J. (2006). Approche quantitative pour la prédiction et la cartographie du drainage des sols par l'intégration de données ASTER et RADARSAT-1. Déc. 2006. Institut national de la recherche scientifique, centre Eau, Terre et Environnement, Québec, Canada. *Rapport de recherche R-905,* ISBN: 2-89146-530-X. 57 p

Niang, M.; Bernier, M.; Nolin, M.C.; Ka, O. & Liu, J. (2007). Apport de l'imagerie radar multipolarisée et polarimétrique à la classification et à la cartographie du drainage des sols du bassin versant du Bras d'Henri. Mars 2007. Institut national de la recherche scientifique, centre Eau, Terre et Environnement, Québec, Canada. *Rapport de recherche R-928,* ISBN: 978-2-89146-540-3. 42 p

Nolin, M.C.; Lamontagne, L. & Dubé, J.C. (1994). Cadre méthodologique d'une étude détaillée des sols et son application en terrain plat. Bulletin technique 1994-4F Direction générale de la recherche. AAC. Ste-Foy, Québec. *CRTRB Contrib.* No 93-41, 78 p.

Jennifer, W. H., Rose, M., Cherie, S., David, K. S. &Anthony, D. Mc. (2001). Soil drainage and its potential for influencing wildfires in Alaska. Studies by U.S. Geological Survey in Alaska, *U.S. Geological Survey Professional*, paper 1678, pp. 139-144.

Smith, A.M.; Eddy, P.R.; Bugden-Storie, J.; Pattey, E.; McNairn, H.; Nolin, M.; Perron, I.; Hinther, M.; Miller, J. & Haboudane, D. (2006). Multi-polarized radar for delineating within-field variability in corn and wheat, *Canadian Journal of Remote Sensing,* Vol. 45 (4), pp. 300-313

Smith, A.M.; Eddy P.; Bugden-Storie J.; Pattey E.; McNairn H.; Nolin, M.C.; Miller J. & Haboudane, D. (2003). Delineating within-field management zones using multi-temporal, multi-polarized airborne SAR imagery. *Proceedings 25th Canadian Remote Sensing Conference*, Montreal, Quebec, October 2003

Touzi, R. & Charbonneau, F.J. (2004). PWS: A friendly and effective tool for polarimetric image analysis. *Can. J. Remote Sens.*, Vol. 30(3), pp. 566–571

Ulaby, F.T.; Moore, R.K. & Fung, A.K. (1986). *Microwave Remote Sensing, Active and Passive, Volume Scattering and Emission Theory – Advanced Systems and Applications 3,* Artech, Dedham, MA

van der Sanden, J.J. (2004). Anticipated applications potential of RADARSAT-2 data. *Can. J. Remote Sensing*, Vol. 30(3), pp. 369-379

van Zyl, J.J. (1989). Unsupervised classification of scattering behavior using radar polarimetry data. *IEEE Trans. GRS*, Vol. GE-27, pp. 36-45

11

Electromagnetic Models for Remote Sensing of Layered Rough Media

Pasquale Imperatore, Antonio Iodice and Daniele Riccio
University of Naples "Federico II",
Department of Biomedical, Electronic and Telecommunication Engineering,
Napoli, Italy

1. Introduction

Each region of the Earth's crust can be morphologically modeled as a suitable layered structure, in which some amount of roughness is presented by every interface. Actually, propagation in stratified soil, sand cover of arid regions, forest canopies, urban buildings, snow blanket, snow cover ice, sea ice and glaciers, oil flood on sea surface, and other natural scenes can be modeled referring to most likely discrete (piecewise-constant) systems, rather than continuous, with some amount of roughness presented by every interface. Moreover, a key issue in remote sensing of other Planets is to reveal the content under the surface illuminated by the sensors: also in this case a layered model is usually employed.
The aim of this chapter is to provide a structured presentation of the main theoretical and conceptual foundations for the problem of the electromagnetic wave interaction with layered rough media. In the first part, special emphasis is on the analytical models obtainable in powerful framework of the perturbation approach. The comprehensive scattering model based on the *Boundary Perturbation Theory* (BPT), which permits to systematically analyze the bi-static scattering patterns of 3D multilayered rough media, is then presented highlighting the formal connections with all the previously existing simplified perturbative models, as well as its wide relevance in the remote sensing applications scenario. The *polarimetric Scattering Matrix* of a multilayered medium with an arbitrary number of rough interfaces is also provided. The second part is devoted to a mathematical description which connects the concepts of *local* scattering and *global* scattering. Consequently, a functional decomposition of the BPT global scattering solution in terms of basic single-scattering local processes is rigorously established. The *scattering decomposition* gives insight into the BPT analytical results, so enabling a relevant physical-revealing interpretation involving ray-series representation. Accordingly, in first-order limit, the way in which the character of the local scattering processes emerges is dictated by the nature of the structural filter action, which is inherently governed by the series of coherent interactions with the medium boundaries. As a result, the phenomenologically successful BPT model opens the way toward new techniques for solving the inverse problem, for designing SAR processing algorithms, and for modelling the time-domain response of layered structures.

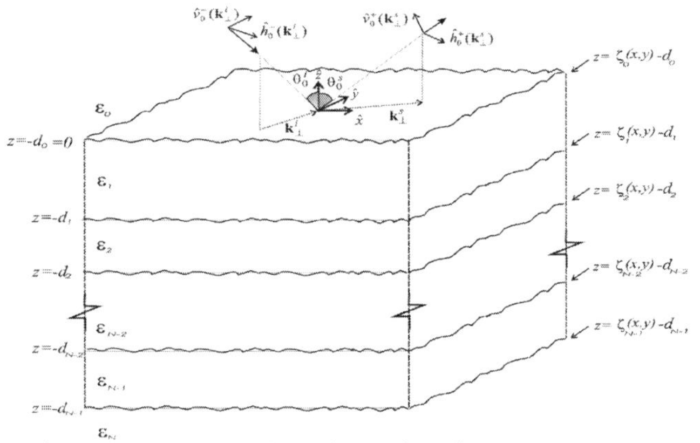

Fig. 1. Geometry for an N-rough boundaries layered medium

2. Problem definition

When stratified media with rough interfaces are concerned, the possible approaches to cope with the EM scattering problem fall within three main categories. First, the numerical approaches do not permit to attain a comprehensive understanding of the general functional dependence of the scattering response on the structure parameters, as well as do not allow capturing the physics of the involved scattering mechanisms. Layered structures with rough interfaces have been also treated resorting to *radiative transfer theory* (RT). However, coherent effects are not accounted for in RT theory and could not be contemplated without employing full wave analysis, which preserves phase information. Another approach relies on the full-wave methods. Although, to deal with the electromagnetic propagation and scattering in complex random layered media, several analytical formulation involving some idealized cases and suitable approximations have been conducted in last decades, the relevant solutions usually turn out to be too complicated to be generally useful in the remote sensing scenario, even if simplified geometries are accounted for. The proliferation of the proposed methods for the simulation of wave propagation and scattering in a natural stratified medium and the continuous interest in this topic are indicative of the need of appropriate modelling and interpretation of the complex physical phenomena that take place in realistic environmental structures. Indeed, the availability of accurate, sound physical and manageable models turns out still to be a strong necessity, in perspective to apply them in retrieving of add-valued information from the data acquired by microwave sensors. For instance, such models are high desirable for dealing with the inversion problem as well as for the effective design of processing algorithms and simulation of Synthetic Aperture Radar signals. Generally speaking, an exact analytical solution of *Maxwell* equations can be found only for a few idealized problems. Subsequently, appropriate approximation methods are needed. Regarding the perturbative approaches, noticeable progress has been attained in the investigation on the extension of the classical SPM (small perturbation method) solution for the scattering from rough surface to specific layered configurations. Most of previous existing works analyze different layered configurations in the first-order limit, using procedures, formalisms and final solutions that can appear of

difficult comparison (Yarovoy et al., 2000), (Azadegan and Sarabandi, 2003), (Fuks, 2001). All these formulations, which refer to the case of a single rough interface, have been recently unified in (Franceschetti et al, 2008). On the other hand, solution for the case of two rough boundaries has also been proposed in (Tabatabaeenejad and Moghaddam, 2006).
Methodologically, we underline that all the previously mentioned existing perturbative approaches, followed by different authors in analyzing scattering from simplified geometry, imply an inherent analytical complexity, which precludes the treatment to structures with more than one (Fuks, 2001) (Azadegan et al., 2003) (Yarovoy et al., 2000) or two (Tabatabaeenejad er al., 2006) rough interfaces.
The general problem we intend to deal with here refers to the analytical evaluation of the electromagnetic scattering by layered structure with an arbitrary number of rough interfaces (see Figure 1). As schematically shown in Figure 1, an arbitrary polarized monochromatic plane wave

$$\mathbf{E}_0^i(\mathbf{r}) = [E_0^{ih}\hat{h}_0^-(\mathbf{k}_\perp^i) + E_0^{iv}\hat{v}_0^-(\mathbf{k}_\perp^i)]\, e^{j(\mathbf{k}_\perp^i \cdot \mathbf{r}_\perp - k_{z0}^i z)} \tag{1}$$

is considered to be incident on the layered medium at an angle θ_0^i relative to the $\hat{z}$ direction from the upper half-space, where in the field expression a time factor *exp(-jωt)* is understood, and where, using a spherical frame representation, the incident vector wave direction is individuated by θ_0^i, φ_0^i:

$$k_0\hat{k}_0^i = \mathbf{k}_0^i = \mathbf{k}_\perp^i - \hat{z}k_{z0}^i = k_0(\hat{x}\sin\theta_0^i\cos\varphi_0^i + \hat{y}\sin\theta_0^i\sin\varphi_0^i - \hat{z}\cos\theta_0^i), \tag{2}$$

with

$$\hat{h}_0^-(\mathbf{k}_\perp^i) = \frac{\hat{k}_0^i \times \hat{z}}{\left|\hat{k}_0^i \times \hat{z}\right|} = \sin\varphi_0^i\hat{x} - \cos\varphi_0^i\hat{y},$$

$$\hat{v}_0^-(\mathbf{k}_\perp^i) = \hat{h}_0^-(\mathbf{k}_\perp^i) \times \hat{k}_0^i = (\hat{x}\cos\varphi_0^i + \hat{y}\sin\varphi_0^i)\cos\theta_0^i + \hat{z}\sin\theta_0^i,$$

where $\mathbf{k}_\perp^i = k_x^i\hat{x} + k_y^i\hat{y}$ is the two-dimensional projection of incident wave-number vector on the plane *z=0*. The parameters pertaining to layer *m* with boundaries $-d_{m-1}$ and $-d_m$ are distinguished by a subscript *m*. Each layer is assumed to be homogeneous and characterized by arbitrary and deterministic parameters: the *dielectric relative permittivity* ε_m, the *magnetic relative permeability* μ_m and the *thickness* $\Delta_m=d_m-d_{m-1}$. With reference to Figure 1, it has been assumed that in particular, $d_0=0$. In the following, the symbol ⊥ denotes the projection of the corresponding vector on the plan *z=0*. Here $\mathbf{r} = (\mathbf{r}_\perp, z)$, so we distinguish the transverse spatial coordinates $\mathbf{r}_\perp = (x, y)$ and the longitudinal coordinate *z*. In addition, each *m*th rough interface is assumed to be characterized by a zero-mean two-dimensional *stochastic process* $\zeta_m = \zeta_m(\mathbf{r}_\perp)$ with normal vector $\hat{n}_m$. No constraints are imposed on the degree to which the rough interfaces are correlated.
A general methodology has been developed by *Imperatore et al.* to analytically treat EM bistatic scattering from this class of layered structures that can be described by small changes with respect to an idealized (unperturbed) structure, whose associated problem is

exactly solvable. A thorough analysis of the results of this theoretical investigation (BPT), which is based on perturbation of the boundary condition, will be presented in the following, methodologically emphasizing the development of the several inherent aspects.

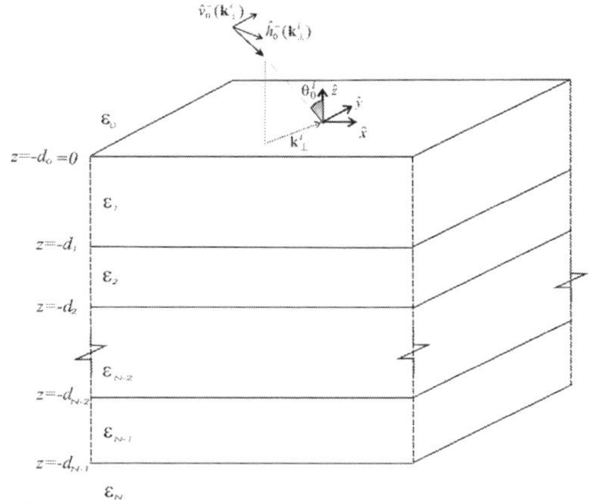

Fig. 2. Geometry for a flat boundaries layered medium

3. Preliminary notation and definitions

This section is devoted preliminary to introduce the formalism used in the following of this chapter. The Flat Boundaries layered medium (*unperturbed structure*) is defined as a stack of parallel slabs (Figure 2), sandwiched in between two half-spaces, whose structure is shift invariant in the direction of x and y (infinite lateral extent in x-y directions). With the notations $T^{p}_{m-1|m}$ and $R^{p}_{m-1|m}$, respectively, we indicate the *ordinary transmission* and *reflection coefficients* at the interface between the regions (m-1) and m, with the superscript $p \in \{v, h\}$ indicating the polarization state for the incident wave and may stand for *horizontal* (h) or *vertical* (v) polarization (Tsang et al., 1985) (Imperatore et al. 2009a). In addition, we stress that:

$$T^{p}_{i|j} = 1 + R^{p}_{i|j} \qquad R^{p}_{i|j} = -R^{p}_{j|i} \qquad i=j\pm 1 \tag{3}$$

The *generalized reflection coefficients* $\Re^{p}_{m-1|m}$, for the p-polarization (TE or TM), at the interface between the regions (m-1) and m are defined as the ratio of the amplitudes of *upward*- and *downward*-propagating waves immediately above the interface, respectively. They can be expressed by recursive relations as in (Chew W. C., 1997) (Imperatore et al. 2009a):

$$\Re^{p}_{m-1|m} = \frac{R^{p}_{m-1|m} + \Re^{p}_{m|m+1} e^{j2k_{zm}\Delta_m}}{1 + R^{p}_{m-1|m} \Re^{p}_{m|m+1} e^{j2k_{zm}\Delta_m}}. \tag{4}$$

Likewise, at the interface between the regions (m+1) and m, $\Re^{p}_{m+1|m}$ is given by:

$$\Re^{p}_{m+1|m} = \frac{R^{p}_{m+1|m} + \Re^{p}_{m|m-1} e^{j2k_{zm}\Delta m}}{1 + R^{p}_{m+1|m} \Re^{p}_{m|m-1} e^{j2k_{zm}\Delta_m}}, \tag{5}$$

where

$$k_{zm} = \sqrt{k_m^2 - |\mathbf{k}_\perp|^2} = k_m \cos\theta_m, \tag{6}$$

where $k_m = k_0\sqrt{\mu_m \varepsilon_m}$ is the *wave number* for the electromagnetic medium in the *m*th layer, with $k_0 = \omega / c = 2\pi / \lambda$, and where $\mathbf{k}_\perp = k_x\hat{x} + k_y\hat{y}$ is the two-dimensional projection of vector wave-number on the plane *z=0*. It should be noted that the factors

$$\vec{M}^{p}_{m}(k_\perp) = 1 - R^{p}_{m|m-1} \Re^{p}_{m|m+1} e^{j2k_{zm}\Delta_m}, \tag{7}$$

$$\vec{M}^{p}_{m}(k_\perp) = 1 - \Re^{p}_{m|m-1} R^{p}_{m|m+1} e^{j2k_{zm}\Delta_m}, \tag{8}$$

$$\vec{M}^{p}_{m}(k_\perp) = 1 - \Re^{p}_{m|m-1} \Re^{p}_{m|m+1} e^{j2k_{zm}\Delta_m}, \tag{9}$$

take into account the multiple reflections in the *m*th layer. On the other hand, the *generalized transmission coefficients* in *downward* direction can be defined as:

$$\Im^{p}_{0|m}(k_\perp) = \exp\left[j\sum_{n=1}^{m-1} k_{zn}\Delta_n \right] \prod_{n=0}^{m-1} T^{p}_{n|n+1} \left[\prod_{n=1}^{m} \vec{M}^{p}_{n} \right]^{-1}, \tag{10}$$

where $p \in \{v, h\}$. The *generalized transmission coefficients* in *upward* direction are then given by:

$$\Im^{p}_{m|0} = \begin{cases} \Im^{p}_{0|m} \dfrac{\mu_0 k_{zm}}{\mu_m k_{z0}} & for\ p = h \\ \Im^{p}_{0|m} \dfrac{\varepsilon_0 k_{zm}}{\varepsilon_m k_{z0}} & for\ p = v \end{cases} \tag{11}$$

which formally express the *reciprocity* of the generalized transmission coefficients for an arbitrary flat-boundaries layered structure (Imperatore et al. 2009b). In addition, with reference to a layered slab sandwiched between two half-space, we consider the *generalized transmission coefficients* in *upward* direction for the layered *slab* between two half-spaces (*m*,0), which are defined as

$$\Im^{p(slab)}_{m|0}(k_\perp) = \exp\left[j\sum_{n=1}^{m-1} k_{zn}\Delta_n \right] \prod_{n=0}^{m-1} T^{p}_{n+1|n} \left[\prod_{n=1}^{m-1} \vec{M}^{p}_{n} \right]^{-1} \tag{12}$$

Note also that

$$\Im^{p}_{m|0}(k_\perp) = \Im^{p(slab)}_{m|0}(k_\perp)\,[\vec{M}^{p}_{m}(k_\perp)]^{-1}. \tag{13}$$

The *generalized transmission coefficients* in *downward* direction for the layered *slab* between two half-spaces (m,0), can be defined as

$$\Im_{0|m}^{p(slab)}(k_\perp) = \exp\left[j\sum_{n=1}^{m-1} k_{zn}\Delta_n \right] \prod_{n=0}^{m-1} T_{n|n+1}^{p} \left[\prod_{n=1}^{m-1} \vec{M}_n^p \right]^{-1} . \tag{14}$$

On the other hand, it should be noted that the $\Im_{0|m}^{p}$ are distinct from the coefficients $\Im_{0|m}^{p(slab)}$, because in the evaluation of $\Im_{0|m}^{p}$ the effect of all the layers under the layer m is taken into account, whereas $\Im_{0|m}^{p(slab)}$ are evaluated referring to a different configuration in which the intermediate layers *1...m* are bounded by the half-spaces *0* and *m*. In the following, we shown how the employing the generalized reflection/transmission coefficient notions not only is crucial in obtaining a compact closed-form perturbation solution, but it also permit us to completely elucidate the obtained analytical expressions from a physical point of view, highlighting the role played by the *equivalent reflecting interfaces* and by *the equivalent slabs*, so providing the inherent connection between local and global scattering responses.

4. Spectral Representation of the Stochastic Geometry Description

In this section, the focus is on stochastic description for the geometry of the investigated structure, and the notion of wide-sense stationary process is detailed. First of all, when the description of a rough interface by means of deterministic function $\zeta_m(\mathbf{r}_\perp)$ is concerned, the corresponding *ordinary* 2-D *Fourier Transform* pair can be defined as

$$\tilde{\zeta}_m(\mathbf{k}_\perp) = (2\pi)^{-2} \iint d\mathbf{r}_\perp \, e^{-j\mathbf{k}_\perp \cdot \mathbf{r}_\perp} \zeta_m(\mathbf{r}_\perp), \tag{15}$$

$$\zeta_m(\mathbf{r}_\perp) = \iint d\mathbf{k}_\perp \, e^{j\mathbf{k}_\perp \cdot \mathbf{r}_\perp} \tilde{\zeta}_m(\mathbf{k}_\perp). \tag{16}$$

Let us assume now that $\zeta_m(\mathbf{r}_\perp)$, which describes the generic (mth) rough interface, is a 2-D stochastic process satisfying the conditions

$$< \zeta_m(\mathbf{r}_\perp) > = 0, \tag{17}$$

$$< \zeta_m(\mathbf{r}_\perp + \boldsymbol{\rho}) \zeta_m(\mathbf{r}_\perp) > = B_{\zeta_m}(\boldsymbol{\rho}), \tag{18}$$

where the *angular bracket* denotes statistical ensemble averaging, and where $B_{\zeta_m}(\boldsymbol{\rho})$ is the interface *autocorrelation* function, which quantifies the similarity of the spatial fluctuations with a displacement ρ. Equations (17)-(18) constitute the basic assumptions defining a *wide sense stationary* (*WSS*) stochastic process: the statistical properties of the process under consideration are invariant to a spatial shift. Similarly, concerning two mutually correlated random rough interfaces ζ_m and ζ_n, we also assume that they are *jointly WSS*, i.e.

$$<\zeta_m(\mathbf{r}_\perp+\boldsymbol{\rho})\zeta_n(\mathbf{r}_\perp)>=B_{\zeta_m\zeta_n}(\boldsymbol{\rho}), \tag{19}$$

where $B_{\zeta_m\zeta_n}(\boldsymbol{\rho})$ is the corresponding *cross-correlation function* of the two random processes. It can be readily derived that

$$B_{\zeta_m\zeta_n}(\boldsymbol{\rho})=B_{\zeta_n\zeta_m}(-\boldsymbol{\rho}). \tag{20}$$

The integral in (15) is a *Riemann* integral representation for $\zeta_m(\mathbf{r}_\perp)$, and it exists if $\zeta_m(\mathbf{r}_\perp)$ is piecewise continuous and *absolutely integrable*. On the other hand, when the spectral analysis of a stationary random process is concerned, the integral (15) does not in general exist in the framework of theory of the ordinary functions. Indeed, a WSS process describing an interface $\zeta_m(\mathbf{r}_\perp)$ of infinite lateral extension, for its proper nature, is not *absolutely integrable*, so the conditions for the existence of the Fourier Transform are not satisfied. In order to obtain a spectral representation for a *WSS* random process, this difficulty can be circumvented by resorting to the more general *Fourier-Stieltjes* integral (Ishimaru, 1978); otherwise one can define space-truncated functions. When a finite patch of the rough interface with area *A* is concerned, the space-truncated version of (15) can be introduced as

$$\tilde{\zeta}_m(\mathbf{k}_\perp;A)=(2\pi)^{-2}\iint_A d\mathbf{r}_\perp\, e^{-j\mathbf{k}_\perp\cdot\mathbf{r}_\perp}\zeta_m(\mathbf{r}_\perp), \tag{21}$$

subsequently, $\tilde{\zeta}_m(\mathbf{k}_\perp)=\lim_{A\to\infty}\tilde{\zeta}_m(\mathbf{k}_\perp;A)$ is not an ordinary function. Nevertheless, we will use again the (15)-(16), regarding them as symbolic formulas, which hold a rigorous mathematical meaning beyond the ordinary function theory (generalized Fourier Transform). We underline that by virtue of the condition (17) directly follows also that $<\tilde{\zeta}_m(\mathbf{k}_\perp)>=0$. Let us consider

$$<\zeta_m(\mathbf{r}'_\perp)\zeta_n^*(\mathbf{r}''_\perp)>=\iint d\mathbf{k}'_\perp\iint d\mathbf{k}''_\perp\, e^{j(\mathbf{k}'_\perp\cdot\mathbf{r}'_\perp-\mathbf{k}''_\perp\cdot\mathbf{r}''_\perp)}<\tilde{\zeta}_m(\mathbf{k}'_\perp)\tilde{\zeta}_n^*(\mathbf{k}''_\perp)>, \tag{22}$$

where the asterisk denotes the complex conjugated, and where the operations of average and integration have been interchanged. When *jointly WSS* processes ζ_m and ζ_n are concerned, accordingly to (19), the RHS of (22) must be a function of $\mathbf{r}'_\perp-\mathbf{r}''_\perp$ only; therefore, it is required that

$$<\tilde{\zeta}_m(\mathbf{k}'_\perp)\tilde{\zeta}_n^*(\mathbf{k}''_\perp)>=W_{mn}(\mathbf{k}'_\perp)\delta(\mathbf{k}'_\perp-\mathbf{k}''_\perp), \tag{23}$$

where $\delta(\cdot)$ is the *Dirac* delta function, and where $W_{mn}(\boldsymbol{\kappa})$ is called the (spatial) *cross power spectral density* of two interfaces ζ_m and ζ_n, for the spatial frequencies of the roughness. Equation (23) states that the different spectral components of the two considered interfaces must be uncorrelated. This is to say that the (generalized) Fourier transform of jointly WSS processes are *jointly non stationary* white noise with average power $W_{mn}(\mathbf{k}'_\perp)$. Indeed, by using (23) into (22), we obtain

$$<\zeta_m(\mathbf{r}'_\perp)\zeta_n(\mathbf{r}''_\perp)> = \iint d\mathbf{k}''_\perp\, e^{j\cdot\mathbf{k}''_\perp\cdot(\mathbf{r}'_\perp-\mathbf{r}''_\perp)} W_{mn}(\mathbf{k}''_\perp), \tag{24}$$

where the RHS of (24) involves an (ordinary) 2D Fourier Transform. Note also that as a direct consequence of the fact that $\zeta_n(\mathbf{r}_\perp)$ is real we have the relation $\tilde{\zeta}_n(\mathbf{k}_\perp) = \tilde{\zeta}_n^*(-\mathbf{k}_\perp)$. Therefore, setting $\boldsymbol{\rho} = \mathbf{r}'_\perp - \mathbf{r}''_\perp$ in (24), we have

$$B_{\zeta_m\zeta_n}(\boldsymbol{\rho}) = \iint d\boldsymbol{\kappa}\, e^{j\boldsymbol{\kappa}\cdot\boldsymbol{\rho}}\, W_{mn}(\boldsymbol{\kappa}). \tag{25}$$

The *cross-correlation function* $B_{\zeta_m\zeta_n}(\boldsymbol{\rho})$ of two interfaces ζ_m and ζ_n, is then given by the (inverse) 2D Fourier Transform of their *(spatial) cross power spectral density*, and Equation (25) together with its Fourier inverse

$$W_{mn}(\boldsymbol{\kappa}) = (2\pi)^{-2} \iint d\boldsymbol{\rho}\, e^{-j\boldsymbol{\kappa}\cdot\boldsymbol{\rho}} B_{\zeta_m\zeta_n}(\boldsymbol{\rho}), \tag{26}$$

may be regarded as the (generalized) *Wiener-Khinchin* theorem. In particular, when $n=m$, (23) reduces to

$$<\tilde{\zeta}_m(\mathbf{k}'_\perp)\tilde{\zeta}_m^*(\mathbf{k}''_\perp)> = W_m(\mathbf{k}'_\perp)\delta(\mathbf{k}'_\perp - \mathbf{k}''_\perp), \tag{27}$$

where $W_m(\boldsymbol{\kappa})$ is called the *(spatial) power spectral density* of nth corrugated interface ζ_m and can be expressed as the *(ordinary)* 2D Fourier transform of n-corrugated interface autocorrelation function, i.e., satisfying the transform pair:

$$W_m(\boldsymbol{\kappa}) = (2\pi)^{-2} \iint d\boldsymbol{\rho}\, e^{-j\boldsymbol{\kappa}\cdot\boldsymbol{\rho}} B_{\zeta_m}(\boldsymbol{\rho}), \tag{28}$$

$$B_{\zeta_m}(\boldsymbol{\rho}) = \iint d\boldsymbol{\kappa}\, e^{j\boldsymbol{\kappa}\cdot\boldsymbol{\rho}}\, W_m(\boldsymbol{\kappa}), \tag{29}$$

which is the statement of the classical *Wiener-Khinchin* theorem. We emphasize the physical meaning of $W_m(\boldsymbol{\kappa})d\boldsymbol{\kappa} = W_m(\kappa_x,\kappa_y)d\kappa_x d\kappa_y$: it represents the power of the spectral components of the mth rough interface having spatial wave number between κ_x and $\kappa_x + d\kappa_x$ and κ_y and $\kappa_y + d\kappa_y$, respectively, in x and y direction. Furthermore, from (20) and (26) it follows that

$$W_{mn}(\boldsymbol{\kappa}) = W_{nm}^*(\boldsymbol{\kappa}). \tag{30}$$

This is to say that, unlike the power spectral density, the cross power spectral density is, in general, neither real nor necessarily positive. Furthermore, it should be noted that the *Dirac*'s delta function can be defined by the integral representation

$$\delta(\boldsymbol{\kappa}) = (2\pi)^{-2} \iint d\boldsymbol{\rho}\, e^{-j\boldsymbol{\kappa}\cdot\boldsymbol{\rho}} = \lim_{A\to\infty} \delta(\boldsymbol{\kappa}; A). \tag{31}$$

By using in (27) and (23) the relation $\delta(0;A) = A/(2\pi)^2$, we have, respectively, that the (spatial) power *spectral density* of nth corrugated interface can be also expressed as

$$W_m(\boldsymbol{\kappa}) = (2\pi)^2 \lim_{A\to\infty} \frac{1}{A} < \left|\tilde{\zeta}_m(\boldsymbol{\kappa}; A)\right|^2 >, \tag{32}$$

and the (spatial) *cross power spectral density* of two interfaces ζ_m and ζ_n is given by

$$W_{mn}(\boldsymbol{\kappa}) = (2\pi)^2 \lim_{A\to\infty} \frac{1}{A} < \tilde{\zeta}_m(\boldsymbol{\kappa}; A)\tilde{\zeta}_n^*(\boldsymbol{\kappa}; A) >. \tag{33}$$

It should be noted that the domain of a rough interface is physically limited by the illumination beamwidth. Note also that the different definitions of the Fourier transform are available and used in the literature: the sign of the complex exponential function are sometimes exchanged and a multiplicative constant $(2\pi)^{-2}$ may appear in front of either integral or its square root in front of each expression (15)-(16). Finally, we recall that the theory of random process predicts only the averages over many realizations.

5. Perturbative Field Formulation

With reference to the geometry of Figure 1, in order to obtain a solution valid in each region of the structure, we have to enforce the continuity of the tangential fields:

$$[\hat{n}_m \times \Delta\mathbf{E}_m]_{z=\zeta_m(\mathbf{r}_\perp)-d_m} = 0, \tag{34}$$

$$[\hat{n}_m \times \Delta\mathbf{H}_m]_{z=\zeta_m(\mathbf{r}_\perp)-d_m} = 0, \tag{35}$$

where $\Delta\mathbf{E}_m = \mathbf{E}_{m+1} - \mathbf{E}_m$, $\Delta\mathbf{H}_m = \mathbf{H}_{m+1} - \mathbf{H}_m$, and the surface normal vector is given by:

$$\hat{n}_m = \frac{\hat{z} - \boldsymbol{\gamma}_m}{\sqrt{1-\gamma_m^2}}, \tag{36}$$

with the slope vector γ_m:

$$\boldsymbol{\gamma}_m = \nabla_\perp \zeta_m = \left[\frac{\partial}{\partial x}\hat{x} + \frac{\partial}{\partial y}\hat{y}\right]\zeta_m, \tag{37}$$

and where $\nabla_\perp$ is the *nabla* operator in the x-y plane. In order to study the fields $\mathbf{E}_m$ and $\mathbf{H}_m$ within the generic mth layer of the structure, we assume then that, for each mth rough interface, the deviations and slopes of surface with respect to the reference mean plane z=-d_m are small enough in the sense of (Ulaby et al, 1982) (Tsang et al., 1985), so that the fields can be expanded about the reference mean plane. The fields expansion can be then injected into the boundary conditions (34)-(35), so that, retaining only up to the first-order terms, the following *nonuniform boundary conditions* can be obtained (Imperatore et al. 2009a)

$$\hat{z}\times\Delta\mathbf{E}_m^{(1)}\Big|_{z=-d_m} = \nabla_\perp\zeta_m \times \Delta\mathbf{E}_m^{(0)}\Big|_{z=-d_m} - \zeta_m\,\hat{z}\times\frac{\partial\Delta\mathbf{E}_m^{(0)}}{\partial z}\Bigg|_{z=-d_m}, \tag{38}$$

$$\hat{z}\times\Delta\mathbf{H}_m^{(1)}\Big|_{z=-d_m} = \nabla_\perp\zeta_m \times \Delta\mathbf{H}_m^{(0)}\Big|_{z=-d_m} - \zeta_m\,\hat{z}\times\frac{\partial\Delta\mathbf{H}_m^{(0)}}{\partial z}\Bigg|_{z=-d_m}, \tag{39}$$

where the field solution has been formally represented as:

$$\mathbf{E}_m(\mathbf{r}_\perp,z)\approx\mathbf{E}_m^{(0)}+\mathbf{E}_m^{(1)}+\mathbf{E}_m^{(2)}+\ldots, \tag{40}$$

$$\mathbf{H}_m(\mathbf{r}_\perp,z)\approx\mathbf{H}_m^{(0)}+\mathbf{H}_m^{(1)}+\mathbf{H}_m^{(2)}+\ldots. \tag{41}$$

Therefore, the boundary conditions from each *m*th rough interface can be transferred to the associated equivalent flat interface. In addition, the right-hand sides of Eqs. (38) and (39) can be interpreted as effective magnetic ($\mathbf{J}_{Hm}^{p(1)}$) and electric ($\mathbf{J}_{Em}^{p(1)}$) surface current densities, respectively, with *p* denoting the incident polarization; so that we can identify the first-order fluctuation fields as being excited by these effective surface current densities imposed on the unperturbed interfaces. Accordingly, the geometry randomness of each corrugated interfaces is then translated in random current sheets imposed on each reference mean plane (z=-d_m), which radiate in an unperturbed (flat boundaries) layered medium. As a result, within the first-order approximation, the field can be than represented as the sum of an unperturbed part $\mathbf{E}_n^{(0)},\mathbf{H}_n^{(0)}$ and a random part, so that $\mathbf{E}_n(\mathbf{r}_\perp,z)\approx\mathbf{E}_n^{(0)}+\mathbf{E}_n^{(1)}$, $\mathbf{H}_n(\mathbf{r}_\perp,z)\approx\mathbf{H}_n^{(0)}+\mathbf{H}_n^{(1)}$. The first is the primary field, which exists in absence of surface boundaries roughness (flat-boundaries stratification), detailed in (Imperatore et al. 2009a); whereas $\mathbf{E}_n^{(1)},\mathbf{H}_n^{(1)}$ can be interpreted as the superposition of single-scatter fields from each rough interface. In order to perform the evaluation of perturbative development, the scattered field is then represented as the sum of *up*- and *down-going* waves, and the first-order scattered field in each region of the layered structure can be represented in the form:

$$\mathbf{E}_m^{(1)}=\mathbf{E}_m^{-(1)}+\mathbf{E}_m^{+(1)}, \tag{42}$$

$$\mathbf{H}_m^{(1)}=\mathbf{H}_m^{-(1)}+\mathbf{H}_m^{+(1)}, \tag{43}$$

With

$$\mathbf{E}_m^{\pm(1)}=\sum_{q=h,v}\iint d\mathbf{k}_\perp e^{j\mathbf{k}_\perp\cdot\mathbf{r}_\perp}\,\hat{q}_m^{\pm}(\mathbf{k}_\perp)S_m^{\pm q(1)}(\mathbf{k}_\perp)e^{\pm jk_{zm}z}, \tag{44}$$

$$\mathbf{H}_m^{\pm(1)}=\sum_{q=h,v}\frac{1}{Z_m}\iint d\mathbf{k}_\perp e^{j\mathbf{k}_\perp\cdot\mathbf{r}_\perp}\,\hat{k}_m^{\pm}\times\hat{q}_m^{\pm}(\mathbf{k}_\perp)S_m^{\pm q(1)}(\mathbf{k}_\perp)e^{\pm jk_{zm}z}, \tag{45}$$

Therefore, a solution valid in each region of the structure can be obtained from (42)-(45) taking into account the non uniform boundary conditions (38)-(39). In order to solve the scattering problem in terms of the unknown expansion coefficients $S_m^{\pm q(1)}(\mathbf{k}_\perp)$, we arrange their amplitudes in a single vector according to the notation:

$$\mathbf{S}_m^{q(1)}(\mathbf{k}_\perp, d_m) = \begin{bmatrix} S_m^{+q(1)}(\mathbf{k}_\perp)e^{-jk_{zm}d_m} \\ S_m^{-q(1)}(\mathbf{k}_\perp)e^{+jk_{zm}d_m} \end{bmatrix}. \tag{46}$$

Subsequently, the *nonuniform boundary conditions* (38)-(39) can be formulated by employing a suitable matrix notation, so that for the (q=h) *horizontal* polarized scattered wave we have (Imperatore et al. 2008a) (Imperatore et al. 2009a):

$$\mathbf{S}_m^{h(1)}(\mathbf{k}_\perp, d_m) + \mathbf{\Theta}_m^p(\mathbf{k}_\perp, \mathbf{k}_\perp^i) = \boldsymbol{N}_{m|m+1}^h(k_\perp)\mathbf{S}_{m+1}^{h(1)}(\mathbf{k}_\perp, d_m), \tag{47}$$

where

$$\mathbf{\Theta}_m^p(\mathbf{k}_\perp, \mathbf{k}_\perp^i) = \begin{bmatrix} -\dfrac{k_0 Z_0 \mu_m}{2k_{zm}}(\hat{k}_\perp \times \hat{z}) \cdot \tilde{\mathbf{J}}_{Em}^{p(1)} + \dfrac{1}{2}\hat{k}_\perp \cdot \tilde{\mathbf{J}}_{Hm}^{p(1)} \\ +\dfrac{k_0 Z_0 \mu_m}{2k_{zm}}(\hat{k}_\perp \times \hat{z}) \cdot \tilde{\mathbf{J}}_{Em}^{p(1)} + \dfrac{1}{2}\hat{k}_\perp \cdot \tilde{\mathbf{J}}_{Hm}^{p(1)} \end{bmatrix} \tag{48}$$

is the term associated with the *effective* source distribution, wherein the expressions of the effective currents $\tilde{\mathbf{J}}_{Em}^{p(1)}$ and $\tilde{\mathbf{J}}_{Hm}^{p(1)}$, imposed on the (flat) unperturbed boundary $z = -d_m$, for an incident polarization $p \in \{v, h\}$ are detailed in (Imperatore et al. 2009a); and where Z_0 is the intrinsic impedance of the vacuum. Furthermore, the fundamental *transfer matrix operator* is given by:

$$\boldsymbol{N}_{m-1|m}^q(k_\perp) = \frac{1}{T_{m-1|m}^q}\begin{bmatrix} 1 & R_{m-1|m}^q \\ R_{m-1|m}^q & 1 \end{bmatrix}, \tag{49}$$

with the superscripts $q \in \{v, h\}$ denoting the polarization. Moreover, it should be noted that on a (kth) flat interface Eq. (47) reduces to the *uniform boundary conditions,* thus getting:

$$\mathbf{S}_k^{h(1)}(\mathbf{k}_\perp, d_k) = \boldsymbol{N}_{k|k+1}^h(k_\perp)\mathbf{S}_{k+1}^{h(1)}(\mathbf{k}_\perp, d_k). \tag{50}$$

We emphasize that Eqs. (47) states in a simpler form the problem originally set by Eqs. (38)-(39): as matter of fact, solving Eq. (47) $\forall m$ implies dealing with the determination of unknown scalar amplitudes $S_m^{\pm q(1)}(\mathbf{k}_\perp)$ instead of working with the corresponding vector unknowns $\mathbf{E}_m^{(1)}, \mathbf{H}_m^{(1)}$. Therefore, the scattering problem in each mth layer is reduced to the algebraic calculation of the unknown expansion scattering coefficients vector (46). As a result, when a structure with rough interfaces is considered, the enforcement of the original non uniform boundary conditions through the stratification (m=0, ..., N-1) can be addressed by writing down a linear system of equations with the aid of the matrix formalism (47)-(48) with m=0, ..., N-1. As a result, the formulation of *non-uniform boundary conditions* in matrix notation (47)-(48) enables a systematic method for solving the scattering problem: For the *N*-layer stratification of Figure 1, we have to find *2N* unknown expansion coefficients, using *N* vectorial equations (47), i.e., *2N* scalar equations. It should be noted that, for the considered configuration, the relevant scattering coefficients $S_N^{+q(1)}(\mathbf{k}_\perp), S_0^{-q(1)}(\mathbf{k}_\perp)$ are obviously supposed to be zero. The scattering problem, therefore, results to be reduced to a formal solution of a linear equation system. We finally emphasize that here we are interested in the scattering from the stratification; therefore, the determination of the unknown expansion coefficients $S_0^{+q(1)}(\mathbf{k}_\perp)$ of the scattered wave into the upper half-space is required only. Full expressions for these coefficients are reported in (Imperatore et al. 2009a).

6. BPT Closed-form Solution

The field scattered upward in the upper half-space in the first-order limit can be written in the form (see (42)-(45)):

$$\mathbf{E}_0^{(1)}(\mathbf{r}) = \sum_{q=h,v} \iint d\mathbf{k}_\perp e^{j\mathbf{k}_\perp \cdot \mathbf{r}_\perp} \hat{q}_0^+(\mathbf{k}_\perp) S_0^{+q(1)}(\mathbf{k}_\perp) e^{jk_{z0}z} \tag{51}$$

By employing the *method of stationary phase,* we evaluate the integral (51) in the *far field* zone, obtaining:

$$\mathbf{E}_0^{(1)}(\mathbf{r}) \cdot \hat{q}_0^+(\mathbf{k}_\perp^s) \cong -j2\pi\, k_0 \cos\theta_0^s \frac{e^{jk_0 r}}{r} S_0^{+q(1)}(\mathbf{k}_\perp^s) \tag{52}$$

with $q \in \{v, h\}$ is the polarization of the scattered field. The scattering *cross section* of a generic (*n*th) rough interface embedded in the layered structure can be then defined as

$$\tilde{\sigma}_{qp,n}^0 = \lim_{r\to\infty} \lim_{A\to\infty} \frac{4\pi r^2}{A} < \left| \mathbf{E}_0^{(1)}(\mathbf{r}) \cdot \hat{q}_0^+(\mathbf{k}_\perp^s) \right|^2 >, \tag{53}$$

where < > denotes ensemble averaging, where $q \in \{v, h\}$ and $p \in \{v, h\}$ denote, respectively, the polarization of scattered field and the polarization of incident field, and where *A* is the illuminated surface area. The estimate of the mean power density can be obtained by averaging over an ensemble of statistically identical interfaces. Therefore, considering that the (spatial) *power spectral density* $W_n(\boldsymbol{\kappa})$ of *n*th corrugated interface is defined as in (32), the *scattering cross section* relative to the contribution of the *n*th corrugated interface, according to the formalism used in [Franceschetti et al. 2008], can be expressed as:

$$\tilde{\sigma}_{qp,n}^0 = \pi k_0^4 \left| \tilde{\alpha}_{qp}^{n,n+1}(\mathbf{k}^s, \mathbf{k}^i) \right|^2 W_n(\mathbf{k}_\perp^s - \mathbf{k}_\perp^i) \,, \tag{54}$$

with p, $q \in \{v, h\}$ denoting, respectively, the incident and the scattered polarization states, which may stand for *horizontal* polarization (*h*) or *vertical* polarization (*v*); the coefficients $\tilde{\alpha}_{qp}^{m,m+1}$ are relative to the *p*-polarized incident wave impinging on the structure from upper half space 0 and to the q-polarized scattering contribution from structure into the upper half space, originated from the rough interface between the layers m, $m+1$:

$$\tilde{\alpha}_{hh}^{m,m+1} = (\varepsilon_{m+1} - \varepsilon_m)\frac{k_{z0}^s}{k_{zm}^s}(\hat{k}_\perp^s \cdot \hat{k}_\perp^i)$$
$$e^{jk_{zm}^s\Delta_m}\mathfrak{I}_{m|0}^h(k_\perp^s)[1+\mathfrak{R}_{m|m+1}^h(k_\perp^s)]\,e^{jk_{zm}^i\Delta_m}\mathfrak{I}_{0|m}^h(k_\perp^i)[1+\mathfrak{R}_{m|m+1}^h(k_\perp^i)], \tag{55}$$

$$\tilde{\alpha}_{vh}^{m,m+1} = (\varepsilon_{m+1} - \varepsilon_m)\frac{k_{z0}^s}{k_0}\hat{z}\cdot(\hat{k}_\perp^i \times \hat{k}_\perp^s)$$
$$e^{jk_{zm}^s\Delta_m}\mathfrak{I}_{m|0}^v(k_\perp^s)[1-\mathfrak{R}_{m|m+1}^v(k_\perp^s)]\,e^{jk_{zm}^i\Delta_m}\mathfrak{I}_{0|m}^h(k_\perp^i)[1+\mathfrak{R}_{m|m+1}^h(k_\perp^i)] \tag{56}$$

$$\tilde{\alpha}_{hv}^{m,m+1} = (\varepsilon_{m+1} - \varepsilon_m)\frac{k_{z0}^s k_{zm}^i}{k_0 k_{zm}^s \varepsilon_m}\hat{z}\cdot(\hat{k}_\perp^i \times \hat{k}_\perp^s)$$
$$e^{jk_{zm}^s\Delta_m}\mathfrak{I}_{m|0}^h(k_\perp^s)[1+\mathfrak{R}_{m|m+1}^h(k_\perp^s)]\,e^{jk_{zm}^i\Delta_m}\mathfrak{I}_{0|m}^v(k_\perp^i)[1-\mathfrak{R}_{m|m+1}^v(k_\perp^i)] \quad , \tag{57}$$

$$\tilde{\alpha}_{vv}^{m,m+1} = (\varepsilon_{m+1} - \varepsilon_m)\frac{k_{z0}^s}{k_0^2 k_{zm}^s \varepsilon_m}e^{jk_{zm}^s\Delta_m}\mathfrak{I}_{m|0}^v(k_\perp^s)e^{jk_{zm}^i\Delta_m}\mathfrak{I}_{0|m}^v(k_\perp^i)$$
$$\{[1+\mathfrak{R}_{m|m+1}^v(k_\perp^s)][1+\mathfrak{R}_{m|m+1}^v(k_\perp^i)]\frac{\varepsilon_m}{\varepsilon_{m+1}}k_\perp^i k_\perp^s \tag{58}$$
$$-[1-\mathfrak{R}_{m|m+1}^v(k_\perp^s)][1-\mathfrak{R}_{m|m+1}^v(k_\perp^i)]k_{zm}^s k_{zm}^i(\hat{k}_\perp^s \cdot \hat{k}_\perp^i)\} \quad ,$$

where $\mathfrak{I}_{m|0}^p(k_\perp)$ are the generalized transmission coefficients in upward direction (see (11)).

Furthermore, we stress when the backscattering case $(\hat{k}_\perp^s \times \hat{k}_\perp^i = 0)$ is concerned, our cross-polarized scattering coefficients (55)-(58) evaluated in the plane of incidence vanish, in full accordance with the classical first-order SPM method for a rough surface between two different media (Ulaby et al, 1982) (Tsang et al., 1985).

We now show that the solution, given by the expression (55)-(58), is susceptible of a straightforward generalization to the case of arbitrary stratification with N-rough boundaries. Taking into account the contribution of each nth corrugated interface, the global *bi-static scattering cross section* of the N-rough interface layered media can be expressed as:

$$\tilde{\sigma}_{qp}^0 = \pi k_0^4 \sum_{n=0}^{N-1} \left|\tilde{\alpha}_{qp}^{n,n+1}(\mathbf{k}^s,\mathbf{k}^i)\right|^2 W_n(\mathbf{k}_\perp^s - \mathbf{k}_\perp^i) + \pi k_0^4 \sum_{i\neq j} \mathrm{Re}\left\{\tilde{\alpha}_{qp}^{i,i+1}\left(\tilde{\alpha}_{qp}^{j,j+1}\right)^*\right\} W_{ij}(\mathbf{k}_\perp^s - \mathbf{k}_\perp^i) \tag{59}$$

with p, $q \in \{v, h\}$, where the asterisk denotes the complex conjugated, where $\tilde{\alpha}_{qp}^{i,i+1}$ are given by (55)-(58), and where the *cross power spectral density* W_{ij}, between the interfaces i and j, for the spatial frequencies of the roughness is given by (33). As a result, the scattering from the rough layered media is sensitive to the correlation between rough profiles of different interfaces. In fact, a real layered structure will have interfaces cross-correlation somewhere between two limiting situations: perfectly correlated and uncorrelated roughness. Consequently, the degree of correlation affects the phase relation between the fields scattered from each rough interface. Obviously, when the interfaces are supposed to be uncorrelated, the second term in (59) vanishes and, in the first-order approximation, the total scattering from the structure arises from the incoherent superposition of radiation scattered from each interface.

As a result, an elegant closed form solution is established, which takes into account parametrically the dependence of scattering properties on structure (geometric and electromagnetic) parameters. In addition, as it will be shown in the following, the proposed global solution turns out to be completely interpretable with basic physical concepts, clearly discerning the physics of the involved scattering mechanisms.

7. Generalized Scattering Matrix

In this section, to emphasize the polarimetric character of the BPT solution, we introduce the *generalized bistatic scattering matrix* of the layered rough media, which can be then formally expressed by:

$$\begin{bmatrix} E_0^{sv} \\ E_0^{sh} \end{bmatrix} = \pi k_0^2 \frac{e^{jk_0 r_0}}{r_0} \sum_{m=0}^{N-1} \tilde{\zeta}_m(\mathbf{k}_\perp^s - \mathbf{k}_\perp^i)\tilde{\boldsymbol{\alpha}}^{m,m+1}(\mathbf{k}^s,\mathbf{k}^i) \begin{bmatrix} E_0^{iv} \\ E_0^{ih} \end{bmatrix}. \tag{60}$$

In particular,

$$\mathcal{S}^{m|m+1}(\mathbf{k}^s,\mathbf{k}^i) = \pi k_0^2 \tilde{\zeta}_m(\mathbf{k}_\perp^s - \mathbf{k}_\perp^i)\,\tilde{\boldsymbol{\alpha}}^{m,m+1}(\mathbf{k}^s,\mathbf{k}^i) \tag{61}$$

characterizes the polarimetric response of the generic (mth) rough interface of the layered structure, for a plane wave incident in direction $\mathbf{k}^i$ and for a given observation direction $\mathbf{k}^s$, with

$$\tilde{\boldsymbol{\alpha}}^{m,m+1}(\mathbf{k}^s,\mathbf{k}^i) = \begin{bmatrix} \tilde{\alpha}_{vv}^{m,m+1}(\mathbf{k}^s,\mathbf{k}^i) & \tilde{\alpha}_{vh}^{m,m+1}(\mathbf{k}^s,\mathbf{k}^i) \\ \tilde{\alpha}_{hv}^{m,m+1}(\mathbf{k}^s,\mathbf{k}^i) & \tilde{\alpha}_{hh}^{m,m+1}(\mathbf{k}^s,\mathbf{k}^i) \end{bmatrix}, \tag{62}$$

and wherein

$$\begin{aligned} \tilde{\alpha}_{vv}^{m,m+1}(\mathbf{k}^s,\mathbf{k}^i) = (\varepsilon_{m+1} - \varepsilon_m)\Bigg[& \frac{\varepsilon_m}{\varepsilon_{m+1}} \frac{k_\perp^s}{k_0\varepsilon_m} \xi_{0\to m}^{+v}(k_\perp^s) \frac{k_\perp^i}{k_0\varepsilon_m} \xi_{0\to m}^{+v}(k_\perp^i) \\ & - \hat{k}_\perp^s \cdot \hat{k}_\perp^i \frac{k_{zm}^s}{k_0\varepsilon_m} \xi_{0\to m}^{-v}(k_\perp^s) \frac{k_{zm}^i}{k_0\varepsilon_m} \xi_{0\to m}^{-v}(k_\perp^i) \Bigg] \end{aligned}, \tag{63}$$

$$\tilde{\alpha}_{vh}^{m,m+1}(\mathbf{k}^s,\mathbf{k}^i) = (\varepsilon_{m+1} - \varepsilon_m)\hat{z} \cdot (\hat{k}_\perp^i \times \hat{k}_\perp^s) \frac{k_{zm}^s}{k_0\varepsilon_m} \xi_{0\to m}^{-v}(k_\perp^s)\xi_{0\to m}^{+h}(k_\perp^i), \tag{64}$$

$$\tilde{\alpha}_{hv}^{m,m+1}(\mathbf{k}^s,\mathbf{k}^i) = (\varepsilon_{m+1} - \varepsilon_m)\,\hat{z} \cdot (\hat{k}_\perp^i \times \hat{k}_\perp^s)\xi_{0\to m}^{+h}(k_\perp^s) \frac{k_{zm}^i}{k_0\varepsilon_m} \xi_{0\to m}^{-v}(k_\perp^i), \tag{65}$$

$$\tilde{\alpha}_{hh}^{m,m+1}(\mathbf{k}^s,\mathbf{k}^i)=(\varepsilon_{m+1}-\varepsilon_m)\hat{k}_\perp^s\cdot\hat{k}_\perp^i\,\xi_{0\to m}^{+h}(k_\perp^s)\,\xi_{0\to m}^{+h}(k_\perp^i),\qquad(66)$$

where we have introduced the notation

$$\xi_{0\to m}^{\pm p}(k_\perp)=\Im_{0|m}^{p}(k_\perp)e^{jk_{zm}\Delta_m}[1\pm\Re_{m|m+1}^{p}(k_\perp)].\qquad(67)$$

It should be noted that (63)-(67) are obtained directly by (55)-(58) by making use of (11). Note also that the coefficients $\tilde{\alpha}_{qp}^{m,m+1}$ are expressible in a direct closed-form and depend parametrically on the unperturbed structure parameters. We also emphasize that the scattering configuration we have considered is compliant with the classical *Forward Scattering Alignment* (FSA) convention adopted in radar polarimetry.

Denoting with the superscript T the transpose, it can be verified that the scattering matrix satisfies the following relationship (Imperatore et al. 2009b)

$$\mathcal{S}^{m|m+1}(\mathbf{k}^s,\mathbf{k}^i)=\left[\mathcal{S}^{m|m+1}(-\mathbf{k}^i,-\mathbf{k}^s)\right]^T,\qquad(68)$$

which concisely expresses the *reciprocity* principle of the electromagnetic theory, as expected. This is to say that the scattering experiment is invariant for interchanging the role of transmitter and receiver. Note that the inversion of the projections on the *z=0* plane $\mathbf{k}_\perp^i$ and $\mathbf{k}_\perp^s$ are directly related to the inversion of the incident and scattered vector wave $\mathbf{k}^i=\mathbf{k}_\perp^i-k_{z0}^i\hat{z}$ and $\mathbf{k}^s=\mathbf{k}_\perp^s+k_{z0}^s\hat{z}$, respectively.

As a result, the presented closed-form solution permits the *polarimetric* evaluation of the scattering for a *bi-static* configuration, once the *three-dimensional* layered structure's parameters (shape of the roughness spectra, layers thickness and complex permittivities), the incident field parameters (frequency, polarization and direction) and the observation direction are been specified. Therefore, our formulation leads to a direct functional dependence (no integral evaluation is required) and, subsequently, allows us to show that the scattered field can be parametrically evaluated considering a set of parameters: some of them refer to an unperturbed structure configuration, i.e. intrinsically the physical parameters of the smooth boundary structure, and others which are determined exclusively by (random) deviations of the corrugated boundaries from their reference position. Procedurally, once the *generalized reflection/transmission coefficients* are recursively evaluated, the (63-67) can be than directly computed, so that the scattering cross section (59) or the generalized scattering matrix (60) of a structure with rough interfaces can be finally predicted. Finally, it should be noted that the method to be applied needs only the classical gently-roughness assumption, without any further approximation.

8. Unifying Perspective on Perturbation Approaches

In this section, we first summarize and discuss the previous existing scattering models introduced to cope with simplified layered geometry with only one (or two) rough interface, whose derivation methods belong to the class of perturbative methods. In the perspective of providing a unifying insight for the different perturbative formulations, the aim is to reconsider the state of art in an organized mathematical framework, analytically demonstrating the formal consistency of *BPT* general scattering solution, which permits to deal with layered media with an arbitrary number of rough interfaces, with the previous existing perturbative models, whose relevant first-order solutions can appear already of difficult mutual comparison (Franceschetti et al 2003) (Franceschetti et a. 2008).

In (Fuks, 2001) a model to calculate scattering from a rough surface on top of a stratified medium (see the geometry of Figure 3a) has been proposed. Expressions for scattering bi-static cross section were obtained by using the plane wave expansion of scattered EM fields and an equivalent current method, without using to the Green's function formalism. With reference to the geometry pictured in Figure 3b, an analytical small-perturbation-based model was developed to deal with a slightly rough interface boundary covered with a homogeneous dielectric layer (Azadegan et al., 2003) (Sarabandi et al, 2000). Starting from a perturbation series expansion and by employing the Green's function formalism, a solution to predict the first order bi-static scattering coefficients was found. On the other hand, the backscattering problem from the two-middle layer structure with one embedded corrugation, as schematized in Figure 3c, was investigated in (Yarovoy et al., 2000) in the first-order approximation, by using the small perturbation method combined with the Green's function approach. This approach leads to some cumbersome analytical expressions for backscattering coefficients.

As matter of fact, all these models, which refer to different simplified geometry, employ different perturbative procedures and different notations in the relative analytical derivation, so that the resulting solutions turn out mutually of difficult comparison. Besides, the finding of the connection between these existing functional forms is not a trivial task. With regard to these models, in (Franceschetti et al., 2008) it was essentially demonstrated the equivalence of the relevant analytical procedures and the consistency of the respective solutions. It should be mentioned that the models in (Fuks, 2001) and (Azadegan et al., 2003) (Sarabandi et al, 2000) are derived for a bi-static configuration. Conversely, the solution derived in (Yarovoy et al., 2000) with reference to the geometry of Figure 3c, which is even relatively more general since contemplates flat-boundaries stratification above and under the roughness, is given only in backscattering case. On the other hand, none of the pertinent configurations of these simplistic considered models is directly applicable to an actual remote sensing scenario. In fact, the natural stratified media are definitely constituted by corrugated interfaces, each one exhibiting a certain amount of roughness, whereas the flatness is an idealization which does not occur in natural media. More specifically, it can occur that, for a given roughness, one might consider an operational EM wavelength for which the roughness itself can be reasonably neglected. However, in principle, there is no defensible motivation, beyond the relevant limitation of the involved analytical difficulties, for considering the effect of only one interfacial roughness, neglecting the other relevant ones. This poses not only a conceptual limitation. In fact, in the applications perspective of retrieving geo-physical parameters by scattering measurements, whether there is a dominant interfacial roughness, and, in case, which the dominant one is, should be established after the remote sensing data are analyzed and, conversely, they cannot constitute *a priori* assumptions.

Each of the existing first-order models referring to a simplified geometry with one (Fuks, 2001) (Azadegan et al., 2003) (Sarabandi et al, 2000) (Yarovoy et al., 2000) or two (Tabatabaeenejad et al., 2006) rough interfaces, can be rigorously regarded as a particular case of our general model. Indeed, it can be analytically demonstrated that when the general geometry reduces to each simplified one, the consistency of the relevant solutions formally holds. In fact, when the (63)-(66) are specialized for the case of Figure 3a, the factors $\Im^{p}_{0|m}(k_{\perp})\exp(jk_{zm}\Delta_{m})$ turn out to be unitary and the general solution formally reduces to the one in (Fuks, 2001). Similarly, by specializing the solution to the configuration of Figure 3b (Azadegan et al., 2003), the computation is reduced to only $\tilde{\alpha}^{1,2}_{qp}$, in which $\xi^{\pm p}_{0\to 1}(k_{\perp}) = T^{p}_{0|1}(k_{\perp})[1 + R^{p}_{0|1}(k_{\perp})R^{p}_{1|2}(k_{\perp})e^{j2k_{z1}\Delta_{1}}]^{-1} e^{jk_{z1}\Delta_{1}}[1 \pm R^{p}_{1|2}(k_{\perp})]$, so the equivalent solution in (Azadegan et al., 2003) (Franceschetti et al., 2008) is formally obtained. Finally, specializing the general solution to the geometry of Figure 3c, and considering the backscattering case ($\mathbf{k}^{i}_{\perp} = -\mathbf{k}^{s}_{\perp}$), the computation is reduced to only $\tilde{\alpha}^{1,2}_{qp}$ in which $\xi^{\pm p}_{0\to 1}(k^{i}_{\perp}) = T^{p}_{0|1}(k^{i}_{\perp})[1 + R^{p}_{0|1}(k^{i}_{\perp})\Re^{p}_{1|2}(k^{i}_{\perp})e^{j2k^{i}_{z1}\Delta_{1}}]^{-1} e^{jk_{z1}\Delta_{1}}[1 \pm \Re^{p}_{1|2}(k^{i}_{\perp})]$, so we formally obtain the equivalent solution in (Yarovoy et al., 2000) (Franceschetti et al., 2008).

Analytically speaking, this allows us to obtain, in a unitary formal framework, a comprehensive insight into the first-order perturbation solutions formalism for scattering from stratified structure with rough interfaces.

Finally, the *Boundary Perturbation Theory* results can be also regarded as a generalization of the classical SPM for rough surface (Ulaby et al., 1982) (Tsang et al., 1985) to layered media with rough interfaces.

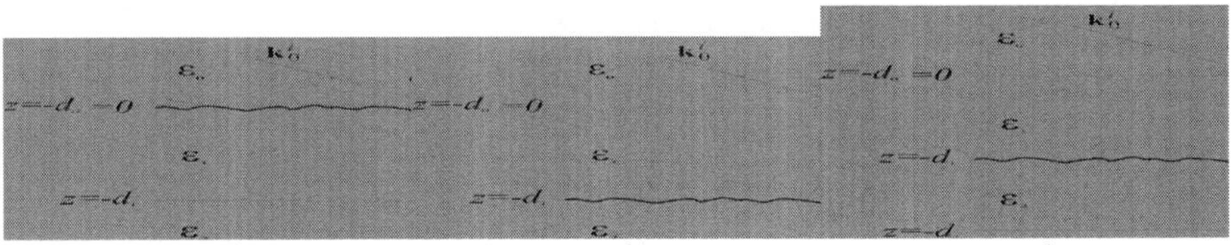

a) b) c)

Fig. 3. Simplified geometry considered by other Authors

9. Wave Scattering Decomposition

In this section, the focus is on the intrinsic significance of the global BPT scattering solution, getting more concrete insight into the physics of the problem of the scattering from rough interfaces of a layered media. In order to be able to express the solution in terms of readable basic physical phenomena, a key point is to exploit the *local scattering* concept, differently from (Yarovoy et al., 2000) and (Fuks, 1998) wherein the authors resort to the *radar contrast*.

It should be noted that the exact *analytic decomposition* of the solution in terms of local interactions is rigorously feasible, since, in the first-order perturbative approximation, the scattering amplitude can be written as a single space integral with a kernel that depends only on the rough interface height and on its first-order derivatives at a given point. Moreover, since in the limit of first-order BPT solution the global response of a structure with all rough interfaces can be directly obtained considering the superposition of the response from each interface, we firstly focus our attention to a generic embedded rough interface. Afterwards, the general interpretation for a layered structure with an arbitrary number of rough interfaces can be addressed.

To focus formally on the relations among *local* and *global* scattering concepts, the identified *Wave Scattering Decomposition* (Imperatore et al 2008c) (Imperatore et al. 2009c), for the global scattering response of the structure in terms of the four types of local interactions, can be expressed with a compact notation as:

$$\tilde{\alpha}_{qp}^{m,m+1} = \mathbf{P}_m^{qp}(k_\perp^s, k_\perp^i) \cdot \mathbf{\Psi}_{qp}^{m,m+1}(\mathbf{k}^s, \mathbf{k}^i) \tag{69}$$

Wherein

$$\mathbf{\Psi}_{qp}^{m,m+1}(\mathbf{k}^s, \mathbf{k}^i) = [\alpha_{qp}^{m,m+1} \quad \beta_{qp}^{m+1,m} \quad \beta_{qp}^{m,m+1} \quad \alpha_{qp}^{m+1,m}]^T \tag{70}$$

captures the local response of the mth rough interface between two layer of permittivity $\varepsilon_m, \varepsilon_{m+1}$ respectively, and the *transfer vector* $\mathbf{P}_m^{qp}$ is related to the coherent propagation inside the stratification (Imperatore et al. 2009c). Specifically, four distinct types of local interaction with an embedded rough interface can be distinguished: two of them identifiable as local scattering through the relevant interfacial roughness and other ones as local scattering from the roughness. We emphasize that the corresponding coefficients $\alpha_{qp}^{m,m+1}$ and $\alpha_{qp}^{m+1,m}$ refer to cases in which both the observation and incidence directions are, respectively, above and under the roughness; whereas $\beta_{qp}^{m,m+1}$ and $\beta_{qp}^{m+1,m}$ concern the local scattering contribution that cross the roughness in opposite directions. In addition, we stress that the local scattering coefficients are formally identical to the classical ones relative to a rough surface between two half-spaces (Ulaby et al., 1982) (Tsang et al., 1985). On the other hand, we emphasize that the *transfer vector,* which measures the influence of the stratification on the local scattering, whatever the roughness is, can be expressed in terms of the *generalized transmission/reflection coefficients* (Imperatore et al. 2009c). Once the local nature has been recognized, the solution can be suitably expanded, so that it can be expressed as a ray series or optical geometric series, where each term of the series is susceptible of a powerful physical interpretation (Imperatore et al. 2008c) as illustrated in the next section.

10. Physical Interpretation

In this section, we show in detailed the physical meaning of the wave scattering decomposition obtained in the previous section. The analytical solution (69), after suitable expansion of the elements of the *transfer vector,* is then susceptible of an expression in terms of an infinite sum of contributions (geometric power series). Consequently, the suitably expanded solution can be expressed as an *optical geometric series,* where each term of the series is susceptible of a direct physical interpretation. In particular, each individual term of the *absolutely summable* infinite series can be physically identified as a wave propagating in the structure that experiences a single-scattering local interaction with the roughness.
To this purpose, we introduce the following useful notations

$$\vec{\Lambda}_m^h(k_\perp) = R^h_{m|m-1}(k_\perp)\Re^h_{m|m+1}(k_\perp)e^{j2k_{zm}\Delta_m}\,, \tag{71}$$

$$\vec{\Lambda}_m^h(k_\perp) = \Re^h_{m|m-1}(k_\perp)\Re^h_{m|m+1}(k_\perp)e^{j2k_{zm}\Delta_m}\,, \tag{72}$$

and recognize that these factors correspond to a complete roundtrip in the intermediate layer with coherent reflections at the layer boundaries. Moreover, in order to provide a symmetrical expansion, it is possible to explicit the factor $\vec{M}_m^h(k_\perp^i)$, which is associated with the multiple round trip in the mth layer and is included in $\Im^h_{0|m}(k_\perp^i)$; so we can write:

$$\Im^h_{0|m}(k_\perp^i) = [\vec{M}_m^h(k_\perp^i)]^{-1}\Im'^h_{0|m}(k_\perp^i) \tag{73}$$

It should be noted that the $\Im'^p_{0|m}$ are distinct from the coefficients $\Im^{p(slab)}_{0|m}$, because in the evaluation of the former the effect of all the layers under the layer m is taken into account, whereas the latter is evaluated referring to a different configuration in which the intermediate layers $1...m$ are bounded by the half-spaces 0 and m.

Furthermore, we focus our attention on the two layers just above (m) and under ($m+1$) the generic roughness. In Figure 4, the remaining part of the structure is visualized condensed in two equivalent slabs constituted, respectively, by the intermediate layers $m+2,...,N-1$ (under the ($m+1$)th layer) and $1,...,m-1$ (above the mth layer). Without loss of generality, since analogous considerations hold for the other polarization combinations, the analysis can be conducted for the hh case only. Consequently, four distinct families of rays can be recognized; each one associated to one type of local interaction, so that each term of the expansion of the (69) can be readily identified as follows (Imperatore et al. 2009c):

a) Local upward scattered waves from rough interface: each of these waves (see Figure 4.a) undergoes a coherent transmission into mth layer ($\Im'^h_{0|m}(k_\perp^i)\exp(jk_{zm}^i\Delta_m)$), through the intermediate layers $1,\dots,m-1$, then j_1 complete round-trips ($[\vec{\Lambda}_m(k_\perp^i)]^{j_1}$) in the mth layer with coherent reflections at the incident angle ($k_\perp^i$), then an incoherent local scattering from the rough interface ($\alpha_{hh}^{m,m+1}(\mathbf{k}_\perp^s,\mathbf{k}_\perp^i)$), upward within the observation plane ($\mathbf{k}_\perp^s$), subsequently j_2 complete round-trips ($[\vec{\Lambda}_m^h(k_\perp^s)]^{j_2}$) in the mth layer with coherent reflections at the scattering angle ($k_\perp^s$), and finally a coherent transmission ($\exp(jk_{zm}^s\Delta_m)\Im^{h(slab)}_{m|0}(k_\perp^s)$) in the upper half-space through the intermediate layers $m-1,\dots,1$.

b) Local upward scattered waves through rough interface: each of these waves (see Figure 4.b) undergoes a coherent transmission into the mth layer ($\Im'^h_{0|m}(k_\perp^i)\exp(jk_{zm}^i\Delta_m)$), through the intermediate layers $1,\dots,m-1$, then j_1 complete round-trips ($[\vec{\Lambda}_m(k_\perp^i)]^{j_1}$) in the mth layer with coherent reflections at the incident angle ($k_\perp^i$), subsequently a coherent transmission ($T^h_{m|m+1}(k_\perp^i)$) followed by n_1 complete round-trips in the ($m+1$)th layer ($[\vec{\Lambda}_{m+1}^h(k_\perp^i)]^{n_1}$) and by further bounce on the (*m+1*)th flat interface ($\Re^h_{m+1|m+2}(k^i_\perp)\exp(j2k^i_{z(m+1)}\Delta_{m+1})$) at the incident angle ($k^i_\perp$), and after that an incoherent local scattering through the rough interface ($\beta^{m+1,m}_{hh}(\mathbf{k}^s_\perp,\mathbf{k}^i_\perp)$), upward within the observation plane ($\mathbf{k}^s_\perp$), subsequently j_2 complete round-trips ($[\vec{\Lambda}^h_m(k^s_\perp)]^{j_2}$) in the *m*th layer with coherent reflections at the scattering angle ($k^s_\perp$), and finally a coherent transmission ($\exp(jk^s_{zm}\Delta_m)\Im^{h(slab)}_{m|0}(k^s_\perp)$) in the upper half-space through the intermediate layers *m-1,…,1*.

c) Local downward scattered waves through rough interface: each of these waves (see Figure 4.c) undergoes a coherent transmission into the *m*th layer ($\Im'^h_{0|m}(k^i_\perp)\exp(jk^i_{zm}\Delta_m)$), through the intermediate layers *1,…,m-1*, then j_1 complete round-trips ($[\Lambda_m(k^i_\perp)]^{j_1}$) in the *m*th layer with coherent reflections at the incident angle ($k^i_\perp$), then an incoherent local scattering through the rough interface ($\beta^{m,m+1}_{hh}(\mathbf{k}^s_\perp,\mathbf{k}^i_\perp)$) downward in the observation plane ($\mathbf{k}^s_\perp$) followed by further bounce on the (*m+1*)th flat interface ($\Re^h_{m+1|m+2}(k^s_\perp)\exp(j2k^s_{z(m+1)}\Delta_{m+1})$) with subsequently n_2 complete round-trips in the (*m+1*)th layer ($[\vec{\Lambda}^h_{m+1}(k^s_\perp)]^{n_2}$) at the scattering angle ($k^s_\perp$), next a coherent transmission ($T^h_{m+1|m}(k^s_\perp)$) followed by subsequently j_2 complete round-trips ($[\vec{\Lambda}^h_m(k^s_\perp)]^{j_2}$) in the *m*th layer with coherent reflections at the scattering angle ($k^s_\perp$), and finally a coherent transmission ($\exp(jk^s_{zm}\Delta_m)\Im^{h(slab)}_{m|0}(k^s_\perp)$) in the upper half-space through the intermediate layers *m-1,…,1*.

d) Local downward scattered waves from rough interface: each of these waves (see Figure 4.d) undergoes a coherent transmission into the *m*th layer ($\Im'^h_{0|m}(k^i_\perp)\exp(jk^i_{zm}\Delta_m)$), through the intermediate layers *1,…,m-1*, then j_1 complete round-trips ($[\vec{\Lambda}_m(k^i_\perp)]^{j_1}$) in the *m*th layer with coherent reflections at the incident angle ($k^i_\perp$), next a coherent transmission ($T^h_{m|m+1}(k^i_\perp)$) followed by n_1 complete round-trips in the (*m+1*)th layer ($[\vec{\Lambda}^h_{m+1}(k^i_\perp)]^{n_1}$) and by further bounce on the (*m+1*)th flat interface ($\Re^h_{m+1|m+2}(k^i_\perp)\exp(j2k^i_{z(m+1)}\Delta_{m+1})$) at the incident angle ($k^i_\perp$), and after that an incoherent local scattering from the rough interface ($\alpha^{m+1,m}_{hh}(\mathbf{k}^s_\perp,\mathbf{k}^i_\perp)$), downward in the observation plane ($\mathbf{k}^s_\perp$), followed by further bounce on the (*m+1*)th flat interface ($\Re^h_{m+1|m+2}(k^s_\perp)\exp(j2k^s_{z(m+1)}\Delta_{m+1})$) with subsequently n_2 complete round-trips in the (*m+1*)th layer ($[\vec{\Lambda}^h_{m+1}(k^s_\perp)]^{n_2}$) at the scattering angle ($k^s_\perp$), then a coherent transmission ($T^h_{m+1|m}(k^s_\perp)$) followed by subsequently j_2 complete round-trips ($[\vec{\Lambda}^h_m(k^s_\perp)]^{j_2}$) in the *m*th layer with coherent reflections at the scattering angle ($k^s_\perp$), and finally a coherent transmission ($\exp(jk^s_{zm}\Delta_m)\Im^{h(slab)}_{m|0}(k^s_\perp)$) in the upper half-space through the intermediate layers *m-1,…,1*.

Note also that when an arbitrary layered structure with all rough interfaces is concerned, since in the first-order limit the multiple scattering contributions are neglected, the relative physical interpretation can be obtained effortless by superposition of the several ray contributions obtained considering separately each rough interface.

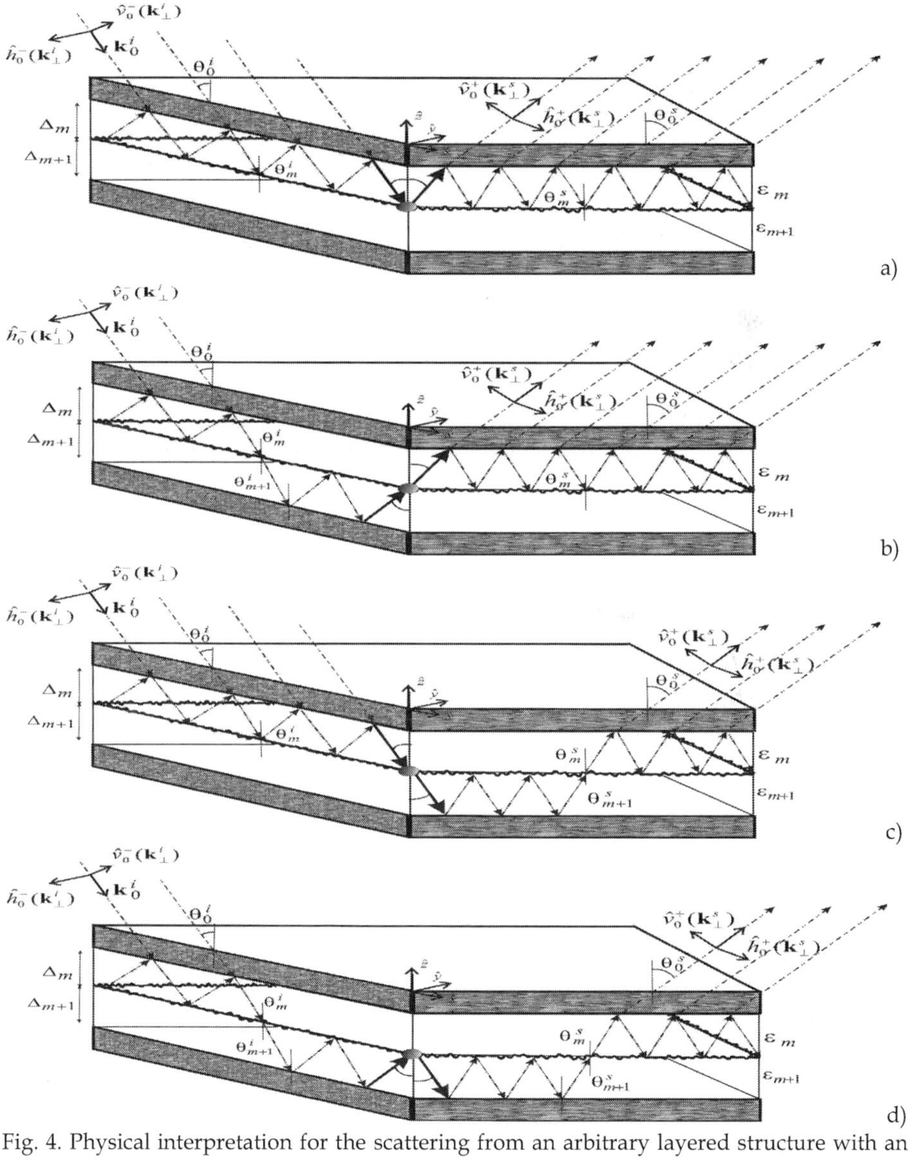

Fig. 4. Physical interpretation for the scattering from an arbitrary layered structure with an embedded rough interface.

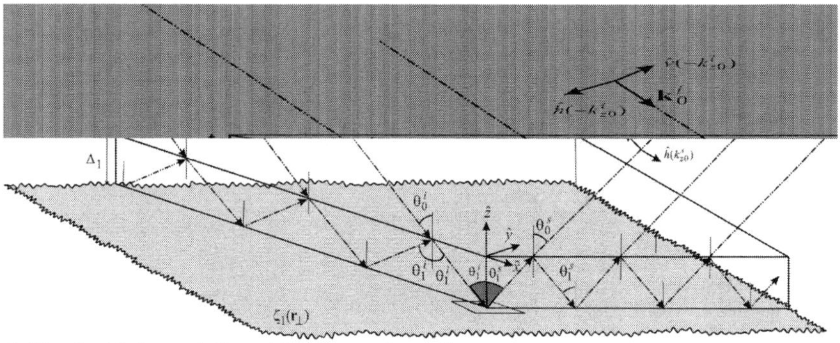

Fig. 5. Physical Interpretation: bistatic configuration

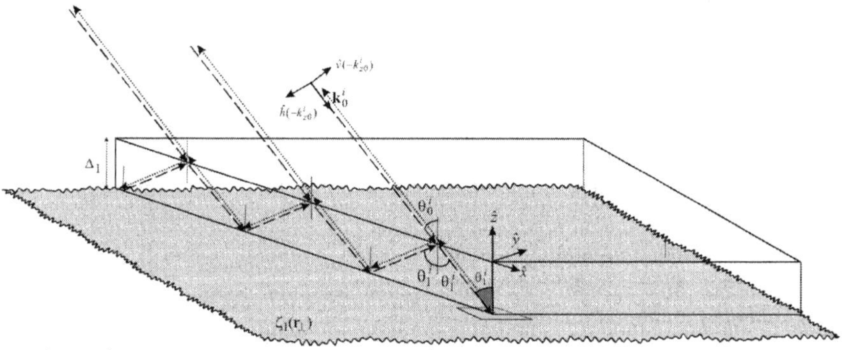

Fig. 6. Physical Interpretation: mono-static configuration

As a result, the obtained interpretation (Figure 4) enables the *global* scattering phenomenon involved to be visualized as a superposition of *local* interactions, emphasizing the role of the interference effects in the structure as well (Imperatore et al. 2009c). It should be also noted that, despite the expansion is attained rigorously without any further approximation with respect to the solution proposed (see (54)-(58)), the resulting interpretation turns out to be extremely intuitive and surprisingly simple. In particular, when the configuration reduces to a rough interface covered by a dielectric layer, as the reader can easily verify, we obtain the interpretation (Franceschetti et al., 2008) for the bistatic and monostatic configuration illustrated in Figure 5 and Figure 6, respectively.

11. Scattering patterns computation

In this section, we present some numerical examples aimed at studying scattering coefficients (59). To this purpose, we consider the canonical layered media with three rough interfaces pictured in Figure 8, which is representative of several situations of interest. In common with classical theoretical studies of the scattering of waves from random surfaces, we assume that the interfaces constitute Gaussian 2D random processes with *Gaussian correlations*, whose spectral representation is given by

$$W_n(\mathbf{\kappa}) = (\sigma_n^2 l_n^2 / 4\pi)\exp(-|\mathbf{\kappa}|^2 l_n^2 / 4) \tag{74}$$

where, with regard to the nth interface, σ_n and l_n are the surface height standard deviation and correlation length, respectively. In order to perform a consistent comparison, we refer to interfaces with the same roughness. In addition, we suppose no correlation between the interfaces. For instance, we analyze the layered medium with three rough interfaces schematized in Figure 7, which can be parametric characterized as follows. We assume $k_0 l_n=1.5$, $k_0\sigma_n=0.15$ for $n=0, 1, 2$. In addition, the considered vertical profile is characterized by the following parameters: $\varepsilon_0=1$, $\varepsilon_1=3.0+j0.0$, $\varepsilon_2=5.5+j0.00055$, $\varepsilon_3=10.5+j1.55$; $\Delta_1/\lambda=1.50$, $\Delta_2/\lambda=2.80$ (see Table 1). Once this reference structure has been characterized, we study the scattering cross section of the structure as a function of the scattering direction in the upper half-space, assuming fixed the incident direction. It should be noted that, also considering a limited number of layers, the number of parameters involved by the model makes difficult the jointly visualization of the multi-variables dependency. As matter of fact, once the structure has been parametrically defined and incident direction has been fixed, it is possible to visualize the scattering cross section of the structure as a function of the scattering direction in the upper half-space (Imperatore et al. 2008c). Therefore, to characterize the re-irradiation pattern of the structure in three-dimensional space, scattering cross-section distributions are represented (Figure 8) as function of zenithal and azimuthal angles and are treated as three-dimensional surfaces. To save space, only the case *hh* is considered. In addition, we assume fixed the incidence angle $\theta_0^i = 45°$ ($\hat{k}_\perp^i = \hat{x}$). Therefore, to evaluate the effect on the global response of each rough interface, the several single contributions are shown in Figure 8a, Figure 8b, and Figure 8c, respectively. In addition, the total contribution is also pictured (Figure 8.d). It should be noted that to visualize the patterns an offset of +40dB has been considered for the radial amplitude, so that scattering coefficients less than -40dB are represented by the axes origin.

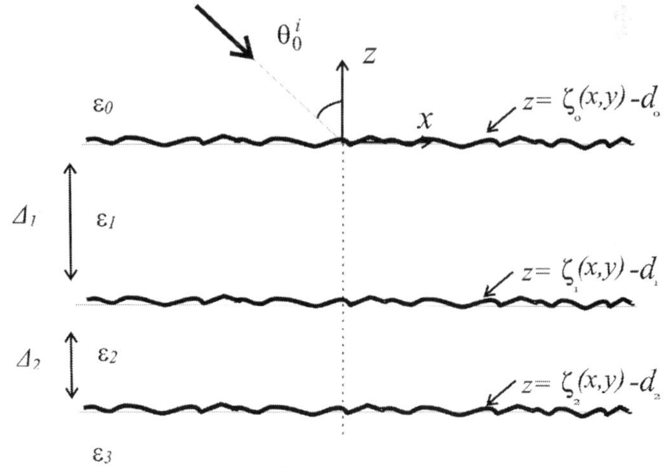

Fig. 7. Three rough interfaces layered media

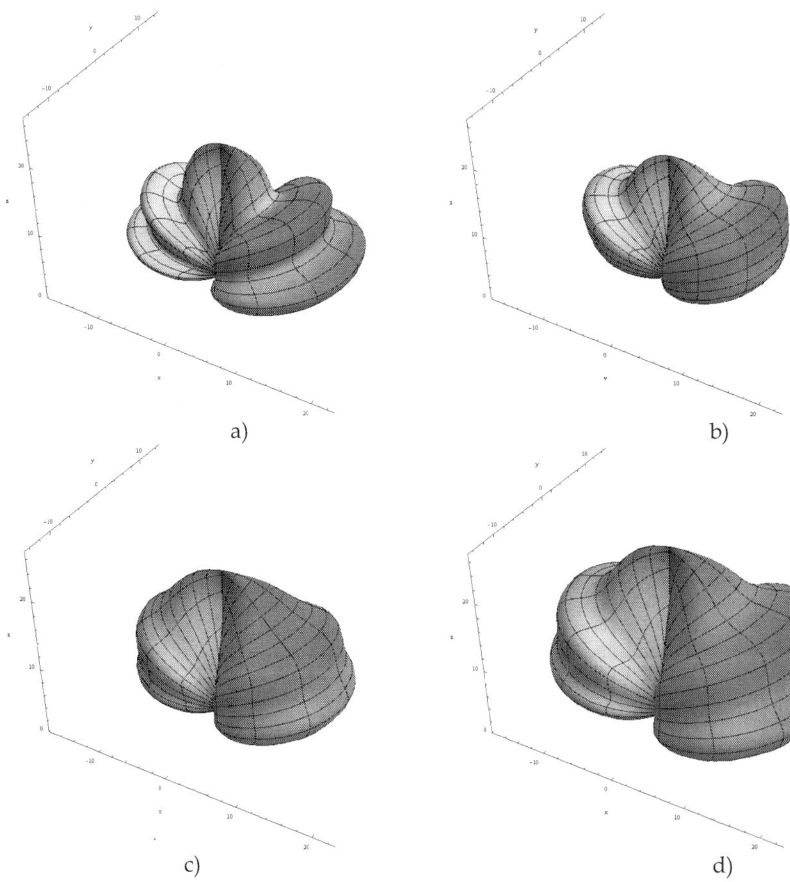

Fig. 8. Bi-static scattering coefficients *hh* for a three rough interfaces layered media: ζ_0 contribution (a), ζ_1 contribution (b), ζ_2 contribution (c), total contribution (d) (note that scattering coefficients values less than -40 dB are represented by the axes origin).

θ_0^i	45°	$k_0\sigma_0$	0.15
Δ_1/λ	1.50	$k_0\sigma_1$	0.15
Δ_2/λ	2.80	$k_0\sigma_2$	0.15
f	1.0 GHz	$k_0 l_0$	1.5
ε_1	3.0	$k_0 l_1$	1.5
ε_2	5.5+j 0.00055	$k_0 l_2$	1.5
ε_3	10.5+j 1.55		

Table 1. Parametric characterization of the layered media of Figure 7

12. Conclusion

A quantitative mathematical analysis of wave propagation in three-dimensional layered rough media is fundamental in understanding intriguing scattering phenomena in such structures, especially in the perspective of remote sensing applications. The results of the *Boundary Perturbation Theory* (BPT), as introduced by P. Imperatore and his coauthors in many different papers, essentially constitutes the content of this chapter in which the theoretical body of results is presented in organized manner, emphasizing the applications perspective. These formally symmetric and physically revealing analytical results are crucial and will contribute to innovatory applications in microwave remote sensing. For instance, they open the way toward new techniques for solving the inverse problem, for designing SAR processing algorithms, and for modelling the time-domain response of complex layered structures.

13. References

Franceschetti G., Imperatore P., Iodice A., Riccio D., and Ruello G. (2003), Scattering from a Layered Medium with One Rough Interface: Comparison and Physical Interpretation of Different Methods, *Proceedings of IEEE International Geoscience and Remote Sensing Symposium (IGARSS 2003), Toulouse, France,* pp. 2912–2914, Jul. 2003.

Franceschetti G., Imperatore P., Iodice A., Riccio D., and Ruello G. (2008). Scattering form Layered Strucutures with One rough Interface: A Unified Formulation of Perturbative Solutions. *IEEE Trans. on Geoscience and Remote Sensing,* vol.46, no.6, June 2008, pp.1634-1643.

Imperatore P., Iodice A., Riccio D. (2008a), Perturbative Solution for the Scattering from Multilayered Structure with Rough Boundaries, *Proceedings of Microwave Radiometry and Remote Sensing of the Environment* (MICRORAD 2008), pp.1-4, 11-14 March 2008, Florence, Italy.

Imperatore P., Iodice A., Riccio D. (2008b), Small Perturbation Method for Scattering From Rough Multilayers, *Proceedings of IEEE International Geoscience and Remote Sensing Symposium,* 2008 (IGARSS 2008), vol.5, pp.V-271-V-274, Boston 7-11 July 2008.

Imperatore P., Iodice A., Riccio D. (2008c), Interpretation of Perturbative Solution for the Scattering From Layered Structure With Rough Interfaces, *Proceedings of IEEE International Geoscience and Remote Sensing Symposium,* 2008 (IGARSS 2008), vol.4, pp. IV -1141-IV-1144, Boston 7-11 July 2008.

Imperatore P., Iodice A., Riccio D. (2009a). Electromagnetic Wave Scattering From Layered Structures with an Arbitrary Number of Rough Interfaces. *IEEE Trans. on Geoscience and Remote Sensing,* vol.47, no.4, pp.1056-1072, April 2009.

Imperatore P., Iodice A., Riccio D. (2009b). Transmission Through Layered Media With Rough Boundaries: First-Order Perturbative Solution. *IEEE Transaction on Antennas and Propagation,* vol.57, no.5, pp.1481-1494, May 2009.

Imperatore P., Iodice A., Riccio D. (2009c). Physical Meaning of Perturbative Solutions for Scattering from and Through Multilayered Structure with Rough Interfaces. *submitted to IEEE Trans. on Antennas and Propagation.*

Sarabandi K., Azadegan R. (2000), Simulation of interferometric SAR response to rough surfaces covered with a dielectric layer, *Proceedings* IEEE *IGARSS,* 24-28 July 2000 pp.1739 - 1741 vol.4.

Azadegan R. and Sarabandi K. (2003), Analytical formulation of the scattering by a slightly rough dielectric boundary covered with a homogeneous dielectric layer, *Proceedings of IEEE AP-S Int. Symp.*, Columbus, OH, Jun. 2003, pp. 420–423.

Fuks I. M.(1998), Radar contrast polarization dependence on subsurface sensing, *Proceedings of IEEE IGARSS'98*, vol. 3, Seattle, WA, USA, July 6–10, 1998, pp. 1455–1459.

Fuks I.M. (2001). Wave diffraction by a rough boundary of an arbitrary plane-layered medium. *IEEE Transaction on Antennas and Propagation,* vol. 49, no.4, Apr. 2001. pp.630–639.

Yarovoy A.G., de Jongh R.V., Ligthard L.P. (2000). Scattering properties of a statistically rough interface inside a multilayered medium. *Radio Science*, vol.35, n.2, 2000, pp. 445–462.

Tabatabaeenejad A., Moghaddam M. (2006). Bistatic scattering from three-dimensional layered rough surfaces. *IEEE Trans. on Geoscience and Remote Sensing*, vol. 44, no. 8, pp. 2102–2114, Aug. 2006.

Chew W. C. (1997). *Waves and Fields in Inhomogeneous Media*. IEEE Press, 1997.

Ishimaru A. (1978). *Wave Propagation and Scattering in Random Media*, vol.2. Academic Press, New York, 1978.

Tsang L., Kong J. A, and Shin R. T. (1985). *Theory of Microwave Remote Sensing*. New York: Wiley, 1985.

Ulaby F. T., Moore R. K., and Fung A. K. (1982). *Microwave Remote Sensing*, vol. I, II, II. Reading, MA: Addison-Wesley, 1982.

12

Highly accurate geometric correction for NOAA AVHRR data

An Ngoc Van
Institute of Industrial Science, University of Tokyo
Japan

Mitsuru Nakazawa and Yoshimitsu Aoki
Graduate School of Science and Technology, Keio University
Japan

1. Introduction

Since 1970, AVHRR (Advanced Very High Resolution Radiometer) on board the NOAA (National Oceanic and Atmospheric Administration) series of satellites has been an ideal observatory for daily global observation of the Earth. NOAA AVHRR data provides very useful information about ecosystems, climate, weather and water from all over the world. It is also widely used for land cover monitoring at global and continental scales.

NOAA AVHRR data are presented in the image coordinate system. Frequently, however, information extracted from AVHRR data is integrated with map data or given to consumers in a map-like form. Therefore, it is necessary to transform NOAA AVHRR data from the image coordinate system into the map coordinate system. In those applications using NOAA AVHRR data, geometric correction with high accuracy plays a very important role to ensure that NOAA AVHRR data is precisely transformed from one coordinate system to another.

Some geometric correction methods for NOAA AVHRR data, or NOAA images, have been proposed. The most popular methods can be divided into two types: orbital geometry model and transformation based on ground control points (GCPs). In the former, the knowledge about the characteristics of the satellite is used to build a physical model that defines the sources of error and the direction as well as the magnitude of their effects. However, this type of method is based on only nominal parameters. It takes into account only selected factors that cause geometric distortion. Variations in the altitude or attitude of the satellite are not considered because the information needed to correct caused by these variations is not generally available (Mather, 2004). The latter looks at the problem from the opposite of view. Rather than attempt to construct the physical model that produces errors, it uses an empirical method to compare the differences between the positions of GCPs, which can be identified both on the image and on the map of the same area. From the differences between the distributions of GCPs on the image and on the map, the errors can be estimated and removed (Mather, 2004). Recently, precise geometric correction method (Ono & Takagi, 2001; Takagi, 2003), which uses GCP template matching and considers elevation effect, has obtained accurate results by considering residual errors and elevation effect when acquiring the errors, as well as applying affine transform to correct them. Although the precision of this method is high, the errors still exist on the rough regions after correction.

In order to improve the precision of the geometric correction for NOAA AVHRR data, this book chapter is dedicated to introduce a novel highly accurate geometric correction method. Section 2 of this chapter will write about NOAA AVHRR data, the steps of NOAA AVHRR data processing (including geometric correction) and the applications relating to NOAA AVHRR data. Section 3 will review the current geometric correction methods, which used orbital geometry model or GCP model, to point out the advantages and disadvantages of each type of those methods. Based on the analysis in section 3, section 4 will introduce a novel highly accurate geometric correction for NOAA AVHRR data, which takes the advantage of the method using GCP and adds more input data to make the result more accurate. This section will also give some discussions on the novel method. The final section will be conclusion and future works.

2. NOAA AVHRR data

2.1 NOAA POES satellites

Eight years from establishment, in October 1978, NOAA launched TIROS-N, their first POES satellite. This series of satellites continued with the satellites from NOAA-6 to NOAA-14, which were launched from to 1979 to 1994. In the spring of 1998, a new series of NOAA POES satellites commenced with the launch of NOAA-K. NOAA-K and its immediate successors, such as NOAA-L and NOAA-M, NOAA-N, represented an improvement over the previous series of satellites. Figure 1 is a NOAA POES satellite (NASA, 2002).

Fig. 1. A NOAA POES satellite

POES satellite's orbit is near-polar orbit, which means the satellite travels from the pole to pole, and the angle between the equatorial plane and the satellite orbital plane is nearly 90 degrees. With the altitude between $830km$ and $870km$, NOAA POES satellites can cover the most difficult-to-cover parts of the world. NOAA POES satellites operate in a sun-synchronous orbit, so they pass the equator and each latitude at the same local solar time each day. This feature enables regular data collection at consistent times as well as long-term comparisons (NOAA, 1998)(NOAA, 2000). NOAA has two polar-orbiting meteorological satellites in orbit at all times (NOAASIS-a, 2008). Together they provide twice-daily global coverage, and ensure that data for any region of the earth are no more than 6 hours old. The swath width, the width of the area on the Earth's surface that the satellite can "see", is approximately $2,500km$ (NOAA, 1998)(NOAA, 2000).

2.2 AVHRR sensor

NOAA POES satellites use Advanced Very High Resolution Radiometer (AVHRR) as the instrument to collect data from the Earth. AVHRR uses passive detection. Each sensor on board an AVHRR instrument is corresponding to a channel, which is designed to record information from a different range of wavelengths, from visible to infrared. The first AVHRR was a 4-channel radiometer, first carried on TIROS-N (launched October 1978). This was subsequently improved to a 5-channel instrument (AVHRR/2) that was initially carried on NOAA-7 (launched June 1981). The latest instrument version is AVHRR/3, with channel 3 is divided into 3A and 3B, first carried on NOAA-K launched in May 1998. The sensors scan line by line as the satellite moves forward along its track. This forward motion of the satellite is used to build up an image by the collection of the scan lines.

2.3 NOAA AVHRR data

After the information of Earth's surface is recorded by the sensors, it is sent from the satellite to the ground receiving stations in a raw digital format. This data is NOAA AVHRR data and is stored in a digital image (NOAA AVHRR image or NOAA image) including the recorded data lines. Figure 2 is the channel 4 of a NOAA image received in Tokyo, Japan in August 1998.

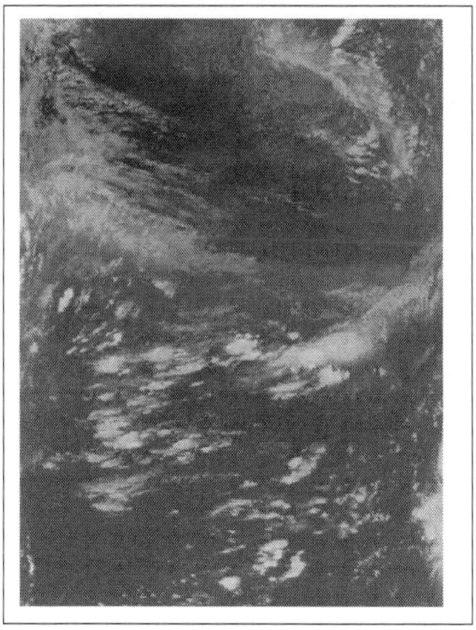

Fig. 2. Channel 4 of a NOAA image received in Tokyo, Japan in August, 1998

A NOAA image contains the lines scanned by the AVHRR sensors. Each line includes the data from all 5 channels of the AVHRR sensors. Every channel in a line contains 2048 pixels. The value of a pixel is coded in 10 bits. The highest spatial resolution of NOAA AVHRR image is

$1.1km$, which means that one pixel at the nadir of the satellite is corresponding with an area of $1.1km$ by $1.1km$. The resolution of the pixels at the edges of the images is lower (Mather, 2004).

2.4 Data processing

Many image processing and analysis techniques have been applied to extract as much information as possible from the NOAA images. These techniques often include three basic steps: preprocessing, image enhancement, and information extraction. The data processing begins with the correction of the errors by some degree of preprocessing including one or more of cosmetic correction, geometric correction, atmospheric correction and radiometric correction. After preprocessing step, image enhancement operations are carried out to improve the interpretability of the image by increasing apparent contrast among various features in the scene. Information Extraction is the last step toward the final output of the remote sensing data processing.

Both image enhancement and information extraction steps are deployed on the output of preprocessing step. Therefore, the accuracy in the data preprocessing, including geometric correction, plays a very important role in the precision of remote sensing data processing.

2.5 The applications of NOAA AVHRR data

Because of the polar orbiting nature of the POES series satellites, these satellites are able to collect global data on a daily basis for a variety of land, ocean, and atmospheric applications. NOAA AVHRR data supports a broad range of environmental monitoring applications including weather analysis and forecasting, climate research and prediction, global sea surface temperature measurements, atmospheric soundings of temperature and humidity, ocean dynamics research, volcanic eruption monitoring, forest fire detection, global vegetation analysis, search and rescue, and many other applications. Nowadays, NOAA AVHRR data is also combined with high-resolution remote sensing data such as MODIS, LANDSAT TM, SPOT HRVÉ to extract very useful information.

3. Review on geometric correction for NOAA AVHRR data

3.1 Geometric distortions

NOAA AVHRR data, as other types of remote sensing data, obtained by the satellite often contains geometric distortions. These distortions can be caused by the Earth's rotation, the velocity variations in the scanning process and forward movement of the satellite during the acquisition of data, the variations of Earth's elevation and the orbit height. Geometric distortions are generally divided into two types: systematic distortion and non-systematic distortion (Gibson & Power, 2000).

Systematic distortions are introduced by the remote sensing system itself or in combination with Earth's rotation or curvature characteristics. These distortions include: skew, variation in ground resolution, and displacement or scale distortion. Most systematic distortions can be corrected using the data obtained from accurate monitoring of the satellite's orbital path and the knowledge of the scanning system's characteristics.

Non-systematic distortions are usually introduced by phenomena that vary in nature through space and time. These distortions result mainly from variations in the satellite's orbit and terrain's features, which usually involve altitude and attitude changes. Geometric non-systematic distortions can be corrected only by using ground control points (Murai, 1980). A ground control point (GCP) is a location on the surface of the Earth (e.g., a road intersection, a river intersection) that can be identified easily on the image and located accurately on a map. The image analyst must be able to obtain two distinct sets of coordinates associated with each GCP, which are image coordinates (e.g., specified in i rows and j columns) and map coordinates (e.g., x, y measured in degrees of latitude and longitude). The paired coordinates (i, j and x, y) from many GCPs can be modeled to derive geometric transformation coefficients to correct the distortions (Mather, 2004).

Figure 3a is an example of the skew distortion caused by the Earth rotation effect. The dotted rectangle is the image received from satellite; the solid parallelogram is the data after correction. Due to the Earth's eastwards rotation, the start of each scan line is displayed slightly westwards. Figure 3b is an example of the distortion caused by altitude change. The dotted rectangle is the image received from satellite; the solid trapezium is the data after correction. Due to the altitude change, a trapezium on the Earth's surface becomes a square in the remote sensing image.

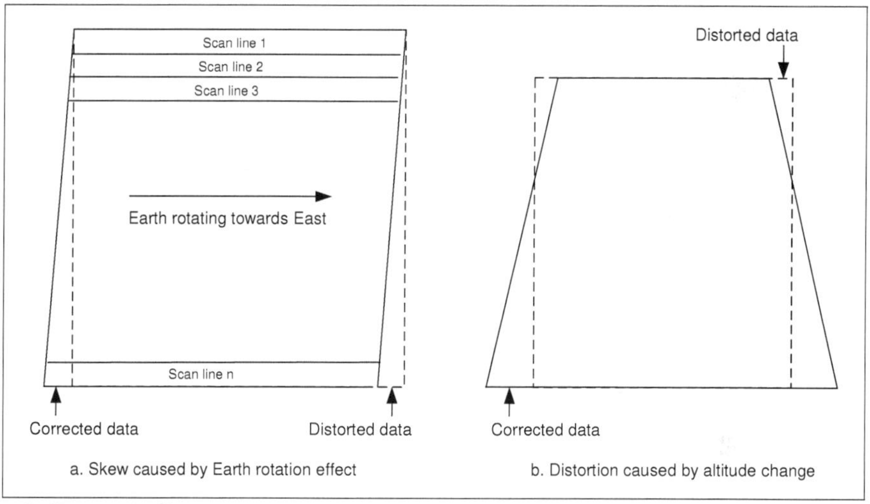

Fig. 3. Geometric distortions

3.2 Geometric corection

Geometric correction process can be considered to include these operations: (1) determination of a relationship between the coordinate system of map and image or image and image, (2) establishment of a set of points defining pixel centers in the corrected image that, when considered as a rectangular grid, define an image with the desired cartographic properties, and (3) estimation of pixel values to be associated with those points (Mather, 2004).

Based on these operations, a lot of geometric correction methods have been proposed. Generally, geometric correction methods can be divided into two types: the methods based on the nominal parameters and the methods based on the map (Mather, 2004).

In the former, the knowledge about the characteristics of the satellite's orbit is used to build a physical model that defines the sources of error and the direction as well as the magnitude of their effects. For example, Bannari (Bannari et al., 1995) used the properties of the satellite orbit and the viewing geometry to relate the image coordinate system to the geographical coordinate system. Another method that corrects the coordinate system of remote sensing images using approximate orbit parameters was proposed by Landgrebe (Landgrebe, 1980). In this method, the orbital parameters are used to express and correct the distortions, such as the skew caused by Earth rotation effect. This type of methods is based upon only the nominal parameters than the actual orbital parameters. Only the selected factors that cause geometric distortions are taken into account. The variations in the altitude or attitude of the satellite are not considered because the information needed to correct for these variations is not generally available.

The latter looks at the problem from the opposite of view. Rather than attempt to construct the physical model that defines the sources of errors and the direction and magnitude of their effects, an empirical method is used which compares the differences between the positions of the GCPs on the image and on the map. From the differences between the distribution of GCPs on the image and the distribution of GCPs on the map, the nature distortions present in the image can be estimated, and an empirical transformation to relate the image and map coordinate systems can be computed and applied. This empirical function should be calibrated (using GCPs), applied to the image, and then validated (using separate test set of GCPs) (Mather, 2004). The correction methods based on map is used to correct the non-systematic distortions, which are caused by attitude and altitude change.

For NOAA AVHRR data, both types of geometric correction methods are described (Brush, 1985; Crawford et al., 1996; Moreno & Melia, 1993; Purevdorj & Yokoyama, 2000; Saitoh et al., 1995; Tozawa, 1983). However, most of them did not consider or considered insufficiently the non-systematic distortions produced by the altitude change, particularly the terrain relief. As mentioned by Mather (Mather, 2004), the altitude change can cause very considerable distortions in the satellite images. If this type of distortion can be corrected, the results of geometric correction will be improved significantly. Therefore, the method to correct this type of distortion in NOAA images is important.

Recently, precise geometric correction method (Ono & Takagi, 2001; Takagi, 2003; Yasukawa et al., 2004), which is based on GCPs and considers the elevation effect, has obtained accurate results. In this method, the variation effect is considered to acquire the distortion more precisely. The next section will explain the precise geometric correction method in more detail.

3.3 Precise geometric correction for NOAA AVHRR data

In the precise geometric correction method (Ono & Takagi, 2001; Takagi, 2003), because the error correction in the image coordinate system is more accurate than the one in the map coordinate system, the errors in NOAA images are corrected in the image coordinate system before the images are transformed into the map coordinate system. Figure 4 shows the steps of this method.

3.4 GCP Template Matching considering elevation effect

3.4.1 Elevation effect

Elevation effect is caused by elevation error. Figure 5 is an example of the schematic illustration of the situation that the elevation error occurs (O is the center of Earth, $Re = 6378.14km$ is the Radius of Earth, $r = 7228km$ is the distance OC). There is a mountain at A, the top of this mountain is D whose height is h. The data the satellite receives from D should be assigned to A with the scanning angle θ_1. However, because the satellite does not take the height h into account, the data it receives from D is assigned to A' with scanning angle θ_2. In this case, AA' is elevation error.

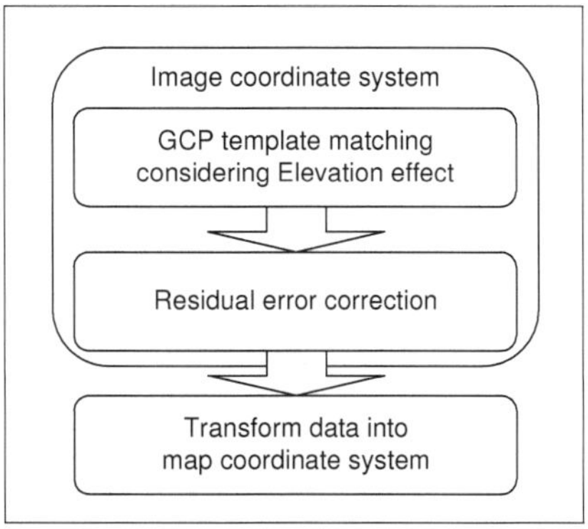

Fig. 4. Steps of the Precise Geometric Correction Method

In the rectangle ODC, DC is found based on θ_1, Re, r and h; and the relationship between θ_2 and the sides is represented by (1). Because $OC = r$ and $OD = Re + h$, θ_2 is found as (2).

$$OD^2 = DC^2 + OC^2 - 2.DC.OC.cos\theta_2 \quad (1)$$

$$\theta_2 = \cos^{-1}\left[(DC^2 + r^2 - (Re + h)^2)/(2.r.DC)\right] \quad (2)$$

As point out in (Takagi, 2003), the difference of the scan angles $(\theta_2 - \theta_1)$ is directly proportional to the elevation error; furthermore, because OC ($7228km$) is much bigger than BC (satellite's height: $830km$), the angles ϕ_1 and ϕ_2 in the figure 2 are very small. Therefore, AA' can be calculated as (3) (unit: pixel) (Takagi, 2003), where 1024 (pixels) is the half width of a NOAA images, and maximum scan angle of AVHRR sensor is 55.4°.

$$AA' = (\theta_2 - \theta_1) * 1024/55.4 \quad (3)$$

The elevation errors are dominant in longitude direction near equator and in latitude direction near polar region in the map coordinate system, which corresponds to the pixel direction in the image coordinate system. The higher the area is, the bigger its elevation error is. The elevation data is read from GTOPO30 dataset, whose resolution is 0.008 degree in both latitude and longitude directions. This elevation data can be used to correct the elevation errors in NOAA images because the resolution of AVHRR mapping system is 0.01 degree.

3.4.2 GCP Template Matching

In the precise geometric correction method, it is very important to accurately acquire the residual errors, which present the displacement of a point from incorrect to correct position. Residual errors are measured by GCP template matching. The information about GCP templates is stored in the Digital Chart of the World. GCP templates are selected from the coastlines (such as the coastline of islands) and the inland objects (such as the coastline of lakes or rivers), which have robust features to specify the displacement (Takagi, 2003).

GCP template matching starts with the generating of GCP templates in the map coordinate system. The size of each template is 33x33 pixels. These templates are then transformed into the image coordinate system. In order to correct the elevation errors for GCP templates, each point in every template is moved in the pixel direction according to its elevation error.

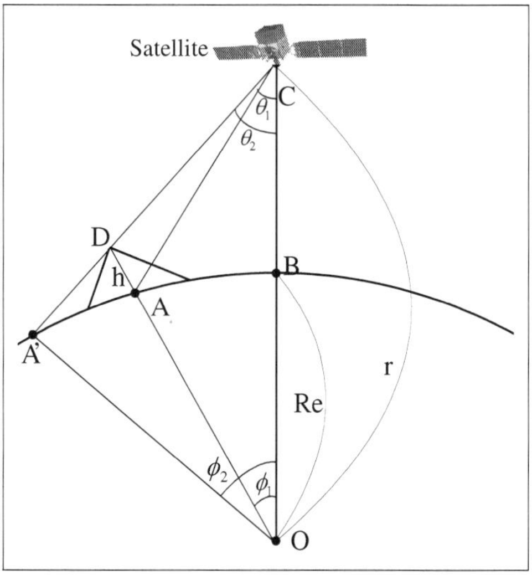

Fig. 5. Schematic illustration of the situation that elevation error occurs

Next, the pieces of 65x65 pixels of the original image corresponding to GCP templates are selected. They are binarized to create the binary and edge images. After binarizing, the piece 65x65 pixels and the GCP template are matched by SSDA (Takagi, 2003) to find out the best matching position.

3.5 Residual Error Correction

In order to specify the residual errors, the positions of the coastlines in the image coordinate system are generated from the coastline data in the map coordinate system considering the elevation effect. The coastlines are then compared with the GCP templates, whose positions are already matched by GCP template matching. The displacements from the positions of GCP templates to the coastlines are residual error vectors, which are used to present residual errors. Finally, the affine transform are used to correct the residual error vectors (Ono & Takagi, 2001; Takagi, 2003).

Affine transform is used to calculate the error $(\Delta l, \Delta p)$ for each point (l, p) in the image coordinate system. The error $(\Delta l, \Delta p)$ is calculated as follow:

$$\begin{pmatrix} \Delta l \\ \Delta p \end{pmatrix} = \begin{pmatrix} \alpha_{11} & \alpha_{12} \\ \alpha_{21} & \alpha_{22} \end{pmatrix} \begin{pmatrix} l \\ p \end{pmatrix} + \begin{pmatrix} \beta_1 \\ \beta_2 \end{pmatrix} \quad (4)$$

where α_{11}, α_{12}, α_{21}, α_{22}, β_1 and β_2 are the coefficients which are specified based on residual error vectors by the precise geometric correction method (Ono & Takagi, 2001).

The correct position of the point (l, p) will be calculated by (5):

$$\begin{cases} l &= l + \Delta l \\ p &= p + \Delta p \end{cases} \quad (5)$$

According to the correction process, because affine transform is applied to calculate error vectors, the variation of elevation is considered as a linear variation on all regions. However, in fact, the variation of elevation on rough regions is complicated; therefore, affine transform will produce incorrect results, especially on the rough regions.

3.6 Transforming Data into the Map Coordinate System

In the final step, NOAA data in the image coordinate system is transformed into the map coordinate system. In order to transform, the data in the map coordinate system is divided into the blocks of 16x16 points and the position in the image coordinate system of each block is calculated. Figure 6 is an example of this transformation. Figure 5a is a block of 16x16

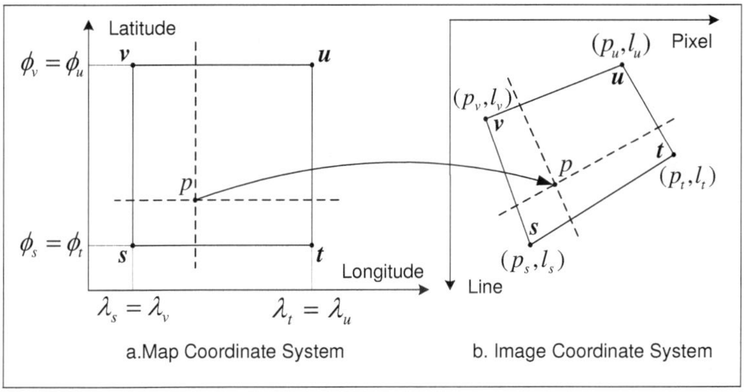

Fig. 6. Bilinear Interpolation

points in the map coordinate system; the position in the image coordinate system of this block is shown in the figure 5b. Firstly, the positions in the image coordinate system of four corner points s, t, u and v in this block are precisely calculated based on the parameters of the satellite and earth's location. Next, the position in the image coordinate system of the point p inside this block is interpolated from four corner points by using bilinear interpolation with (6) (Ono & Takagi, 2001), where E_s, E_t, E_u, E_v are the elevation errors of s, t, u, v, correspondingly. This process is repeated for all blocks. Finally, the data in the image are assigned to the correct positions in the map.

$$\begin{cases} l_p = \begin{pmatrix} (\phi_v - \phi_p)(\lambda_t - \lambda_p)l_s \\ +(\phi_p - \phi_s)(\lambda_t - \lambda_p)l_v \\ +(\phi_v - \phi_p)(\lambda_p - \lambda_s)l_t \\ +(\phi_p - \phi_s)(\lambda_p - \lambda_s)l_u \end{pmatrix} \\ \qquad /(\phi_v - \phi_s)(\lambda_t - \lambda_s) \\ p_p = \begin{pmatrix} (\phi_v - \phi_p)(\lambda_t - \lambda_p)(p_s + E_s) \\ +(\phi_p - \phi_s)(\lambda_t - \lambda_p)(p_v + E_v) \\ +(\phi_v - \phi_p)(\lambda_p - \lambda_s)(p_t + E_t) \\ +(\phi_p - \phi_s)(\lambda_p - \lambda_s)(p_u + E_u) \end{pmatrix} \\ \qquad /(\phi_v - \phi_s)(\lambda_t - \lambda_s) \end{cases} \tag{6}$$

Referred to (6), the elevation errors of the points inside a block are interpolated from the elevation errors of only four corner points. For this reason, (6) will produce correct result on only the flat blocks; it will be incorrect on the rough blocks, where the variation of elevation is complicated.

4. Highly accurate geometric correction for NOAA AVHRR data

By considering elevation effect, the precise geometric correction method (Ono & Takagi, 2001)(Takagi, 2003) has improved the correction result. After correction, the residual errors are considerably reduced. However, as shown in the previous chapter, the precise geometric correction method still has some problems:

- Because affine transform is applied to calculate the residual errors, the variation of elevation error is considered as a linear variation on all regions of the image. However, the variation of elevation error is complicated, especially on the rough regions, it cannot be considered as a linear variation on all regions. Therefore, affine transform will produce incorrect residual error correction result.
- By using bilinear interpolation to transform data from the image coordinate system into the map coordinate system, the elevation errors of the points inside a block are interpolated from the elevation errors of the corner points of that block. For this reason, bilinear interpolation will have correct result only on the flat blocks, where the difference of the elevation errors between the corner points and the rest points inside the block is small; it will be wrong on the rough blocks, where the variation of elevation is big.

In order to increase the precision of geometric correction for NOAA AVHRR data, this section proposes a new geometric correction method. The proposed method solves the problems of the precise method by considering the variation of elevation. Depending on the variation of elevation, residual errors are corrected by different transforms on different regions; data is also transformed into the map coordinate system by different methods on different regions. Additionally, to make GCP template matching to be more precise, more GCP templates are automatically generated from the feature of the coastline.

4.1 Steps of the proposed method

Figure 7 shows the steps of the proposed method.

4.2 Finding flat and rough blocks

As mentioned above, when using affine transform to calculate the residual errors, the results on the rough regions are not good because the variation of elevation is complicated on these regions and affine transform is not able to express it; therefore, affine transform should be applied only on the flat regions, where the difference of elevation between to neighboring points is small. On the other hand, when transforming data from the image coordinate system into the map coordinate system by using bilinear interpolation, because the elevation error of a point inside a block is interpolated from the elevation errors of four corner points, bilinear interpolation will produce errors on rough regions. For these reasons, to identify flat and rough regions is necessary to reduce the error.

In order to find out the flat and rough block, the data in the map coordinate system is divided into blocks of 16x16 points. The elevation data of every point in each block are read from GTOPO30 dataset. The highest and lowest points in each block are compared to decide if it is flat or rough block. If a block is rough, it will be divided into four same-size smaller blocks. The dividing process is repeated on the rough blocks until the sizes of all rough blocks are 4x4 points.

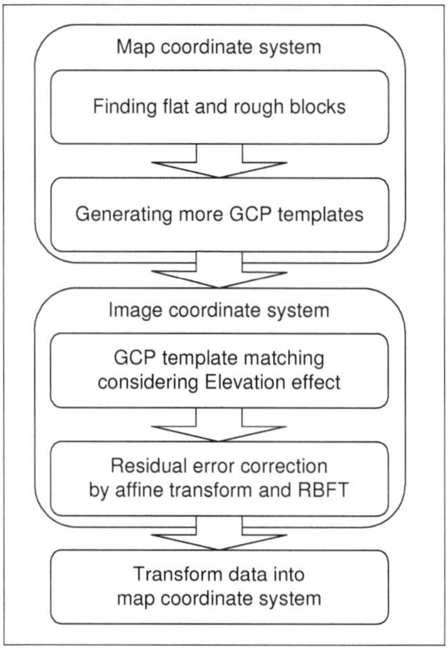

Fig. 7. Steps of the proposed method

Figure 8 is an example of the dividing a block into smaller flat and rough blocks. In figure 8a, an original block of 16x16 points is identified as a rough block, and it is divided into 4 smaller blocks of 8x8 points. The highest and lowest points in each smaller block are compared to find out one flat block (block 1) and three rough blocks (block 2, 3, 4). Each rough block in figure 8a continues to be divided into 4 smaller blocks of 4x4 points in figure 8b (from block 5 to block 16 in figure). The highest and lowest points in these 4x4 points blocks are compared to identify flat blocks (block 5, 7, 10, 14, 15, 16) and rough blocks (block 6, 8, 9, 11, 12, 13).
In this study, a block is considered as a flat block if the difference in the elevation between the lowest and the highest point is smaller than $50m$; otherwise, the block is considered as a rough block. The value $50m$ is selected as the threshold for a rough block because if the difference in elevation is greater than $50m$, the elevation error on the high altitude regions at the left or right side of the image may be greater than 0.5 pixel, and when this value is rounded, it will become 1 pixel displacement.

4.3 Generating more GCP templates

In order to acquire residual errors precisely, the number of GCP templates using in GCP template matching is very important. The greater number of GCP templates is used in GCP template matching, the more residual errors are calculated precisely by GCP template matching. On some regions, because of the cloud effect or the feature of the terrain, GCP template cannot be found or the number of GCP templates is not enough for GCP template matching. Therefore, in addition to the templates in the GCP template database, a method which automatically generates GCP templates based on the feature of the coastline described by Nakano

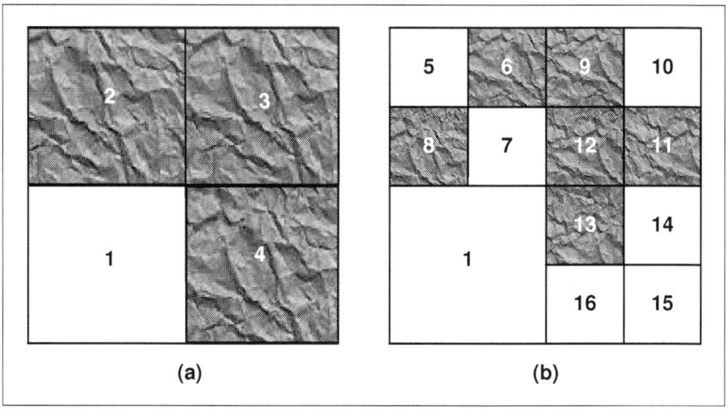

Fig. 8. Dividing data into flat and rough blocks

et al. (Nakano et al., 2004) is considered. According to this method, GCP templates are generated based on the feature of the coastline. The complexity in the shape of the coastline is taken into account to decide whether or not that region should become a GCP template.
GCP templates are generated automatically as following:
Firstly, the coastline image is created from the coastline database in the map coordinate system. A square of 33x33 points whose center is a point belonging to coastline is specified. In figure 9, a square of 33x33 points with center C_1 (belongs to the coastline) is created from the coastline image.

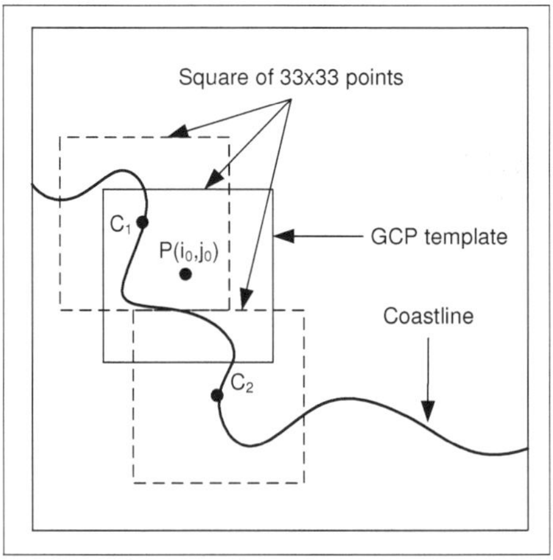

Fig. 9. Automatically generating GCP template

With every point p in this square, P is calculated as:

$$P(i,j) = \sum_{k=-r}^{r} \sum_{l=-r}^{r} c(i+k, j+l) \tag{7}$$

where (i, j) is the position of the point p, $r = 16$, and c is a function described by (8).

$$c(x,y) = \begin{cases} 1 & \text{if } (x,y) \text{ belongs to coastline} \\ 0 & \text{otherwise} \end{cases} \tag{8}$$

The values of P corresponding to the different positions of $p(i, j)$ are compared to find out the maximal value P_{max}. P_{max} is then compared with a threshold. If P_{max} is greater than that threshold, a square of 33x33 points, whose center is the point corresponding to P_{max}, is generated. And if this template is a flat region, it will be considered as a GCP template. In figure 9, $p(i_0, j_0)$ is supposed to correspond to P_{max}, and the square of 33x33 points whose center is $p(i_0, j_0)$ is generated. If P_{max} is greater than the threshold and this square locates on a flat region, it will be a GCP template.

Next, the square of 33x33 points, which locates next to the current square and whose center belongs to the coastline, is taken into account. In the figure 9, after finding GCP template in the square whose center is C_1, the square whose center is C_2 will be considered.

This process is repeated until no more squares can be found. Figure 10 (Nakano et al., 2004) is an example of the GCP templates automatically generated by this method.

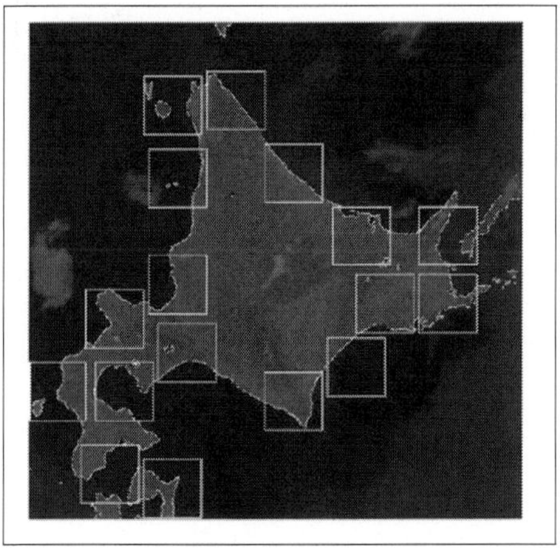

Fig. 10. A result of automatically generating GCP template

4.4 GCP template matching using reference data

Similar to the precise geometric correction method explained in the previous section, the result of GCP template matching is used to specify the residual errors. In the proposed method, both the GCP templates in the template database and the GCP templates generated automatically based on the features of the coastline are used for GCP template matching. GCP template matching is applied as it is described in the section 3.4.2.

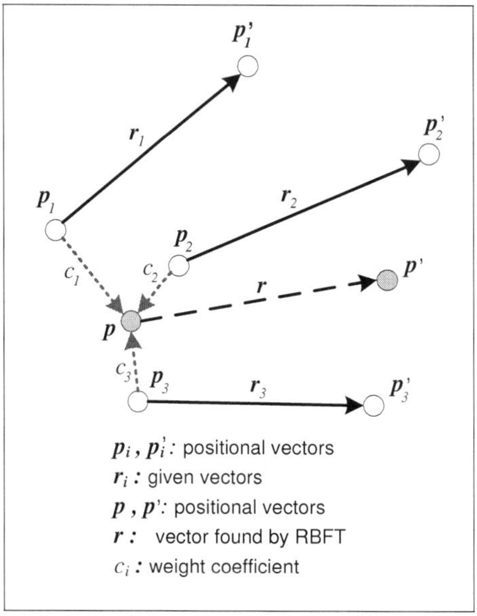

Fig. 11. Radial Basis Fucntion Transform

4.5 Residual error correction considering affine transform and Radial Basis Function Transform

After GCP template matching, the residual errors of the GCP template's center points are found. In order to calculate the residual errors for the rest points in the NOAA image, some mathematical transformation is applied.

In a flat block, because the difference of the height between the highest and lowest points are quite small, the elevations of all points in this flat block are similar, and the variation of the elevation therefore can be considered as a linear variation. For this reason, in the proposed method, affine transform is applied to calculate the residual errors for only the points in the flat blocks.

In the rough blocks, the variation of elevation is more complicated than that in the flat blocks. Thus, rather than applying affine transform, another mathematical transformation should be used to calculate the residual errors in the rough blocks. Based on the fact that the elevation error of a point is affected more strongly by the elevation errors of its neighboring points than by the elevation errors of the further points, Radial Basis Function Transform (RBFT) (Pighin et al., 1998) is used to calculate the residual errors in the rough blocks.

Figure 11 shows an example of using RBFT. Supposed that N vectors $\boldsymbol{r}_i$ are already known, $i = \overline{1,N}$. $\boldsymbol{p}_i$ and $\boldsymbol{p}'_i$ is the positional vector of the start point and end point of vector $\boldsymbol{r}_i$ respectively. Based on the vectors $\boldsymbol{r}_i$, $\boldsymbol{p}_i$ and $\boldsymbol{p}'_i$, RBFT is applied to find the vector $\boldsymbol{r}$ whose start point is expressed by the positional vector $\boldsymbol{p}$. In the formulas (9) and (10), $g(t)$ is the Radial Basis Function. Firstly, from the vectors $\boldsymbol{r}_i$ and the positional vectors $\boldsymbol{p}_i$, the weight coefficients c_j are found by (9):

$$\boldsymbol{r}_i = \sum_{j=1}^{N} c_j g(\|\boldsymbol{p}_i - \boldsymbol{p}_j\|) \tag{9}$$

Based on these weight coefficients, vector $\boldsymbol{r}$ is calculated as (10):

$$\boldsymbol{r} = \sum_{i=1}^{N} c_i g(\|\boldsymbol{p} - \boldsymbol{p}_i\|) \tag{10}$$

By using RBFT, the smaller the distance from $\boldsymbol{p}$ to $\boldsymbol{p}_i$ is, the greater the vector $\boldsymbol{r}_i$ affects to the vector $\boldsymbol{r}$.

In the proposed method, affine transform and RBFT are used to find the residual errors in the images as the following steps:

1. Use bilinear interpolation to find the positions in the image coordinate system for the points in the flat blocks. With the rough blocks, because the variation of elevation is complicated, the positions in the image coordinate system of all points in these blocks are found from the satellite's parameters and Earth's location.
2. Apply affine transform to calculate the residual errors for the points in the flat blocks.
3. With the rough blocks:

   (a) Mark the center points of GCP templates as the calculated points. All the points in the rough blocks are marked as un-calculated points.

   (b) With each un-calculated point, find its nearest GCP template.

   (c) Find the un-calculated points whose distances from them to their nearest GCP templates are the shortest.

   (d) With the points found in step (c), select the point whose elevation is the most similar to the elevation of its nearest GCP template.

   (e) Apply RBFT to the residual error vectors of the calculated points to find the residual error vector for the point found in step (d). After the residual error is found, mark this point as a calculated point.

   (f) Repeat from step (c) to step (e) until all the points in the rough blocks are calculated points.

4. All points in the image are moved to their corrected positions according to their residual error vectors.

By applying RBFT, the residual error vector of a point in a rough block is specified based on the residual error vectors of other points as well as the distances from the current point to those points. The shorter the distance between points is, the more strongly these points affect to each other. It reflects the real situation that the elevation error of a point is affected more strongly by its neighboring points' elevation errors than by the further points'.

4.6 Transforming data into map coordinate system

In the final step, data is transformed into the map coordinate system. By transforming data from the map coordinate system into the image coordinate system in the previous step, each point in the image coordinate system has its corresponding point in the map coordinate system. After moving all points in the image coordinate system to their correct positions according to their residual errors, the values of these points are assigned to their corresponding positions in map coordinate system.

As described in the previous step, because bilinear interpolation is applied only on the flat blocks, where the elevation errors of all points are fairly same, the error of bilinear interpolation is reduced. Furthermore, since satellite's parameters and Earth's location are used to transform data on the rough regions, the precision of transformation on the rough regions is high.

4.7 Result and Evaluation

The proposed method was applied to correct the errors for the NOAA images receiving in Tokyo (Japan), Bangkok (Thailand) and Ulaanbaatar (Mongolia). In order to evaluate the precision of the proposed method, the correction results of the precise geometric correction method (precise method) - described by Ono and Takagi (Ono & Takagi, 2001), and some different versions of the proposed method are compared.

In order to evaluate the correction results, some tests are performed. In each test, a number of NOAA images are corrected. With each NOAA image, 90% number of GCP templates are used for GCP template matching. After correction, the residual errors at the positions corresponding to the rest 10% number of GCP templates (which are used as the checking points) are acquired by using GCP template matching in the map coordinate system.

4.7.1 Selecting the radial basic function for RBFT

There are several types of radial basic function, which can be used for RBFT. In the proposed method, to express the fact that the elevation error of a point is affected by more strongly by the elevation errors of its neighboring points than by the elevation errors of the further points, the non-linear radial basis function whose shape is similar to the graph in the figure 12 is used. In this figure, the x axis is the distance between two points; the bigger the distance between two points is, the more weakly their errors affect to each other.

From the above shape, the most common radial basis function types including Gaussian function $g(t) = e^{-kt}$, Inverse-Multiquadric function $(1 + kt)^{-1/2}$ and Cauchy function $(1 + kt)^{-1}$ are considered. In order to select the best radial basic function for RBFT, these functions are applied to correct the distortions and the correction results are evaluated.

The value of k in each function is specified so that the residual error vectors of all the points in a NOAA image are used for RBFT. Because each NOAA image often includes 5000 lines, each line contains 2048 pixels, the maximum distance between two points in the image is around $\sqrt{(5000^2 + 2048^2)} \approx 5400$ pixels. This maximum distance should correspond to the minimum value of the radial basic function. Therefore, in the proposed method, the output of radial basic function should be nearly zero when the value of its variable is the maximum distance. In calculating, the minimum value of the radial basic function is selected to be equal to 10^{-6}.

Based on this idea, the best value of k for each radial basis function will be found by comparing the average error after correction of the tests on 100 NOAA images when the maximum distance is changed from 4000 pixels to 6000 pixels.

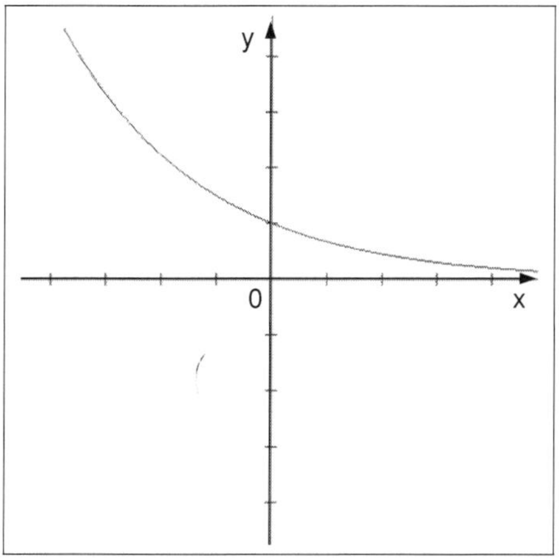

Fig. 12. The shape of the radial basic function

Table 1 is the result of this test. In this table, the average errors after correction in latitude and longitude directions are calculated for each type of radial basis function; G is Gaussian, I.M is Inverse Multiquadric and C is Cauchy; Lat is latitude and Lon is longitude.

Maximum Distance	G		I.M.		C.	
(pixels)	Lat	Lon	Lat	Lon	Lat	Lon
4000	0.25	0.28	0.31	0.35	0.33	0.40
4200	0.21	0.26	0.25	0.29	0.27	0.32
4400	0.16	0.20	0.23	0.26	0.23	0.27
4600	0.12	0.16	0.19	0.24	0.22	0.25
4800	0.15	0.17	0.22	0.25	0.23	0.26
5000	0.16	0.19	0.23	0.27	0.25	0.29
5200	0.17	0.21	0.23	0.29	0.27	0.31
5400	0.19	0.23	0.25	0.30	0.28	0.33
5600	0.21	0.25	0.25	0.32	0.30	0.34
5800	0.23	0.28	0.28	0.33	0.32	0.36
6000	0.25	0.30	0.29	0.35	0.34	0.39

Table 1. Average residual errors after correction with different maximum distances (unit: pixel)

This table shows that with each maximum distance, the Gaussian function is the best radial basic function because the average error of after correction when applying RBFT with this function is the smallest. This table also points out that the best maximum distance should be selected at around 4600 pixels, which corresponds to the value 0.003 of k in the Gaussian function.

4.7.2 Comparing affine transform and RBFT

The effectiveness of the affine transform and RBFT is compared by applying them to correct the same NOAA images and then finding the difference in the errors after correction.
In the first test, only the flat regions in 100 NOAA images are corrected by affine transform and then by RBFT. Table 2 is the results this test.

Image (Checking Points)	Value	Affine		RBFT	
		Latitude	Longitude	Latitude	Longitude
AH14060402222944 (45)	Average	0.12	0.15	0.11	0.13
	Maximum	1.00	1.00	1.00	1.00
	Minimum	0.00	0.00	0.00	0.00
AH16110602044118 (41)	Average	0.10	0.11	0.10	0.12
	Maximum	1.00	1.00	1.00	1.00
	Minimum	0.00	0.00	0.00	0.00
AH16110702183721 (46)	Average	0.12	0.14	0.11	0.14
	Maximum	1.00	2.00	1.00	1.00
	Minimum	0.00	0.00	0.00	0.00
100 NOAA images (44)	Average	0.11	0.12	0.10	0.12

Table 2. Residual errors after correction the flat regions by using affine transform and RBFT (unit: pixel)

Table 2 shows that, the average error after correction by affine transform and RBFT on the flat regions are not so different. Although the result of RBFT is a little better, affine transform will be selected for the proposed method because its computing cost is smaller than the one of RBFT.
In the second test, 100 NOAA images are corrected by the first method, which uses only affine transform, and the second method, which uses affine transform on the flat regions and RBFT on the rough regions. Table 3 is the results this test.
Table 3 shows that, compared with the correction results when applying only affine transform, the average error after correction is reduced 50% in latitude direction (0.22/0.44) and 49% in longitude direction (0.25/0.50) when applying affine transform on the flat regions and RBFT on the rough regions. It means that, compared with affine transform, RBFT reduces about 50% of the errors after correction on the rough regions.

4.7.3 Comparing the correction result of the precise method and the proposed method

In this test, the correction results of the proposed method are compared with those of the precise method. 100 NOAA images are corrected by the proposed method and the precise method. Table 4 shows the result of acquiring residual errors after correction. The average, maximum and minimum values of the residual errors in latitude and longitude direction are recorded. It is seen that both methods give highly accurate correction results, but the residual errors after correction of the proposed method are much smaller than the one of the precise method. The smaller residual errors after correction proved that the proposed method is more accurate than the precise method.
Figure 13 shows the correction results (in the map coordinate system) of the precise method (left pictures) and the proposed method (right pictures) on two regions of China where the

Image (Checking Points)	Value	Affine only		Affine and RBFT	
		Latitude	Longitude	Latitude	Longitude
AH14060402222944	Average	0.53	0.64	0.24	0.27
	Maximum	1.00	1.00	1.00	1.00
(45)	Minimum	0.00	0.00	0.00	0.00
AH16110602044118	Average	0.37	0.62	0.22	0.27
	Maximum	1.00	2.00	1.00	1.00
(41)	Minimum	0.00	0.00	0.00	0.00
AH16110702183721	Average	0.41	0.45	0.21	0.23
	Maximum	1.00	2.00	1.00	1.00
(46)	Minimum	0.00	0.00	0.00	0.00
100 NOAA images (44)	Average	0.44	0.51	0.22	0.25

Table 3. Residual errors after correction by using affine transform and by using both affine transform and RBFT (unit: pixel)

elevation are $4392m$ and $4831m$. In each picture, the black object is generated in the map coordinate system from the coastline database; the white curve is the result of geometric correction, it is the border of the black object. The black objects in all pictures should fit their white borders. The better geometric correction is, the better the black objects fit their white borders.

From the pictures in the figure 13, it is easy to see that, at both elevations ($4392m$ and $4831m$), the black objects fit their white borders in the right pictures (which are the result of the proposed method) better than those in the left pictures (which are the result of the precise method).

Figures 14 and 15 show the residual error vectors after correction (in the map coordinate system) of the precise method and the proposed method for the NOAA images AH16110602044118 and AH16110702183721, respectively. The vectors in these pictures are the residual error vectors after correction. It is clear that the residual error vectors after correction are improved by the proposed method. On some regions where the residual error vectors are big in the result of the precise method, the residual error vectors in the result of the proposed method are smaller.

4.7.4 Comparing the correction result of the proposed method and other methods

In this test, the correction results of the proposed method (method P) are compared with those of other tow methods. Method S is systematic method, which just transforms data from the image coordinate system into map coordinate system by using bilinear interpolation (Ono & Takagi, 2001). Method C is conventional method, which correct data using GCP template matching but does not consider the elevation effect (Ono & Takagi, 2001). 02 NOAA images are corrected by these methods. Table 5 shows the result of acquiring residual errors after correction. The average, maximum and minimum values of the residual errors in latitude (Lat) and longitude (Lon) direction are recorded. It is seen that the residual errors after correction of the proposed method are much smaller than other two methods.

Image (Checking points)	Value	Precise Method		Proposed Method	
		Latitude	Longitude	Latitude	Longitude
AH14060402222944 (Precise:37/Proposed:45)	Average	0.53	0.64	0.17	0.20
	Maximum	1.00	1.00	1.00	1.00
	Minimum	0.00	0.00	0.00	0.00
AH16110602044118 (Precise:36/Proposed:41)	Average	0.37	0.62	0.12	0.17
	Maximum	1.00	2.00	1.00	1.00
	Minimum	0.00	0.00	0.00	0.00
AH16110702183721 (Precise:38/Proposed:46)	Average	0.41	0.45	0.11	0.14
	Maximum	1.00	2.00	1.00	1.00
	Minimum	0.00	0.00	0.00	0.00
100 NOAA images (Precise:38/Proposed:44)	Average	0.44	0.51	0.12	0.16

Table 4. Residual errors after correction by the precise method and proposed method (unit: pixel)

Image	Value	S		C		P	
		Lat	Lon	Lat	Lon	Lat	Lon
AH12102499232725	Average	3.00	3.40	0.50	0.90	0.18	0.22
	Maximum	5.00	12.00	2.00	3.00	1.00	1.00
	Minimum	1.00	1.00	0.00	0.00	0.00	0.00
AH14092199212611	Average	1.27	3.00	0.60	0.90	0.16	0.19
	Maximum	2.00	6.00	2.00	4.00	1.00	1.00
	Minimum	1.00	1.00	0.00	0.00	0.00	0.00

Table 5. Residual errors after correction by the proposed method and other methods (unit: pixel)

4.7.5 Evaluating the contribution of the steps in the proposed method

In this test, 100 NOAA images are corrected by three methods:

- M1: this is the proposed method without generating more GCP templates
- M2: this is the proposed method

Table 6 shows the average residual errors after correction by M1 and M2. When comparing the results of M1 and the precise method (in table 4), it can be seen that M1 reduces the errors considerably by dividing data into flat and rough blocks as well as applying both affine transform and RBFT to correct residual errors. After generating more GCP templates based on the feature of the coastline, the errors after correction are considerably reduced with M2.

Image	Value	M1		M2	
		Lat	Lon	Lat	Lon
100 NOAA images	Average	0.22	0.25	0.12	0.16

Table 6. Average residual errors after correction by M1 and M2 (unit: pixel)

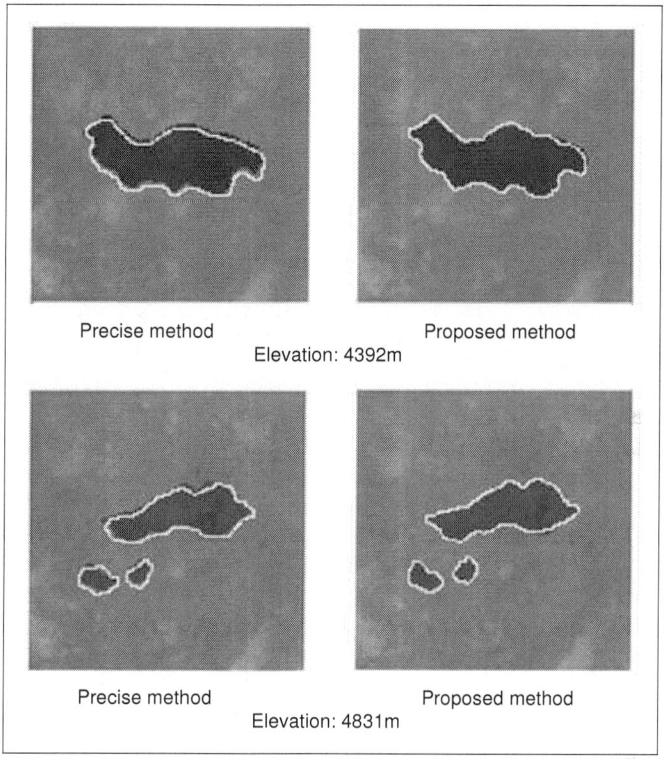

Fig. 13. Correction result of precise method and proposed method

4.7.6 Evaluating the processing time

Table 7 shows the processing time of the precise method and the proposed method. 100 NOAA images are processed by both methods on the computer Sun Ultra 45 Workstation with 1.6*GHz* Sun UltraSPARC IIIi processor and 1GB RAM. In these tests, all 5 channels of each image are corrected and the mapping unit is 0.01 degree. The maximum, minimum and average processing time of all channels are calculated. From the table 7, it is possible to say that, though the proposed method takes longer processing time, the difference between two methods is acceptable.

4.8 Discussion

This section proposed a novel geometric correction method for NOAA AVHRR data. The correction results have shown that the precision of the proposed method is high. Compared with the precise geometric correction method, which is described by Ono and Takagi (Ono & Takagi, 2001), the proposed method has some advantages:

- Data is divided into flat and rough blocks by identifying the variation of elevation. This dividing allows the errors in the different regions to be corrected in different ways.

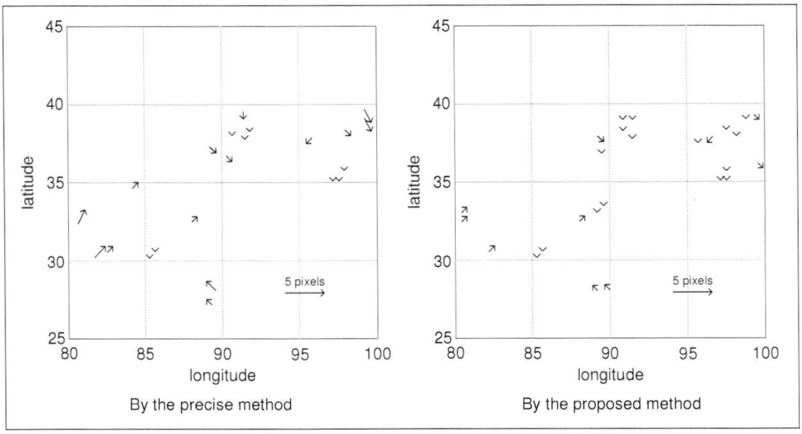

Fig. 14. Residual error vectors after correction the image AH16110602044118 by the precise method and the proposed method

Time	Precise Method	Proposed Method
Maximum	377.16 sec	390.17 sec
Minimum	359.49 sec	375.42 sec
Average	368.27 sec	381.48 sec

Table 7. Average processing time of 100 images

- Missing lines and noise pixels are corrected before applying GCP template matching. This improves the result of GCP template matching.
- The number of GCP templates is increased by automatically generating more GCP templates based on the feature of the coastline. With more GCP templates, more residual errors are acquired precisely by GCP template matching.
- Residual errors on the flat regions are corrected by affine transform; residual errors on the rough regions are corrected by RBFT with Gaussian radial basis function. With RBFT, the residual errors on the rough regions are corrected more accurately.
- Bilinear interpolation is applied only on the flat regions. This reduces the errors of bilinear interpolation on the rough regions, where the variation of elevation is big.

Besides the advantages, the proposed method also has some problems:

- If the interest region is a flat region, the precise method and the proposed method may produce the same correction results; however, the proposed method will takes longer processing time because it needs time to identify the flat and rough regions.
- There is no relationship in the residual error correction results between flat and rough regions. The residual error correction results on the flat regions are not considered when correcting residual errors on the rough regions. Therefore, there is a case that two neighboring points with the same elevation error, one belongs to a flat region, another

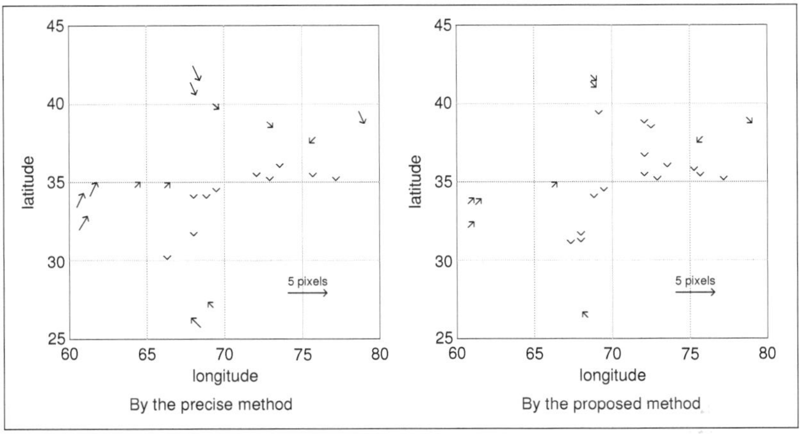

Fig. 15. Residual error vectors after correction the image AH16110702183721 by the precise method and the proposed method

belongs to a rough region, will have quite different residual errors because they are corrected by different methods.

4.9 Conclusion of future works

This book chapter proposed a novel highly accurate geometric correction method considering the variation of elevation effect.

With the proposed method, the data in the map coordinate system is divided into flat and rough blocks in order to identify the variation of elevation. More GCPs are generated based on the features of the coastline. Residual errors are specified in the image coordinate system by GCP template matching and they are corrected based on the variation of elevation. Affine and RBFT is used to correct the residual errors on flat rough blocks, respectively. For this reason, residual errors are corrected more precisely. When transforming data into the map coordinate system, bilinear interpolation is applied only on the flat blocks; satellite's parameters and earth's location are used to transform data on rough blocks. Therefore, the precision of this transformation process is higher.

With higher precision and the processing time is not too different compared to the precise geometric correction method, it is possible to say that the proposed method has higher efficiency than the precise geometric correction method. It can be integrated into some NOAA AVHRR processing applications such as PaNDA (Yasuoka and Takeuchi Laboratory, 2006) to improve the processing results. NOAA images corrected by this algorithm can also be easier to combine with other high-resolution types satellite image to complete other processing operations.

The correction results of proposed method show that the errors still exist after correction. In order to improve the correction result, some future works should be taken into account. Firstly, the number of GCPs and their precision can be increased by combination of the proposed method in this book chapter with the characteristics of the terrain, not only the feature of the coastline but also the information from the better database or from the higher resolution satellite images. Next, some new mathematical transformations should be considered to select the appropriate transformations between the image coordinate system and the map coordinate system. Finally, other geometric correction methods, which have been applied for the high resolution satellite images can be a good references for NOAA images.

5. References

Bannari, A.; Morin, D.; Benie, G. B.; Bonn, F. J. (1995). A theorical review of different mathematical models of geometric corrections applied to remote sensing images, *Remote sensing reviews*, pp.27-47, Vol.13.

Brush, R. J. H. (1985). A method for real-time navigation of AVHRR imagery, *IEEE Transactions on Geoscience and Remote Sensing*, pp.876-887, Vol.23.

Crawford, P. S.; Brooks, A. R.; Brush, R. J. H. (1996). Fast navigation of AVHRR images using complex oribtal models, *International Journal of Remote Sensing*, pp.197-212, Vol. 17.

Gibson, P. G. & Power, C. H. (200). *Introductory Remote Sensing, Digital image processing and applications*, Routledge, England.

Landgrebe, D. A. (1980). The development of a spectral spatial classifier for Earth oabservation data, *Journal of Pattern recognition*, pp.165-175, Vol.12, Issue 3.

Mather, P. M. (2004). *Computer Processing of Remotely-Sensed Images*, Third Ed., John Widley & Sons, ISBN, England.

Moreno, J. F. & Melia, J. (1993). A method for accurate geometric correction for NOAA AVHRR HRPT data, *IEEE Transactions on Geoscience and Remote Sensing*, pp.204-226, Vol.31, No. 3.

Murai, S. (1980). Geometric and Geographic Correction for Remote Sensing Data, *Proceedings of Asian Conference on Remote Sensing*, http://www.gisdevelopment.net/aars/acrs/1980/techsa/techs002pf.htm.

Nakano, M.; Kalpoma, K.; Kudoh, J. (2004). Automatic GCP creation for NOAA AHVRR geometric correction, *Proceedings of IEEE International Geoscience and Remote Sensing Symposium*, pp.3733-3735.

NASA, (2002). *NOAA-17 (M) Environmental Satellite Successfully Launched*, http://earthobservatory.nasa.gov/Newsroom/view.php?old=200206249448.

NOAA, (1998). *NOAA Polar Orbiter Data User's Guide*, http://www2.ncdc.noaa.gov/docs/podug/index.htm.

NOAA, (2000). *NOAA KLM User's guide*, http://www.ncdc.noaa.gov/oa/podguide/ncdc/docs/klm/index.htm.

NOAA Satellites and Information Service, (2008). *NOAA's Geostationary and Polar-Orbiting Weather Satellites*, http://noaasis.noaa.gov/NOAASIS/ml/genlsatl.html.

NOAA Satellites and Information Service, (2008). *Advanced Very High Resolution Radiometer - AVHRR*, http://noaasis.noaa.gov/NOAASIS/ml/avhrr.html.

Ono, T. & Takagi, M. (2001). Precise Geometric Correction of NOAA Satellite imagery Considering Elevation effects, *Proceedings of International Symposium on Remote Sensing*, pp.378-388, Vol.31, No.3.

Purevdorj, T., Yokoyama, R. (2000). Geometric Registration Method For 10-Day Composite AVHRR Data For Asian Region, *Proceedings of Asian conference on remote sensing*, http://www.gisdevelopment.net/aars/acrs/2000/ts9/imgp0015.asp.

Pighin, F.; Hecker, J.; Lichinski, D. (1998). Synthesizing realisitic facial expressions form photographs, *Proceedings of SIGGRAPH98*, pp.75-84.

Saitoh, K; Hashimoto, T.; Kajiwara, K.; Tateishi, R. (1995). Geometric Correction of NOAA AVHRR GAC Data, *Proceedings of Asian conference on remote sensing*, http://www.gisdevelopment.net/aars/acrs/1995/ts5/ts5006.asp.

Takagi, M. (2003). Precise geometric correction for NOAA and GMS images considering elevation effects using GCP template matching and affine transform, *Proceedings of SPIE Conference on Remote Sensing, Image and Signal Processing for Remote Sensing IX*, pp.132-141, Vol. 5238, Barcelona, Spain.

Tozawa, Y. (1983). Fast Geometric correction of NOAA AVHRR, *Proceedings of Symposium on machine processing of remotely-sensed data*, pp.46-53.

Yasukawa, M.; Takagi, M.; Kitsuregawa, M. (2004). High-speed and precise geometric correction for GMS S-VISSR data, *Proceedings of IEEE International Geoscience and Remote Sensing Symposium*, pp.3970-3973. Vol. 6.

Yasuoka and Takeuchi Laboratory (2006). WebPaNDA - NOAA AVHRR data processing, http://webpanda.iis.u-tokyo.ac.jp/.

13

C-band Scatterometers and Their Applications

Vahid Naeimi and Wolfgang Wagner
Vienna University of Technology
Austria

1. Introduction

Scatterometers are non-imaging active sensors used to measure the intensity of microwave backscatter while scanning the surface of the earth from an aircraft or a satellite. Active microwave sensors are radars providing their own illumination and do not depend upon ambient radiation like passive microwave sensors. They transmit microwave electromagnetic pulses toward the surface and measure how much of that signals return after interacting with the target. Scatterometer is a form of radar that is used to investigate different geophysical properties of the surface and few centimeters beneath. Spaceborne scatterometers have the advantage of providing global coverage on a continuous basis, which cannot be achieved through airborne or ground measurements. They have the capability of providing day and night time measurements unaffected by cloud cover. Scatterometers were originally designed to study ocean winds but have been also used to study of cryosphere, vegetation, and soil surface properties.

A number of scatterometers have been flown on space missions since the early 1970s. The first scatterometer in space was a Ku-band instrument on Skylab mission. Investigations on the potential use of scatterometers in geosciences achieved a major technical milestone with the launch of Seasat, carrying a Ku-band scatterometer (SASS), in 1978. Other missions have followed SASS; C-band scatterometers onboard the European Space Agency's (ESA) Earth Remote Sensing (ERS 1 & ERS-2) satellites in 1991 and 1995, the NASA's Ku-band scatterometer (NSCAT) in 1996, SeaWinds on QuikSCAT in 1999, SeaWinds on ADEOS-II in 2002, and Advanced Scatterometer (ASCAT) onboard Metop-A launched in 2006.

In this study we focus on spaceborne C-band scatterometers and present an overview of their applications in geoscience.

2. C-band Scatterometers

2.1 SCAT onboard ERS satellites

The first spaceborne C-band scatterometer was flown on ERS-1, the European Earth observation mission. ERS-1, launched in July 1991, was aimed to provide environmental monitoring particularly in the microwave spectrum. ERS-1 has been placed in a near-polar orbit at a mean altitude of about 780km with an instrument payload comprising active and passive microwave sensors and a thermal infra-red radiometer. ERS-2 the follow-up ESA mission of ERS-1 was launched in 1995. The ERS-2 satellite is a copy of ERS-1 except that it includes a number of enhancements and new payload instruments. Both scatterometers onboard ERS-1 and ERS-2 are part of an Active Microwave Instrument (AMI) operating in C-band (5.3 GHz). The AMI incorporates two separate radar systems; Synthetic Aperture Radar (SAR) and scatterometer (SCAT) operating in three different modes. SAR for Image and Wave mode operations, and scatterometer for Wind mode operation. The Wind and Wave modes are capable of interleaved operation, i.e. so-called Wind/Wave mode, but the operation in Image mode excludes the operation of the other two modes (Attema, 1991).

2.2 ASCAT onboard Metop satellites

The Advanced Scatterometer (ASCAT) is the new generation and successor of the ERS SCATs onboard the Meteorological Operational (Metop) series of satellites. Metop-A, launched on 19 October 2006, is the first satellite in the series foreseen in EUMETSAT Polar System (EPS) program (Klaes et al., 2007). Like SCAT, ASCAT system uses a fan-beam antenna technology and transmits vertically polarized pulses at frequency of 5.255 GHz with high radiometric stability. Contrary to SCAT it uses two sets of three antennas instead

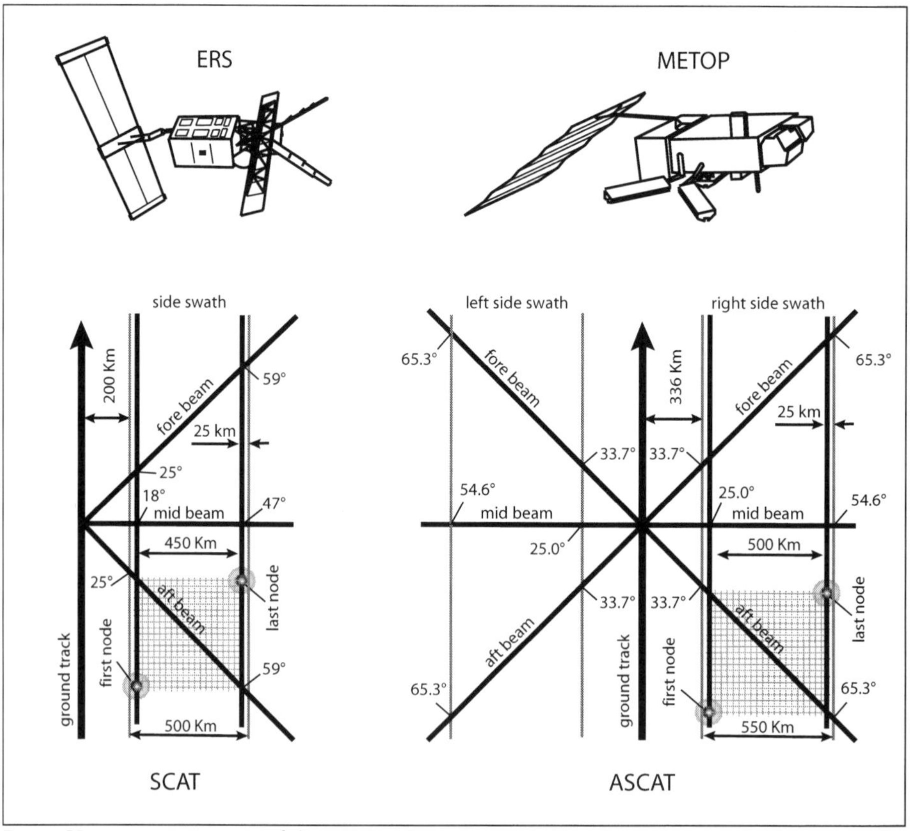

Fig. 1. Viewing geometries of the scatterometers onboard ERS and Metop satellites.

of one. For ASCAT the incidence angle range has been extended from 25° to 65°. Hence ASCAT covers two 550 km swaths to the left and right of the satellite ground track which are separated from the satellite ground track by about 336 km. This results in over twice faster global-coverage capability than its predecessor SCAT. Beside an optimized viewing geometry, ASCAT also features a number of technical improvements. The improved instrument design and radiometric performance results in higher stability and reliability of ASCAT measurements. Additionally EUMETSAT foresees to generate a research product at a resolution of 25km (Figa-Saldana et al., 2002). Figure 1 illustrates the viewing geometries of SCAT and ASCAT. Specifications of the C-band scatterometers and their carrier satellites are given in table-1.

Satellite Specifications	**ERS-1**	**ERS-2**	**Metop-A**
Launch Time	17 July 1991	21 April 1995	19 October 2006
Launch Mass	2354 kg	2516 kg	4093 kg
launcher	Ariane 4		Soyuz/ST
Spacecraft Altitude	770 to 785 km		800 to 850 km
Inclination	98.52°		98.7°
Local Solar Time	10:30 am*		9:30 am*
Orbit Period	100 minutes		101 minutes
Orbit	Near-circular, polar, Sun-synchronous		
Scatterometer Specifications	**SCAT**		**ASCAT**
Frequency	5.3 GHz. (C-Band)		5.255 GHz. (C-Band)
Polarization	VV		VV
Swath Width	500 km		550 km (double swath)
Swath Stand-off	200 km to the right of sub-satellite track		336 km
Localization Accuracy	5 km		4.4 km
Spatial Resolution	50 km		50 km, 25 km
Sampling Interval	25 km		25 km, 12.5 km
* equatorial crossing time at the descending node			

Table 1. Specifications of the European C-band scatterometers

3. Wind Speed and Direction Measurement

The primary application of the spaceborne scatterometry has been the measurement of near-surface winds over the ocean. The concept of retrieving wind speed at sea surface from the radar backscatter goes back to the Second World War. During the World War II, marine radar operators observed disturbing noises, called *"clutter"*, on their radar screens, which made them difficult detecting targets on the ocean surface (Moore et al., 1979). The clutters were the backscatter of the radar pulses from the small waves on the sea surface. Since that time many theoretical studies and experiments have been carried out to find the relationship between the microwave backscatter and the surface wind speed (Liu, 2002). The idea of remote sensing of the wind relies on the fact that winds over the sea cause small-scale disturbances of the sea surface which modify the radar backscattering characteristics. The backscatter from oceans is largely due to these small centimeter ripples, capillary waves, which is in equilibrium with the local wind stress. The backscatter depends not only on the magnitude of the wind stress but also the wind direction relative to the direction of the radar beam. By combining backscatter measurements from different azimuth angles, the near-surface wind vector over the ocean's surface can be determined using a Geophysical Model Function (GMF). The first operational GMF used for ERS-1 scatterometer data by ESA was a prelaunch transfer function denoted CMOD2, derived from aircraft-mounted instrument data (Long, 1985). An improved transfer function, CMOD4 was presented by Stoffelen et al. (1997) with full specification. CMOD4 adopted by ESA since March 1993 for wind retrieval. The latest C-band GMF used for wind retrieval is CMOD5, which is derived on the basis of measurements from the ERS-2 scatterometer. The CMOD5 algorithm corrects some shortcomings in the earlier models and result in a better wind retrieval at high wind speed and more uniform performance across the scatterometer swath (Hersbach et al., 2007). The estimated accuracy of the ASCAT 50-km wind product is 2 m/s RMS difference in wind vector components and 0.5 m/s bias in wind speed (ASCAT product guide). The wind observations at sea surface are essential to describe the atmospheric flow and therefore have many meteorological and oceanographic applications. Wind information is useful for weather forecasting, prediction of extreme events, and climate studies. Figure 2 indicates two examples of the ASCAT 25- and 12.5-km wind products (Verhoef et al., 2009). Processing of the wind product is done in near-real time at EUMETSAT's processing facility. From the sensing time, it takes approximately 2 hours to get the corresponding wind product ready at KNMI. The wind data are disseminated through the EUMETCast system (EUMETCast).

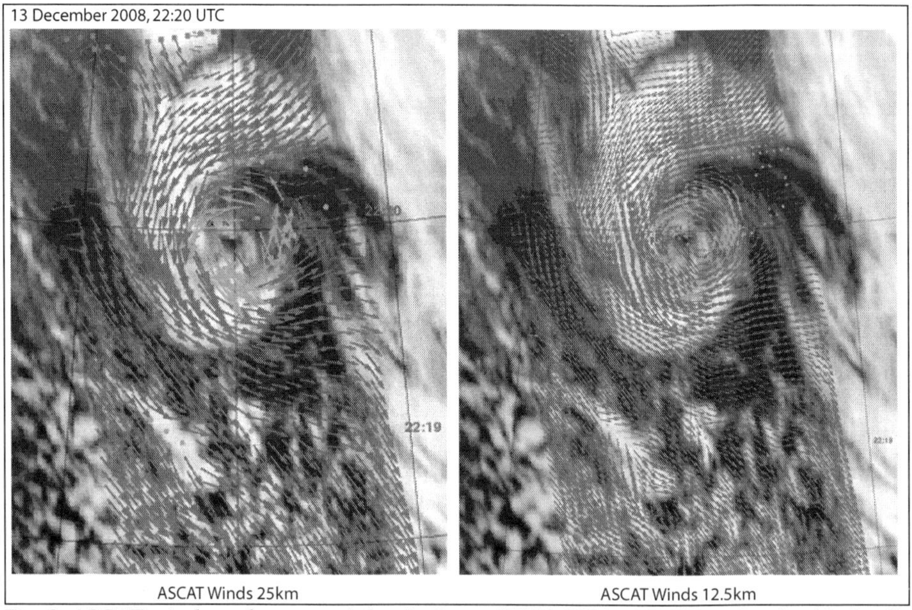

Fig. 2. ASCAT wind product over Atlantic Ocean (55°N-65°N, ~15° West, South of Iceland). Background image shows the infrared cloud image of the METEOSAT9 geostationary satellite. Images are adopted from (Verhoef et al., 2009).

4. Monitoring Seasonal Dynamics of Vegetation

The intensity of the backscattered signal over land is affected by roughness, vegetation structure, vegetation water content, and soil moisture. These factors influence the backscattering coefficient σ^0 on different time scales. At the resolution of the ERS and Metop scatterometers, surface roughness can be in general considered as a temporally invariant parameter. Surface soil moisture changes rapidly within hours to days, contrary to the vegetation canopy and vegetation water content, which vary within several days to weeks. Scattering from the vegetated surface is a complex phenomenon and difficult to model as the volume scattering contributes in total backscattering. Preliminary studies indicated the potential of the C-band scatterometer data for monitoring the seasonal variation of vegetation using multi-temporal analysis (Wismann et al., 1994; Mougin et al., 1995; Frison et al., 1996a; Frison et al., 1996b). Many studies used semi-empirical models to model vegetation effect on backscatter (Magagi et al., 1997; Woodhouse et al., 2000; Jarlan et al., 2003). There have been several canopy scattering models developed to describe σ^0 in terms of vegetation and soil surface parameters based on a solution of the radiative transfer equation (Attema et al., 1978; Ulaby et al., 1990; Karam et al., 1992; Saatchi et al., 1994). Radiative transfer theory describes the propagation of radiation through a medium affected by absorption, emission and scattering processes (Fung, 1994). But the problem with all complex theoretical scattering models is that their input data requirements are very challenging and for solving the equations many parameters are needed such as leaf diameter, branch length, trunk moisture, and probability functions representing the orientational distribution of leaves, branches, and trunks.

The incidence angle of scatterometer observations varies from acquisition to acquisition. Since the intensity of backscatter signal strongly depends on the incidence angle, in the most of the multi-temporal vegetation studies using scatterometer data, σ^0 measurements are averaged over longer periods (e.g. one month) to make σ^0 measurements comparable. But the averaging procedure does not allow us to distinguish the impact of the soil moisture and vegetation cover on backscatter. Wagner et al. (1999a) used a simple model fitted to scatterometer observations to model the incidence angle dependency of backscatter:

$$\sigma'(\theta) = \frac{d\sigma^0(\theta)}{d\theta} \tag{1}$$

σ' shows the incidence angle dependency of σ^0. Knowing the incidence angle dependency, σ^0 can be normalized at a reference incidence angle. In this approach σ' is calculated for each triplet, which contains concurrent measurements representing the same soil moisture condition. Therefore the effect of soil moisture on incidence angle behavior of σ^0 is negligible, if not completely removed from the backscattered signal:

$$\sigma'\left(\frac{\theta_m - \theta_{a/f}}{2}\right) = \frac{\sigma^0_m(\theta_m) - \sigma^0_{a/f}(\theta_{a/f})}{\theta_m - \theta_{a/f}} \tag{2}$$

where the index m stands for the mid-beam and the indices a and f for the aft and fore beam measurements.

The backscattered energy received by the scatterometer sensor increases with decreasing incidence angle. The rate of backscatter change due to incidence angle variation depends on the surface roughness. Bare soil roughness is basically constant in time but vegetation can have a seasonal influence on the incidence angle dependency behavior of backscatter. With increasing vegetation density, the shape of incidence angle dependency of backscatter changes depending on the type and density of vegetation as well as the orientation of vegetation elements. Having multi-year scatterometer data, the seasonal variation of slope can be extracted for a reference incidence angle (e.g. 40°). Slope function at 40°, $\sigma'(40)$ correlates pretty well with the seasonal vegetation change (Naeimi et al., 2009a). Figure 3-top shows slope values globally calculated for the mid of July. Figure 3-bottom illustrates three examples of $\sigma'(40)$ from different regions compared with the Normalized Vegetation index (NDVI). The vegetation index data have been derived from a 16-day Moderate Resolution Imaging Spectroradiometer (MODIS) NDVI product (Huete et al., 2002). NDVI

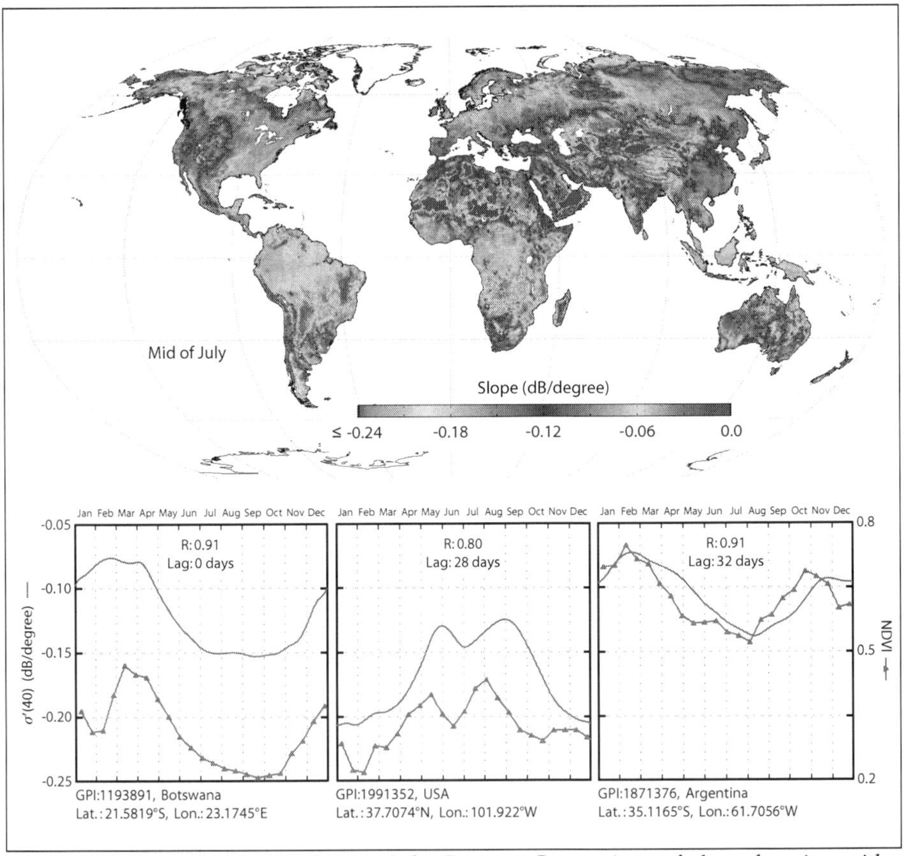

Fig. 3. Above: Global slope values in July. Bottom: Comparison of slope function with NDVI in three different areas.

values are averaged over three years (2000–2002) to estimate the yearly vegetation variation. Depending on land cover type there is a time lag between NDVI and $\sigma'(40)$ in most regions (Doubkova et al., 2009). This implies the fact that the σ' derived from C-band backscatter observations corresponds to vegetation structure development whereas NDVI represents only greenness of vegetation canopy.

5. Soil Moisture Change Detection

As it mentioned in section 4, σ^0 is affected over land by surface roughness, vegetation, and soil moisture. The major challenge of extracting soil moisture from scatterometer data is the presence of the other additional factors influencing the signal. Most studies have introduced physical inversion methods describing scattering process to model roughness and vegetation contributions on backscatter signal (Frison et al., 1997; Pulliainen et al., 1998; Woodhouse et al., 2000; Magagi et al., 2001; Jarlan et al., 2002; Zine et al., 2005). Although theoretical models are useful for understanding and interpreting scattering behavior of natural surfaces, the major problems of these retrieval concepts appear to be their complexity and physical validity at large scales. A promising solution to the problems of physically based inversion models is using change detection method rather than using a complex model to describe the full range of parameters influencing the scattering process. Availability of several years of backscatter data, multi-viewing capability, and high temporal sampling rate of scatterometers make them appropriate instruments for change detection methods. The potential of using change detection techniques for active sensors has been demonstrated in several studies (Wagner, 1998, Moeremans et al., 1998, Quesney et al., 2000; Moran et al., 2000; Le Hegarat-Mascle et al., 2002; De Ridder, 2000).

5.1 TUWien change detection method

Wagner et al., (1999b) presented a change detection method for soil moisture retrieval from ERS scatterometers. A processing algorithm for soil moisture retrieval based on change detection technique has been developed at the Institute of Photogrammetry and Remote Sensing (IPF) of the Vienna University of Technology (TUWien) which will further be referred to as the TUWien method. In the TUWien method soil moisture dynamics are extracted after modeling the behavior of σ^0 with respect to the surface roughness and the local variability of vegetation and eventually subtracting them from the backscatter signal. In the retrieval algorithm, multi-looking direction ability of scatterometer is used to describe the incidence angle behavior of the backscatter signal as a seasonal function, $\sigma'(\theta)$. The incidence angle dependency of backscatter can be described by the derivatives of σ^0 at a reference incidence angle (set to 40°) according to the Taylor series expansion:

$$\sigma'(\theta) = \sigma'(40) + \sigma''(40)(\theta - 40) \tag{3}$$

$\sigma'(40)$ and $\sigma''(40)$, called slope and curvature at 40°, are calculated by fitting a regression line to the obtained local slope values in equation-2 during a certain period of the year. After determination of slope and curvature for each day of year and using the following second-order polynomial equation based on Taylor series, $\sigma^0(\theta)$ measurements are extrapolated to 40° incidence angle:

$$\sigma^0(40) = \sigma^0(\theta) - \sigma'(40)(\theta - 40) - \frac{1}{2}\sigma''(40)(\theta - 40)^2 \tag{4}$$

Eventually the normalized backscatter $\sigma^0(40)$ is scaled between the lowest and highest values ever measured within the long-term $\sigma^0(40)$ observations, $\sigma^0_{wet}(40)$ and $\sigma^0_{dry}(40)$, which represent the driest and wettest conditions:

$$\Theta_s = \frac{\sigma^0(40) - \sigma^0_{dry}(40)}{\sigma^0_{wet}(40) - \sigma^0_{dry}(40)} \times 100 \tag{5}$$

Θ_s corresponds to the normalized volumetric water content at topmost 2 cm soil surface ranging between 0% and 100% with presumption of linear relationship between $\sigma^0(40)$ and the surface soil moisture (Ulaby et al., 1982). In addition the TUWien retrieval algorithm includes processing modules for vegetation correction, wet reference correction and soil moisture uncertainty analysis (Naeimi et al., 2009a). An operational processing system based on the TUWien retrieval algorithm is implemented at EUMETSAT to provide near-real time ASCAT soil moisture data (Hasenauer et al., 2006). The data have been made available through the EUMETCast system (EUMETCast). Figure 4 shows SCAT/ASCAT soil moisture time series compared with precipitation data at a grid point located in Lower Austria. An example of global distribution of the mean soil moisture values retrieved from long-term SCAT time series is shown in Figure 5. The spatial variability of the estimated mean of soil moisture is connected to atmospheric-forcing related soil moisture signal. Soil moisture retrieval from scatterometer data has also limitations when the soil is frozen or

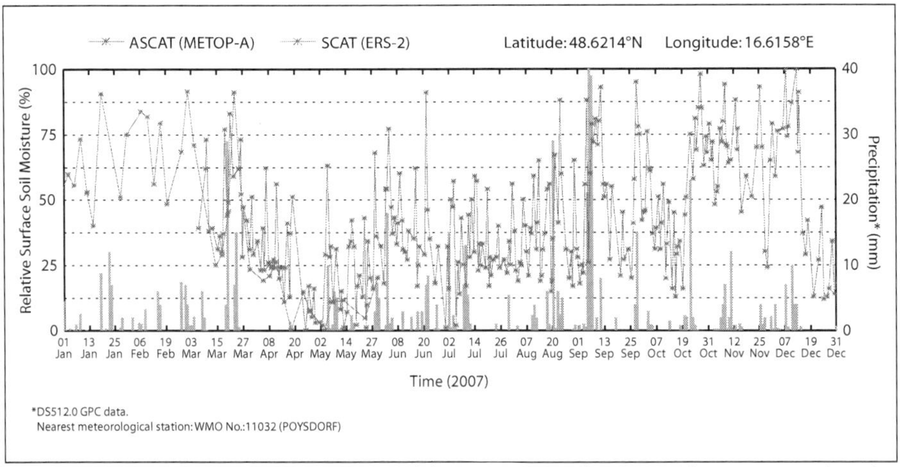

Fig. 4. Soil moisture time series retrieved from SCAT and ASCAT data compared with precipitation data in lower Austria.

covered with snow. As soon as the soil freezes the dielectric constant of the soil drops drastically and results in low backscatter. Therefore the backscattering behaviors of dry and frozen soil are similar. The scattering behavior of snow is more complex and depends on the dielectric properties of the ice particles and on their distribution and density. Furthermore, land cover has also impacts on the quality of soil moisture retrieval from scatterometer data. There is a strong response of the azimuthal noise level of backscatter to different land cover types like rainforests, lakes, rivers, floodplains, coastal areas, urban areas, and sand deserts as well as areas with complex topography (Naeimi et al., 2008). An uncertainty analysis module using Monte Carlo error propagation (Naeimi, 2009b) is implemented within the TUWien algorithm which identifies such problematic areas for soil moisture retrieval from scatterometer data.

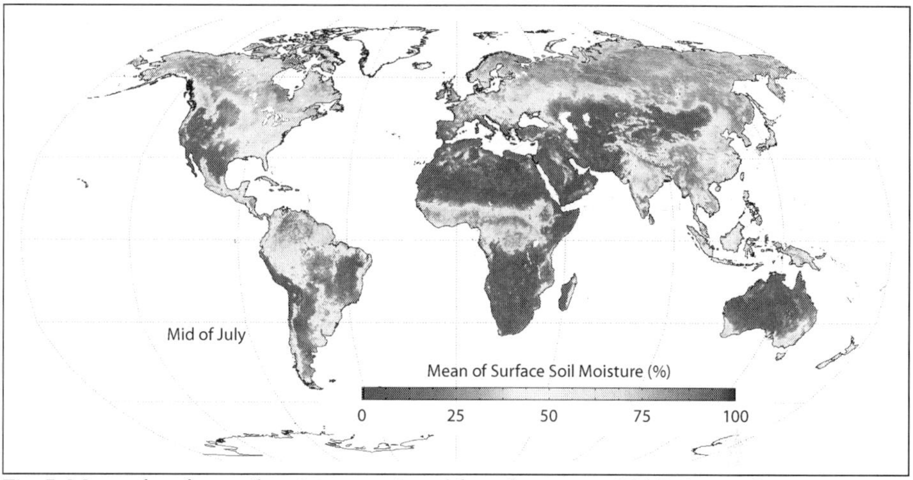

Fig. 5. Mean of surface soil moisture retrieved from long-term SCAT time series.

5.2 Surface soil moisture anomalies

Anomalies of soil moisture, precipitation, temperature, and vegetation indices are parameters that are used as indicator of extreme weather conditions. Scatterometer soil moisture anomalies can be calculated by comparing the current values with mean and standard deviation values in the same time of year over the long-term ERS/Metop scatterometer time series. Figure 6 illustrates monthly anomalies of ASCAT soil moisture compared with the NDVI anomaly images derived from MODIS data (NASA-EO). The extremely dry conditions are visible in parts of Europe during July 2007 (Figure 6-a). As reported by the authorities the 2007 drought in Moldova was the most severe in living memory. The World Food Program compared its severity to the drought of 1946 during which many Moldovans starved. The Cereal production at that year was down by 63% compared to the year before, and was about 70% lower than the average of the five years before (FAO news). Figure 6-b shows another example of extreme condition, which is evident in ASCAT soil moisture anomalies. The anomalous wet soil in March 2008 in parts of India provided a suitable condition for vegetation growth. By early April 2008, plants throughout the country were responding to the plentiful water supply that led to record of harvest yield in April (NASA-EO).

5.3 Soil Water index (SWI)

The C-band scatterometer derived soil moisture represent only top few centimeter of soil. Nevertheless, thanks to the high temporal sampling of scatterometers (about 80% global daily coverage for ASCAT), soil moisture in plant root zone can be estimated by using an infiltration model. Wagner et al. (1999b) proposed a simple two-layer water balance model to estimate profile soil moisture. The remotely sensed topsoil represents the first layer and the second layer extends downwards from the bottom of the surface layer. In this model, the water content of the reservoir layer is described in terms of a Soil Water Index (SWI), which is controlled only by the past soil moisture conditions in the surface layer in a way that the influence of measurements decreases by increasing the time:

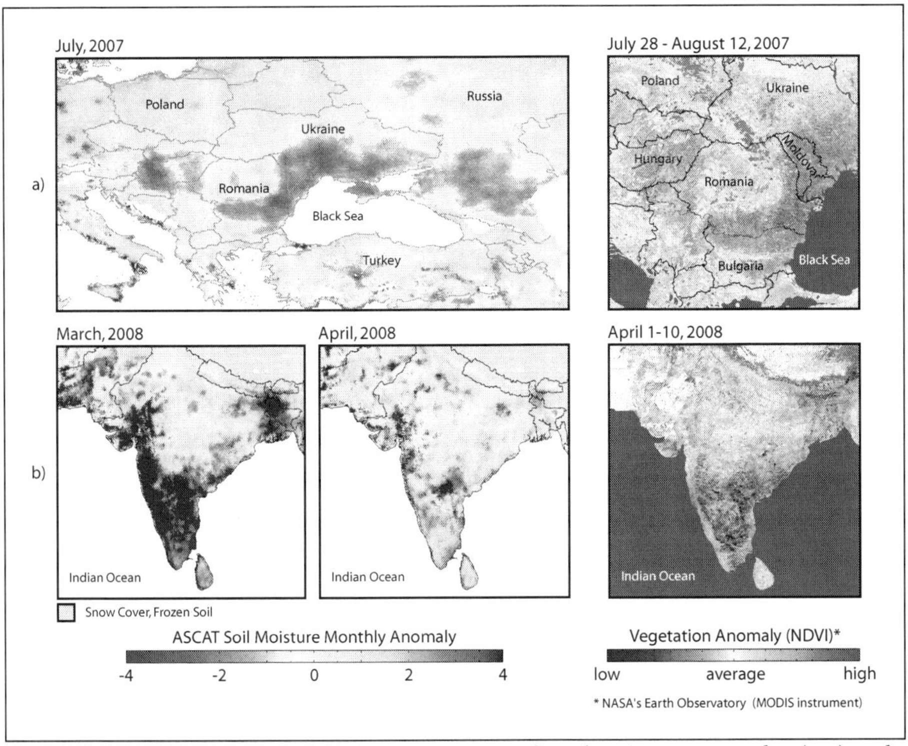

Fig. 6. Examples of the ASCAT soil moisture anomalies showing extreme dry (top) and wet conditions (bottom) compared with NDVI anomalies extracted from MODIS data.

$$SWI(t_n) = \frac{\sum_i^n \Theta_s(t_i) e^{-\frac{t_n - t_i}{T}}}{\sum_i^n e^{-\frac{t_n - t_i}{T}}} \quad for\, t_i \leq t_n \tag{6}$$

$\Theta_s(t_i)$ is the surface soil moisture measured at time t_i and T is the characteristic time length connected to the depth of reservoir which describes the linkage between the surface layer and the reservoir by:

$$T = L/C \,, \quad L\frac{d\Theta(t)}{dt} = C \cdot [\Theta_s(t) - \Theta_r(t)] \tag{7}$$

where L is the depth of the reservoir layer and C is a pseudo-diffusivity coefficient that depends on soil properties. Θ_s and Θ_r are the volumetric moisture content of the surface and reservoir respectively.
Daily images of SWI calculated at five different T values (10, 20, 40, 60, 100) retrieved from ASCAT-25km observations using a near-real time recursive processor will be available through the geoland project (geoland-II). Figure 7 indicates the global ASCAT-50km SWI image calculated for T=10 as an example.

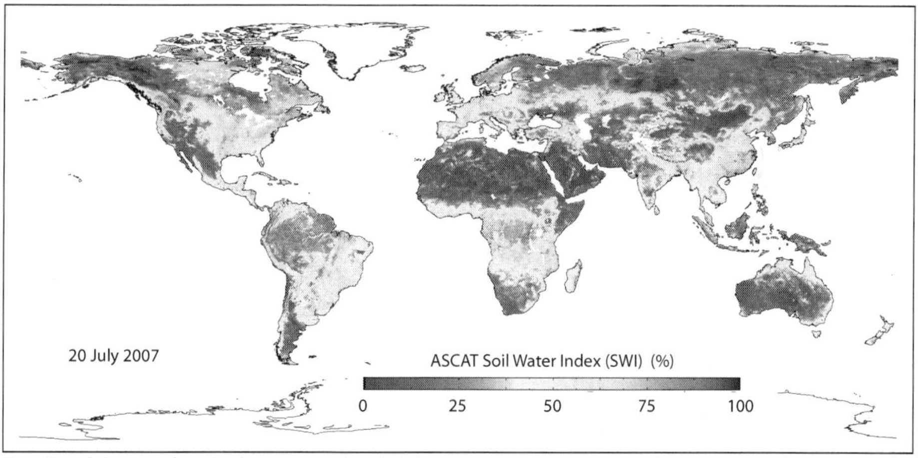

Fig. 7. ASCAT-50km Soil Water Index (SWI) at T=10.

6. Monitoring Cryosphere

The cryosphere consists of the parts of the Earth's surface where water exists in solid form, including snow cover, frozen ground, glacier, see ice, ice sheets and any other form of ice on land or in ocean. The cryosphere plays an important role in the global climate system and therefore impacts significantly human life. More than about 70% of the Earth's freshwater is frozen in ocean ice sheets, glaciers or permafrost areas (UNESCO report, 2006). Permafrost regions are of major interest in climate studies as several hundred gigatons of carbon are stored in frozen soils in high latitudes. Thawing of permafrost could supercharge the global warming process. There is also a major concern about the possibility of shrinking the Earth's ice sheets due to the global warming which could raise the global sea level by several meters. There are many cryosphere-climate feedback mechanisms in the global climate system over a wide range of spatial and temporal scales. Snow and ice have a remarkable effect on climate as they modulate energy exchanges between the surface and the atmosphere because of their physical properties. One of the most important properties is the surface reflectance (albedo). Non-melting snow and ice can reflect between ~80-90% of incident solar energy whereas vegetation and soil surface reflect as little as 20-30%. The reflected sunlight into space does not get absorbed by the Earth as heat. Therefore the high albedo plays as a cooling factor in the global climate system. The thermal properties of cryospheric elements have also major consequences for the climate and hydrological cycle. Snow and ice have much lower thermal diffusivities than air and build an insulating layer over land and ocean surfaces decoupling the surface-atmosphere interface with respect to both heat and moisture fluxes. High latent heat is another thermal property of snow and ice that act to moderate temperature in warm seasons because of the large amount of energy required to melt ice.

Scatterometry has been proven to be useful for monitoring and understanding the cryosphere. Several studies have investigated the applicability of scatterometer data in various cryosphere research areas for instance; mapping snowmelt extent (Wismann et al., 1997; Wismann, 2000), snow accumulation in Greenland (Drinkwater et al., 2001), snow cover over the Northern Hemisphere (Nghiem et al., 2001), frozen terrain in Alaska (Kimball et al., 2001). Other studies have used scatterometer data for determination of freeze/thaw cycles in Northern Latitudes (Bartsch et al., 2007), spatial and temporal variability of sea ice (Drinkwater et al., 2000), classification of sea ice in Polar Regions (Remund et al., 2000), deriving the surface wind-induced patterns over Antarctica by measuring the azimuthal modulation of backscatter (Long et al., 2000).

In winter when soil surface freezes, dielectric properties of the soil changes significantly which results in low backscatter values. As snow begins to fall and accumulates over the surface, due to volume scattering, backscatter signals increase depending on microwave frequency. The response of dry snow volume to microwaves is rather complex and depends on snow properties like snow depth, density, and average grain size as well as the age of snowpack. With increasing temperature in spring, snow begins to melt and water covers the surface of snow pack which causes a sudden drop in backscatter. After snow melting period, soil and vegetation begin to thaw and consequently backscatter arise again. Figure 8 shows a typical example of freeze/thaw process as described above observable in ASCAT normalized backscatter at 40°. High diurnal difference of backscatter (green bars) implies frozen condition in the morning and thawing in the evening which can be used as an indicator of the transition between different phases.

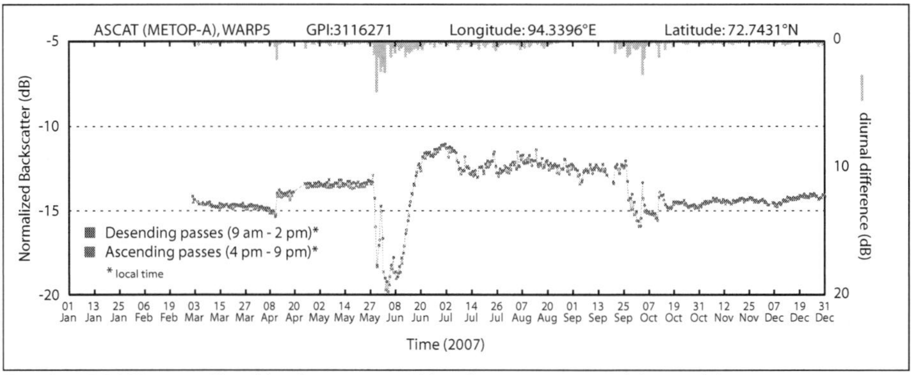

Fig. 8. ASCAT normalized backscatter at 40° indicating seasonal freeze/thaw process.

The high temporal sampling of the scatterometers in Polar Regions despite the frequent cloud cover and poor sunlight make them valuable instruments for sea ice observations. The sea ice imaging is based on the sensitivity of scatterometer to ice roughness and relatively high difference between the backscatter from open water and sea ice.

Long et al. (1999) used a simple linear function to approximate the backscatter at 40° (reference incidence angle):

$$\sigma^0(\theta) = A + B(\theta - 40) \tag{8}$$

where A is the σ^0 at 40° incidence and B describes the incidence angle dependency of backscatter.

The A and B parameters are calculated after combining the scatterometer observations from multiple passes from several days and using the Scatterometer Image Reconstruction (SIR) algorithm to enhance the resolution (Early et al., 2001). The A and B images represent the backscatter properties of the surface and are related to ice and snow characteristics over the imaging period (Long et al., 2001). Figure 9 illustrates examples of the normalized backscatter retrieved from ERS-1/2 scatterometer data available through the Scatterometer Climate Record Pathfinder (SCP) project (NASA-SCP).

7. Conclusion

C-band scatterometers have demonstrated to be valuable sensors for large-scale observation of the Earth's surface in a variety of disciplines. High temporal sampling in all weather conditions, multi-viewing capability and availability of long-term measurements make the European C-band scatterometers excellent Earth observation tools. Scatterometer data are used to extract geophysical parameters such as wind speed and direction, surface soil moisture, seasonal dynamics of vegetation, spatial and temporal variability of frozen train in high latitudes, snowmelt and sea ice. Furthermore the scatterometer data are utilized in hydrological modeling, observation of extreme events, flood and drought monitoring, and also used for climate change studies. The observations of the ERS-1/2 scatterometers

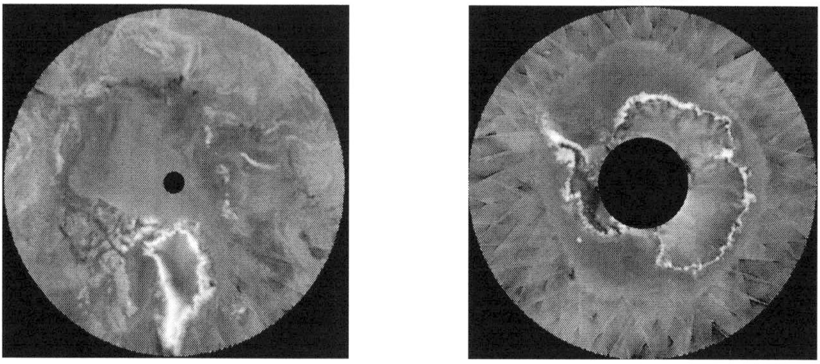

Fig. 9. Images of parameter A in equation-8 (the normalized backscatter at 40°) retrieved from ERS-2 scatterometer data after resolution enhancement. Adopted from (NSIDC).

(SCATs) together with the new series of advanced scatterometers (ASCAT) onboard Metop satellites ensure long-term global observation (from 1991 until at least 2020).

Acknowledgements

This work was supported by the geoland-II project in frame of the Global Monitoring for the Environment and Security (GMES), a joint initiative of European Commission (EC) and European Space Agency (ESA), and the Austrian Research Promotion Agency (FFG) through the Global Monitoring of Soil Moisture (GMSM) project.

8. References

ASCAT Product Guide v4 (2009), EUM/OPS-EPS/MAN/04/0028, EUMETSAT, Darmstadt, Germany, http://oiswww.eumetsat.org/WEBOPS/eps-pg/ASCAT/ASCAT-PG-0TOC.htm.

Attema E. P. W. and F. T. Ulaby (1978), Vegetation Modeled as a Water Cloud, Radio Science, Vol. 13, pp. 357.

Bartsch A., R. A. Kidd, W. Wagner and Z. Bartalis (2007), Temporal and spatial variability of the beginning and end of daily spring freeze/thaw cycles derived from scatterometer data, Remote Sensing of Environment, Vol. 106, pp. 360.

De Ridder K. (2000), Quantitative estimation of skin soil moisture with the Special Sensor Microwave/Imager, Boundary-Layer Meteorology, Vol. 96, pp. 421-432.

Doubkova M., V. Naeimi, W. Wagner and G. Henebry (2009), On the ability of the ERS scatterometer to detect vegetation properties, IEEE International Geoscience and Remote Sensing (IGARSS), Cape Town, South Africa, 12-17 July 2009.

Drinkwater M. R. and X. Liu (2000), Seasonal to interannual variability in Antarctic sea-ice surface melt, IEEE Transactions on Geoscience and Remote Sensing, Vol. 38, pp. 1827.

Drinkwater M. R., D. G. Long and A. W. Bingham (2001), Greenland snow accumulation estimates from satellite radar scatterometer data, Journal of Geophysical Research D: Atmospheres, Vol. 106, pp. 33935.

Early D. S. and D. G. Long (2001), Image reconstruction and enhanced resolution imaging from irregular samples, IEEE Transactions on Geoscience and Remote Sensing, Vol. 39, pp. 291.

EUMETCast, (accessed 2009), http://www.eumetsat.int/home/Main/What_We_Do/EUMETCast/index.htm.

FAO news report, (Accessed July 2009), http://www.fao.org/newsroom/en/news/2007/1000667.

Figa-Saldana J., J. W. Wilson, E. Attema, R. Gelsthorpe, M. R. Drinkwater and A. Stoffelen (2002), The advanced scatterometer (ASCAT) on the meteorological operational (MetOp) platform: A follow on for European wind scatterometers, Canadian Journal of Remote Sensing, Vol. 28, pp. 404-412.

Frison P. L. and E. Mougin (1996a), Monitoring global vegetation dynamics with ERS-1 wind scatterometer data, International Journal of Remote Sensing, Vol. 17, pp. 3201.

Frison P. L. and E. Mougin (1996b), Use of ERS-1 wind scatterometer data over land surfaces, IEEE Transactions on Geoscience and Remote Sensing, Vol. 34, pp. 550.

Frison P. L., E. Mougin and P. Hiernaux (1997), Observations and simulations of the ERS wind scatterometer response over a sahelian region, International Geoscience And Remote Sensing Symposium (IGARSS), pp. 1832-1834.

Geoland-II project, Integrated GMES project on land cover and vegetation, http://www.gmes-geoland.info/PROJ/index.php.

Hasenauer S., W. Wagner, K. Scipal, V. Naeimi and Z. Bartalis (2006), Implemetation of near real time soil moisture products in the SAF network based on METOP ASCAT data, paper presented at EUMETSAT Meteorological Satellite Conference, Helsinki, Finland, 12-16 June 2006.

Hersbach H., A. Stoffelen and S. De Haan (2007), An improved C-band scatterometer ocean geophysical model function: CMOD5, Journal of Geophysical Research C: Oceans, Vol. 112.

Huete A., K. Didan, T. Miura, E. P. Rodriguez, X. Gao and L. G. Ferreira (2002), Overview of the radiometric and biophysical performance of the MODIS vegetation indices, Remote Sensing of Environment, Vol. 83, pp. 195.

Jarlan L., P. Mazzega and E. Mougin (2002), Retrieval of land surface parameters in the Sahel from ERS wind scatterometer data: A "brute force" method, IEEE Transactions on Geoscience and Remote Sensing, Vol. 40, pp. 2056-2062.

Jarlan L., P. Mazzega, E. Mougin, F. Lavenu, G. Marty, P. L. Frison and P. Hiernaux (2003), Mapping of Sahelian vegetation parameters from ERS scatterometer data with an evolution strategies algorithm, Remote Sensing of Environment, Vol. 87, pp. 72-84.

Kimball J. S., K. C. McDonald, A. R. Keyser, S. Frolking and S. W. Running (2001), Application of the NASA scatterometer (NSCAT) for determining the daily frozen and nonfrozen landscape of Alaska, Remote Sensing of Environment, Vol. 75, pp. 113.

Klaes D. and K. Holmlund (2007), The EPS/Metop system: overview and first results, paper presented at Joint 2007 EUMETSAT Meteorological Satellite Conference and the 15th Satellite Meteorology & Oceanography Conference of the American Meteorological Society, Amsterdam, The Netherlands, 24-28 September 2007.

Le Hegarat-Mascle S., M. Zribi, F. Alem, A. Weisse and C. Loumagne (2002), Soil moisture estimation from ERS/SAR data: Toward an operational methodology, IEEE Transactions on Geoscience and Remote Sensing, Vol. 40, pp. 2647.

Liu W. T. (2002), Progress in scatterometer application, Journal of Oceanography, Vol. 58, pp. 121-136.

Long A. E. (1985), Towards a C-Band radar sea echo model for the ERS-1 scatterometer, Proceedings Conference on Spectral Signatures, Les Arcs, France, December 1985. European Space Agency Special Publication, ESA SP-247, 29-34.

Long D. G. and M. R. Drinkwater (1999), Cryosphere applications of NSCAT data, IEEE Transactions on Geoscience and Remote Sensing, Vol. 37, pp. 1671.

Long D. G. and M. R. Drinkwater (2000), Azimuth variation in microwave scatterometer and radiometer data over Antarctica, IEEE Transactions on Geoscience and Remote Sensing, Vol. 38, pp. 1857-1870.

Long D. G., M. R. drinkwater, B. Holt, S. Saatchi and C. Bertoia (2001), Global Ice and Land Climate Studies Using Scatterometer Image Data, EOS, Transaction of the American Geophysical Union, Vol. 82, pp. 503.

Magagi R. D. and Y. H. Kerr (1997), Retrieval of soil moisture and vegetation characteristics by use of ERS-1 wind scatterometer over arid and semi-arid areas, Journal of Hydrology, Vol. 188-189, pp. 361-384.

Magagi R. D. and Y. H. Kerr (2001), Estimating surface soil moisture and soil roughness over semiarid areas from the use of the copolarization ratio, Remote Sensing of Environment, Vol. 75, pp. 432-445.

Moeremans B. and S. Dautrebande (1998), Use of ERS SAR interferometric coherence and PRI images to evaluate crop height and soil moisture and to identify crops, Remote Sensing For Agriculture, Ecosystems, And Hydrology, Vol. 3499, pp. 9-19.

Moore R. K. and A. K. Fung (1979), Radar Determination Of Winds At Sea, Proceedings Of The Ieee, Vol. 67, pp. 1504-1521.

Moran M. S., D. C. Hymer, J. Qi and E. E. Sano (2000), Soil moisture evaluation using multi-temporal synthetic aperture radar (SAR) in semiarid rangeland, Agricultural and Forest Meteorology, Vol. 105, pp. 69.

Mougin E., A. Lopes, P. L. Frison and C. Proisy (1995), Preliminary analysis of ERS-1 wind scatterometer data over land surfaces, International Journal of Remote Sensing, Vol. 16, pp. 391.

NASA-EO, NASA's Earth observatory, (Accessed July 2009), http://earthobservatory. nasa.gov/NaturalHazards/category.php?cat_id=6.

NASA-SCP, The NASA Scatteromter Climate Record Pathfinder (SCP), http://www.scp.byu.edu/.

Naeimi V., C. Kuenzer, S. Hasenauer, Z. Bartalis and W. Wagner (2008), Evaluation of the influence of land cover on the noise level of ERS-scatterometer backscatter, International Geoscience and Remote Sensing Symposium (IGARSS), Barcelona, Spain. pp. 3685-3688.

Naeimi V., K. Scipal, Z. Bartalis, S. Hasenauer and W. Wagner (2009a), An Improved Soil Moisture Retrieval Algorithm for ERS and METOP Scatterometer Observations, IEEE Transactions on Geoscience and Remote Sensing, Vol. 47, pp. 555-563.

Naeimi V., (2009), Model improvements and error characterization for global ERS and METOP scatterometer soil moisture data, dissertation, pp. 111, Vienna University of Technology, Vienna.

Nghiem S. V. and W. Y. Tsai (2001), Global snow cover monitoring with spaceborne Ku-band scatterometer, IEEE Transactions on Geoscience and Remote Sensing, Vol. 39, pp. 2118.

NSIDC, National Snow and Ice Data Center, (Accessed July 2009), http://nsidc.org/data/nsidc-0260.html

Pulliainen J. T., T. Manninen and M. T. Hallikainen (1998), Application of ERS-1 wind scatterometer data to soil frost and soil moisture monitoring in boreal forest zone, IEEE Transactions On Geoscience And Remote Sensing, Vol. 36, pp. 849-863.

Quesney A., S. Le Hegarat-Mascle, O. Taconet, D. Vidal-Madjar, J. P. Wigneron, C. Loumagne and M. Normand (2000), Estimation of watershed soil moisture index from ERS/SAR data, Remote Sensing of Environment, Vol. 72, pp. 290.

Remund Q. P., D. G. Long and M. R. Drinkwater (2000), An iterative approach to multisensor sea ice classification, IEEE Transactions on Geoscience and Remote Sensing, Vol. 38, pp. 1843.

Stoffelen A. and D. Anderson (1997), Scatterometer data interpretation: Estimation and validation of the transfer function CMOD4, Journal of Geophysical Research C: Oceans, Vol. 102, pp. 5767-5780.

Ulaby F. T., R. K. Moore and A. K. Fung (1982), Microwave remote sensing: active and passive. Volume II. Radar remote sensing and surface scattering and emission theory.

UNESCO report (2006), United Nations World Water Developement Report 2, Water, a Shared Responsibility, p.121

Verhoef A. and Ad Stoffelen (2009), Validation of ASCAT 12.5-km winds, version 1.2, SAF/OSI/CDOP/KNMI/TEC/RP/147, EUMETSAT.

Wagner W. (1998), Soil Moisture Retrieval from ERS Scatterometer Data, dissertation, pp. 111, Vienna University of Technology, Vienna.

Wagner W., G. Lemoine, M. Borgeaud and H. Rott (1999a), A study of vegetation cover effects on ERS scatterometer data, IEEE Transactions on Geoscience and Remote Sensing, Vol. 37, pp. 938-948.

Wagner W., G. Lemoine and H. Rott (1999b), A method for estimating soil moisture from ERS scatterometer and soil data, Remote Sensing of Environment, Vol. 70, pp. 191-207.

Wismann V. (2000), Monitoring of seasonal snowmelt on Greenland with ERS scatterometer data, IEEE Transactions on Geoscience and Remote Sensing, Vol. 38, pp. 1821.

Wismann V. and K. Boehnke (1997), Monitoring snow properties on Greenland with ERS scatterometer and SAR, European Space Agency, (Special Publication) ESA SP. pp. 857-861.

Wismann V. R., K. Boehnke and C. Schmullius (1994), Global land surface monitoring using the ERS-1 scatterometer, International Geoscience and Remote Sensing Symposium (IGARSS), pp.1488.

Woodhouse I. H. and D. H. Hoekman (2000), Determining land-surface parameters from the ERS wind scatterometer, IEEE Transactions on Geoscience and Remote Sensing, Vol. 38, pp. 126-140.

Zine S., L. Jarlan, P. L. Frison, E. Mougin, P. Hiernaux and J. P. Rudant (2005), Land surface parameter monitoring with ERS scatterometer data over the Sahel: A comparison between agro-pastoral and pastoral areas, Remote Sensing of Environment, Vol. 96, pp. 438.

14

Monitoring of terrestrial hydrology at high latitudes with scatterometer data

Annett Bartsch
Vienna University of Technology
Austria

1. Introduction

The mission of this chapter is to provide insight into the capabilities of scatterometer data for climate change relevant monitoring at high latitudes of the terrestrial hydrosphere (excluding large ice caps). Scatterometer are active microwave instruments. Spaceborne sensors have been developed for operational ocean wind monitoring but they have also been proven of high value for applications over land area within especially the last decade (Wagner et al., 2007). The applications cover a wide range of subjects from snowmelt to phenology. What all have in common is the focus on monitoring of dynamic processes.

Scatterometer are non-imaging radars. Currently operational sensors which are used for land applications operate in Ku- ($\approx$ 2.1 cm wavelength) and C-band ($\approx$ 5.6 cm wavelength). The spatial resolution is coarse compared to most optical and SAR sensors and ranges between 25 km and 50 km. The temporal resolution, however, can be better than daily especially at high latitudes as they are mounted on polar orbiting platforms what results in overlapping orbits. Especially issues in hydrology require a high sampling rate. The interaction of the electromagnetic waves at the earth surface is influenced by dielectric properties, roughness and land cover parameters such as vegetation (Ulaby et al., 1982). Additionally the properties of the waves (frequency, polarization, incidence angle) determine the sensitivity to certain changes (spatially and temporally) at the earth surface. For the derivation of most land surface parameters single sensor approaches have been developed which exploit the specific sensor properties. As the backscatter return is complex to model, they are largely based on change detection approaches. These parameters and thus results from scatterometers with different configurations can be jointly used in order to get an advanced insight to earth surface processes and long term changes (Bartsch et al., 2007a).

The focus of this chapter is on applicability of scatterometer products for investigation of basin hydrology. High latitudes are of special interest for climate change monitoring (Hinzman et al., 2005; IPCC, 2007). Predicted and observed changes affect the hydrosphere, especially snowmelt timing (Dye & Tucker, 2003; Smith et al., 2004) and permafrost (e.g. Callaghan et al. (2004)). Scatterometers provide valuable data for the monitoring of these changes on regional to global scale. River runoff maxima occur in conjunction with snowmelt (e.g.Khan et al. (2008); Yang et al. (2007)). These melt patterns can be determined using scatterometer products (Bartsch et al., 2007b; Frolking et al., 1999; Kimball et al., 2004a;

Wismann, 2000). Frozen ground impedes drainage (French, 1996; Lilly et al., 2008; Williams & Smith, 1989) and may thus impact the relationship between snowmelt patterns and river discharge (Kane, 1997; Kane et al., 2003). When snowmelt has ceased changes in discharge result from e.g. precipitation and subsurface melt. Near surface soil water content can be also captured with active microwave data and therefore allow an assessment of the hydrological status of entire basins (Scipal et al., 2005).

In the following sections, available sensors and change detection approaches relevant to basin hydrology at high latitudes are reviewed and eventually results for selected basins presented and discussed.

2. Sensors and data

ERS-1 has been launched in 1991 and ERS-2 in 1996. The wind scatterometer on these platforms provide 50 km resolution datasets in C-band. A global coverage can be theoretically achieved within 3-4 days (Wagner et al., 1999a). A similar sensor has been launched with METOP ASCAT (Advanced Scatterometer) in October 2006. The ground coverage is considerably improved compared to ERS due to a second swath and spatial resolution is 25km (Klaes et al., 2007). Measurements are consistent with the preceding sensors and allow continuation of products developed for ERS (Bartalis et al., 2007; Naeimi et al., 2009).

First Ku-band scatterometer studies are based on NSCAT onboard the Advanced Earth Observation Satellite (ADEOS). It was launched on August 1996 and operated until June 1997. The spatial resolution was 25km and a 90% global coverage has been achieved within two days allowing for twice daily acquisitions at high latitudes (Frolking et al., 1999). The later Seawinds instruments (on QuikScat and ADEOS2) cover 90% of the Earth's surface daily and provide up to 10 measurements towards 75 °N (Bartsch et al., 2007b). Seawinds QuikScat is also a Ku-Band sensor with 25km resolution and is to date in operation.

Important steps in the preprocessing of scatterometer data are normalization and gridding as they are non-imaging radars. Normalization is required since the incidence angle varies from acquisition to acquisition what causes differences in backscatter. Both NSCAT and ERS measurements are usually normalized to 40° (Kimball et al., 2001; Wagner et al., 1999a; Wismann, 2000; Zhribi et al., 2008). Nghiem & Tsai (2001) used 45° for NSCAT. Often scatterometer data have been gridded into rectangular cells of e.g. 0.5° x 1° (Prigent et al., 2007; Wismann, 2000), 0.5° x 0.5° (Abdel-Messeh & Quegan, 2000) or 0.25° x 0.25° (Zhribi et al., 2008) for ERS, 25 km x 25 km for NSCAT (Kimball et al., 2004a; 2001; Nghiem & Tsai, 2001) and 12.5 km x 12.5 km for QuikScat (Kidd et al., 2003). The TU Wien ERS product (Scipal, 2002) uses a Discrete Global Grid (DGG) which is a sinusoidal global grid generated by an adapted partitioning of the globe with orininally a 25 km grid spacing for ERS and recently a 12.5 km grid incorporating also ASCAT (Bartalis et al., 2006; Naeimi et al., in press). The spatial interpolation of the data in each grid point is performed after the incidence angle normalization of the backscatter measurements in dB (σ^0), by using the Hamming window function (Scipal, 2002). A resolution enhanced QuikScat product has been developed based on multiple measurements available during short time intervals (Early & Long, 2001). QuikScat σ^0 data are made available as "eggs" or "slices" depending on the processing method. Egg-based QuikScat images have a nominal pixel spacing of 4.45 km and an estimated effective resolution of 8-10 km (Long & Hicks, 2005, BYU product). Daily "eggs" data have been used by Hardin & Jackson (2003), Frolking et al. (2005), Brown et al. (2007) and Wang et al. (2008) for land applications outside glaciated areas. On a global level those are assembled from four days of data and for polar regions for separated day times due to increased revisit intervals (applied in e.g. Wang et al. (2008)).

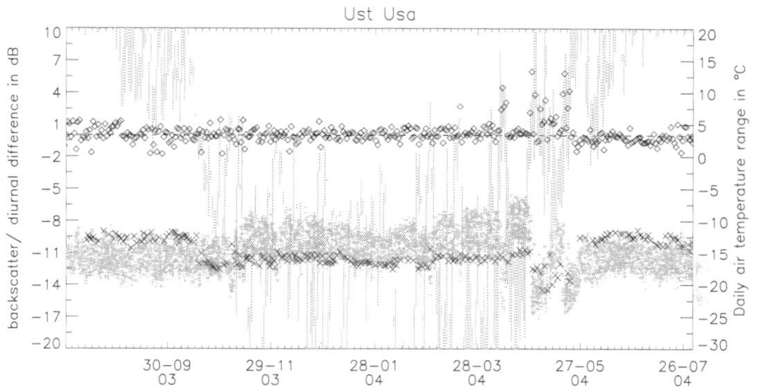

Fig. 1. Typical backscatter time series (in dB) for C-Band (blue crosses, source: ERS) and Ku-band (grey points, source: QuikScat) at Ust Usa (56.92 ° E, 65.97 ° N) for August 2003 to July 2004. Daily air temperature range in °C extracted from the WMO512 dataset is shown as yellow bars. Diamonds represent Ku-band diurnal backscatter difference in dB.

A typical backscatter time series of C-band (ERS) and Ku-band (QuikScat) for high latitude environment is shown in Figure 1. Although there are similarities in surface interaction, seasonal backscatter behaviour differs between Ku- and C-Band. This is especially pronounced if a snow cover is present. Microwave backscatter differs significiantly due to changing dielectric properties between frozen and unfrozen ground (e.g. Ulaby et al. (1982); Way et al. (1997); Wegmüller (1990)). In case of Ku-band, the backscatter is low before snow arrival, it gradually increases with snow accumulation, then rapidly decreases when the snow starts melting and eventually increases again when all snow has melted (Nghiem & Tsai, 2001). The level of summer backscatter is lower than winter backscatter. In C-band, the summer backscatter is higher than when snow is present or the ground is frozen. When the snow surfaces recyristallize after a midwinter short-term melt event, backscatter can increase up to summer levels in C-band (Wismann, 2000). The formation of ice crust after mid-winter thaw and subsequent backscatter increase is also strongly visibly in Ku-band (Kimball et al., 2001). QuikScat also allows the investigation of diurnal differences during the snowmelt period (Bartsch et al., 2007b; Nghiem & Tsai, 2001). The snow is then often frozen in the morning and the surface is undergoing melt in the evening due to air temperatures increase above 0 °C during the day. This results in strong differences between morning and evening backscatter (Figure 1).

Microwave backscatter during freeze/snow free conditions increases with increasing soil moisture (Ulaby et al., 1982). This has been demonstrated for C-band (e.g. Wagner et al. (1999b); Zhribi et al. (2008)) and Ku-Band (Mladenova et al., in press) scatterometer.

Additionally, variation in summer can be caused by phenology. Backscatter increases with vegetation growth (Frolking et al., 2005; Hardin & Jackson, 2003). The magnitude of contribution at C-band is, however, low compared to soil water changes (Wagner et al., 1999c).

Tundra regions are often characterized by a high number of small lakes and ponds which can be easily identified with higher resolution microwave satellite data (Synthetic aperture radars - SARs) due to the specific low backscatter of smooth water surfaces (e.g. Bartsch et al. (2008a). For coarse resolution data such as from the ERS scatterometer, however, it has been found that contributions of lakes and rivers to the overall backscatter is very small and can be neglected (Wismann, 2000).

Variations in backscatter are also introduced by instrument noise, speckle and azimuthal effects (Wagner et al., 1999b). In C-band especially azimuthal effects add to noise for different land cover types (Bartalis et al., 2006; Naeimi et al., in press). Seawinds QuikScat data exhibit also strong noise which varies over differing land cover (Bartsch et al., 2007b). Figure 2 demonstrates the typical noise at Ku-Band of un-glaciated terrain in high latitudes (Estimated standard deviation of noise - ESD). It is much higher than the ESD determined for ASCAT in those environments (Naeimi et al., in press). It is typically below 0.3 dB for ASCAT. It usually exceeds 0.5 dB (mean of 0.57 above 60 ° north) for QuikScat. This needs to be accounted for especially when change detection methods which use thresholds are applied.

C-band scatterometer (ERS-1, Metop ASCAT) find mostly application for detection of soil moisture changes (Bartalis et al., 2007; Wagner et al., 1999b; Zhribi et al., 2008). The higher sensitivity to changes in snow properties of the shorter Ku-band (from Seawinds/QuikScat and NSCAT) is employed for mostly glaciological and seasonal snow cover monitoring applications. These sensors have been also investigated outside the high latitudes for phenology (Frolking et al., 2005; Hardin & Jackson, 2003; Oza & Parihar, 2007; Prigent et al., 2001), urban mapping (Nghiem et al., in press) and soil moisture (Mladenova et al., in press) applications.

3. Change Detection Approaches

3.1 Freeze/thaw and snow monitoring

First analyses of scatterometer for seasonal thaw are based on ERS-1 data as complete coverage of seasonal cycles from this sensor are already available since 1992. Boehnke & Wismann (1996) calculated the typical summer (July) and Winter (February) backscatter level in order to determine the thaw timing. When a minimum of 50% of the winter summer difference is exceeded for at least two consecutive measurements ground thaw is detected. However, since re-crystallization of snow can cause similar backscatter levels as during summer in C-band an enhanced method has been developed (Wismann, 2000) which applies additionally a maximum likelihood classification over neighbouring pixels in cases when the initial detection fails.

Although the available record of NSCAT (Ku-band) is rather short (eleven months) it provided a first dataset covering an entire northern hemisphere winter and spring period at this wavelength. Its suitability for detection of freeze/thaw was tested by Frolking et al. (1999). They introduced a change detection algorithm which considers differences between

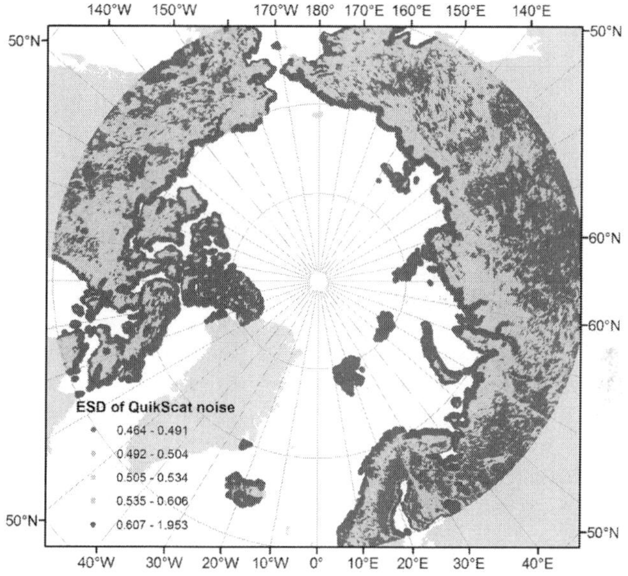

Fig. 2. Estimated Standard Deviation (ESD) of QuikScat long-term noise above 60 ° N; oceans and ice caps are excluded

five day averages and location specific differences from the overall mean value. This approach has been build on and extended within a number of follow-up studies with Ku-Band scatterometer. Kimball et al. (2001) transferred the C-band approach of Boehnke & Wismann (1996) to NSCAT data. The five day average method (Frolking et al., 1999) has been extended for NSCAT by extraction of three instead of one specific date of thaw: the start, the end and the primary thaw date which is the day with the highest backscatter difference (Kimball et al., 2004a). The five day moving average approach has been subsequently transferred to QuikScat (launched 1999) for final thaw date extraction and also applied in a similar way to autumn refreeze (Kimball et al., 2004b). A further method which is taking winter (February) mean backscatter into consideration is applying fixed thresholds for daily mean values in order to determine the onset of snowmelt (Brown et al., 2007). As QuikScat provides sufficient morning and evening measurement, a new adaptive approach based on diurnal thaw and refreeze of snow cover could be developed (Bartsch et al., 2007b; Kidd et al., 2003, TU Wien method). Thresholds are defined for each single grid cell depended on the estimated standard deviation of long-term noise (Figure 2) and the actual number of measurements available during each 12 hour period. Significant diurnal backscatter changes occur throughout the snowmelt period several times but not necessarily on subsequent days. This occurrence of multiple events has been solved with a clustering method. In case of multiple melt periods (several clusters of at minimum two days with diurnal thaw and refreeze), the last one is identified as the major melt period. An analyses limited to evening measurements using the five day average approach (Frolking et al., 1999) plus the summer mean backscatter (August) has been carried out by Wang et al. (2008). The evening values are taken from the BYU "egg" product (Long & Hicks, 2005). Static thresholds are used for thaw day definitions and it is assumed that relevant melt periods are longer than two days. If multiple events occur, the longest has been selected. This does not account for short term interruptions and thus supplies the end of melt only with respect to the entire spring melt.

Advanced products are snow covered area and melt area. The snow covered area can be determined with all above mentioned approaches as well as with optical data (Scherer et al., 2005). The melt area can be derived with the methods of Kimball et al. (2001) or Bartsch et al. (2007b) as they consider beginning and end of spring thaw. This differs from glaciological applications where single days or conseqtive days with surface melt need to be identified for melt season length determination (e.g. Tedesco (2007)) as any interruption of melt impacts the mass balance. Surface melt of seasonal snow cover, especially in relation to rain-on-snow events, affects thermal properties of the snow pack and the soil beneath (Putkonen & Roe, 2003). Even, single days of thaw during spring can cause an increase in heterotrophic soil respiration (Bartsch et al., 2007b). The primary thaw day (Kimball et al., 2001) extracted for the year 2000 in the circum-boreal and -arctic regions showed good correlations (R=0.75) with modelled timing of water content increase in the snowpack (Rawlins et al., 2005).

Current approaches are not applicable in regions where no continuous snow cover/frozen ground conditions during the winter time exists as they are designed to identify one seasonal thaw event or period only. The presence of snow itself is not considered in all approaches. The presence of melting snow causes decreased backscatter similar to water in both C-band and Ku-Band. Independent whether ground thaw or snow thaw is sought for multiple thaw periods within one winter season need to be accounted for at all latitudes. This has been so far considered in two mapping approaches only (Bartsch et al., 2007b; Wang et al., 2008). The for QuikScat typical variations in noise are only accounted for by the TU Wien method (Bartsch et al., 2007b; Kidd et al., 2003).

3.2 Soil moisture monitoring

The only change detection method for the determination of surface soil moisture from scatterometer data has been introduced by Wagner et al. (1999a). It is based on the assumption that most backscatter variation within the freeze free period is caused by changes in soil water content. The minimum (dry reference) and maximum (wet reference) values are site specific. Once they have been determined from a sufficiently long enough record each measurement can be scaled between those boundary values and a relative near surface soil moisture content determined. These datasets are available globally (since 2002, Scipal (2002)) and in case of ASCAT in near-real-time (Naeimi et al., in press). They can be thus applied e.g. for operational applications such as assimilation into weather forecasts (Scipal et al., 2008).

4. Examples

4.1 Snow melt

The TU Wien product which is based on an adaptive diurnal difference approach introduced by Bartsch et al. (2007b) can be applied to QuikScat data for the extraction of the beginning and the end of thaw. It has therefore been chosen for investigation of the melt area and river discharge behaviour over selected Russian basins. It considers the varying noise levels and captures the final thaw period with respect to multiple thaw events before the final snowmelt period and short term variations during spring thaw. Typical duration of final spring melt in central Siberia above 60 ° N is two weeks to one month.

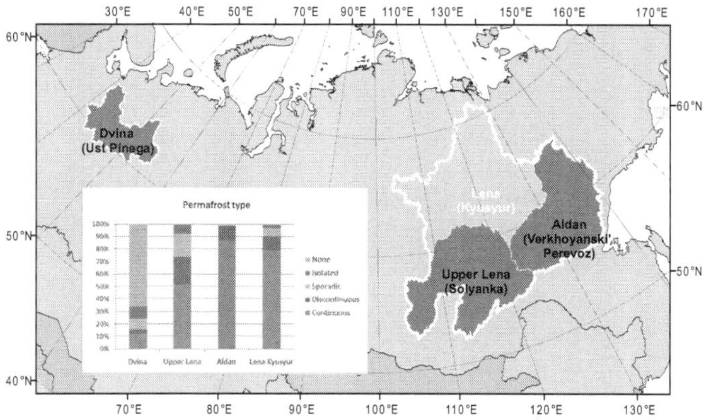

Fig. 3. Overview map of selected basins in Russia and proportion of permafrost types (source: NSDIC, Brown et al. (1998))

The area which undergoes snowmelt at a certain day has been extracted for three basins in Russia for the years 2001 to 2008. Those are the Dvina upstream of Ust Pinega ($\approx$270.000 km^2), the Lena river upstream of Kyusyur ($\approx$2.440.000 km^2) and two of its subbasins: the upper Lena upstream of Solyanka ($\approx$770.000 km^2) and the Lena river tributary Aldan (Verkhoyanski' Perevoz, $\approx$695.000 km^2). These basins show varying Permafrost characteristics (Figure 3, source: NSDIC, Brown et al. (1998)). Dvina has only 12.5% continuous permafrost. This proportion is higher for all other selected basins, 50% for upper Lena and 80% for Aldan and the entire Lena basin. The upper Lena basin constitutes most of the none-continuous permafrost of the Lena basin. Most of it, however, is also characterized by discontinuous and sporadic permafrost.

Figure 4 and Figure 5 show time series of melt area and discharge for the years 2001-2008. River runoff measurements are provided through ArcticRIMS (Regional, integrated Hydrological Monitoring System)/ R-ArcticNET (www.russia-arcticnet.sr.unh.edu). All basins are characterized by a pronounced runoff peak in spring.

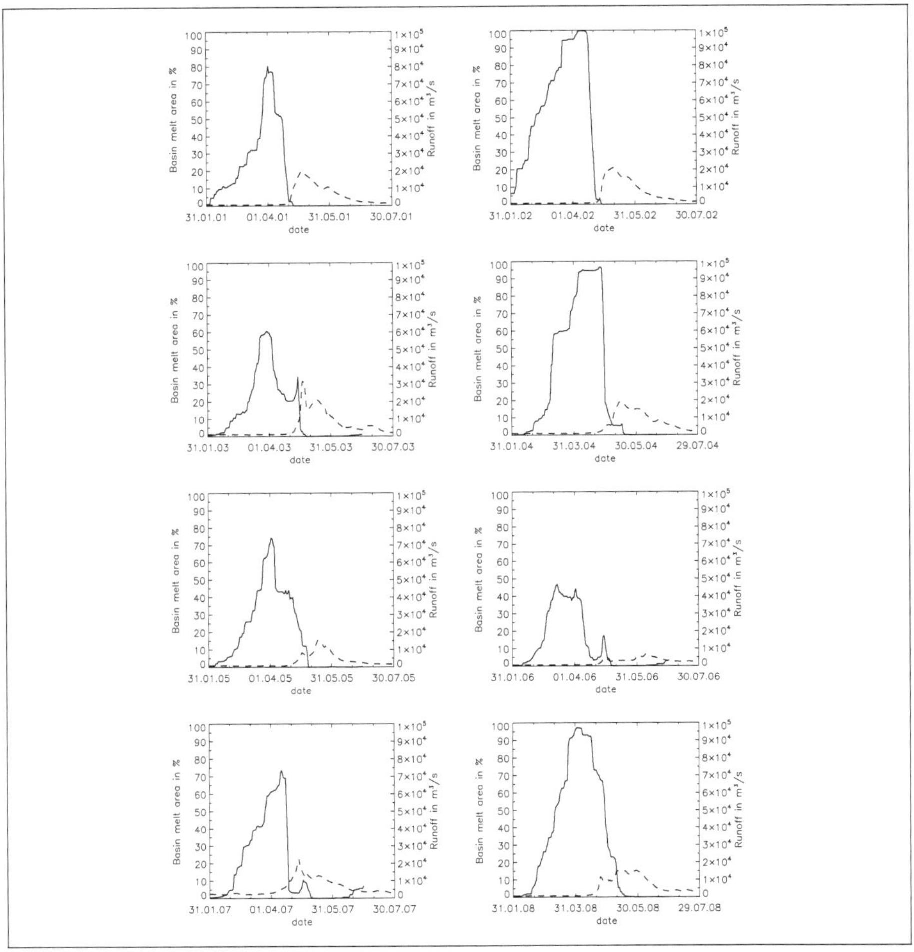

Fig. 4. QuikScat derived daily basin melt area in % of the Dvina basin (solid line) and river discharge in m^3/s at Ust Pinega (dashed line), 2001-2008

Dvina melt area and discharge measurements vary considerably over the eight analysed years. This applies to the magnitude as well as timing. The timing for the Lena is more constant and also does not vary much between the subbasins. Although the start of the snowmelt in the upper Lena basin is often earlier, it only slightly deviates from Aldan or entire Lena. This overlap may contribute to the distinct peak discharge observed at Kyusyur. The duration of spring snow smelt of Dvina is longer than for all Lena subbasins. This could be related to the higher average snow depth in the Eurasian Arctic than over the Lena basin (Khan et al., 2008).

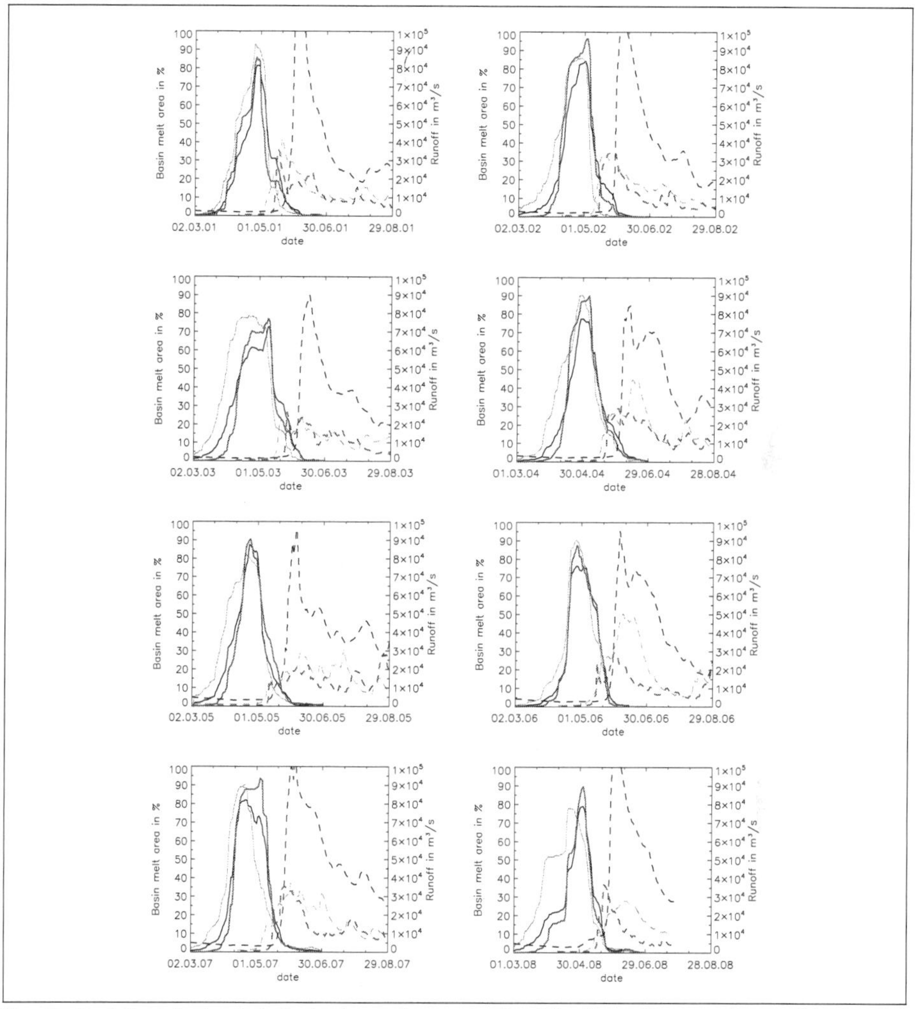

Fig. 5. QuikScat derived daily basin melt area in % of the Lena Kyusyur basin (black solid thick line), Uppler Lena Solyanka basin (solid thin grey line), Aldan basin (thick grey solid line) and river discharge in m^3/s at corresponding stations (dashed lines), 2001-2008

The magnitude of the melt area maximum and the river discharge spring maximum shows only a high correlation (R^2=0.79) for the upper Lena basin (Figure 6). This relationship is also partly visible for Aldan, but no distinct discharge peak could be observed in 2005 (Figure 5). The overall Lena basin spans over several degrees latitude and includes mountain ranges and therefore does not show a direct relationship.

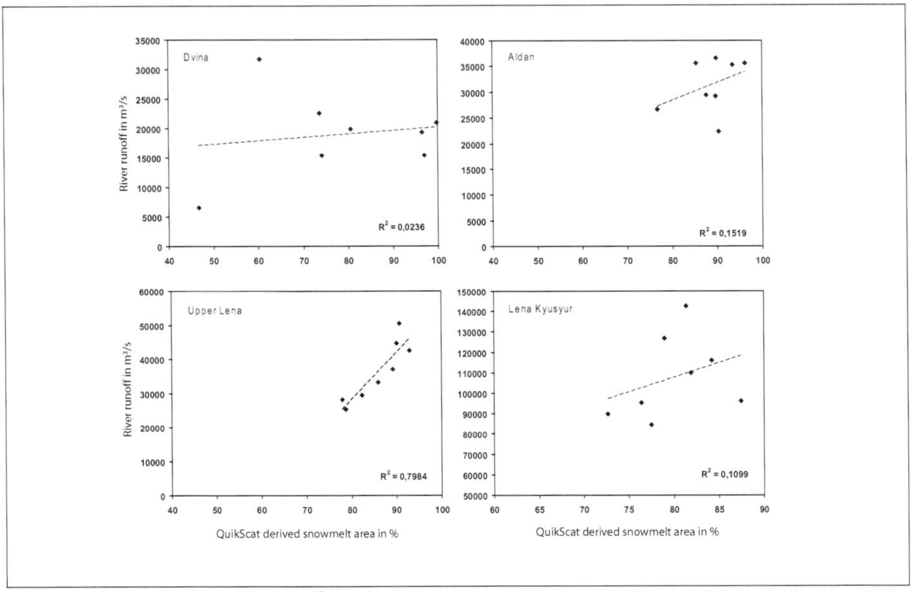

Fig. 6. Spring peak discharge (m^3/s) and QuikScat derived annual peak basin melt area (in%) comparison, 2001-2008

In order to compare the actual temporal offset between the melt area and discharge maxima for the different basins, the basin size needs to be taken into consideration. Therefore, the offset (in days) has been divided by basin area (100.000 km^2). The normalized offset is shortest for Aldan and Lena (Kyusyur), somewhat longer in most years for the upper Lena and clearly longer (and more variable) for the Dvina basin (Figure 7). The higher the extent of continuous permafrost the shorter the temporal offset between melt area maximum and spring discharge peak.

4.2 Near surface soil moisture

River discharge measurements from subtropic environment have already shown to have high correlation with ERS estimated soil water index (SWI, Scipal et al. (2005)). The soil water index is derived from the original surface soil moisture product using an exponential function in order to model infiltration (Wagner et al., 1999a). The advantage is that moisture estimates for larger depths become available. The original measurements only represent the upper 2-5 cm. The surface values are available in irregular intervals. The model output on the other hand supplies a regular 10-day dataset with respect to the varying global coverage. The percolation depth is static for the specified analyses region. However, if permafrost is present, the depth of unthawed ground (active layer) varies throughout the season. Therefore, 10-day means have been extracted from the original surface soil moisture calculations for the high latitude analyses. These values are averaged over the entire basins. In case of the Lena basin upstream of Kyusyur (Figure 3) a Pearson correlation of R^2=0.62 between the basin mean

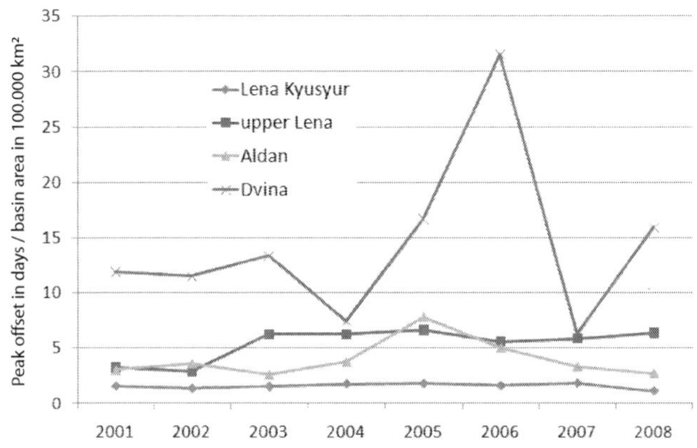

Fig. 7. Peak offset between spring discharge and QuikScat derived annual maximum melt area in days by basin area (in 100.000 km^2), 2001-2008

surface soil moisture and the river discharge measured at Kyusyur can be determined for the summer periods (Mid-June to end of August) of 1992-2000 (Figure 8).

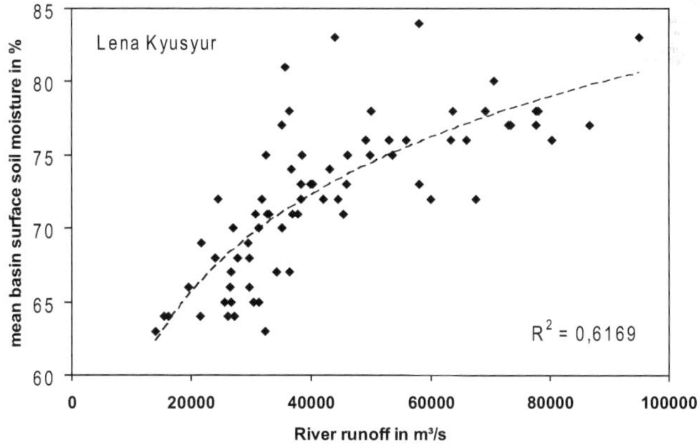

Fig. 8. ERS scatterometer derived 10-day mean basin surface soil moisture in % and River discharge in m^3/s for the Lena Kyusyur basin, Mid-June to end of August 1992-2000

5. Discussion

Variation in melt area and soil moisture can be introduced by variations in terrain and latitude. This has been show especially for the Mackenzie previously (Bartsch et al., 2007a). The use of the surface wetness in order to monitor runoff is only applicable when there are no other sources than rain or local ground thaw. This limits the application to regions without significant contribution by glacier melt water. Intermediate storage in lakes and wetlands can also decrease this relationship (Bartsch et al., 2007a). Surface wetness can be also derived from ScanSAR data as the ENVISAT ASAR operating in Global Monitoring Mode (Pathe et al., 2009). This provides more detailed (1km) although less frequent measurements and has been also demonstrated to be correlated to river runoff in subtropical environments (Bartsch et al., 2008b).

Although drainage can be also impeded by other ground characteristics, a decrease in permafrost extent may impact the for those basins currently determined relationship between river discharge and both soil moisture and snow melt patterns. Additionally a change in greening-up dates has been observed during the recent decades (Myneni et al., 1997). A possible related change in snowmelt timing can be monitored with the currently available scatterometers complementary to optical satellite data which are impacted by cloud coverage. Continuation of C-band scatterometer is ensured until 2020 within the Metop series of satellites (Naeimi et al., in press). Both C-band and Ku-band scatterometer are widely used for ocean applications and therefore a need for continuation exists for different purposes. So far, the joint use of both bands was limited due to the unavailability of ERS data after 2000 for many parts of the globe. A synergistic use became now possible due to the launch of ASCAT. This will allow a comprehensive monitoring of catchment hydrology in regions with seasonal snow cover.

6. Conclusion

Scatterometer are capable of providing a range of climate change relevant land surface parameters. They are especially sensitive to changes in the hydrological cycle. Products cover freeze/thaw status, snowmelt patterns and soil moisture variations. C-band data have been especially proven valuable for soil moisture monitoring. The variation of surface wetness over Lena River basin with more than $\approx$80% continuous permafrost captured during the snow/freeze free period highly correlates to measured river runoff without any offset. This significantly differs from basins in subtropic environments with similar size where water can percolate deeper into the ground. This delays the transport to the river courses and offsets can be several months. An impact of impeded drainage over permafrost can be also observed for the peak runoff associated with spring snowmelt. The temporal offset between melt area maximum and river discharge maximum decreases with increasing proportion of continuous permafrost in the basin. The maximum melt area reached over basins with high proportion of continuous permafrost can correlate in cases with the magnitude of peak discharge (R^2= 0.8 for upper Lena). In spite of the coarse resolution of scatterometer data, they provide valuable operational monitoring tools of terrestrial hydrology at high latitudes on regional to circumpolar scale.

Acknowledgments

This study has been funded by the Austrian Science Fund through the Hertha Firnberg program.

7. References

Abdel-Messeh, M. & Quegan, S. (2000). Variability in ERS scatterometer measurements over land, *IEEE Transactions on Geoscience and Remote Sensing* **38**(4): 1767–1776.

Bartalis, Z., Scipal, K. & Wagner, W. (2006). Azimuthal anisotropy of scatterometer measurements over land, *IEEE Transactions on Geoscience and Remote Sensing* **44**(8): 2083–2092.

Bartalis, Z., Wagner, W., Naeimi, V., Hasenauer, S., Scipal, K., Bonekamp, H., Figa, J. & Anderson, C. (2007). Initial soil moisture retrievals from the METOP-a advanced scatterometer (ASCAT), *Geophysical Research Letters* **34**: L20401.

Bartsch, A., Doubkova, M., Pathe, C., Sabel, D., Wolski, P. & Wagner, W. (2008b). River flow & wetland monitoring with ENVISAT ASAR global mode in the Okavango basin and delta, *Proceedings of the Second IASTED Africa Conference, September 8-10, 2008 Gaborone, Botswana, Water Resource Management (AfricaWRM 2008)*, IASTED, pp. 152–156.

Bartsch, A., Kidd, R. A., Wagner, W. & Bartalis, Z. (2007b). Temporal and spatial variability of the beginning and end of daily spring freeze/thaw cycles derived from scatterometer data, *Remote Sensing of Environment* **106**: 360–374.

Bartsch, A., Pathe, C., Wagner, W. & Scipal, K. (2008a). Detection of permanent open water surfaces in central Siberia with ENVISAT ASAR wide swath data with special emphasis on the estimation of methane fluxes from tundra wetlands, *Hydrology Research* **39**(2): 89–100.

Bartsch, A., Wagner, W., Rupp, K. & Kidd, R. (2007a). Application of C and Ku-band scatterometer dara for catchment hydrology in northern latitudes, *Proceedings of the 2007 IEEE International Geoscience and Remote Sensing Symposium. 23-27 July, Barcelona, Spain.*, IEEE.

Boehnke, K. & Wismann, V. R. (1996). ERS scatterometer land applications: Detecting the thawing of soils in Siberia, *Earth Observation Quarterly 52*, ESA Publication Division.

Brown, J., Ferrians Jr., O. J., Heginbottom, J. A. & Melnikov, E. S. (1998). Circum-arctic map of permafrost and ground-ice conditions, Boulder, CO: National Snow and Ice Data Center/World Data Center for Glaciology. Digital Media. revised February 2001.

Brown, R., Derksen, C. & Wang, L. (2007). Assessment of spring snow cover duration variability over Northern Canada from satellite dataset, *Remote Sensing of Environment* **111**: 367–381.

Callaghan, T. V., Björn, L. O., Chernov, Y., Chapin, T., Christensen, T. R., Huntley, B., Ims, R. A., Johansson, M., Jolly, D., Jonasson, S., Matveyeva, N., Panikov, N., Oechel, W., Shaver, G., Schaphoff, S. & Sitch, S. (2004). Effects of changes in climate on landscape and regional processes, and feedbacks to the climate system, *Ambio* **33**(7): 459–468.

Dye, G. D. & Tucker, C. J. (2003). Seasonality and trends of snow cover, vegetation index and temperature in northern Eurasia, *Geophysical Research Letters* **30**(7): 1405.

Early, D. S. & Long, D. G. (2001). Image reconstruction and enhanced resolution imaging from irregular samples, *IEEE Transactions on Geoscience and Remote Sensing* **39**(2): 291–302.

French, H. M. (1996). *The Periglacial Environment*, 2nd edn, Longman, Harlow.

Frolking, S., Fahnestock, M., Milliman, T., McDonald, K. & Kimball, J. (2005). Interannual variability in North American grassland biomass/productivity detected by SeaWinds scatterometer backscatter, *Geophysical Research Letters* **32**: L21409.

Frolking, S., McDonald, K. C., Kimball, J. S., Way, J. B., Zimmermann, R. & Running, S. W. (1999). Using the space-borne NASA scatterometer (NSCAT) to determine the frozen and thawed seasons, *Journal of Gephysical Research* **104**(D22): 27895–27907.

Hardin, P. J. & Jackson, M. W. (2003). Investigating SeaWinds terrestrial backscatter: Equatorial savannas of South America, *Photogrammetric Engineering & Remote Sensing* **69**(11): 1243–1254.

Hinzman, L. D., Bettez, N. D., Bolton, W. R., Chapin, F. S., Dyurgerov, M. B., Fastie, C. L., Griffith, B., Hollister, R. D., Hope, A., Huntington, H. P., Jensen, A. M., Jia, G. J., Jorgenson, T., Kane, D. L., Klein, D. R., Kofinas, G., Lynch, A. H., Lloyd, A. H., McGuire, A. D., Nelson, F. E., Oechel, W. C., Osterkamp, T. E., Racine, C. H., Romanovsky, V. E., Stone, R. S., Stow, D. A., Sturm, M., Tweedie, C. E., Vourlitis, G. L., Walker, M. D., Walker, D. A., Webber, P. J., Welker, J. M., Winker, K. S. & Yoshikawa, K. (2005). Evidence and implications of recent climate change in northern alaska and other arctic regions, *Climatic Change* **72**: 251–298.

IPCC (2007). *Climate Change 2007: The Physical Science Basis. Contribution of Working Group I to the Fourth Assessment Report of the Intergovernmental Panel on Climate Change*, Cambridge University Press, Cambridge, United Kingdom and New York, NY, USA.

Kane, D. L. (1997). The impact of arctic hydrologic perturbations on arctic ecosystems induced by climate change, *in* W. C. Oechel (ed.), *Global Change and Arctic Terrestrial Ecosystems*, Vol. 124 of *Ecological Studies*, Springer, pp. 63–81.

Kane, D. L., McNamara, J. P., Yang, D., Olsson, P. Q. & Gieck, R. E. (2003). An extreme rainfall/runoff event in Arctic Alaska, *Journal of Hydrometeorology* **4**: 1220–1228.

Khan, V., Holko, L., Rubinstein, K. & Breiling, M. (2008). Snow cover characteristics over the main Russian river basins as represented by reanalyses and measured data, *Journal of Applied Meteorology and Climatology* **47**(6): 1819–1833.

Kidd, R. A., Trommler, M. & Wagner, W. (2003). The development of a processing environment for time-series analyses of SeaWinds scatterometer data, *IGARSS Proceedings*, IEEE, pp. 4110–4112.

Kimball, J. S., McDonald, K. C., Frolking, S. & Running, S. W. (2004a). Radar remote sensing of the spring thaw transition across a boreal landscape, *Remote Sensing of Environment* **89**: 163–175.

Kimball, J. S., McDonald, K. C., Keyser, A. R., Frolking, S. & Running, S. W. (2001). Application of the NASA scatterometer (NSCAT) for determining the daily frozen and nonfrozen landscape of Alaska, *Remote Sensing of Environment* **75**: 113–126.

Kimball, J. S., McDonald, K. C., Running, S. W. & Frolking, S. E. (2004b). Satellite radar remote sensing of seasonal growing seasons for boreal and subalpine evergreen forests, *Remote Sensing of Environment* **90**: 243–258.

Klaes, K. D., Cohen, M., Buhler, Y., Schlüssel, P., Munro, R., Luntama, J.-P., Engeln, A. V., Clerigh, E. O., Bonekamp, H., Ackermann, J. & Schmetz, J. (2007). An introduction to the EUMETSAT polar system, *Bulletin of the American Meteorological Society* **88**(7): 1085–1096.

Lilly, M. R., Paetzold, R. F. & Kane, D. L. (2008). Tundra soil-water content and temperature data in support of winter tundra travel, *Proceedings of the Ninth International Symposium on Permafrost*, Fairbanks, Alaska, pp. 1067–71.

Long, D. G. & Hicks, B. R. (2005). Standard BYU QuikSCAT/SeaWinds land/ice image products, *MERS Technical Report 05-04*, Brigham Young University.

Mladenova, I., Lakshmi, V., Walker, J. P., Long, D. G. & De Jeu, R. (in press). An assessment of QuikSCAT Ku-band scatterometer data for soil moisture sensitivity, *IEEE Geoscience and Remote Sensing Letters* .

Myneni, R. B., Keeling, C. D., Tucker, C. J., Asrar, G. & Nemani, R. R. (1997). Increased plant growth in the northern high latitudes from 1981-1991, *Nature* **386**: 698–702.

Naeimi, V., Bartalis, Z. & Wagner, W. (2009). ASCAT soil moisture: An assessment of the data quality and consistency with the ERS scatterometer heritage, *Journal of Hydrometeorology* **10**: 555–563.

Naeimi, V., Scipal, K., Bartalis, Z., Hasenauer, S. & Wagner, W. (in press). An improved soil moisture retrieval algorithm for ERS and METOP scatterometer observations, *IEEE Transactions on Geoscience and Remote Sensing* .

Nghiem, S., Balk, D., Rodriguez, E., Neumann, G., Sorichetta, A., Small, C. & Elvidge, C. (in press). Observations of urban and suburban environments with global satellite scatterometer data, *ISPRS Journal of Photogrammetry and Remote Sensing* .

Nghiem, S. V. & Tsai, W.-Y. (2001). Global snow cover monitoring with spaceborne ku-band scatterometer, *IEEE Transactions on Geoscience and Remote Sensing* **39**(10): 2118–2134.

Oza, S. R. & Parihar, J. S. (2007). Evaluation of Ku-band QuikSCAT scatterometer data for rice crop growth stage assessment, *International Journal of Remote Sensing* **28**(16): 3447–3456.

Pathe, C., Wagner, W., Sabel, D., Doubkova, M. & Basara, J. (2009). Using ENVISAT ASAR global mode data for surface soil moisture retrieval over Oklahoma, USA, *IEEE Transactions on Geoscience and Remote Sensing* **47**(2): 468–480.

Prigent, C., Matthews, E., Aires, F. & Rossow, W. B. (2001). Remote sensing of global wetland dynamics with multiple satellite data sets, *Geophysical Research Letters* **28**: 4631–4634.

Prigent, C., Papa, F., Aires, F., Rossow, W. B. & Matthews, E. (2007). Global inundation dynamics inferred from multiple satellite observations, 1993–2000, *Journal of Geophysical Research* **112**: D12107.

Putkonen, J. & Roe, G. (2003). Rain-on-snow events impact soil temperatures and affect ungulate survival, *Geophysical Research Letters* **30**(4): 1188.

Rawlins, M. A., McDonald, K. C., Frolking, S., Lammers, R. B., Fahnestock, M., Kimball, J. S. & V&Oumlr&Oumlsmarty, C. J. (2005). Remote sensing of snow thaw at the pan-Arctic scale using the SeaWinds scatterometer, *Journal of Hydrology* **312**: 294–311.

Scherer, D., Hall, D. K., Hochschild, V., König, M., Winther, J.-G., Duguay, C. R., Pivot, F., Mätzler, C., Rau, F., Seidel, K., Solberg, R. & Walker, A. E. (2005). Remote sensing of snow cover, *in* C. R. Duguay & A. Pietroniro (eds), *Remote Sensing in Northern Hydrology: Measuring Environmental Change*, Vol. 163 of *Geophysical Monograph Series*, American Geophysical Union, Washington, pp. 7–38.

Scipal, K. (2002). *Global Soil Moisture Monitoring Using ERS Scatterometer Data*, PhD thesis, Vienna University of Technology.

Scipal, K., Drusch, M. & Wagner, W. (2008). Assimilation of a ERS scatterometer derived soil moisture index in the ECMWF numerical weather prediction system, *Advances in Water Resources* **31**: 1101–1112.

Scipal, K., Scheffler, C. & Wagner, W. (2005). Soil moisture-runoff relation at the catchment scale as observed with coarse resolution microwave remote sensing, *Hydrology and Earth System Sciences* **9**: 173–183.

Smith, N. V., Saatchi, S. S. & Randerson, J. T. (2004). Trends in high northern latitude soil freeze thaw cycles from 1988 to 2002, *Journal of Geophysical Research* **109**: D12101.

Tedesco, M. (2007). Snowmelt detection over the Greenland ice sheet from SSM/I brightness temperature daily variations, *Geophysical Research Letters* **34**: L02504.

Ulaby, F. T., Moore, R. K. & Fung, A. (1982). *Microwave Remote Sensing–Active and Passive*, Vol. II, Artech House, Norwood, Mass.

Wagner, W., Blöschl, G., Pampaloni, P., Calvet, J.-C., Bizzarri, B., Wigneron, J.-P. & Kerr, Y. (2007). Operational readiness of microwave remote sensing of soil moisture for hydrologic applications, *Nordic Hydrology* **38**: 1–20.

Wagner, W., Lemoine, G., Borgeaud, M. & Rott, H. (1999c). A study of vegetation cover effects on ERS scatterometer data, *IEEE Transactions on Geoscience and Remote Sensing* **37**(2): 938–948.

Wagner, W., Lemoine, G. & Rott, H. (1999a). A method for estimating soil moisture from ERS scatterometer and soil data, *Remote Sensing of Environment* **70**: 191–207.

Wagner, W., Noll, J., Borgeaud, M. & Rott, H. (1999b). Monitoring soil moisture over the Canadian prairies with the ERS scatterometer, *IEEE Transactions on Geoscience and Remote Sensing* **37**(1): 206–216.

Wang, L., Derksen, C. & Brown, R. (2008). Detection of pan-Arctic terrerstrial snowmelt from QuikSCAT, 2000-2005, *Remote Sensing of Environment* **112**(10): 3794–3805.

Way, J. B., Zimmermann, R., Rignot, E., McDonald, K. & Oren, R. (1997). Winter and spring thaw as observed with imaging radar at BOREAS, *Journal of Geophysical Research* **102**: 29673–29684.

Wegmüller, U. (1990). The effect of freezing and thawing on the microwave signatures of bare soil, *Remote Sensing of Environment* **33**: 123–135.

Williams, P. J. & Smith, M. W. (1989). *The Frozen Earth: Fundamentals of Geocryology*, Cambridge University Press, New York.

Wismann, V. (2000). Monitoring of seasonal thawing in Siberia with ERS scatterometer data, *IEEE Transactions on Geoscience and Remote Sensing* **38**(4): 1804–1809.

Yang, D., Zhao, Y., Armstrong, R., Robinson, D. & Brodzik, M.-J. (2007). Streamflow response to seasonal snow cover mass changes over large Siberian watersheds, *Journal of Geophysical Research* **112**: F02S22.

Zhribi, M., André, C. & Decharme, B. (2008). A method for soil moisture estimation in Western Africa based on the ERS scatterometer, *IEEE Transactions on Geoscience and Remote Sensing* **46**(2): 438–448.

15

Ocean wind fields from satellite active microwave sensors

S. Zecchetto
Istituto Scienze dell'Atmosfera e del Clima
Padova, Italy
s.zecchetto@isac.cnr.it

1. The Marine Atmospheric Boundary Layer

"In the Earth's atmosphere, the planetary boundary layer is the air layer near the ground affected by diurnal heat, moisture or momentum transfer to or from the surface". This definition, obtained from [1], may introduce the Marine Atmospheric Boundary Layer (MABL) as the planetary boundary layer over the sea surface. In this layer, important exchanges of sensible and latent heat and momentum take place over a large spectrum of time and spatial scales, driving the sea waves, the drift ocean currents and the storage of CO_2 by the sea due to the wind and the breaking waves. In this context, the leading quantity is the wind vector **U**. Its assessment is of paramount importance in the evaluation of the wind stress $\tau = C_d(T_a, T_s, T_d) \cdot |\mathbf{U}|^2$, (the drag coefficient C_d is a function depending, in a first approximation, on the air T_a, the sea T_s and the dew T_d temperatures), and of the gas transfer velocity $k = 2.8310^{-2} \cdot |\mathbf{U}|^3$ (Monahan, 2002), for instance.
One of the major problems in understanding the dynamics of the wind in the surface layer, the bottom layer inside the MABL where the turbulent fluxes exhibit a variability smaller than 10%, is the difficulty to get experimental data at spatial scales from few meters to few kilometers.
The satellite sensors discussed in this chapter measure the backscatter from the sea surface, providing maps directly related to the characteristics of the surface layer and to the wind blowing inside this layer. Satellite active microwave sensors are the only instruments able to provide information about the spatial structure of the wind in the marine surface layer over large areas.

2. Satellite active microwave sensors

The active microwave sensors (Campbell, 2002; CCRS, 2009; Elachi, 1988) are radars operating in the microwave region (1 to 30 GHz in frequency, 1 to 30 cm in wavelength) at different polarizations and incidence angles. Over the sea, the radar return depends, besides the geometry of the radar illumination, from the degree of development of the sea surface roughness (Valenzuela, 1978), composed by centimeter sea waves produced by the wind. Since the wind field has its own spatial pattern, which depends on its strength, on the thermodynamic characteristics at the air-sea interface and on the interaction between the wind

[1] `http://en.wikipedia.org/wiki/Boundary_layer`

flow and the orography, the sea surface roughness it generates its spatial features. The radar backscatter does reproduce, in turns, the sea surface roughness. Therefore, the study of the characteristics of the radar backscatter provides information on the characteristics of the wind and of the MABL.

The sea surface roughness is also modulated by some pre-existing oceanographic phenomena, like sea surface gravity waves, internal waves and ocean currents, or by the presence of oil slicks on the sea surface, which muffle the roughness. These modulations permit the detection of these oceanographic phenomena, besides the wind field.

This section introduces the two most popular radar sensors: the scatterometer, used to measure the wind field over the ocean, and the Synthetic Aperture Radar (SAR), used for a variety of applications, from land (forestry, geology, agriculture) to ocean (ocean surface waves, currents, ocean wind).

2.1 Spaceborne scatterometers

At present, the two most important satellites carrying scatterometers are the NASA *QuikSCAT* (JPL, 2006) and the Eumetsat *Metop* (Eumetsat, 2007). Both fly on a polar sun-synchronous orbit of about 100 minute of period. *QuikSCAT* has a repetition cycle of 4 days, whereas *Metop* of 29 days. This means that every 4 (29) days the scatterometers cover exactly the same areas of the Earth. The scatterometer winds are referenced to 10 m of height above the sea surface and to equivalent neutral air-sea stability conditions.

Scatterometer data are widely used by the scientific meteorologic community: they are assimilated into the global atmospheric models (Isaksen & Janssen, 2008; Isaksen & Stoffelen, 2000), used operationally for coastal (Lislie et al., 2008; Milliff & Stamus, 2008) and tropical cyclone (Brennan et al., 2008; Singh et al., 2008) wind forecasting, in global scale and mesoscale meteorology studies (Chelton et al., 2004; Liu et al., 1998; Zecchetto & Cappa, 2001), in climatological studies (Kolstad, 2008; Risien & Chelton, 2008; Zecchetto & De Biasio, 2007), in the assessment of the performances of the global (Chelton & Freilich, 2005) and regional atmospheric models (Accadia et al., 2007), in the oceanic simulations (Millif et al., 2001; Ruti et al., 2008).

2.1.1 SeaWinds on board of QuikSCAT satellite

QuikSCAT is a NASA satellite launched in June 1999. It provides, by means of the on board scatterometer *SeaWinds* working at Ku band (13.4 GHz), wind fields with spatial resolution of 25 km × 25 km and 12.5 km × 12.5 km at neutral air-sea stability conditions. *SeaWinds* is a scatterometer with a rotating antenna, measuring the wind in swaths 1800 km wide. Because of the operating frequency, QuikSCAT data can be seriously contaminated by rain (Jones et al., 1999; Portabella & Stoffelen, 2001). For this reason the wind data are provided with the probability that the columnar rate of rain exceedes 2 km mm h^{-1}, (Huddleston & Stiles, 2000), which can be used to discard the contaminated data. Figure 1 reports a *SeaWinds* swath over the European waters.

The data used here are the level L2B data set, available at PODAAC[2]. According to the sensor specifications, the QuikSCAT winds have an accuracy of 2 ms^{-1} in speed and 20° in direction in the wind speed range 3-20 ms^{-1}, but the actual accuracies are generally better (1 m s^{-1} and 23° (Ebuchi et al., 2002), 1.3 m s^{-1} and 27° (Pickett et al., 2003), 1.7 m s^{-1} and 14° (Chelton & Freilich, 2005)).

[2] available at podaac.jpl.nasa.gov/pub/ocean_wind/quikscat

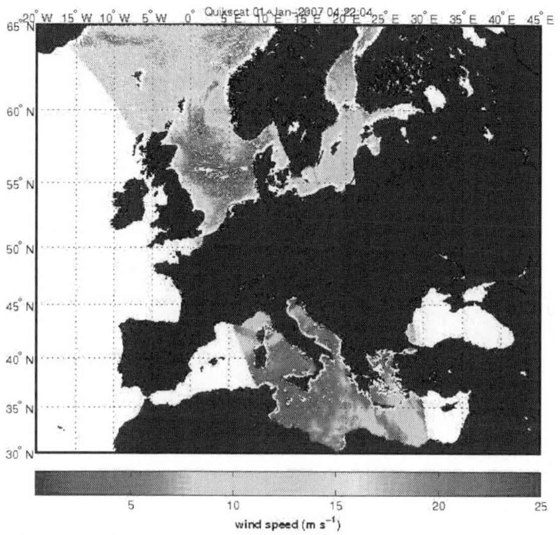

Fig. 1. A swath of *SeaWinds* over the European waters. 1 January 2007 at 04:22 GMT. Ascending orbit.

2.1.2 ASCAT on board of Metop satellite

Since May 2007, the European satellite *Metop* is operational: among other instruments, it carries the scatterometer *ASCAT*. Differently from *QuikSCAT*, *ASCAT* has fixed antennas in the two sides of the satellite, producing a swath composed by two sub-swaths 500 km wide, 768 km apart. The available spatial resolutions are 25 km by 25 km and 12.5 km by 12.5 km. Working at C-band (5.255 GHz), ASCAT data are only slightly affected by rain. Figure 2 reports a swath of *ASCAT* over the European waters at spatial resolution of 25 km by 25 km. *ASCAT* wind data are available at Eumetsat[3] or in near real time from the Dutch Met Office (`www.knmi.nl`), disseminating the data on behalf of the Ocean & Sea Ice Satellite Application Facility (`www.osi-saf.org`) of EUMETSAT (`www.eumetsat.org`).

2.2 The Synthetic Aperture Radar

At present, several Synthetic Aperture Radar (SAR) instruments are flying above us: the Advanced SAR instrument of Envisat (March 2002) (ESA, 2002), the German TerraSAR-X (June 2007) (DLR, 2003), the Italian Cosmo-Skymed programme (from June 2007) (ASI, 2007), the Canadian commercial satellite RADARSAT-2 (December 2007) (CSA, 2001; Morena et al., 2004). Table 1 reports the main characteristics of the mentioned SARs.

The term polarization refers to the polarization of the transmitted Tx and received Rx electromagnetic waves. Single polarization can be (TxRx) VV or HH or VH or HV; dual polarization comprises HH and HV or VV and VH; quad (fully) polarization is when all the possible polarization combinations are acquired, i. e. HH, HV, VV, VH. Terrasar-X, CosmoSkyMed and RADARSAT-2 are fully polarimetric SARs.

[3] http://archive.eumetsat.int/umarf/

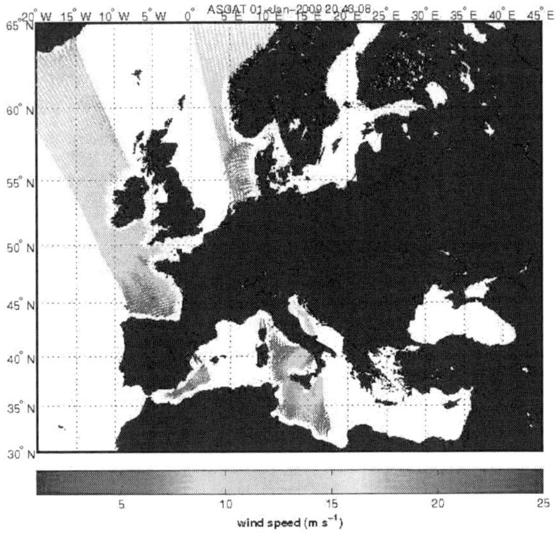

Fig. 2. A swath of *ASCAT* over the European waters. 1 January 2009 at 20:43 GMT. Ascending orbit.

Satellite	Polarization	Frequency	Spatial resolution (width x length) m	Swath width (km)
Envisat[1]	Single, dual	C-band (5.3 GHz)	Polarization mode: ~ 30 x 30 Wide Swath mode: ~ 150 x 150 Global Monitoring mode: ~ 1000 x 1000	 up to 100 400 > 400
TerraSAR-X[2]	Single, dual, quad	X-band (9.6 GHz)	SpotLight: up to 1 StripMap: up to 3 ScanSAR: up to 18	10 30 100
Radarsat-2[3]	Single, dual, quad	C-band (5.4 GHz)	Ultra-Fine: 3 x 3 Multi-Look Fine: 8 x 8 Standard: 25 x 26 Wide: 30 x 26 ScanSAR narrow: 50 x 50 ScanSAR wide: 100 x 100 Standard Quad-pol: 12 x 8 Fine Quad-pol: 25 x 8	20 50 100 150 300 500 25 25
CosmoSkyMed[4]	Single, dual, quad	X-band (9.6 GHz)	Spotlight-2: 1 x 1 Stripmap: 3 x 3 5 x 5 Scansar: 30 x 30 100 x 100	10 30 40 100 200

Table 1. The main characteristics of the operational SAR instruments. From: [1]envisat.esa.int; [2]www.infoterra.de/terrasar-x; [3]www.radarsat2.info;[4]www.e-geos.it/docs/asi.pdf

3. Mesoscale wind meteorology from scatterometer data

The mesoscale may be defined, according to Orlanski (1975), as composed by three subranges: the mesoscale γ, from 2 km to 20 km, β, from 20 km to 200 km and α, from 200 km to 2000 km. This range is of uttermost importance, since in this range the wind controls the atmosphere's dynamics. This range is also sensitive to local modulations of the wind field, especially in regions where steep orography surrounds the various basins. Indeed, this is the range where global models have a decreased ability in reproducing the surface wind field. One of the regions where the atmospheric phenomena frequently occur in the mesoscale range is the Mediterranean Basin, which is chosen here as the area of interest.

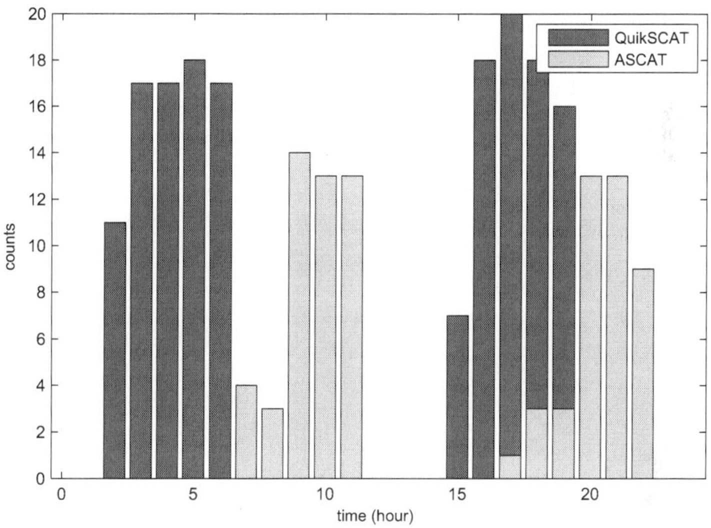

Fig. 3. Frequency distribution of QuikSCAT (red) and ASCAT (yellow) passes over the Mediterranean Basin as a function of the day time (GMT). January 2008.

The Mediterranean Basin is a semi-enclosed basin, having maximum extent of about 4000 km east-west and of about 1200 km north-south. It is almost entirely surrounded by mountain chains (with the exception of the east coast of Tunisia), which often raise nearby the coastline. The complexity of the coastal orography and the presence of mountainous islands deeply influence the local scale atmospheric circulation in the MABL, producing local effects at spatial scales down to a few kilometers. In the Mediterranean Basin, many regional wind systems, local cyclogeneses and wind flow disturbances induced by orography have a spatial variability at the mesoscale β. Up to now, the atmospheric phenomena at this scale in the Mediterranean Basin have not been extensively studied, mainly due to the lack of high spatial resolution data.

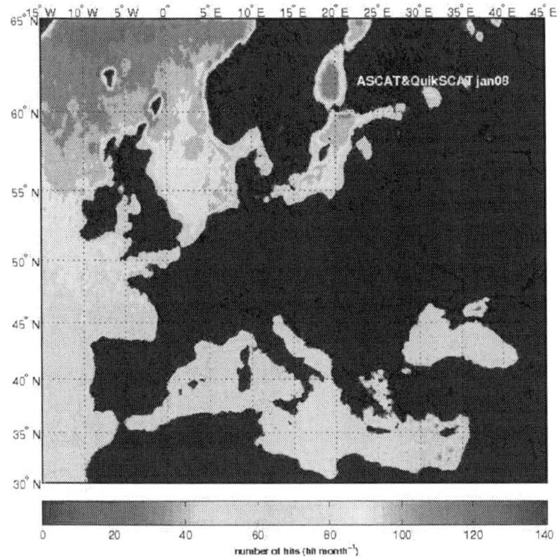

Fig. 4. Monthly coverage of QuikSCAT and ASCAT scatterometers over the European waters. January 2008.

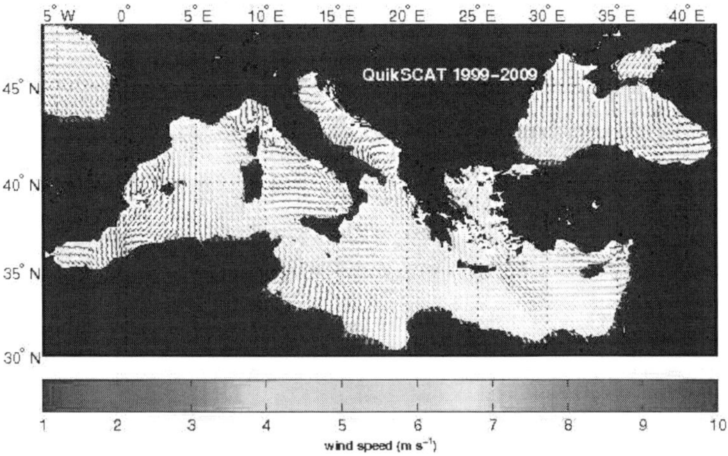

Fig. 5. Mean wind speed field over the Mediterranean Basin (July 1999 to May 2009) derived from QuikSCAT data. The vectors are plotted at one third of their original space resolution for readability

3.1 Parameters describing the wind field characteristics

Besides the mean wind field, several other parameters are important in defining the wind spatial structure, i. e. the wind gustiness, the wind steadiness and the wind speed variability.

The wind speed gustiness G is defined as:

$$G = (\sum_{i=1}^{n}(U_i - \overline{U})^2)^{1/2}/\overline{U} \tag{1}$$

where U_i is the wind speed modulus at time i, and $\overline{U}$ the mean value of ensemble $\{U_i\}$. The gustiness is a non-dimensional parameter providing information on the turbulent intensity of the wind. All the previous quantities are related to wind speed, neglecting wind direction. The wind steadiness coefficient S, expressed as

$$S = 100\frac{[(\sum_{i=1}^{n}(u_i)^2) + (\sum_{i=1}^{n}(v_i)^2)]^{1/2}}{\sum_{i=1}^{n}(u_i{}^2 + v_i{}^2)^{1/2}} \tag{2}$$

where u_i and v_i are the zonal and meridional wind components, provides insights into the variability of the wind direction. This non-dimensional parameter, which expresses the ratio between the mean vector and the mean scalar wind speed, ranges from 0 (wind direction randomly changing) to 100 (constant wind direction). It permits the identification of persistent wind regimes. The wind speed variability σ_w is the wind standard deviation computed over the period considered, and may be considered to integrate the information provided by the non-dimensional gustiness.

3.2 Temporal sampling and spatial coverage

To study the climatological spatial properties of a field, it is important to know how it has been produced and the temporal sampling of the area of interest.

One of the important aspects concerns the scatterometer pass time over a region of interest. Considering the Mediterranean Basin, the pass time of QuikSCAT and ASCAT, regardless the number of data per passage, may be inferred from Fig. 3, which reports the frequency distribution of the pass time as a function of the day time for January 2008. QuikSCAT swaths the Mediterranean Basin in the early morning and early afternoon, while ASCAT in the middle morning and evening. Figure 4 reports the map of the number of hits provided by QuikSCAT and ASCAT together over the European waters for one month (January 2008). The sampling roughly increases with latitude, from the $\approx$ 50 hit month^{-1} of the eastern Mediterranean to the $\approx$ 140 hit month^{-1} above 60°. In the Mediterranean Basin there are about two measurements per day: this permits to represent the temporal evolution of the wind only at scales longer than one day, but prevents to study of the wind associated to phenomena like fronts or cyclogeneses.

With the present coverage provided jointly by QuikSCAT and ASCAT in the Mediterranean Basin, it is possible to study the spatial structure of the winds in the mesoscale α and β, while their temporal evolution only in the mesoscale α.

3.3 Climatological spatial structure of the wind

The short time climatology of the spatial structure of the wind has been built over the ten years of QuikSCAT data available (July 1999 to May 2009), and presented here in terms of seasonal fields. To illustrate some aspect of the climatological spatial structure of the wind over the Mediterranean Basin, the winter and summer maps of wind speed variability σ_w and of wind steadiness S are presented. Before to present them, it is useful to sketch the general characteristics of the large scale wind circulation over the Mediterranean basin.

The Mediterranean weather is mainly driven by the Atlantic weather which, locally modified over the Mediterranean area (HMSO, 1962), produces a secondary weather system of spatial scales smaller than 500 km or so, frequent in all seasons. This is due both to the basin size and its geographic configuration: the coastal steep orography and the presence of large islands may modify the synoptic scale air flows over the region producing well distinct circulations on regional scale (Zecchetto & Cappa, 2001; Zecchetto & De Biasio, 2007). Many of the most important regional winds in this basin are indeed the product of the interaction between the synoptic scale flow and the local orography. However, the Mediterranean Sea, due to its longitudinal extension, can be subject to various (sometimes co-exiting) circulation regimes as, for instance, the etesian wind circulation in the Eastern Mediterranean during summer, mainly linked to the Asian Monsoon (Ziv et al., 2004).

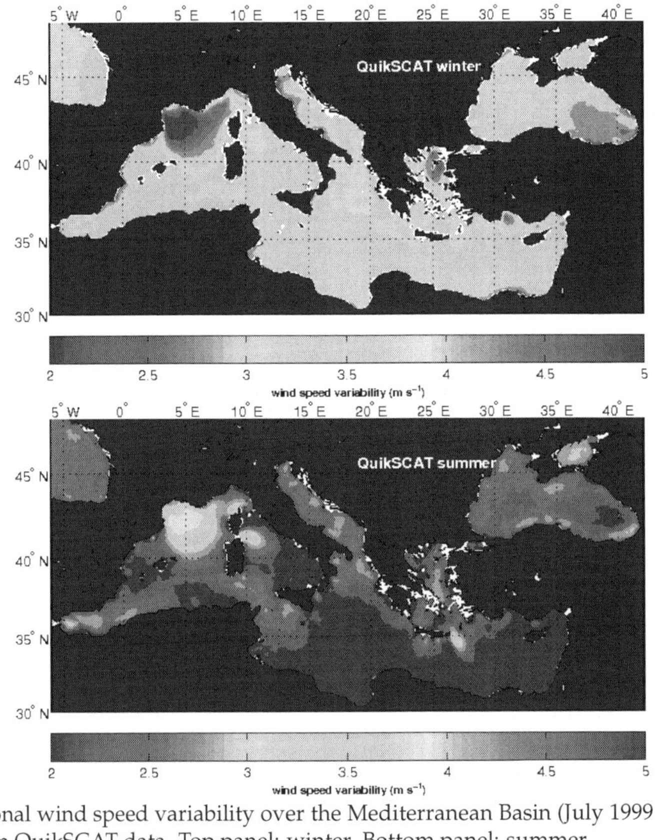

Fig. 6. Seasonal wind speed variability over the Mediterranean Basin (July 1999 to May 2009) derived from QuikSCAT data. Top panel: winter. Bottom panel: summer.

Figure 5 reports the QuikSCAT mean wind field over ten years (July 1999 to May 2009), plotted at one third of its original space resolution for readability. The basin scale wind circulation is mainly from northwest, with a strong signature of mistral in the Gulf of Lion and of the etesian in the eastern part of the basin. However, several regional wind circulations are co-existing, such as the northeastern bora wind (Pandžić & Likso, 2005; Yoshino, 1976) affecting the Adriatic Sea, and the northern winds of Black Sea, where a cyclonic circulation dominates the eastern part of the basin and an anticyclonic one prevails in its western side (Efimov & Shokurov, 2002).

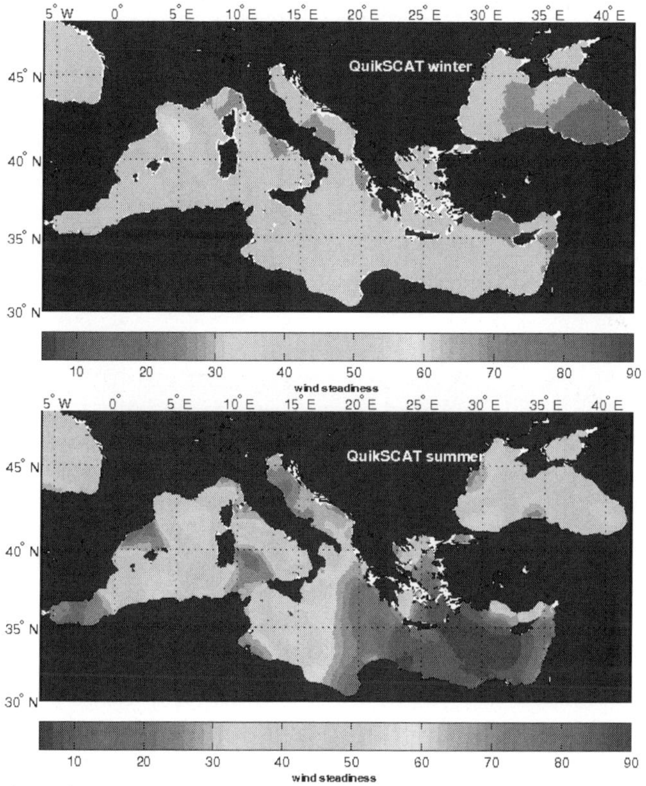

Fig. 7. Seasonal wind steadiness over the Mediterranean Basin (July 1999 to May 2009) derived from QuikSCAT data. Top panel: winter. Bottom panel: summer.

Figure 6 reports the maps of wind speed variability (top panel) for winter (December to February) and summer (June to August, bottom panel). Keeping in mind the general circulation shown in Fig. 5, it is easy to attribute the highest wind variability, found in areas at the wind lee side, to the influence of the interaction between the wind flow and the orography. This is particularly evident between Corsica and Sardinia and between Crete and Rhodos islands in the eastern basin in winter and summer, both due to the funneling and island shielding effects. The orographically induced effects may be seen also in Adriatic and Aegean seas in winter and in the southern Turkey. Figure 7 reports the maps of wind steadiness for winter (top panel) and summer (bottom panel). The highest steadiness is found in the eastern Mediterranean in summer, as effect of the etesian wind circulation. In winter, the moderate steadiness is highest in the area interested by the mistral (Gulf of Lion up to the Sicily Channel) and south of the Crete-Rhodos islands. In this season, also the steadiness pattern in the Adriatic Sea reveals the signature of the northeastern bora.

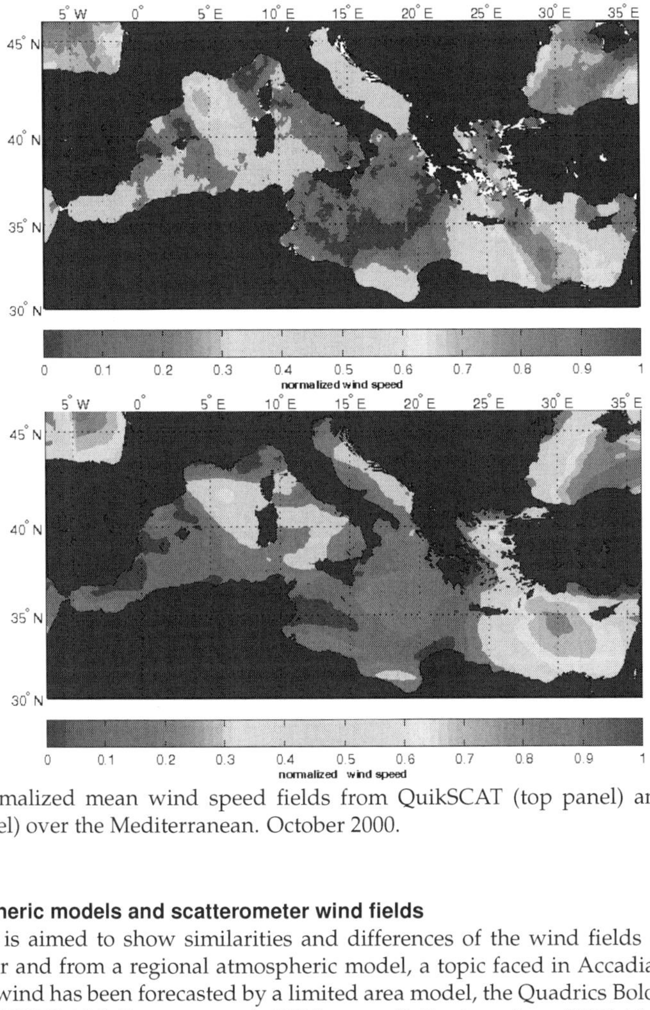

Fig. 8. Normalized mean wind speed fields from QuikSCAT (top panel) and QBOLAM (bottom panel) over the Mediterranean. October 2000.

3.4 Atmospheric models and scatterometer wind fields

This section is aimed to show similarities and differences of the wind fields derived from scatterometer and from a regional atmospheric model, a topic faced in Accadia et al. (2007). The surface wind has been forecasted by a limited area model, the Quadrics Bologna Limited-Area Model (QBOLAM) (Speranza et al., 2004), a parallelized version of BOLAM (Buzzi et al., 1994), covering the whole Mediterranean area with 0.1° by 0.1° grid resolution. Figure 8 reports the fields of the normalized mean wind speed for October 2000, as derived from QuikSCAT (top panel) and QBOLAM (bottom panel). The fields are normalized to make them more comparable. Apart from the differences on the spatial structure of the wind (the model winds are more westerlies than those measured by scatterometer), note the different representation of details provided by the two fields. Despite QBOLAM has a nominal spatial resolution of 0.1° (about 10 km in latitude and 7 km in longitude), higher than that of scatterometer (12.5 km by 12.5 km), its fields result smoother. This common feature of the model fields, discussed in Chèruy et al. (2004) and Skamarock (2004), stresses the importance of the satellite winds in studying the mesoscale spatial features of the wind.

4. Small-scale structure of the MABL from SAR images

The increased availability of satellite SAR images offers to scientists many opportunities to investigate the structure of the MABL over the sea and coastal areas. Scientific literature about SAR images over the ocean has shown a variety of geophysical phenomena detectable by SAR (Alpers & Brümmer, 1994; Kravtsov et al., 1999; Mitnik et al., 1996; Mityagina et al., 1998; Mourad, 1996; Sikora et al., 1997; Zecchetto et al., 1998), including the multiscale structure in the atmospheric turbulence under high winds and the structure of the convective turbulence under low wind. More recently, some effort has been devoted to evaluate the wind direction, using the backscatter signatures produced by the atmospheric wind rolls or those occurring at the lee side of islands (Vachon & Dobson, 2000) as effect of wind shielding, by computing the local gradient of the image backscatter (Horstmann et al., 2002; Koch, 2004) or by using the two dimensional Continuous Wavelet Transform (CWT2) (Zecchetto & De Biasio, 2002; Zecchetto & De Biasio, 2008).
This section illustrates the ability of the CWT2 in detecting and quantifying the backscatter pattern linked to the spatial structure of the MABL. It summarizes the CWT2 methodology applied to SAR images, providing the results obtainable by showing a case study chosen among the hundreds of images analyzed. The extraction of the wind field from SAR images, a follow up of the CWT2 analysis, is then illustrated at the end.

4.1 The methodology

The Continuous Wavelet Transform (Beylkin et al., 1992; Foufoula-Georgiou & Kumar, 1994) $\tilde{f}$ of a function $f(u)$ is a local transform, dependent on the parameters s and τ, defined as

$$\tilde{f}(s,\tau) = \langle \psi_{(s,\tau)}, f \rangle = \int_{-\infty}^{+\infty} du\; \psi^*_{(s,\tau)}(u) f(u) \tag{3}$$

where $\psi_{(s,\tau)}(u) = \psi\left(\frac{u-\tau}{s}\right)$ is the mother wavelet at a given scale (or dilation) s and location τ (the asterisk denotes complex conjugation). The quantity $|\tilde{f}(s,\tau)|^2$ plays the role of local energy density at given (s,τ). The Continuous Wavelet Transform in two dimensions (CWT2) is then,

$$\tilde{f}(s_x,\tau_x;s_y,\tau_y) = \iint_{-\infty}^{+\infty} du\, dv\; \psi^*_{(s_x,\tau_x)}(u)\; \psi^*_{(s_y,\tau_y)}(v)\; f(u,v).$$

The CWT2 has been computed using the Mexican Hat as mother wavelet, able to capture the fine scale structure of the data and suitable for the continuous wavelet transform because it is non-orthogonal.
The images must be preprocessed before the CWT2 analysis, to mask the land and to mitigate the effects introduced by the variation in range of the radar incidence angle. This avoids that structures on the inner part of the image, where the radar incidence angle is smaller and the radar backscatter higher, prevail on the outer ones.
The choice of the scales is very important because it defines the geophysical phenomena to investigate: if the wind field retrieval is of interest, the spatial range is set from 300 m to 4 km; if phenomena such as the atmospheric gravity waves are the object of study, the spatial range has to be set from 4 km up to 20 km.

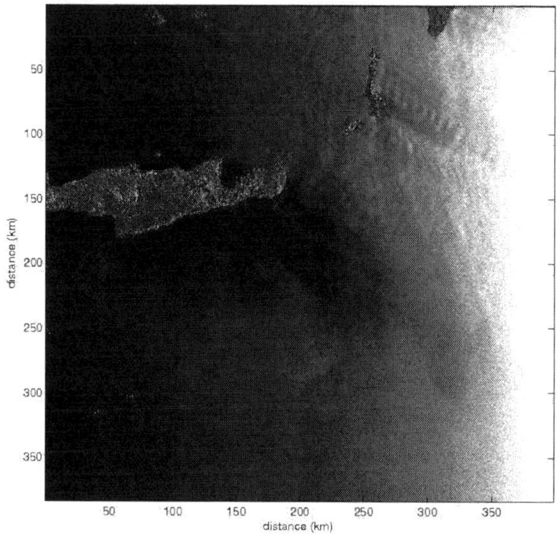

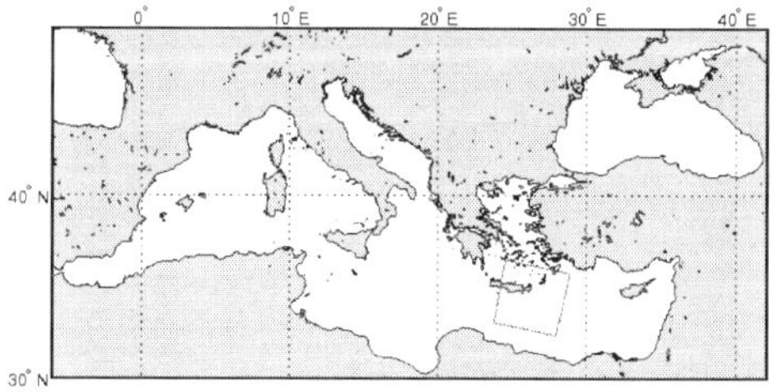

Fig. 9. The Envisat ASAR Wide Swath image selected for the case study. Top panel: the ASAR image (15-May-2008 at 08:20:47 GMT). Bottom panel: the image location in the Mediterranean Sea.

A basic quantity yielded by the CWT2 is the wavelet variance map, derived from the wavelet coefficients. Providing information about the energy distribution as a function of (s_r, s_c), in the same way as the two dimensional Fourier spectrum does as a function of wavenumbers, it is used to select the scales, taken around the maximum of the wavelet variance map, to build a SAR-like map (reconstructed map). This is obtained adding the wavelet coefficient maps at the selected scales: a SAR-like image is thus obtained, representing a spatial pattern due to the most energetic spatial scales present in the original SAR image.
The reconstructed map undergoes a threshold process to isolate the structures from the background. The result of this procedure is a map of backscatter cells, then used as a mask on the original SAR image to get the values of the radar backscatter inside the detected cells, as well as to estimate their shape and size. The reconstructed map depends on the range of

scales chosen in the analysis. As used here, the CWT2 methodology acts as a filtering based on energetic considerations.

4.2 A case study

The image selected for the case study (Fig. 9, top panel) is an Envisat ASAR Wide Swath image taken in the Crete island area (eastern Mediterranean Sea, Fig. 9, bottom panel). This image covers about 400 km by 400 km, with a pixel resolution of 75 m by 75 m. It has been downloaded from the ESA site[4].

The tilting effect due to the change of the radar incidence angle - from 16° on the right side to 43° on the left side, hinders to see the fine structure of the radar backscatter, however well visible in the image blow-up reported in Fig. 10: the wind rolls may be seen in many parts of this image, especially in its top right part.

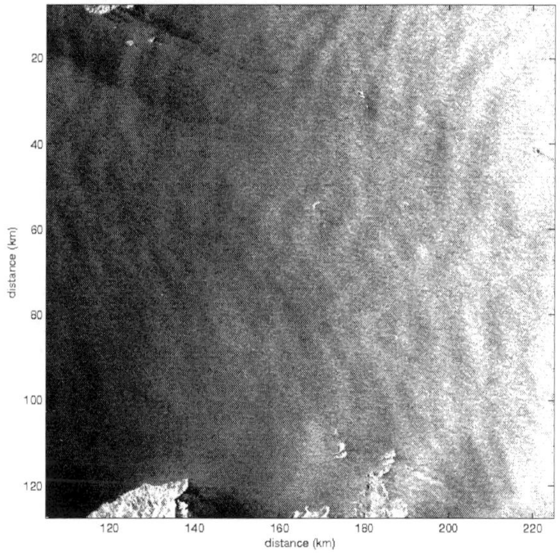

Fig. 10. A blow-up of the ASAR Wide Swath image shown in Fig. 9, roughly corresponding to the area at north-east of Crete.

The larger backscatter structures, as those due to the atmospheric gravity waves east of Karpathos and to the wind sheltering by islands, at the islands lee side) (the wind blowed from northwest) are easily detectable.

[4] *http://oa-ip.eo.esa.int/ra/asa*

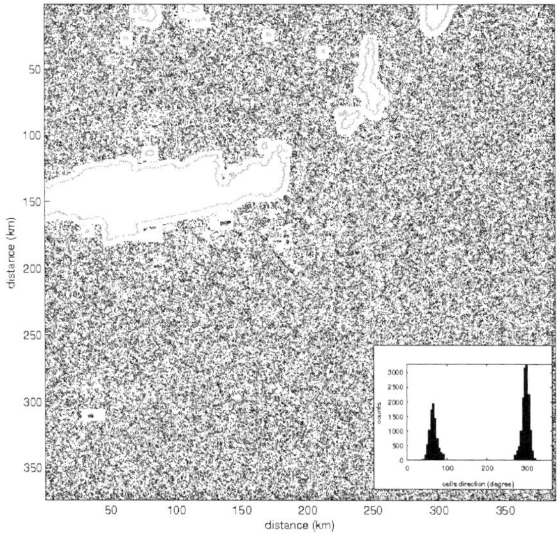

Fig. 11. Map reconstruction in the spatial range 0.3 km ÷ 4 km. Inside panel: the distribution of the orientation of cells' major axis as a function of the angle RGN.

The map reconstructed in the range 0.3 km ÷ 4 km, shown in the left panel of Fig. 11, evidences the small scale structure of the radar backscatter, formed by elliptic cells with major axis orientation falling into two classes, as evidenced by their distribution reported in the inset. The existence of these two classes is due to the texture of the SAR images, and does not represent the geophysical pattern of the backscatter cells excited by the turbulent wind, which may be singled out taking those with directions close to the most probable one, in this case $\theta = 300^\circ$. Thus a reconstructed map with only the cells produced by the wind can be obtained. Figure 12 reports it for the whole image of Fig. 9 (left panel) and for a portion of it (right panel). Note the uneven spatial distribution of the cells but also the high spatial resolution of information obtained. From this map, used as a mask over the original one, it is then possible to retrieve the wind field (Zecchetto & De Biasio, 2008) and to produce a statistics of the cell's size, which may have important implications of the study of the air-sea interaction because it can be linked to the structure of the MABL.

The map reconstructed in the range 4 km ÷ 20 km, reported in the left panel of Fig. 13, clearly shows the pattern of the atmospheric gravity waves in its upper right part. The two dimensional spectral analysis of this map yields the 2D spectrum shown in the right panel of Fig. 13, where two directions are evidenced: that of the maximum energy, occurring at a peak wavelength of 8350 m and an aliased direction of propagation of 296°, and a secondary one, due to the presence of different atmospheric gravity wave trains in the image, with a peak wavelength of 16.7 km and a direction of 63°.

These information may be used, as in Sikora et al. (1997), to estimate the vertical thickness of the MABL.

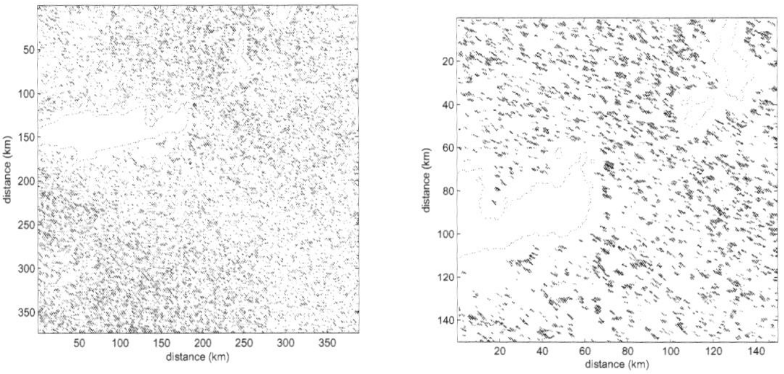

Fig. 12. Reconstructed map with only the cells produced by the wind. Left panel: whole map, corresponding to Fig. 9. Right panel: a blow up of it.

4.3 Wind field extraction: choice of the wind aliased direction

The aliased wind orientation is taken as that corresponding to the most frequent mode of the distribution of cell's direction: in the example reported, the directions around $\Phi = 300°$ have a frequency of 54%, whereas those around $\Phi = 60°$ a frequency of 46 %. These frequencies may differ more in some case (70% to 30% or so), while in some other they can result very similar making difficult the choice of the aliased direction. Their variability across the SAR data set likely depends on the characteristics of the images.

4.4 Wind field extraction: dealiasing

The dealiasing technique takes advantage of the idea, formulated by Zecchetto et al. (1998) in a case of convective turbulence, that the wind gustiness, modulating the mean wind speed, produces patches of roughness characterized by an asymmetric distribution of energy along the wind direction. The speed modulation acts inside the cells: higher backscatter is expected at the leading edge of the patches, then decreasing towards the trailing edge, allowing the wind direction dealiasing. This figure is coherent with the layout of the wind cells, organized like "pearls on a string" (Etling & Brown, 1993), as well as with their inner structure (Zecchetto & De Biasio, 2002).

4.5 Wind field extraction: wind speed computation

Once assessed the wind direction, the speed has been computed from the mean radar backscatter of the selected cells using the CMOD5 model (Hersbach et al., 2007), an empirical model converting the radar cross section at C-band to the wind speed, once the radar incidence angle and the wind direction are known.

4.6 The resulting wind field

The wind field derived from the ASAR image of Fig. 9 is shown in the left panel of Fig. 14, along with the contour plot of the wind speed in the right panel. The wind field is spatially uneven because it has been computed over the detected cells. Where the wind is low, as at

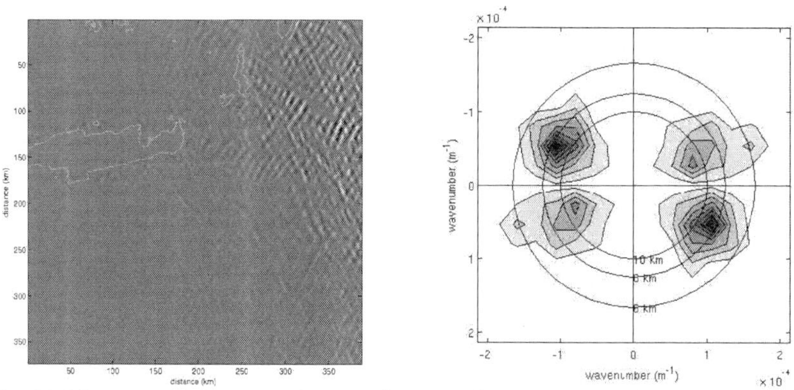

Fig. 13. Map reconstruction in the spatial range 4 km ÷ 20 km. Left panel: the reconstructed map. Right panel: the 2D power spectrum of the reconstructed map.

the lee side of eastern Crete, the spatial density of cells is low too and the wind vectors result more sparse.

The SAR derived wind field provides very detailed information about the spatial structure of the wind and an estimate of the wind much closer to coast than scatterometer, as the Fig. 15, which reports the QuikSCAT wind field at 12.5 km of resolution (left panel) and the contour plot of the wind speed (right panel) suggests.

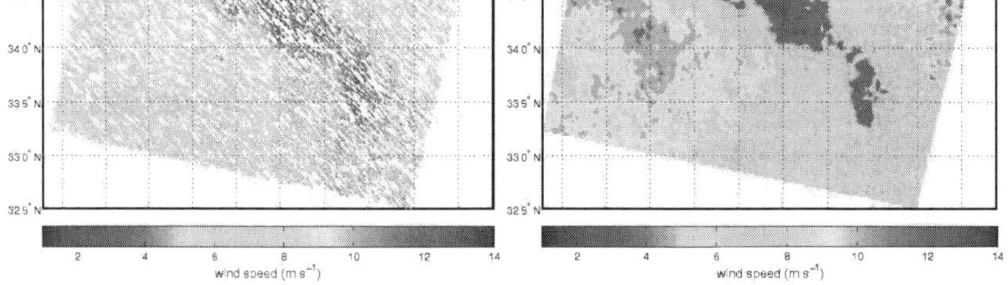

Fig. 14. The wind field derived from the processing with CWT2 method of the ASAR image of Fig. 9. Left panel: the wind field. Right panel: contour map of the wind speed.

Thus, SAR derived winds are an unique experimental tool for coastal wind study in the mesoscale β and γ.

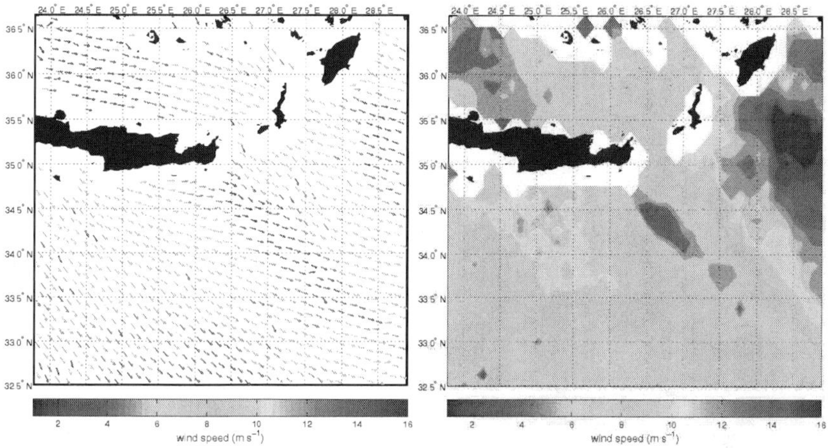

Fig. 15. The wind field from QuikSCAT in the area imaged by ASAR, taken 8 hours and 57 minutes later the ASAR pass time. Left panel: the wind field. Right panel: contour map of the wind speed.

5. Conclusions

This chapter has introduced the satellite scatterometer and SAR, the two satellite radar sensors which may be used to evaluate the wind fields over the sea. A third instrument, the radar altimeter, able to provide only the wind speed over the satellite track, has not be treated because it hardly can be used for mesoscale wind study.

Scatterometer is the most experienced sensor for the measure of the wind field, and its ability to detect detailed features of the wind in the mesoscale is well known. The spatial resolution it provides is sufficient for open sea applications, but insufficient for coastal wind studies, since the data closest to coast are at least 25 km away. Furthermore, the temporal sampling at middle latitudes, roughly two samples per day, is still insufficient for a suitable description of the time evolution of the winds associated to the frontal passage or local cyclogenesis.

The SAR derived wind fields solve the problem of coverage close to coasts, providing very resolute wind fields and permitting to infer the wind speed variability in these areas, as done by Young et al. (2008). Concerning the time sampling provided by operative SARs, this is an open question, the answer depending on many factors: the imaging capabilities of satellites (Radarsat2 has an imaging capability of 28 minutes/orbit, Envisat ASAR 30 minute/orbit for imaging modes and all orbit in the Global Monitoring Mode), the priorities of the different SAR missions (SAR is used over land and over sea), the spatial resolution required. To provide some number, the Envisat ASAR has an average revisit of seven days at the equator, improving to nearly every five days at 45°. With such a revisit time, only research applications can be envisaged, as a monitoring of whatever atmospheric phenomenon would suffer for the unsuitable time sampling. However, the constellations of satellites like the CosmoSkyMed mission, will offer, in principle, a revisiting period of < 12 hours, approaching the threshold of six hours considered the minimum time sampling to describe the evolution of the winds.

Acknowledgments

Scatterometer QuikSCAT data have been downloaded from the Physical Oceanography Distributed Active Archive Center (PODAAC) of the Jet Propulsion Laboratory, Pasadena, USA. The *ASCAT* data have been obtained from the Koninklijk Nederlands Meteorologisch Instituut (Dutch Meteorological Service *KNMI*, `www.knmi.nl`) operating in the framework of the Ocean & Sea Ice Satellite Application Facility (`www.osi-saf.org`) of EUMETSAT. The Envisat ASAR Wide Swath image has been downloaded from the ESA web server `http://oa-ip.eo.esa.int/ra/asa` on the framework of the Project Start Up C1P.5404 of the European Space Agency.

6. References

Accadia, C., Zecchetto, S., Lavagnini, A. & Speranza, A. (2007). Comparison of 10-m wind forecasts from a regional area model and QuikSCAT scatterometer wind observations over the Mediterranean Sea, *Monthly Weather Review* **135**: 1946–1960.

Alpers, W. & Brümmer, B. (1994). Atmospheric boundary layer rolls observed by the synthetic aperture radar aboard the ERS-1 satellite, *Journal of Geophysical Research* **99**(C6): 12613 – 12621.

ASI (2007). COSMO-SkyMed system description and user guide, *Technical Report ASI-CSM-ENG-RS-093-A*, Agenzia Spaziale Italiana, Rome, Italy.

Beylkin, G., Coifman, R., Daubechies, I., Mallat, S., Meyer, Y. & Raphael, L. (1992). *Wavelets and their applications*, Jones and Barlett Publishers, Boston.

Brennan, M. J., Hennon, C. C. & Knabb, R. D. (2008). The operational use of QuikSCAT ocean surface vector winds at the National Hurricane Center, *Weather and Forecasting* **23**: 168182. (doi: 10.1175/2008WAF2222188.1).

Buzzi, A., Fantini, M., Malguzzi, P. & Nerozzi, F. (1994). Validation of a limited area model in cases of Mediterranean cyclogenesis: Surface fields and precipitation scores, *Meteor. Atmos. Phys.* **53**: 53–67.

Campbell, A. B. (2002). *Radar remote sensing of planetary surfaces*, Cambridge University Press, Cambridge.

CCRS (2009). Tutorial: Fundamentals of Remote Sensing, *Technical report*, Canada Centre for Remote Sensing. `http://www.ccrs.nrcan.gc.ca/resource/tutor/gsarcd.`

Chelton, D. & Freilich, M. (2005). Scatterometer-based assessment of 10-m wind analyses from the operational ECMWF and NCEP numerical weather prediction models, *J. of Geophys. Res.* **133**: 409–429.

Chelton, D., Schlax, M. G., Freilich, M. & Millif, R. F. (2004). Satellite radar measurements reveal short-scale features in the wind stress field over the world ocean, *Science* **303**: 978–983.

Chèruy, F., Speranza, A., Sutera, A. & Tartaglione, N. (2004). Surface winds in the Euro-Mediterranean area: the real resolution of numerical grids, *Annales Geophysicae* **22**: 4043–4048.

CSA (2001). Applications potential of RADARSAT-2 - A Preview, *Technical report*, Canadian Space Agency. (J. J. Van der Sanden and S. G. Ross eds.).

DLR (2003). TerraSAR-X Science Plan, *Technical Report TX-PGS-PL-4001*, Deutschen Zentrums fur Luft- und Raumfahrt (DLR), Oberpfaffenhofen, Germany.

Ebuchi, N., Graber, H. & Caruso, M. J. (2002). Evaluation of wind vectors observed by QuikSCAT/SeaWinds using ocean buoy data, *J. of Atmosph. and Oceanic Tech.* **19**: 2049–2062.

Efimov, V. V. & Shokurov, M. V. (2002). Spatiotemporal structure of the surface wind field over the Black Sea, *Izvestiya Atmospheric and Oceanic Physics* **38**(4): 421–430.

Elachi, C. (1988). *Spaceborne Radar Remote Sensing: Applications and Techniques*, IEEE Press, New York, USA.

ESA (2002). ASAR Product Handbook, *Technical report*, European Space Agency, Paris, France.

Etling, D. & Brown, R. A. (1993). Roll vortices in the planetary boundary layer: a review, *Boundary Layer Meteorol.* **65**: 15–248.

Eumetsat (2007). ASCAT Wind Product User Manual, *Technical Report SAF/OSI/CDOP/KNMI/TEC/MA/126*, Eumetsat, Darmstadt, Germany.

Foufoula-Georgiou, E. & Kumar, P. (eds) (1994). *Wavelet in Geophysics*, Vol. 4 of *Wavelet analysis and its applications*, Academic Press, San Diego, CA.

Hersbach, H., Stoffelen, A. & de Haan, S. (2007). An improved scatterometer ocean geophysical model function: CMOD5, *Journal of Geophysical Research* **112**: 5767–5780. doi:10.1029/2006jc003743.

HMSO (1962). Weather in the Mediterranean, *Volume 1 (second ed.)*, Her Majesty's Stationary Office, London, UK.

Horstmann, J., Koch, W. & Lehner, S. (2002). High resolution wind fields retrieved from SAR in comparison to numerical models, *Proceedings of the IEEE International Geoscience and Remote Sensing Symposium*, Toronto, Canada.

Huddleston, J. M. & Stiles, B. W. (2000). Multidimensional histogram (mudh) rain flag. Product description ver. 2.1, *Technical report*, Jet Propulsion Laboratory,Pasadena, USA.

Isaksen, L. & Janssen, P. A. E. M. (2008). Impact of ERS scatterometer winds in ECMWF's assimilation, *Q. J. R. Meteorol. Soc.* **130**: 1793–1814.

Isaksen, L. & Stoffelen, A. (2000). ERS scatterometer wind data impact on ECMWF's tropical cyclone forecasts, *IEEE Trans. Geosci. Remote Sens.* **138**: 1885–1892.

Jones, W. L., Mladen, S., Park, J. & Mehershadi, R. (1999). A quality control rain flag using QuikSCAT radiometric observations, *Proc. of QuikSCAT Cal/Val Workshop*, Pasadena, USA.

JPL (2006). QuikSCAT Science Data Product User's Manual (Ver. 3), *Technical Report Publ. D-18053*, Jet Propulsion Laboratory, Pasadena, USA.

Koch, W. (2004). Directional analysis of SAR images aiming at wind direction, *IEEE Trans. Geosci. Remote Sens.* **42**(4): 702 – 710.

Kolstad, E. W. (2008). A QuikSCAT climatology of ocean surface winds in the nordic seas: identification of features and comparison with the NCEP/NCAR reanalysis, *Journal of Geophysical Research* **113**. D11106, doi:10.1029/2007JD008918.

Kravtsov, Y., Mityagina, M., Pungin, V. & Sabinin, K. (1999). Manifestation of the wind-field fine structure ahead of an atmospheric cold front on radar imagery of the sea surface, *Earth Obs. Rem. Sens.* **5**: 513–525.

Lislie, L. M., Buckley, B. W. & Leplastier, M. (2008). The operational impact of QuikSCAT winds in Perth, Australia: examples and limitations, *Weather and Forecasting* **23**(1): 183–193. (doi: 10.1175/2007WAF2007027.1).

Liu, T. W., Tang, W. & Polito, P. S. (1998). NASA scatterometer provides global ocean-surface wind fields with more structures than numerical weather prediction, *Geophysical Research Letters* **25**: 761–764.

Millif, R. F., Morzel, J., Danabasoglu, G. & Chin, T. M. (2001). Ocean general circulation model sensitivity to forcing from scatterometer winds, *J. Geophys. Res.* **104C**: 11337–11358.

Milliff, R. F. & Stamus, P. A. (2008). QuikSCAT impacts on coastal forecasts and warnings: operational utility of satellite ocean surface vector wind data, *Weather and Forecasting* **23**(5): 878890. (doi: 10.1175/2008WAF2007081.1).

Mitnik, L., Hsu, M. & Mitnik, M. (1996). Sharp gradients and organised structures in sea surface wind field in the region of polar vortex formation, *Global Atmos. Ocean Syst.* **4**: 335–361.

Mityagina, M., Pungin, V. & Yakovlev, V. (1998). Two-polarization K_u-band radar imagery of the sea surface in presence of atmospheric boundary layer motions, *Waves Random Media* **8**: 111–118.

Monahan, E. C. (2002). The physical and practical implications of CO_2 gas transfer coefficient that varies as the cube of the wind speed, *in* E. S. S. M. A. Donelan, W. M. Drennan & R. Wanninkhof (eds), *Gas transfer at water surface*, American Geophysical Union, Washington DC, USA, pp. 193–197.

Morena, L. C., James, K. V. & Beck, J. (2004). An introduction to the RADARSAT-2 mission, *Canadian J. Remote Sensing* **30**(3): 221–234.

Mourad, P. (1996). Inferring multiscale structure in the atmospheric turbulence using satellite-based synthetic aperture radar imagery, *J. Geophys. Res.* **101**: 18433–18449.

Orlanski, I. (1975). A rational subdivision of scales for atmospheric processes, *Bull. Amer. Meteor. Soc.* **56**: 527–530.

Pandžić, K. & Likso, T. (2005). Eastern Adriatic typical wind field patterns and large-scale atmospheric conditions, *Int. J. of Climatology* **25**: 81–98.

Pickett, M. H., Tang, W., Rosenfeld, L. K. & Wash, C. H. (2003). QuikSCAT satellite comparisons with nearshore buoy wind data off the U.S. west coast, *J. of Atmos. and Ocean. Tech.* **20**: 1869–1879.

Portabella, M. & Stoffelen, A. (2001). Rain detection and quality control of SeaWinds, *J. of Atmos. and Ocean. Tech.* **18**: 1171–1183.

Risien, C. M. & Chelton, D. B. (2008). A global climatology of surface wind and wind stress fields from eight years of QuikSCAT scatterometer data, *Journal of Phys. Ocean.* **38**(11): 2379–2413. (doi: 10.1175/2008JPO3881.1).

Ruti, P. M., Marullo, S., D'Ortenzio, F. & Tremant, M. (2008). Comparison of analyzed and measured wind speeds in the perspective of oceanic simulation over the Mediterranean basin: analysed, QuikSCAT and buoy data, *J. of Marine Systems* **70**: 33–48. (doi: 10,1016/j.jmarsys.2007.02.026).

Sikora, T., Young, G., Shirer, H. & Chapman, R. (1997). Estimating convective atmospheric boundary layer depth from microwave radar imagery of the sea surface, *J. Appl. Meteorol.* **36**: 833–845.

Singh, R., Pal, P. K., Kishtawal, C. M. & Joshi, P. C. (2008). The impact of variational assimilation of SSM/I and QuikSCAT satellite observations on the numerical simulation of Indian ocean tropical cyclones, *Weather and Forecasting* **23**: 460–476. (doi: 10.1175/2007WAF2007014.1).

Skamarock, W. C. (2004). Evaluating mesoscale NWP models using kinetic energy spectra, *Monthly Weather Review* **132**: 3019–3032.

Speranza, A., Accadia, C., Casaioli, M., Mariani, S., Monacelli, G., Inghilesi, R., Tartaglione, N., Ruti, P. M., Carillo, A., Bargagli, A., Pisacane, G., Valentinotti, F. & Lavagnini, A. (2004). POSEIDON: an integrated system for analysis and forecast of hydrological, meteorological and surface marine fields in the Mediterranean area, *Il Nuovo Cimento* **27C**: 329–345.

Vachon, P. & Dobson, F. (2000). Wind retrieval from Radarsat SAR images: selection of a suitable C-Band HH polarization wind retrieval model, *Canadian Journal of Remote Sensing* **24**(4): 306 – 313.

Valenzuela, G. (1978). Theories for the interaction of electromagnetic and oceanic waves- A review, *Boundary Layer Meteorology* **13**: 61–85.

Yoshino, M. (1976). *Local wind Bora*, University of Tokio Press, Tokio, Japan.

Young, G., Sikora, T. & Winstead, N. (2008). Mesoscale near-surface wind speed variability mapping with synthetic aperture radar, *Sensors* **8(11)**: 7012–7034. (doi: 10.3390/s8117012).

Zecchetto, S. & Cappa, C. (2001). The spatial structure of the Mediterranean Sea winds revealed by ERS-1 scatterometer, *Int. J. Remote Sensing* **22**(1): 45–70.

Zecchetto, S. & De Biasio, F. (2002). On shape, orientation and structure of atmospheric cells inside wind rolls in SAR images, *IEEE Trans. of Geoscience and Remote Sensing* **40**(10): 2257 – 2262.

Zecchetto, S. & De Biasio, F. (2007). Sea surface winds over the Mediterranean Basin from satellite data (2000-04): meso- and local-scale features on annual and seasonal time scales, *J. Applied Meteor. and Climatology* **46**: 814–827.

Zecchetto, S. & De Biasio, F. (2008). A wavelet based technique for sea wind extraction from SAR images, *IEEE Trans. of Geoscience and Remote Sensing* **46(10)**: 2983–2989. (doi:10.1109/TGRS.2008.920967).

Zecchetto, S., Trivero, P., Fiscella, B. & Pavese, P. (1998). Wind stress structure in the unstable marine surface layer detected by SAR, *Boundary Layer Meteorol.* **86**: 1–28.

Ziv, B., Saaroni, H. & Alpert, P. (2004). The factors governing the summer regime of the eastern Mediterranean, *Int. J. of Climatology* **24**: 1859–1971.

16

Optical and Infrared Modeling

Abdelaziz Kallel
Tartu Observatory
Estonia

1. Introduction

In order to understand the relationships between the vegetation features (namely amount and structure) and the amount of sunlight reflected in the visible and near- to middle-infrared spectral domains many empirical methods based on various vegetation indices (e.g., NDVI, EVI) (Kallel et al., 2007), and physical approach namely based on radiative transfer (RT) theory have been developed. In RT, two model types can be distinguished: (i) one-Dimensional (1-D) models providing a (semi)analytical expression of the Bidirectional Reflectance Distribution Function (BRDF) of canopy architecture, its scattering parameters, and scene geometry (Gobron et al., 1997; Verhoef, 1984; 1998); (ii) 3-D models based on Monte Carlo simulations of a large number of photons randomly propagating through a canopy (Gastellu-Etchegorry et al., 1996; Lewis, 1999; North, 1996). Compared to 1-D models, such 3-D methods allow to take into account canopy heterogeneity with high accuracy. However, they suffer from long running times making their inversion difficult.

The RT theory was first proposed by Chandrasekhar (1950) to study radiation scattering in conventional (i.e. rotationally invariant) media. Such an assumption could be sufficient to model, for example, light scattering in the atmosphere, but appears rudimentary for modeling the reflectance of leaves, or shoots, in a vegetation canopy. To extend the formulation to such a case, many models are proposed. Among the 1-D model, one can cite SAIL (Verhoef, 1984) that is among the most widely used in case of turbid (null size components) crops canopies. The SAIL model allows to derive a non-isotropic BRDF considering two diffuse fluxes (upward/downward flux) to model the multiple scattering of the radiant flux by the vegetation elements. These fluxes are assumed to be semi-isotropic, which is only an approximation that lead to reflectance underestimation (Pinty et al., 2004). As a solution, Verhoef (1998) developed SAIL++ which is a 1-D model providing accurate reflectance estimation in the turbid case. Indeed, this model divides the diffuses fluxes into 72-subfluxes, and turns the SAIL equation system into a matrix-vector equation. Compared to 3-D models of RAMI 2 database in the turbid case (Pinty et al., 2004), SAIL++ gives accurate results.

Another solution to overcome the semi-isotropy assumption in the turbid case will be presented in this chapter, it is based on the coupling between SAIL and Adding method (Cooper et al., 1982; Van de Hulst, 1980). For such a method, optical characteristics of canopy layers such that reflectance and transmittance are directly defined and handled at the scale of the vegetation layer (as operators). Their physical interpretation is hence easier. However, the vegetation description is rather simplistic and the canopy internal geometry is represented with low accuracy. Indeed, in order to retrieve the adding operators for each layer, Cooper et al. (1982) did not take into account the high order interactions between light and vegetation

which are very important as shown in (Pinty et al., 2004). In order to adapt the Adding method to such a configuration, we need a more accurate estimation of the Adding scattering parameters. Since the Adding method operators are derived from the bidirectional reflectance and transmittance of the considered layer, in this study we propose to introduce the SAIL canopy description into the Adding formulation. The developed model is called AddingS.
Now, since the size of vegetation elements cannot be assumed null. Among others, Kuusk (1985) proposed a correction allowing the extension of the RT models like SAIL and SAIL++ to the discrete case (non-null-size components) (Verhoef, 1998). This approach allows to take into account the hot spot effect representing the bright area in the direction opposite to the direction of a pointlike the light source. This effect is caused by the high probability of backscattering which is proportional to the mean size of medium elements. Such an approach suffers from a severe shortcoming: compared to the turbid case, it increases only the reflectance created by the first collision of the radiation by leaves. As this increasing is not followed by the decreasing of other fluxes, it leads to a violation of the energy conservation law (Kallel, 2007). Therefore, based on the Kuusk (1985) approach, we propose the adaptation of AddingS to the discrete case. The extended model is called AddingSD. This model allows both to conserve the energy and to take into account the hot spot effect between diffuse fluxes. As AddingS/AddingSD are based on adding method then they need a long running time for that in this study, we benefit from both the rapidity of the SAIL++ as well as the hot spot modeling in the AddingSD and we propose a new other approach. This approach is based on the traking of the flux created by the first photons collison by leaves. The analysis of this flux will be done using AddingSD and the RT problem resolution will be based on SAIL++.
The chapter is divided up as follows. First, we present the theoretical background of our models (Section 2). Then, we show model implementation (Section 3), and some validation results (Section 4). Finally, we present our main conclusions and perspectives (Section 5).

2. Theoretical background

In this section, we will first present the models AddingS/AddingSD then we expose our model based on flux decomposition.

2.1 AddingS/AddinSD modeling

The Adding method is based on the assumption that a vegetation layer receiving a radiation flux from bottom or top, partially absorbs it and partially scatters it upward or downward, independently of the other layers (Cooper et al., 1982; Van de Hulst, 1980). Thus, the relationships between fluxes are given by operators which allow the calculation of the output flux density distribution as a function of the input flux density distribution. As the Adding method vegetation layer operators depend on the bidirectional reflectance and transmittance, we propose to derive them both in the turbid and the discrete case based on respectively SAIL and the Kuusk definition of the Hot Spot.
In this section, we first present the Adding operator definition, and secondly the derivation of the bidirectional reflectance and transmittance of a vegetation layer in both turbid and discrete cases corresponding respectively to the operators of the models AddingS and AddingSD.

2.1.1 Adding operators reformulation in the continuous case

In this paragraph, we present a generalization of the Adding operators presented in (Cooper et al., 1982) in the continuous case, dealing with radiance hemispherical distribution.

For a given medium having two parallel sides (top and bottom) receiving a source radiation flux $dE_i(\Omega_i = (\theta_i, \varphi_i))$ (θ_i the zenithal angle and φ_i the azimuthal angle) provided within a cone of solid angle $d\Omega_i = \sin(\theta_i)d\theta_i d\varphi_i$, produces elementary radiances at the top and the bottom of the medium called respectively $\mathrm{d}L_e(\Omega_i, \Omega_e)$ and $\mathrm{d}L'_e(\Omega_i, \Omega'_e)$ in the directions $\Omega_e = (\theta_e, \varphi_e)$ and $\Omega'_e = (\theta_{e'}, \varphi_{e'})$, respectively.
So the BRDF, r, and the bidirectional transmittance distribution function (BTDF), t, are defined respectively as follows:

$$\begin{aligned} r(\Omega_i \to \Omega_e) &= \frac{\pi \mathrm{d}L_e(\Omega_i, \Omega_e)}{\mathrm{d}E_i(\Omega_e)} = \frac{\pi \mathrm{d}L_e(\Omega_i, \Omega_e)}{L_i(\Omega_i)\cos(\theta_i)\mathrm{d}\Omega_i}, \\ t(\Omega_i \to \Omega_{e'}) &= \frac{\pi \mathrm{d}L'_e(\Omega_i, \Omega_{e'})}{\mathrm{d}E_i(\Omega_i)} = \frac{\pi \mathrm{d}L'_e(\Omega_i, \Omega'_e)}{L_i(\Omega_i)\cos(\theta_i)\mathrm{d}\Omega_i}. \end{aligned} \tag{1}$$

where L_i is the radiance provided by the source.
So, we define the two scattering operators $\mathcal{R}$ and $\mathcal{T}$, that give the outward radiance L_e from an incident radiance defined over the whole hemisphere L_i:

$$\mathcal{R}[L_i](.) = \frac{1}{\pi} \overbrace{\int_{\Pi}}^{\text{Over hemisphere}} r(\Omega_i \to .)L_i(\Omega_i)\cos(\theta_i)\mathrm{d}\Omega_i, \tag{2}$$

$$\mathcal{T}[L_i](.) = \frac{1}{\pi}\int_{\Pi} t(\Omega_i \to .)L_i(\Omega_i)\cos(\theta_i)\mathrm{d}\Omega_i. \tag{3}$$

For two medium 1 and 2 such that the second one is above the first one, the top reflectance operator for the canopy is given by (Verhoef, 1985):

$$\mathcal{R}_t = \mathcal{R}_{t,2} + \mathcal{T}_{u,2} \circ (I - \mathcal{R}_{t,1} \circ \mathcal{R}_{b,2})^{-1} \circ \mathcal{R}_{t,1} \circ \mathcal{T}_{d,2}. \tag{4}$$

where $\mathcal{T}_{u,2}$, $\mathcal{T}_{d,2}$ are respectively the upward and downward transmittances of the layer 2, $\mathcal{R}_{t,1}$ and $\mathcal{R}_{b,1}$ are the reflectances of respectively the top of layer 1 and the bottom of layer 2, and I is the identity operator.
Finally, to be implemented such operators have to discretized. Thus, Kallel et al. (2008) propose a regular discretization of the zenithal angle θ and azimuthal angle φ into 20 and 10 intervals respectively. In this case, the reflectance and transmittance operators become matrices and the '$\circ$' operator becomes matrix multiplication.

2.1.2 Turbid case: AddingS

For one vegetation layer, the top and bottom reflectance operators and the downward and upward transmittance operators require the estimation of top and bottom bidirectional reflectances, the downward and upward bidirectional transmittance respectively, r_t, r_b, t_d and t_u. Now, assuming that the vegetation layer is formed by small and flat leaves with uniform azimuthal distribution, the layer has the same response when observed from the top or the bottom. $r_b = r_t$ and $t_u = t_d$. Moreover, two kinds of transmittances can be distinguished: those provided from the extinction of the incident flux, and those provided by the scattering of the incident flux by the vegetation components. So, we called them respectively $t_{.,s}$ and $t_{.,d}$, where . equals d (downward) or u (upward).
The SAIL model allows the BRDF (r_t) and the BTDF by scattering ($t_{d,d}$) derivation of a vegetation layer. Moreover, Kallel et al. (2008) showed that

$$t_{d,s}(\Omega_i \to \Omega_{e'}) = \frac{\tau_{ss}\delta(\theta'_e = \theta_i)\delta(\varphi'_e = \varphi_i)}{\cos(\theta_i)\sin(\theta_i)}, \tag{5}$$

with τ_{ss} the direct transmittance given by SAIL.
As such a model is based on SAIL which assumes that the diffuse fluxes are semi-isotropic, then it is only correct for thin layers (LAI $< 10^{-2}$) where the diffuse fluxes contribution to the BRDF/BTDF are small. Therefore, to estimate the reflectance of a thick layer and overcome the semi-isotropy assumption, we propose to divide the thick layer into thin sublayers with LAI value, $L_{min} = 10^{-3}$. The whole layer reflectance operator is then derived with good accuracy using the adding method Eq. (4) as it allows to model the diffuse flux anisotropy.

2.1.3 Discrete case: AddingSD

In the discrete case, the size of the leaves is no longer assumed null and there is a non-negligible correlation between the incident flux path and the diffused flux: the hot spot effect Kuusk (1985); Suits (1972). Until now, such an effect was taken into account in 1-D model only for the single scattering contribution from soil and foliage that is increased. Now, as the diffuse fluxes are not decreased consequently, the radiative budget is not checked. Now, the hot spot effect occurs also for diffuse fluxes (whose contribution increases with the vegetation depth). We call such a phenomena the multi hot spot effect. In this section, having recall Kuusk' model Kuusk (1985), we present our approach.

2.1.3.1 Kuusk' model

For a layer located at in altitude between -1 and 0, the single scattering reflectance ($\rho_{HS}^{(1)}$) by a leaf M at depth z, for the source and observation directions being respectively Ω_s and Ω_o, is (Verhoef (1998), pp 150-159):

$$\rho_{HS}^{(1)}(z) = P_{so}(\Omega_s,\Omega_o,z)\frac{w(\Omega_s,\Omega_o)}{\pi}, \tag{6}$$

where w is the bidirectional scattering parameter under the vegetation (Verhoef, 1984) and $P_{so}(\Omega_s,\Omega_o,z)$ is the conjoint probability that the incident flux reaches M without any collision with other canopy components and that, after scattering by M, it also reaches the top of the canopy without collisions Kuusk (1985):

$$\begin{aligned} P_o(\Omega_s,\Omega_o,z) &= \exp\left[-\int_z^0 \{k+K-\sqrt{Kk}\exp[(z-x)b]\}\mathrm{d}x\right], \\ &= \exp[(K+k)z]C_{HS}(\Omega_s,\Omega_o,z), \end{aligned} \tag{7}$$

with k, K the extinction respectively in source and observation directions and C_{HS} the correction factor:

$$C_{HS}(\Omega_s,\Omega_o,z) = \exp\left[\frac{\sqrt{kK}}{b}[1-\exp(bz)]\right], \tag{8}$$

where b is a function of the vegetation features, the different solid angles and the hot spot factor d_l defined as the ratio between the leaf radius and the layer height Kuusk (1985); Pinty et al. (2004).

2.1.3.2 Multi hot spot model

Firstly recall that the energy conservation is insured by adding model whatever be the foliage area volume density (FAVD), u_l (cf. Appendix B) or the probability of finding foliage P_χ. In this subsection, we first show that the first order hot spot corresponds to the use of a fictive equivalent P_χ, called $P_{\chi,HS}$.

For a vegetation layer composed of two layers: a thin layer 2 above a layer 1, located respectively in $[z_0,0]$ and $[-1,z_0]$, let $P_{so}(\Omega_s,\Omega_o,z_0,z)$ denotes the joint probability that the two fluxes do not collide with leaves for $z' \in [z_0,0]$ (only in the layer 2). Its expression is obtained from Eq. (7) by changing the integral endpoints $[z,0]$ by $[z_0,0]$:

$$P_{so}(\Omega_s,\Omega_o,z_0,z) = \exp[(K+k)z_0]\mathcal{C}_{HS}(\Omega_s,\Omega_o,z_0,z),$$

with $\mathcal{C}_{HS}$ the generalized correction factor:

$$\mathcal{C}_{HS}(\Omega_s,\Omega_o,z_0,z) = \exp\left[\frac{\sqrt{kK}}{b}\Big(\exp[b(z-z_0)] - \exp[bz]\Big)\right].$$

The conditional probability definition that the flux in the direction Ω_o does not collide leaves given the same property for the incident flux is:

$$P_o(\Omega_o|\Omega_s,z_0,z) = \frac{P_{so}(\Omega_s,\Omega_o,z_0,z)}{P_s(\Omega_s,z_0)},$$

where $P_s(\Omega_s,z_0)$ represents the prior probability of gap in the direction Ω_s. Since $P_s(\Omega_s,z_0) = \exp[kz_0]$, then:

$$P_o(\Omega_o|\Omega_s,z_0,z) = \exp[Kz_0]\mathcal{C}_{HS}(\Omega_s,\Omega_o,z_0,z).$$

In the case of the direct flux, the first order contribution of a leaf $M(z)$ in the layer 1 to the BRDF is:

$$\rho^{(1)}_{HS}(z) = \underbrace{\exp[kz_0]}_{P_s(\Omega_s,z_0)}\underbrace{\rho^{(1)}_{HS}(z-z_0)}_{\text{layer 1}}\underbrace{\exp\{\overbrace{Kz_0 + \log[\mathcal{C}_{HS}(\Omega_s,\Omega_o,z_0,z)]}^{K_{HS}(\Omega_o|\Omega_s,z_0,z)z_0}\}}_{P_o(\Omega_o|\Omega_s,z_0,z)}. \quad (9)$$

In Eq. (9), $\rho^{(1)}_{HS}(z)$ can be interpreted as follows: reaching the top of the canopy the direct flux is partially extinguished in the layer 2 by the factor $P_s(\Omega_s,z_0)$. Then, reaching the interface between the two layers, its amplitude will be determined according to $\rho^{(1)}_{HS}(z-z_0)$ that depends on the layer 1 features. Finally, $K_{HS}(\Omega_o|\Omega_s,z_0,z)$ can be viewed as the 'effective' extinction related to the conditional probability of gap $P_o(\Omega_o|\Omega_s,z_0,z)$ of the layer 2. Indeed, $K_{HS} < K$ means that the probability of collision with leaves (or probability of finding leaves, P_χ) for the exiting flux that it will be noted $L^{(1)}_{o,HS}$, is decreased. Since the extinction depends linearly on P_χ, one can deem that P_χ is locally decreased by the factor $\gamma = \frac{K_{HS}}{K}$:

$$P_{\chi,HS}(\Omega_o|\Omega_s,z_0,z) = \frac{K_{HS}(\Omega_o|\Omega_s,z_0,z)}{K}P_\chi. \quad (10)$$

The physical interpretation of $P_{\chi,HS}$ is as follows. Assume that the probability of gap (for a given flux) is increased in the layer 2. For this flux, the 'effective' probability of being collided by vegetation when crossing the layer is reduced accordingly. Obviously, the fist collision between the flux and the vegetation is reduced according to the same probabilty of finding vegetation or similarly the same density of vegetation. Now, since the layer 2 is thin, its corresponding reflectance and diffuse transmittance depend mainly on the first interaction. So, just an approximation of the multiple scattered fluxes is sufficient to derive the layer 2 scattering terms with good accuracy. For that, the derivation of all diffuse fluxes can be done using this 'effective' probability of finding foliage ($P_{\chi,HS}$ in our case). Moreover, for such a modeling, the interactions of the considered flux and the layer 2 components (transmittance by extinction, reflectance and diffuse transmittance) are derived using exactly the same probability value ($P_{\chi,HS}$), which is physically consistent and thus leads to the conservation of the energy of this flux. Furthermore, by doing the same processing for all fluxes exiting the layer 1 in direction of the layer 2, the energy of all fluxes is conserved and so the energy is conserved in the system composed by the two vegetation layers.

The layer 2 reflectance and diffuse transmittance of the flux $L^{(1)}_{o,HS}$, respectively called $r_{b,2,HS}(z,\Omega_o \to .)$ and $t_{d,2,HS}(z,\Omega_o \to .)$, have therefore to be estimated using $P_{\chi,HS}$ rather than the initial P_χ. The first order hot spot effect can then be viewed as a local reduction of the layer 2 probability of finding leaves. The layer 2 operators are derived accordingly, and the two layer reflectance operator is obtained using Eq. (4). In summary, given a vegetation layer, its corresponding reflectance is computed dividing it into N_{HS} thin sublayers with a value of LAI, $\mathrm{L}_{HS} = 3 \times 10^{-2}$ (L_{HS} is higher than the elementary sublayer LAI corresponding to AddingS model concatenation, $\mathrm{L}_{\min} = 10^{-3}$) and iteratively adding a new sublayer to the current 'stack' of sublayers (from 1 to N_{HS}).

More precisely, beginning from a thin layer, where the neglecting of the hot spot effect appears reasonable, thin layers are added, one after one, to build up a 'system' taking into account the whole hot spot effect (as well as conserving the energy). The contribution of each new sublayer 2 to the high order hot spot effect is computed as follows. The flux reaching the top of the layer 2 is scattered many times before reaching the interface between the two layers where it is considered again as a direct flux (according to the adding method). In layer 1, the first order (direct flux case) hot spot computation is therefore valid. Adding iteratively the thin layers and the contribution of their diffuse fluxes, the hot spot effect between all the diffuse fluxes is taken into account.

Finally, for more information about the implementation of the models AddingS/AddingSD, readers are invited to read the article (Kallel et al., 2008).

2.2 Virtual flux decomposition

In this section, we propose an alternative to AddingSD that is simpler, conserves the energy and based on effective vegetation density too but does not take into account the high order hot spot effect. Moreover, the proposed approach is an extension to the discrete case of SAIL++, that we provide an overview in Appendix A. To do the extension, we study the collision of direct fluxes with vegetation in the discrete homogeneous medium case. The energy will be conserved by increasing the flux created by first collision and decreasing the flux created by this flux scattering.

2.2.1 Derivation of $L_+^{1,n}$

Figure 1 shows two points $M(x,y,z)$ and $N(x',y',t)$ in a vegetation layer assumed be a homogeneous discrete medium such that $t < z$. The elementary volume at M is viewed from N at an elementary solid angle $\mathrm{d}\Omega$ with $\Omega = (\theta,\varphi)$. A direct flux ($E_s(0)$) present above the vegetation layer having direction $\Omega_s = (\theta_s,0)$ passes through the vegetation from the top to N without a collision. By assuming a constant extinction k along the path, E_s at altitude t is

$$E_s(t) = E_s(0)\exp(kt). \tag{11}$$

Then the light is scattered in an elementary volume at N with an elementary thickness $\mathrm{d}t$. Thus scattered radiance in the direction $\mathrm{d}\Omega$ called ($\mathrm{d}L_+^1(N,\Omega)$) is

$$\mathrm{d}L_+^1(N,\Omega) = E_s(t)\pi^{-1}w(\Omega_s \to \Omega)\mathrm{d}t. \tag{12}$$

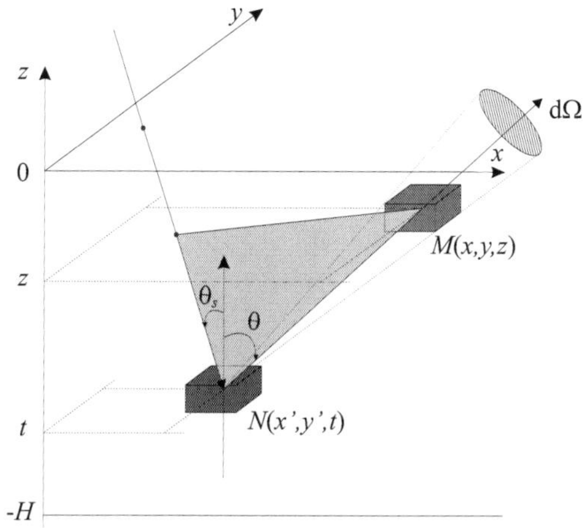

Fig. 1. A vegetation layer located from altitude 0 to $-H$ and assumed a discrete medium. Two point $M(x,y,z)$ and $N(x',y',z)$ are located in the layer. The elementary volume at M is viewed from N under an elementary solid angle (dΩ) with a polar angle θ. A direct flux with zenith angle θ_s collides with vegetation in an elementary volume at point N, is then reflected in the solid angle dΩ and reaches point M without collision. The downward and upward paths are correlated from altitude z to t as shown by the gray triangle linking the two paths.

$\mathrm{d}L_+^1(N,\Omega)$ travels from N to M without collision. Therefore, by assuming a constant extinction κ along the path and without taking into account the dependency between paths, the radiance reaching M called $\mathrm{d}L_+^{1*}(N \rightarrow M,\Omega)$ is

$$\begin{aligned} \mathrm{d}L_+^{1*}(N \rightarrow M,\Omega) &= \mathrm{d}L_+^1(N,\Omega)\exp[\kappa(t-z)], \\ &= E_s(0)\exp[(k+\kappa)(t-z)]\exp(kz)\pi^{-1}w(\Omega_s \rightarrow \Omega)\mathrm{d}t. \end{aligned} \tag{13}$$

Since the medium is assumed discrete, the hot spot effect representing the dependency between downward direct fluxes and diffuse fluxes at N has to be taken into account from depth t to z. Using Kuusk's model [1985], the radiance reaching M called $\mathrm{d}L_+^1(N \rightarrow M,\Omega)$ is

$$\begin{aligned} \mathrm{d}L_+^1(N \rightarrow M,\Omega) &= \mathrm{d}L_+^{1*}(N \rightarrow M,\Omega)\exp\left[\frac{\sqrt{k\kappa}}{b}(1-\exp[-b(z-t)])\right], \\ &= E_s(0)\exp[(k+\kappa)(t-z)]\exp\left[\frac{\sqrt{k\kappa}}{b}(1-\exp[-b(z-t)])\right] \\ &\quad \times \exp(kz)\pi^{-1}w(\Omega_s \rightarrow \Omega)\mathrm{d}t. \end{aligned} \tag{14}$$

Eq. (14) is the foundation of our model. However, since it has a complex expression, in particular in the exponential term corresponding to the hot spot correction, there is no linearity versus z and t enabling a simple solution based on differentiel equations as those of SAIL++ [cf. Eqs. 85]. For that, we propose to apply the Taylor series decomposition to this term

$$\exp\left[\frac{\sqrt{k\kappa}}{b}(1-\exp[-b(z-t)])\right] = \exp\left[\frac{\sqrt{k\kappa}}{b}\right]\sum_{n=0}^{\infty}\frac{(-1)^n(k\kappa)^{n/2}}{n!b^n}\exp[nb(t-z)]. \tag{15}$$

Therefore, Eq. (14) can be written as follows,

$$\mathrm{d}L_{+}^{1}(N \rightarrow M,\Omega) = \sum_{n=0}^{\infty}(-1)^{n}A_{n}\mathrm{d}L_{+}^{1,n}(N \rightarrow M,\Omega), \tag{16}$$

where

$$\begin{aligned} \mathrm{d}L_{+}^{1,n}(N \rightarrow M,\Omega) &= E_{s}(0)\exp[(k+\kappa_{n})(t-z)] \times \exp(kz)\pi^{-1}w(\Omega_{s} \rightarrow \Omega)\mathrm{d}t, \\ A_{n} &= \frac{(k\kappa)^{n/2}}{n!b^{n}}\exp\left[\frac{\sqrt{k\kappa}}{b}\right], \\ \kappa_{n} &= \kappa + nb. \end{aligned} \tag{17}$$

As the vegetation is homogeneous, then $\mathrm{d}L_{+}^{1}(N \rightarrow M,\Omega)$ can be written simply as $\mathrm{d}L_{+}^{1}(t \rightarrow z,\Omega)$. Thus, $L_{+}^{1}(z,\Omega)$ is obtained by integration of $\mathrm{d}L_{+}^{1}$ over the depth $[-H,z]$

$$L_{+}^{1}(z,\Omega) = \int_{t=-H}^{z}\mathrm{d}L_{+}^{1}(t \rightarrow z,\Omega). \tag{18}$$

Based on (16), L_{+}^{1} can be written as

$$L_{+}^{1}(z,\Omega) = \sum_{n=0}^{+\infty}(-1)^{n}A_{n}L_{+}^{1,n}(z,\Omega), \tag{19}$$

where

$$\begin{aligned} L_{+}^{1,n}(z,\Omega) &= \int_{-H}^{z}E_{s}(0)\exp[(k+\kappa+nb)(t-z)]\exp(kz)\pi^{-1}w(\Omega_{s} \rightarrow \Omega)\mathrm{d}t, \\ &= E_{s}(0)\frac{1-\exp[-(k+\kappa+nb)(H+z)]}{k+\kappa+nb}\exp(kz)\pi^{-1}w(\Omega_{s} \rightarrow \Omega). \end{aligned} \tag{20}$$

2.2.2 Application of the effective vegetation density approach

Here, we will try to extend the reformulated SAIL++ equation (cf. Appendix A.3) to the discrete case. Thus, as shown in Section 2.1.3, the hot spot effect will be treated as an increased posterior probability of gap which, in turn, results from a reduction in vegetation density. Then, it was suggested the use of the concept 'effective vegetation density' to describe the phenomenon. In this subsection, we propose to derive this density for $L_{+}^{1,n}, \forall n \in \mathbb{N}$, and to use it further to derive the equations of fluxes created by $L_{+}^{1,n}$ scattering. Moreover, the same effective density using leads to conserve energy (as explained in Section 2.1.3).

In Eqs. (13) (17), the difference between $\mathrm{d}L_{+}^{1*}$ and $\mathrm{d}L_{+}^{1,n}$ is the value of the extinction in the direction Ω (κ and κ_{n} respectively). Note that $\forall n > 0, \kappa_{n} > \kappa$, then $\mathrm{d}L_{+}^{1,n}$ decreases faster than $\mathrm{d}L_{+}^{1}$.

According to our approach described in Section A, the variation in the extinction factor is linked to the variation of the collision probability locally around M. In other words, a decrease in the probability of finding foliage at M decreases P_{χ}, accordingly (cf. Appendix B). Now, according to (77) and (99)

$$\left.\begin{aligned} \kappa &= d_{L}P_{\chi}(M)\kappa_{o} \\ \kappa_{n} &= d_{L}P_{\chi,n}(M)\kappa_{o} \end{aligned}\right\} \Rightarrow P_{\chi,n}(M) = \frac{\kappa_{n}}{\kappa}P_{\chi}(M), \tag{21}$$

with $P_{\chi,n}(M)$ the a posteriori probability of finding vegetation at M for the virtual radiance $\mathrm{d}L_{+}^{1,n}$, and κ_0 the normalized extinction factor [as explained in Eq. (77), it is independent on vegetation density]. We will use this notation in the following for SAIL++ scattering parameters. For each scattering parameter X, one can define the corresponding normalized one X_0 according to Eq. (77).
As we can see in Eqs. (21), $P_{\chi,n}(M)$ does not depend on M. Thus, it will be simply called $P_{\chi,n}$. Then, based on L_{+}^{1} differential equation derivation [cf. Eq. (89)] and replacing κ by κ_n, we obtain,

$$\begin{aligned}\frac{\mathrm{d}L_{+}^{1,n}(z,\Omega)}{\mathrm{d}z} &= [\mathfrak{s}\circ E_s(z,\Omega_s)](\Omega) - \kappa_n L_{+}^{1,n}(z,\Omega) = [\mathfrak{s}\circ E_s(z,\Omega_s)](\Omega) - d_L P_{\chi,n}\kappa_0 L_{+}^{1,n}(z,\Omega),\\ &= [\mathfrak{s}\circ E_s(z,\Omega_s)](\Omega) - d_L P_{\chi,n}[\mathfrak{k}_0\circ L_{+}^{1,n}(z)](\Omega),\end{aligned} \tag{22}$$

where $\mathfrak{k}_0$ is the normalized scattering term corresponding to $\mathfrak{k}$ [cf. Eq. (80)].
It leads to the following important result linking the differentiation of L_{+}^{1} to $(L_{+}^{1,n})_{n\in\mathbb{N}}$:

$$\begin{aligned}\frac{\mathrm{d}L_{+}^{1}(z,\Omega)}{\mathrm{d}z} &= \frac{\mathrm{d}\left\{\sum_{n=0}^{+\infty}(-1)^n A_n L_{+}^{1,n}(z,\Omega)\right\}}{\mathrm{d}z},\\ &= \sum_{n=0}^{+\infty}(-1)^n A_n\{[\mathfrak{s}\circ E_s(z,\Omega_s)](\Omega) - d_L P_{\chi,n}[\mathfrak{k}_0\circ L_{+}^{1,n}(z)](\Omega)\},\\ &= [\mathfrak{s}\circ E_s(z,\Omega_s)](\Omega)\overbrace{\sum_{n=0}^{+\infty}(-1)^n A_n}^{=1} - d_L\sum_{n=0}^{+\infty}(-1)^n A_n P_{\chi,n}[\mathfrak{k}_0\circ L_{+}^{1,n}(z)](\Omega),\\ &= [\mathfrak{s}\circ E_s(z,\Omega_s)](\Omega) - d_L\sum_{n=0}^{+\infty}(-1)^n A_n P_{\chi,n}[\mathfrak{k}_0\circ L_{+}^{1,n}(z)](\Omega).\end{aligned} \tag{23}$$

Thus, the radiance distributions created by $\mathrm{d}L_{+}^{1,n}$ scattering depend on $P_{\chi,n}$ rather than P_{χ}. As explained in Appendix A.3, these radiances are the downward diffuse radiance distribution (L_{-}), upward higher order diffuse radiance distribution (L_{+}^{∞}), upward radiance in observation direction (E_o^{+}) and downward radiance in observation direction (E_o^{-}). Note that, the mathematical validation, in term of global flux estimation, is explained in Subsection 2.2.3 and then shown in Appendix C.
Note that, similar to L_{+}^{1}, the differentiation of E_o^{+} that depends only on E_s is

$$\frac{\mathrm{d}E_o^{+}(z,\Omega_o)}{\mathrm{d}z} = wE_s(z,\Omega_s) - d_L\sum_{n=0}^{+\infty}(-1)^n A_n P_{\chi,n} K_0 E_0^{+,n}(z,\Omega_o), \tag{24}$$

with Ω_o the E_s direction, K the extinction factor in the direction Ω_o and

$$E_0^{+,n}(z,\Omega_o) = E_s(0)\frac{1-\exp[-(k+K+nb)(H+z)]}{k+K+nb}\exp(kz)w(\Omega_s\rightarrow\Omega_o). \tag{25}$$

As in classical models, there is no need to use Eq. (24). We merely assume, as in the turbid case, that

$$\frac{\mathrm{d}E_0^{+}(z,\Omega_o)}{\mathrm{d}z} = wE_s(z,\Omega_s) - KE_0^{+}(z,\Omega_o), \tag{26}$$

and the reflectance provided from the first order collision ($\rho_{so}^{(0),HS}$) will be corrected using the tradional formula (Kuusk, 1985)

$$\rho_{so}^{(0),HS} = w \int_{-H}^{0} \exp\left[(k+K)z + \frac{\sqrt{kK}}{b}[1-\exp(bz)]\right] \mathrm{d}z. \tag{27}$$

2.2.3 Dependency on $L_+^{1,n}$

In this subsection, we propose a modification to the reformulated SAIL++ equation set, presented in Appendix A.3, in order to take into account the effective vegetation density values in the expressions of L_-, L_+^∞, E_o^+ and E_o^- that depend on $L_+^{1,n}$ scattering.

First, let us derive the angular differentiation of E_o^+ ($\mathrm{d}^2E_o^+(z,\Omega \to \Omega_o)$) that depends only on $L_+^{1,n}$. Compared to the dependency on L_+ in classical SAIL++ equations, P_χ has to be replaced by $P_{\chi,n}$. Thus,

$$\frac{\mathrm{d}[\mathrm{d}^2E_o^+(z,\Omega\to\Omega_o)]}{\mathrm{d}z} = w_n'(\Omega\to\Omega_o)L_+^{1,n}(z,\Omega)\cos(\theta)\mathrm{d}\Omega, \tag{28}$$

where

$$w_n'(\Omega\to\Omega_o) = d_L P_{\chi,n} w_0'(\Omega\to\Omega_o), \tag{29}$$

with w_0' the normalized scattering parameter corresponding to w' [cf. Eq. (76)].

Then, the angular differentiation of E_o^+ ($\mathrm{d}^2E_o^+(z,\Omega\to\Omega_o)$) which depends only on L_+^1 is obtained by summing the contribution of the set $(L_+^{1,n})_{n\in\mathbb{N}}$

$$\frac{\mathrm{d}[\mathrm{d}^2E_o^+(z,\Omega\to\Omega_o)]}{\mathrm{d}z} = \sum_{n=0}^{+\infty}(-1)^n A_n w_n'(\Omega\to\Omega_o)L_+^{1,n}(z,\Omega)\cos(\theta)\mathrm{d}\Omega. \tag{30}$$

Note that, based on AddingSD formalism, the validity of our decomposition in this derivation of $P_{\chi,n}$ is shown in Appendix C.

By integration of Ω over the upper-hemisphere [cf. Eqs. (75) (84)], Eq. (30) becomes

$$\begin{aligned}\frac{\mathrm{d}E_o^+(z,\Omega_o)}{\mathrm{d}z} &= d_L\sum_{n=0}^{+\infty}(-1)^n A_n P_{\chi,n}\int_\Pi w_0'(\Omega\to\Omega_o)L_+^{1,n}(z,\Omega)\cos(\theta)\mathrm{d}\Omega,\\ &= d_L\sum_{n=0}^{+\infty}(-1)^n A_n P_{\chi,n}[\mathfrak{v}_0'\circ L_+^{1,n}(z)],\end{aligned} \tag{31}$$

with $\mathfrak{v}_0'$ the normalized scattering parameter corresponding to $\mathfrak{v}'$ [cf. Eq. (74)].

Next, by integrating the dependency on E_s, L_- and L_+^∞, the original reformulated SAIL++ Eq. (92) becomes

$$\frac{\mathrm{d}E_o^+}{\mathrm{d}z} = wE_s + \mathfrak{v}\circ L_- + \mathfrak{v}'\circ L_+^\infty + d_L\sum_{n=0}^{+\infty}(-1)^n A_n P_{\chi,n}[\mathfrak{v}_0'\circ L_+^{1,n}(z)] - KE_o^+. \tag{32}$$

Similarly, Eqs. (91), (90) and (93) become respectively

$$\begin{aligned}\frac{\mathrm{d}L_-}{\mathrm{d}z} &= -\mathfrak{s}'\circ E_s + \mathfrak{A}\circ L_- - \mathfrak{B}\circ L_+^\infty - d_L\sum_{n=0}^{+\infty}(-1)^n A_n P_{\chi,n}[\mathfrak{B}_0\circ L_+^{1,n}(z)],\\ \frac{\mathrm{d}L_+^\infty}{\mathrm{d}z} &= \mathfrak{o}\circ E_s + \mathfrak{B}\circ L_- - \mathfrak{A}\circ L_+^\infty + d_L\sum_{n=0}^{+\infty}(-1)^n A_n P_{\chi,n}[\mathfrak{B}_0'\circ L_+^{1,n}(z)],\\ \frac{\mathrm{d}E_o^-}{\mathrm{d}z} &= -w'E_s - \mathfrak{v}'\circ L_- - \mathfrak{v}\circ L_+^\infty - d_L\sum_{n=0}^{+\infty}(-1)^n A_n P_{\chi,n}[\mathfrak{v}_0\circ L_+^{1,n}(z)] + KE_o^-,\end{aligned} \tag{33}$$

with $\mathfrak{o}$ the vacuum operator, $\mathfrak{B}_0$, $\mathfrak{B}'_0$ and $\mathfrak{v}_0$ the normalized scattering parameters corresponding to $\mathfrak{B}$, $\mathfrak{B}'$ and $\mathfrak{v}$ [cf. Eqs. (73) (74) (75)], respectively.

3. Virtual flux decomposition implementation

As in SAIL++ (cf. Appendix A.2), the implementation needs the discretization of the diffuses fluxes over the hemispheres. These diffuse fluxes correspond to the diffuse radiances $(L_+^{1,n})_{n\in\mathbb{N}}$, L_+^∞ and L_- when only a vegetation layer is considered (cf. Subsection 2.2). The corresponding discrete fluxes will be called $(E_+^{1,n})_{n\in\mathbb{N}}$, E_+^∞ and E_-, respectively. The reflectances created by scattering of $(E_+^{1,n})_{n\in\mathbb{N}}$ and $(E_+^{0,n})_{n\in\mathbb{N}}$ will be separated to the one created by E_s. The separation enables the solution of the RT problem based on SAIL++ formalism.
First, we present the processing of the vegetation layer. Second, we show the soil vegetation coupling.

3.1 Vegetation layer

3.1.1 $E_+^{1,n}$ estimation

As reformulated in Appendix A.3, the difference between SAIL ++ and our model occurs in the calculation of L_+^1. In our model it is decomposed into the sequence $(L_+^{1,n})_{n\in\mathbb{N}}$ thus modifying the expressions of L_-, L_+^∞, E_o^+ and E_o^-. Therefore, in this section, we propose the derivation of a new expression for the discrete fluxes E_- and E_+^∞ as well as the radiances E_o^+ and E_o^- versus $(E_+^{1,n})_{n\in\mathbb{N}}$.
Now, $\forall n \in \mathbb{N}, L_+^{1,n}$ is given by Eq. (20). Let us consider the Verhoef (1998) sphere tessellation into N segments, then the irradiance $E_{+,i}^{1,n}$ of each segment i is

$$\begin{aligned} E_{+,i}^{1,n}(z) &= \int_{\Delta\Omega_i} L_+^{1,n}(z,\Omega)\cos(\theta)\mathrm{d}\Omega, \\ &\approx E_s(0)\frac{1-\exp[-(k+\langle\kappa\rangle_{\Delta\Omega_i}+n\langle b\rangle_{\Delta\Omega_i})(H+z)]}{k+\langle\kappa\rangle_{\Delta\Omega_i}+n\langle b\rangle_{\Delta\Omega_i}}\exp(kz) \\ &\times\int_{\Delta\Omega_i}\pi^{-1}w(\Omega_s\to\Omega)\cos(\theta)\mathrm{d}\Omega, \end{aligned} \tag{34}$$

where $\langle . \rangle_{\Delta\Omega_i}$ is the mean value operator defined for a given function f as follows

$$\langle f(\Omega)\rangle_{\Delta\Omega_i} = \frac{\int_{\Omega\in\Delta\Omega_i} f(\Omega)\cos(\Omega)\mathrm{d}\Omega}{\int_{\Omega\in\Delta\Omega_i}\cos(\Omega)\mathrm{d}\Omega}. \tag{35}$$

Following Verhoef (1998) terminology,

$$\begin{aligned} &\langle\kappa\rangle_{\Delta\Omega_i} = \boldsymbol{\kappa}(i), \\ &\int_{\Delta\Omega_i}\pi^{-1}w(\Omega_s\to\Omega)\cos(\theta)\mathrm{d}\Omega = \mathbf{s}(i), \end{aligned} \tag{36}$$

similarly, we adopt the following notation

$$\langle b\rangle_{\Delta\Omega_i} = \boldsymbol{b}(i), \tag{37}$$

thus κ_n [cf. Eq. (17)] will be extended in the discrete case as follows

$$\boldsymbol{\kappa}_n(i) = \boldsymbol{\kappa}(i) + n\boldsymbol{b}(i). \tag{38}$$

3.1.2 $E_{+,i}^{1,n}$ dependency

Being scattered, $E_{+,i}^{1,n}$ can create both diffuse fluxes E_{+}^{∞} and E_{-} as well as radiances E_{o}^{+} and E_{o}^{-}. The scattering parameters will be called respectively $\mathbf{s}_{i,n}$, $\mathbf{s}'_{i,n}$, $w'_{i,n}$ and $w_{i,n}$. Now,

$$w'_{i,n}(\Omega_o) = d_L P_{\chi,n} \langle w'_0(\Omega \to \Omega_o) \rangle_{\Omega \in \Delta\Omega_i} = P_{\chi,n} \mathbf{v}'_0(i), \tag{39}$$

where $\mathbf{v}'_0$ is the normalized SAIL++ scattering parameter corresponding to $\mathbf{v}'$ [cf. Eq. (85)]. Similarly, one can define $w_{i,n}$ the analogue of $w'_{i,n}$ when $\Delta\Omega_i$ and Ω_o are in the same hemisphere

$$w_{i,n}(\Omega_o) = d_L P_{\chi,n} \langle w_0(\Omega \to \Omega_o) \rangle_{\Omega \in \Delta\Omega_i} = P_{\chi,n} \mathbf{v}_0(i), \tag{40}$$

where $\mathbf{v}_0$ is the normalized scattering parameter corresponding to $\mathbf{v}$ [cf. Eq. (85)].
As in the SAIL++ model (Verhoef, 1998), $\mathbf{s}_{i,n}$ and $\mathbf{s}'_{i,n}$ are integrated values of $w_{i,n}$ and $w'_{i,n}$ over the output solid angle. So, for $m \in \{1, \ldots, N\}$ a given discrete solid angle index

$$\begin{aligned} \mathbf{s}_{i,n}(m) &= \int_{\Delta\Omega_m} w_{i,n}(\Omega_m) \mathrm{d}\Omega_m = d_L P_{\chi,n} \pi^{-1} \langle\langle w_0(\Omega \to \Omega_+) \rangle\rangle_{(\Omega,\Omega_+) \in (\Delta\Omega_i, \Delta\Omega_m)} \frac{2\pi}{N}, \\ &= d_L P_{\chi,n} \mathbf{B}'_0(i \to m), \end{aligned} \tag{41}$$

where $\mathbf{B}'_0$ is the normalized SAIL++ scattering matrix corresponding to $\mathbf{B}'$ [cf. Eq. (86)]. Similarly,

$$\mathbf{s}_{i,n}(m) = d_L P_{\chi,n} \mathbf{B}_0(i \to m), \tag{42}$$

where $\mathbf{B}_0$ is the normalized scattering matrix corresponding to $\mathbf{B}$ [cf. Eq. (85)].

3.1.3 $E_{+,i}^{1,n}$ decomposition

From Eq. (34), one has

$$\begin{aligned} E_{+,i}^{1,n}(z) &= E_s(0) \frac{1 - \exp[-(k + \boldsymbol{\kappa}_n(i))(H + z)]}{k + \boldsymbol{\kappa}_n(i)} \exp(kz) \mathbf{s}(i), \\ &= X_i^n E_{+,i,1}^{1,n}(z) + Y_i^n E_{+,i,2}^{1,n}(z). \end{aligned} \tag{43}$$

with

$$\begin{aligned} X_i^n &= \frac{\mathbf{s}(i)}{k + \boldsymbol{\kappa}_n(i)}, \\ Y_i^n &= -\frac{\mathbf{s}(i)\exp(-kH)}{k + \boldsymbol{\kappa}_n(i)}, \\ E_{+,i,1}^{1,n}(z) &= E_s(0)\exp(kz) = E_s(z), \\ E_{+,i,2}^{1,n}(z) &= E_s(0)\exp[-\boldsymbol{\kappa}_n(i)(H + z)]. \end{aligned} \tag{44}$$

Therefore, $E_{+,i,1}^{1,n}$ and $E_{+,i,2}^{1,n}$ can be viewed as the direct downward and upward fluxes with an extinction factor under the vegetation equal to k and $\boldsymbol{\kappa}_n(i)$, respectively.

Thus, the corresponding RT discrete equation set to the continuous Eqs. (32) (33) presented in the last section is

$$\begin{cases} E^{1,n}_{+,i,1}(0) = E^{1,n}_{+,i,2}(-H) = E_s(0), \forall\{i,n\} \in \{1,\ldots,N\} \times \mathbb{N}, \\ \dfrac{dE_s}{dz} = kE_s, \\ \dfrac{dE^{1,n}_{+,i,1}}{dz} = kE^{1,n}_{+,i,1}, \forall\{i,n\} \in \{1,\ldots,N\} \times \mathbb{N}, \\ \dfrac{dE^{1,n}_{+,i,2}}{dz} = -\boldsymbol{\kappa}_n(i)E^{1,n}_{+,i,2}, \forall\{i,n\} \in \{1,\ldots,N\} \times \mathbb{N}, \\ \dfrac{dE_-}{dz} = -\mathbf{s}'E_s + \mathbf{A}E_- - \mathbf{B}E_+ - \sum\limits_{n=0}^{\infty}(-1)^n \sum\limits_{i=1}^{N} A_i^n (X_i^n \mathbf{s}_{i,n} E^{1,n}_{+,i,1} + Y_i^n \mathbf{s}_{i,n} E^{1,n}_{+,i,2}), \\ \dfrac{dE^{\infty}_{+}}{dz} = \mathbf{B}E_- - \mathbf{A}E_+ + \sum\limits_{n=0}^{\infty}(-1)^n \sum\limits_{i=1}^{N} A_i^n (X_i^n \mathbf{s}'_{i,n} E^{1,n}_{+,i,1} + Y_i^n \mathbf{s}'_{i,n} E^{1,n}_{+,i,2}), \\ \dfrac{dE_o^+}{dz} = wE_s + \mathbf{v}E_- + \mathbf{v}'E_+ + \sum\limits_{n=0}^{\infty}(-1)^n \sum\limits_{i=1}^{N} A_i^n (X_i^n w'_{i,n} E^{1,n}_{+,i,1} + Y_i^n w'_{i,n} E^{1,n}_{+,i,2}) - KE_o^+, \\ \dfrac{dE_o^-}{dz} = -w'E_s - \mathbf{v}'E_- - \mathbf{v}E_+ - \sum\limits_{n=0}^{\infty}(-1)^n \sum\limits_{i=1}^{N} A_i^n (X_i^n w_{i,n} E^{1,n}_{+,i,1} + Y_i^n w_{i,n} E^{1,n}_{+,i,2}) + KE_o^-, \end{cases} \tag{45}$$

with A_i^n the extension of A_n to the discrete case Eq. (17)

$$A_i^n = \frac{(k\boldsymbol{\kappa}(i))^{n/2}}{n!\boldsymbol{b}(i)^n} \exp\left[\frac{\sqrt{k\boldsymbol{\kappa}(i)}}{\boldsymbol{b}(i)}\right]. \tag{46}$$

From a mathematical perspective, System 45 could be viewed as follows. The unknowns are E_-, E^{∞}_{+}, E_o^+ and E_o^-. They have to be solved using three differential equations linking them (three last Equations in Set 45). In addition to the unknown functions, the differential equations contain additive terms composed of linear combinations of known functions which are E_s and $E^{1,n}_{+,i,j}, \forall\{i,j,n\} \in \{1,\ldots,N\} \times \{1,2\} \times \mathbb{N}$. Therefore, solutions to the global differential equation set (E_-, E^{∞}_{+}, E_o^+ and E_o^-) can be written as linear combinations (the same as the combination of the additive terms in the initial set) of the same differential equation set solutions with only one additive term among the set E_s, $E^{1,n}_{+,i,j}, \forall\{i,j,n\} \in \{1,\ldots,N\} \times \{1,2\} \times \mathbb{N}$.
Therefore, we propose the following solution. E_-, E^{∞}_{+}, E_o^+ and E_o^- have to be derived for different sources: $E_s(0)$, $E^{1,n}_{+,i,1}(0)$ and $E^{1,n}_{+,i,2}(-H)$, $\forall\{i,n\} \in \{1,\ldots,N\} \times \mathbb{N}$. For that, one can define the corresponding sub-solutions which are E^s_-, $E^{\infty,s}_{+}$, $E_o^{+,s}$, $E_o^{-,s}$, $\forall\{i,j,n\} \in \{1,\ldots,N\} \times \{1,2\} \times \mathbb{N}$, $E^n_{-,i,j}$, $E^{\infty,n}_{+,i,j}$, $E^{+,n}_{o,i,j}$ and $E^{-,n}_{o,i,j}$, respectively.
According to Eqs. (45), the global solution for $E \in \{E_-, E^{\infty}_{+}, E_o^+, E_o^-\}$ is written as follows

$$E = E^s + \sum_{n=0}^{\infty}(-1)^n \sum_{i=1}^{N} A_i^n (X_i^n E^n_{i,1} + Y_i^n E^n_{i,2}), \tag{47}$$

Now, compared to SAIL++ in terms of boundary conditions (cf. Appendix A.2), each term x of the boundary condition matrix [cf. Eq. (87)] that depends on the direct source flux [cf. Eq.

(88)], i.e.

$$x \in \left\{ \tau_{ss} = \frac{E_s(-H)}{E_s(0)}, \boldsymbol{\tau}_{sd} = \frac{E_-(-H)}{E_s(0)}, \boldsymbol{\rho}_{sd} = \frac{E_+^0(0) + E_+^\infty(0)}{E_s(0)}, \rho_{so} = \frac{E_o^+(0)}{E_s(0)}, \tau_{so} = \frac{E_o^-(0)}{E_s(0)} \right\} \quad (48)$$

has to be modified. The other boundary matrix terms (T, R, $\boldsymbol{\tau}_{do}$, $\boldsymbol{\rho}_{do}$ and τ_{oo}) remain equivalent to SAIL++.
Moreover, $\boldsymbol{\rho}_{sd}$ is divided into two terms

$$\begin{aligned} \boldsymbol{\rho}_{sd} &= \boldsymbol{\rho}_{sd}^1 + \boldsymbol{\rho}_{sd}^\infty, \\ \boldsymbol{\rho}_{sd}^1 &= \frac{E_+^0(0)}{E_s(0)}, \\ \boldsymbol{\rho}_{sd}^\infty &= \frac{E_+^\infty(0)}{E_s(0)}. \end{aligned} \quad (49)$$

In the case of $x \in \{\tau_{ss}, \boldsymbol{\tau}_{sd}, \boldsymbol{\rho}_{sd}^\infty, \rho_{so}, \tau_{so}\}$ and according to Eqs. (47)

$$x = x^s + \sum_{n=0}^{\infty} (-1)^n \sum_{i=1}^{N} A_i^n (X_i^n x_{i,1}^n + Y_i^n x_{i,2}^n), \quad (50)$$

with x^s the value corresponding to the source E_s, and $\forall \{i,j,n\} \in \{1,\ldots,N\} \times \{1,2\} \times \mathbb{N}, x_{i,j}^n$ the value corresponding to the source $E_{+,i,j}^{n,1}$. Based on Eqs. (43) (44)

$$\forall i \in \{1,\ldots,N\}, \boldsymbol{\rho}_{sd}^1(i) = \sum_{n=0}^{\infty} (-1)^n A_i^n (X_i^n + Y_i^n \exp[-\boldsymbol{\kappa}_n(i) H]). \quad (51)$$

Note that τ_{ss}, the direct transmittance, does not change, it is equal to $\exp(-kH)$. Therefore, we have to derive only $\boldsymbol{\tau}_{sd}$, $\boldsymbol{\rho}_{sd}^\infty$, ρ_{so} and τ_{so}.

3.1.4 Sub-solution derivation

Here, we try to find the sub-solution scattering term expressions ($\boldsymbol{\tau}_{sd}$, $\boldsymbol{\rho}_{sd}^\infty$, ρ_{so} and τ_{so}) based on SAIL++ formalism and versus SAIL++ boundary matrix terms. To achieve it, the irradiance E_+^∞, E_- and radiances E_o have to be estimated. The latter terms have first to be estimated for each source among E_s and $\forall \{i,j,n\} \in \{1,\ldots,N\} \times \{1,2\} \times \mathbb{N}$, $E_{-,i,j}^n$ and second combined using Eq. (47).

3.1.4.1 Source E_s

E_-^s, $E_+^{\infty,s}$, $E_o^{+,s}$ and $E_o^{-,s}$ verify

$$\frac{\mathrm{d}}{\mathrm{d}z} \begin{pmatrix} E_s \\ E_-^s \\ E_+^{\infty,s} \\ E_o^{+,s} \\ E_o^{-,s} \end{pmatrix} = \begin{pmatrix} k & 0 & 0 & 0 & 0 \\ -\mathbf{s}' & \mathbf{A} & -\mathbf{B} & 0 & 0 \\ 0 & \mathbf{B} & -\mathbf{A} & 0 & 0 \\ w & \mathbf{v} & \mathbf{v}' & -K & 0 \\ -w' & -\mathbf{v}' & -\mathbf{v} & 0 & K \end{pmatrix} \begin{pmatrix} E_s \\ E_-^s \\ E_+^{\infty,s} \\ E_o^{+,s} \\ E_o^{-,s} \end{pmatrix}, \quad (52)$$

Thus based on Eq. (88) notation, it follows

$$\begin{aligned} \boldsymbol{\tau}_{sd}^s &= \boldsymbol{\tau}_{sd}^{++}(k, \mathbf{s}', 0), \\ \boldsymbol{\rho}_{sd}^{s,\infty} &= \boldsymbol{\rho}_{sd}^{++}(k, \mathbf{s}', 0), \\ \rho_{so}^s &= \rho_{so}^{HS,++}(k, \mathbf{s}', 0, w), \\ \tau_{so}^s &= \tau_{so}^{++}(k, \mathbf{s}', 0, w). \end{aligned} \quad (53)$$

3.1.4.2 Source $E^{1,n}_{+,i,1}$
As for E_s [cf. Eq. (53)], it is straightforward to show that

$$\begin{aligned} \boldsymbol{\tau}^{n}_{sd,i,1} &= \boldsymbol{\tau}^{++}_{sd}(k,\mathbf{s}_{i,n},\mathbf{s}'_{i,n}), \\ \boldsymbol{\rho}^{n,\infty}_{sd,i,1} &= \boldsymbol{\rho}^{++}_{sd}(k,\mathbf{s}_{i,n},\mathbf{s}'_{i,n}), \\ \rho^{n}_{so,i,1} &= \rho^{++}_{so}(k,\mathbf{s}_{i,n},\mathbf{s}'_{i,n},w'_{i,n}), \\ \tau^{n}_{so,i,1} &= \tau^{++}_{so}(k,\mathbf{s}_{i,n},\mathbf{s}'_{i,n},w'_{i,n}). \end{aligned} \tag{54}$$

3.1.4.3 Source $E^{1,n}_{+,i,2}$
As for E_s [cf. Eq. (53)], it is straightforward to show that

$$\begin{aligned} \boldsymbol{\tau}^{n}_{sd,i,2} &= \boldsymbol{\rho}^{++}_{sd}(\boldsymbol{\kappa}_n(i),\mathbf{s}'_{i,n},\mathbf{s}_{i,n}), \\ \boldsymbol{\rho}^{n,\infty}_{sd,i,2} &= \boldsymbol{\tau}^{++}_{sd}(\boldsymbol{\kappa}_n(i),\mathbf{s}'_{i,n},\mathbf{s}_{i,n}), \\ \rho^{n}_{so,i,2} &= \tau^{++}_{so}(\boldsymbol{\kappa}_n(i),\mathbf{s}'_{i,n},\mathbf{s}_{i,n},w'_{i,n}), \\ \tau^{n}_{so,i,2} &= \rho^{++}_{so}(\boldsymbol{\kappa}_n(i),\mathbf{s}'_{i,n},\mathbf{s}_{i,n},w'_{i,n}). \end{aligned} \tag{55}$$

Finally, according to Eqs. (50) (53) (54) (55)

$$\begin{aligned} \boldsymbol{\tau}_{sd} &= \boldsymbol{\tau}^{++}_{sd}(k,\mathbf{s}',0) + \sum_{n=0}^{\infty}(-1)^n \sum_{i=1}^{N} A^n_i\left(X^n_i \boldsymbol{\tau}^{++}_{sd}(k,\mathbf{s}_{i,n},\mathbf{s}'_{i,n}) + Y^n_i \boldsymbol{\rho}^{++}_{sd}(\boldsymbol{\kappa}_n(i),\mathbf{s}'_{i,n},\mathbf{s}_{i,n})\right), \\ \boldsymbol{\rho}^{\infty}_{sd} &= \boldsymbol{\rho}^{++}_{sd}(k,\mathbf{s}',0) + \sum_{n=0}^{\infty}(-1)^n \sum_{i=1}^{N} A^n_i\left(X^n_i \boldsymbol{\rho}^{++}_{sd}(k,\mathbf{s}_{i,n},\mathbf{s}'_{i,n}) + Y^n_i \boldsymbol{\tau}^{++}_{sd}(\boldsymbol{\kappa}_n(i),\mathbf{s}'_{i,n},\mathbf{s}_{i,n})\right), \end{aligned} \tag{56}$$

$$\begin{aligned} x_{so} &= \rho^{HS,++}_{so}(k,\mathbf{s}',0,w) + \sum_{n=0}^{\infty}(-1)^n \sum_{i=1}^{N} A^n_i\big(X^n_i \rho^{++}_{so}(k,\mathbf{s}_{i,n},\mathbf{s}'_{i,n},w'_{i,n}) + \\ &\qquad Y^n_i \tau^{++}_{so}(\boldsymbol{\kappa}_n(i),\mathbf{s}'_{i,n},\mathbf{s}_{i,n},w_{i,n})\big), \\ \tau_{so} &= \tau^{++}_{so}(k,\mathbf{s}',0,w) + \sum_{n=0}^{\infty}(-1)^n \sum_{i=1}^{N} A^n_i\big(X^n_i \tau^{++}_{so}(k,\mathbf{s}_{i,n},\mathbf{s}'_{i,n},w'_{i,n}) + \\ &\qquad Y^n_i \rho^{++}_{so}(\boldsymbol{\kappa}_n(i),\mathbf{s}'_{i,n},\mathbf{s}_{i,n},w_{i,n})\big). \end{aligned} \tag{57}$$

3.2 Concatenation vegetation layer and soil background

Similarly to vegetation, one can define the directional-hemispherical reflectance ($\boldsymbol{r}_{sd}$), hemispherical-directional reflectance ($\boldsymbol{r}_{do}$) and hemispherical-hemispherical reflectance ($\boldsymbol{R}_{dd}$) fore soil which are two vectors and a matrix, respectively.
Based on Adding principle (Van de Hulst, 1980), Verhoef (1998) defines the bidirectional reflectance of the couple soil+vegetation (ρ^t_{so}) as

$$\rho^t_{so} = \rho_{so} + \tau_{oo} r_{so} \tau_{ss} + (\tau_{oo}\boldsymbol{r}^T_{do} + \boldsymbol{\tau}^T_{do}\boldsymbol{R}_{dd})(1 - \mathrm{R}\boldsymbol{R}_{dd})^{-1}\boldsymbol{\tau}_{sd} + (\boldsymbol{\tau}^T_{do} + \tau_{oo}\boldsymbol{r}^T_{do}\mathrm{R})(I - \boldsymbol{R}_{dd}\mathrm{R})^{-1}\boldsymbol{r}_{sd}\tau_{ss}, \tag{58}$$

with I the identity matrix.
Inspired from AddingSD (e.g. Kallel et al., 2008, p. 3647), we propose the following transformation of Eq. (58)

$$\rho^t_{so} = \rho_{so} + \overbrace{\tau_{oo} r_{so} \tau_{ss}}^{r_{sso}} + \overbrace{\boldsymbol{\tau}^T_{do}\boldsymbol{r}_{sd}\tau_{ss}}^{r_{sdo}} + (\boldsymbol{\tau}^T_{do}\boldsymbol{R}_{dd} + \tau_{oo}\boldsymbol{r}^T_{do})(I - \mathrm{R}\boldsymbol{R}_{dd})^{-1}(\overbrace{\mathrm{R}\boldsymbol{r}_{sd}\tau_{ss} + \boldsymbol{\tau}_{sd}}^{\tau_{sdd}}). \tag{59}$$

As rigourously explained in (Kallel et al., 2008), to pass from a turbid to a discrete case and take into account the hot spot effect as well as maintain energy conservation, we have to modify the expression of r_{sso}, r_{sdo} and τ_{sdd} by considering the actual local vegetation density:

- r_{sso} corresponds to the flux passing through the vegetation layer from top to bottom without collision, scattered by the soil and reaching the top of the vegetation without other collisions. For this flux, the classical hot spot effect should be computed as

$$r_{sso} = r_{so} \exp\left[-(k+K)H + \frac{\sqrt{kK}}{b}[1 - \exp(-bH)] \right]; \quad (60)$$

- τ_{sdd} corresponds to the flux passing through the vegetation layer from top to bottom without collisions, scattered by the soil, colliding with the vegetation and reaching the soil again.
- r_{sdo} corresponds to the flux passing through the vegetation layer from top to bottom without collisions, scattered by the soil and reaching the top of the vegetation after multiple collisions.

Using the same principle that for E_+^1 scattering derivation, it is straightforward to show that

$$\begin{aligned} \tau_{sdd} &= \exp(-kH) \sum_{n=0}^{\infty} (-1)^n \sum_{i=1}^{N} A_i^n \mathbf{a}(i) \boldsymbol{\rho}_{sd}^{++}(\kappa_n(i), \mathbf{s}'_{i,n}, \mathbf{s}_{i,n}), \\ r_{sdo} &= \exp(-kH) \sum_{n=0}^{\infty} (-1)^n \sum_{i=1}^{N} A_i^n \mathbf{a}(i) \tau_{so}^{++}(\kappa_n(i), \mathbf{s}'_{i,n}, \mathbf{s}_{i,n}, w_{i,n}). \end{aligned} \quad (61)$$

4. Virtual flux decomposition validation

This section is dedicated to the validation of our virtual flux decomposition. The corresponding will be called the Flux Decomposition Model (FDM). First, model convergence and running time are presented. Second, energy conservation is shown. Third, a comparison between our approach and SAIL/SAIL++ models is presented. Finally, our model is compared to the 3-D models of the RAMI 2 database assumed 'most credible'.

Among the most commonly used models to describe the distribution of leaf zenithal angles is the method assuming an elliptic leaf distribution where the distribution is parameterized by the mean leaf inclination angle, ALA, ranging between 0 and 90° (Campbell, 1990). We will use this distribution in our model simulations. Note that small ALA values correspond to planophile distributions, high values to erectophile distributions, and medium values to extremophile distributions. Moreover, to be compatible with RAMI database simulations, Bunnik's [1978] parametrization will be used in the fourth subsection.

Since FDM is equivalent to SAIL++ in the turbid case. In this paper, we will deal only with the discrete case.

4.1 Running time

Among the strengths of our model is its low running time. The decomposition of L_+^1 into virtual sub-fluxes allowed the use of SAIL++ formalism to solve the RT problem.

Although, according to Eqs. (56) (57) (61), an infinite number of SAIL++ simulations is needed to derive the reflectance, only few first ranked terms are used to achieve accurate results. The sum is provided by Taylor series decomposition. Next, we will present a study on the accuracy of the approximation.

Here, we opt to use fluxes (E_+^∞ in our case) and the corresponding hemispherical scattering (ρ_{sd}^∞) term rather than radiances E_o^+ or E_o^-. For energy conservation, it is more significant to deal with fluxes than their densities.
Recall that

$$\begin{aligned} E_+^\infty &= E_+^{\infty,s} + \sum_{n=0}^{\infty}(-1)^n \sum_{i=1}^{N} A_i^n (X_i^n E_{+,i,j}^{\infty,n} + Y_i^n E_{+,i,j}^{\infty,n}) = E_+^{\infty,s} + \sum_{n=0}^{\infty}(-1)^n \sum_{i=1}^{N} A_i^n E_{+,i}^{\infty,n}, \\ &= E_+^{\infty,s} + \sum_{i=1}^{N}\sum_{n=0}^{\infty}(-1)^n A_i^n E_{+,i}^{\infty,n}, \end{aligned} \tag{62}$$

where $E_{+,i}^{\infty,n}$ is the high order diffuse flux created by $E_{+,i}^{1,n}$. Now, according to Eq. (22), $E_{+,i}^{1,n}$ is created by E_S scattering. Then, due to the energy conservation law

$$\forall (z,i) \in [-H,0] \times \{1,\ldots,N\}, ||E_{+,i}^{\infty,n}(z)|| \leq E_s(0), \tag{63}$$

where $||.||$ of a given discrete flux over a hemisphere is the sum of the sub-fluxes' values in each segment. It corresponds to the integrate radiance distribution over the hemisphere.
Let us assume that, $\forall i \in \{1,\ldots,N\}$, the series $\sum_{n=0}^{\infty}(-1)^n A_i^n E_{+,i}^{\infty,n}$ is truncated to the order u_i. Let us define a vector $\boldsymbol{u}$ by

$$\boldsymbol{u} = [u_1,\ldots,u_N], \tag{64}$$

and the corresponding flux $E_+^{\infty,\boldsymbol{u}}$

$$E_+^{\infty,\boldsymbol{u}} = E_+^{\infty,s} + \sum_{i=1}^{N}\sum_{n=0}^{u_i}(-1)^n A_i^n E_{+,i}^{\infty,n}, \tag{65}$$

then

$$\begin{aligned} ||E_+^\infty - E_+^{\infty,\boldsymbol{u}}|| &= ||\sum_{i=1}^{N}\sum_{n=u_i+1}^{\infty}(-1)^n A_i^n E_{+,i}^{\infty,n}|| \leq \sum_{i=1}^{N}\sum_{n=u_i+1}^{\infty} A_i^n ||E_{+,i}^{\infty,n}|| \leq \sum_{i=1}^{N}\sum_{n=u_i+1}^{\infty} A_i^n E_s(0), \\ &\leq E_s(0)\sum_{i=1}^{N}\exp\left[\frac{\sqrt{k\boldsymbol{\kappa}(i)}}{\boldsymbol{b}(i)}\right]\sum_{n=u_i+1}^{\infty}\frac{(k\boldsymbol{\kappa}(i))^{n/2}}{n!\boldsymbol{b}(i)^n}, \\ &\leq E_s(0)\sum_{i=1}^{N}\exp\left[\frac{\sqrt{k\boldsymbol{\kappa}(i)}}{\boldsymbol{b}(i)}\right]\frac{(k\boldsymbol{\kappa}(i))^{(u_i+1)/2}}{(u_i+1)!\boldsymbol{b}(i)^{u_i+1}}\sum_{n=0}^{\infty}\frac{(k\boldsymbol{\kappa}(i))^{n/2}(u_i+1)!}{(n+u_i+1)!\boldsymbol{b}(i)^n}, \\ &\leq E_s(0)\sum_{i=1}^{N}\exp\left[\frac{\sqrt{k\boldsymbol{\kappa}(i)}}{\boldsymbol{b}(i)}\right]\frac{(k\boldsymbol{\kappa}(i))^{(u_i+1)/2}}{(u_i+1)!\boldsymbol{b}(i)^{u_i+1}}\sum_{n=0}^{\infty}\frac{(k\boldsymbol{\kappa}(i))^{n/2}}{n!\boldsymbol{b}(i)^n}, \\ &\leq E_s(0)\sum_{i=1}^{N}\underbrace{\exp\left[\frac{\sqrt{k\boldsymbol{\kappa}(i)}}{\boldsymbol{b}(i)}\right]A_i^{u_i+1}}_{B_i^{u_i+1}}. \end{aligned} \tag{66}$$

It is clear that, $\lim_{u_i\to+\infty} B_i^{u_i+1} = 0$, then

$$\lim_{u_1\to+\infty}\lim_{u_2\to+\infty}\ldots\lim_{u_N\to+\infty}||E_+^\infty - E_+^{\infty,\boldsymbol{u}}|| = 0. \tag{67}$$

Moreover, let us define $\boldsymbol{\rho}_{sd}^{\infty,\boldsymbol{u}}$ by

$$\boldsymbol{\rho}_{sd}^{\infty,\boldsymbol{u}} = \frac{E_+^{\infty,\boldsymbol{u}}}{E_s(0)}, \tag{68}$$

then according to Eq. (66),

$$||\boldsymbol{\rho}_{sd}^{\infty} - \boldsymbol{\rho}_{sd}^{\infty,\boldsymbol{u}}|| \leq \sum_{i=1}^{N} B_i^{u_i+1}. \tag{69}$$

Therefore, to ensure that $||\boldsymbol{\rho}_{sd}^{\infty} - \boldsymbol{\rho}_{sd}^{\infty,\boldsymbol{u}}|| \leq \varepsilon$ with $\varepsilon \in \mathbb{R}_+^*$ such that $\varepsilon \ll 1$, it is sufficient to choose $\boldsymbol{u}$ as follows,

$$\forall i \in \{1,\ldots,N\}, u_i = \text{argmin}_{n\in\mathbb{N}} B_i^{n+1} \leq \frac{\varepsilon}{N}. \tag{70}$$

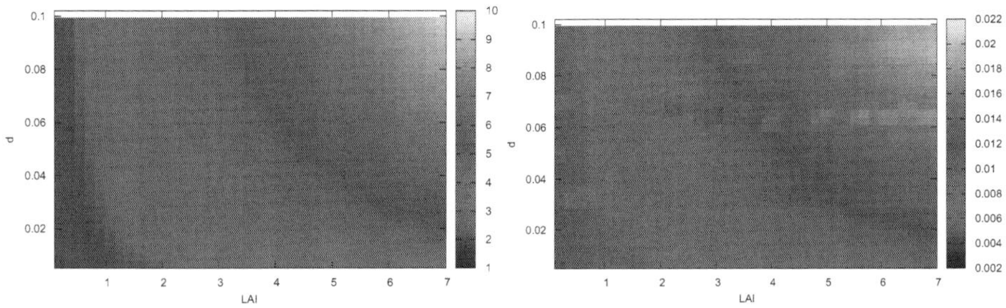

(a) Mean $\boldsymbol{u}$ value (b) Running time

Fig. 2. Variation of a) the mean $\boldsymbol{u}$ value and b) the corresponding model running time vas a function of LAI and the hot spot parameter d for a source zenith angle $\theta_s = 25°$, $ALA = 63°$, $\rho = 0.5$, $\tau = 0.5$ and a model error threshold $\varepsilon = 10^{-4}$.

Figure 2 shows the variation of the average $\boldsymbol{u}$ value ($\langle \boldsymbol{u} \rangle$) as well as the corresponding model running time, T, versus the Leaf Area Index value, LAI, and the hot spot parameter (ratio between the leaf mean radius and the vegetation layer thickness), d, for $\varepsilon = 10^{-4}$. As expected, $\langle \boldsymbol{u} \rangle$ and the corresponding T value increases as a function of LAI and d. $\langle \boldsymbol{u} \rangle$ ranges between 1 and 10 which is a relatively high speed convergence. For medium LAI and d values (respectively equal 3 and 0.05), $\langle \boldsymbol{u} \rangle$ is about 4.2. This speed convergence explains the small running time for such a complex model: it ranges between 2 and 22 milliseconds (ms). In particular, for LAI=3 and $d = 0.05$, it is about $10ms$.

4.2 Energy conservation

Compared to SAIL++, the advantage of our model is energy conservation in the discrete case. To check conservation, we propose to use the same procedure as Kallel et al. (2008). In the purist corner case ($\rho + \tau = 1$), the energy conservation law is

$$B(\Omega_s) = \tau_{ss} + \int_{\Pi} \{\rho_{so}(\Omega_o) + \tau_{so}(\Omega_o)\} \cos(\theta_o) \mathrm{d}\Omega_0 = 1. \tag{71}$$

Figure 3 shows the variation of B versus LAI, for extremophile and erectophile vegetation (ALA equals respectively 45° and 63°) and for different hot spot parameters. In both cases,

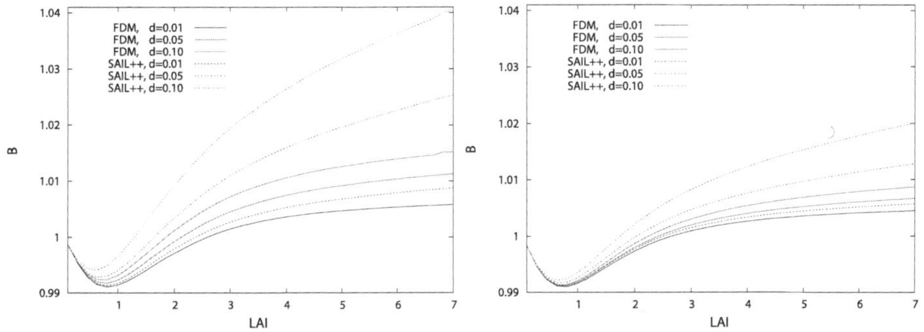

(a) ALA = 45⊠ value   (b) ALA = 63⊠ value

(a) $ALA = 45^\circ$ value   (b) $ALA = 63^\circ$ value

Fig. 3. Variation of energy conservation (B) versus LAI for different values of the hot spot parameter (d), $\theta_s = 25^\circ$, $\rho = 0.5$ and $\tau = 0.5$.

for small LAI values, the energy is well conserved by both models with an error lower than 0.1%. However, for values of LAI ranging from 0.5 to 1, B decreases to around 0.99 for both cases. The decrease of accuracy is due to the sampling of the hemisphere in only 36 segments and assuming a constant radiance distribution over each segment. The decrease is more pronounced for erectophile leaf distributions since the validity of the diffuse flux isotropy assumption weakens as ALA increases (Kallel et al., 2008). An increase in the segment number extends the running time. According to Verhoef (1998), 36 segments is a trade-off between accuracy and running time. Reaching its minimum values, B increases versus LAI in different ways for different models. First, for $d = 0.01$, the increase is relatively small and energy is accurately conserved by both models. Second, for $d = 0.05$ and $d = 0.1$, Figure 3 shows large differences between models. SAIL++ conserves energy less. For example for LAI=7 and $ALA = 45^\circ$, the variation in energy conservation (B) was 1.04 whereas for our model it was less then 1.015. This proves our main objective for the new model: energy conservation. Moreover, comparing extremophile and erectophile cases, Figure 3 shows that energy is batter conserved in the erectophile case. Indeed, mutual shadowing between leaves decreases as a function of ALA, and thus it is higher in the extremophile case.

4.3 Model comparison: SAIL/SAIL++

In this subsection, BRDF and BTDF produced by our model are compared to ones produced by the SAIL and SAIL++ models in the discrete case.

Figure 4 shows the case of a hot spot parameter ($d = 0.02$) for extremophile ($ALA = 45^\circ$) and erectophile vegetation ($ALA = 63^\circ$) in the principle plane. First, figures show the hot spot peaks for $\theta_o = \theta_s$ and $\varphi = 0^\circ$. Second, when a soil background is added, the model reflectances increase but the curve dynamics decrease due to soil lambertianity. Third, since SAIL underestimates the reflectance due to the diffuse flux semi-isotropy assumption (Kallel, 2007), its BRDF curves are always below SAIL++ curves. Fourth, compared to SAIL++, our model's BRDF and BTDF curves are always lower: our model decreases SAIL++ diffuse fluxes E_+^∞ and

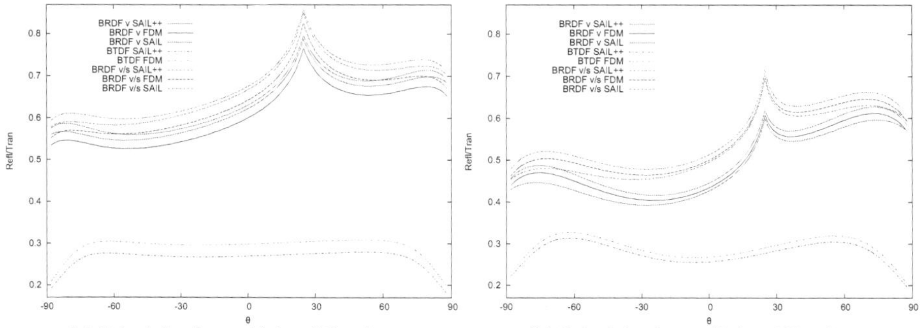

(a) Principle plane, $ALA = 45°$ value

(b) Principle plane, $ALA = 63°$ value

Fig. 4. A comparison between BRDF of SAIL, SAIL++ and FDM versus the observation zenith angle for (i) only a vegetation layer (noted in the legend BRDF v); (ii) for a vegetation layer+lambertian soil (noted in the legend BRDF v/s), and BTDF of SAIL++ and FDM for a vegetation layer. LAI=3, $d = 0.1$, $\theta_s = 25°$, $\rho = 0.5$, $\tau = 0.5$ and $r_{so} = 0.2$.

E_- in order to conserve the energy. Now, let us compare the extremophile and erectophile cases. (i) Extremophile vegetation has a higher reflectance (and a lower transmittance) than the erectophile one. Indeed, interception is higher for more vertical vegetation; (ii) SAIL and SAIL++ curves are closer to each other in the extremophile case because the semi-isotropic diffuse flux assumption is more conserved in this case (Kallel et al., 2008); (iii) FDM curves are closer to SAIL++ in the erectophile vegetation case then the extremophile case because the mutual shadowing decreases as ALA increases. Thus, the hot spot effect is less pronounced for erectophile vegetation. This was also seen in Subsection 4.2: for SAIL++ the energy is less conserved in the extremophile case similarly to Figure 3 where the hot spot peak is more narrow in the erectophile case; (iv) Finally, although SAIL underestimates the reflectance, due to the high hot spot effect, FDM curves are lower than SAIL curves in the extremophile case.

4.4 Model comparison: RAMI database

The RAdiation transfer Model Intercomparison (RAMI) exercise (Pinty et al., 2001; 2004) proposes some protocols to benchmark radiative transfer models applied to plant canopies covering soil surfaces. The object of RAMI is to validate the reliability and accuracy of different models in simulating RT in or near a vegetation canopy for the benefit of remote sensing data interpretation.

The present study only deals with homogenous vegetation assumed a discrete medium. Also, we only present the BRDF relative to the near-infrared domain, since, in this case, the leaf albedo is higher than the other wavelength cases and, thus, the corresponding simulation results have larger contrasts. In the RAMI exercise second phase (Pinty et al., 2004), the two types of RT models were considered: 1-D models, namely SAIL, SAIL++, 1/2 Discrete (Gobron et al., 1997), and 3-D models, namely Flight (North, 1996), DART (Gastellu-Etchegorry et al., 1996), Sprint-2 (Thompson & Goel, 1998), Raytran (Govaerts & Verstraete, 1998), RGM (Qin & Sig, 2000) and Drat (Lewis, 1999). In addition to these models, AddingSD and FDM will be shown in this Section to be compared with the others. RAMI 2 recommended using simulation results of 3-D models as a reference for homogeneous canopies when ground truth data is not available. Moreover, simulation results of Flight, Raytran and Sprint-2 were assumed the 'most credible models'. Therefore, only the 'most credible' 3-D models will be shown in the figures.

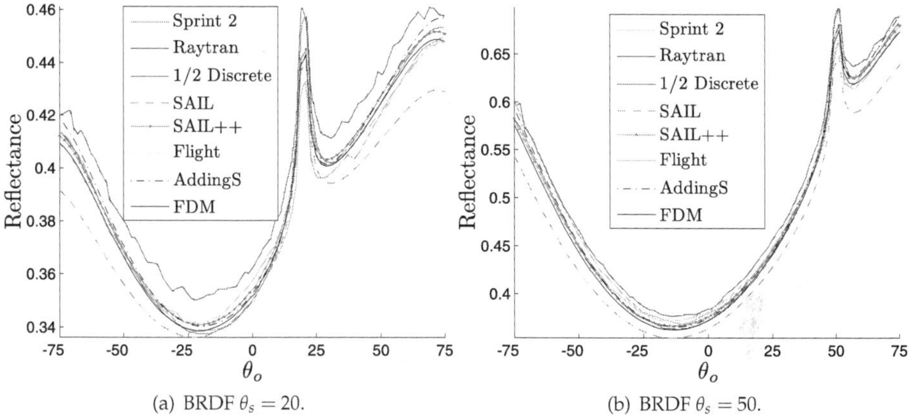

Fig. 5. Canopy BRDF simulations for a discrete medium, at the principal plane. The vegetation features are LAI=3, $H = 2$, leaf radius $= 0.05$, erectophile leaf distribution, $\rho = 0.4957$ and $\tau = 0.4409$. The soil is assumed lambertian with reflectance equal to 0.159.

Figure 5 shows the BRDF simulations in the principal plane. In all cases, the FDM curves are close to the 3-D model ones. This proves the validity of our approach. Moreover, SAIL++ gives results close to FDM because the hot spot parameter is small ($d =$ leaf radius$/H = 0.25$). FDM and AddingSD perform similarly since both models overcome the isotropy assumption and conserve energy. However, AddingSD curves are slightly higher than the FDM ones. This can be explained by two phenomena: (i) AddingSD describes better the multiple scattering under the vegetation since the hemispheres were decomposed into 400 segments, whereas in our case they were decomposed into only 36 segments; (ii) AddingSD takes into account the multi hot spot effect which increases the high order reflectances [cf. Fig 13, p. 3652 of (Kallel et al., 2008)]. Finally, as already shown in the previous subsection, SAIL underestimates the reflectance due to the semi-isotropy assumption.
Furthermore, for quantitative comparison with 3-D models, we already submitted our simulations to RAMI administrator to participate to the forth phase.

5. Conclusion

The goal of this chapter was to derive new methods for computing canopy reflectance so that it both conserves energy and surmounts the assumption that diffuse fluxes E_+/E_- are isotropically distributed over hemispheres. To achieve this object, we first proposed the AddingS/AddingSD models based on SAIL and Adding method and allowing to conserve energy based on the effective vegetation density approach. Second, due to the long running time of such an approach we proposed to benefited from two models: (i) SAIL++ that overcomes

the isotropy assumption and is very fast since it is based on Discrete Cosine Transformation. However, this model does not conserve energy in the hot spot; (ii) AddingSD which also overcomes the isotropy assumption and allows to conserve energy. Our new model was based on injecting the effective vegetation density approach in SAIL++, and therefore, benefited from both energy conservation and a small running time. The procedure was as follows. First, the flux created by direct solar light scattering upwards from vegetation, E_+^1, was computed by taking into account the hot spot effect. Second, according to the effective density approach, the hot spot effect corresponded to a local vegetation density variation. Therefore, the diffuse fluxes and radiances in the observation direction (created by E_+^1) scattering have had to be estimated using the same density achieving consequently energy conservation. Third, since the computation in the latter step was too laborious, E_+^1 was decomposed into virtual subfluxes using Taylor series. Such subfluxes have a simpler expression, and were interpreted each one as a virtual direct solar flux. The provided fluxes, radiances and BRDF/BTDF by the virtual flux scattering were estimated using the SAIL++ formalism. Finally, the total model BRDF/BTDF were estimated by summing the contributions of the subfluxes.
The convergence of the Taylor series decomposition was studied. We showed that only the few first elements were needed to obtain accurate results and a small running time (ranging from 2 to 20 milliseconds). In addition, energy conservation was checked; our model showed good results with an error lower than 2%. Compared to SAIL++, our model's BRDF/BTDF curves were always lower since it decreases high order diffuse fluxes. Finally, compared to 3-D models of RAMI II database, our model gave close results.
Future research will deal with: (i) extending our model to the heterogenous medium case using the Forest Reflectance and Transmittance (FRT) Model (Kuusk & Nilson, 2000); (ii) then validating our model using the multiangular forest reflectance ground truth database described by Kuusk et al. (2008).

A. SAIL++ model

A.1 Continuous case

The formalism shown in this appendix is not presented in the original SAIL++ reference (Verhoef, 1998). However, it is needed in our study to derive the equations of our model.
The SAIL++ equations are written in the continuous case as

$$\frac{dE_s(z,\Omega_s)}{dz} = kE_s(z,\Omega_s), \tag{72}$$

$$\frac{dL_-(z,\Omega_-)}{dz} = -[\mathfrak{s}' \circ E_s(z,\Omega_s)](\Omega_-) + [\mathfrak{A} \circ L_-(z)](\Omega_-) - [\mathfrak{B} \circ L_+(z)](\Omega_-), \tag{73}$$

$$\frac{dL_+(z,\Omega_+)}{dz} = [\mathfrak{s} \circ E_s(z,\Omega_s)](\Omega_+) + [\mathfrak{B} \circ L_-(z)](\Omega_+) - [\mathfrak{A} \circ L_+(z)](\Omega_+), \tag{74}$$

$$\frac{dE_o^+(z,\Omega_o)}{dz} = wE_s(z,\Omega_s) + [\mathfrak{v} \circ L_-(z)] + [\mathfrak{v}' \circ L_+(z)] - KE_o^+(z,\Omega_o), \tag{75}$$

$$\frac{dE_o^-(z,\Omega_o)}{dz} = -w'E_s(z,\Omega_s) - [\mathfrak{v}' \circ L_-(z)] - [\mathfrak{v} \circ L_+(z)] + KE_o^-(z,\Omega_o), \tag{76}$$

where E_s is the solar incident flux, L_-, L_+ are the downward and upward hemispherical distributions of diffuse radiance, E_o^+, E_o^- are the upward and downward radiances in the source direction times k, $\mathfrak{s}$, $\mathfrak{s}'$, $\mathfrak{A}$, $\mathfrak{B}$, w, w', $\mathfrak{v}$, $\mathfrak{v}'$ and K are the generalized Suits scattering terms (Suits, 1972). These parameters are estimated for a given solar and sensor orientation, leaf distribution and foliage density. The foliage area volume density (FAVD) will be called $\bar{u}_L$ (cf. Appendix B) and each Suits parameter (X) is written as follows

$$X = \bar{u}_L X_o, \tag{77}$$

with X_o a scattering parameter depending only on the leaf distribution and the solar and sensor orientations, but not the foliage density. X_o will be called the normalized parameter corresponding to X.
Applied to E_s, $\mathfrak{s}$ and $\mathfrak{s}'$ give respectively

$$\begin{aligned} [\mathfrak{s} \circ E_s(\Omega_s)](.) &= \pi^{-1} w(\Omega_s \to .) E_s, \\ [\mathfrak{s}' \circ E_s(\Omega_s)](.) &= \pi^{-1} w'(\Omega_s \to .) E_s, \end{aligned} \tag{78}$$

where w and w' are respectively the bidirectional scattering terms when the input and the output flux directions are in the same and opposite hemispheres.
Applied to a radiance distribution (L), the operator $\mathfrak{B}$ gives

$$[\mathfrak{B} \circ L](.) = \pi^{-1} \int_{\Pi} w'(\Omega \to .) L(\Omega) \cos(\Omega) \mathrm{d}\Omega. \tag{79}$$

$\mathfrak{A}$ can be divided into two terms

$$\mathfrak{A} = \mathfrak{k} - \mathfrak{B}', \tag{80}$$

where $\mathfrak{k}$ and $\mathfrak{B}'$ are respectively the extinction and the scattering terms.
Since, the extinction of L_- is given by

$$\frac{\mathrm{d}L_-(z, \Omega_-)}{\mathrm{d}z} = \kappa L_-(z, \Omega_-). \tag{81}$$

Therefore, applied to a radiance distribution L, the operator $\mathfrak{k}$ can be defined as

$$[\mathfrak{k} \circ L](\Omega' = (\theta', \varphi')) = \int_{\Pi} \frac{\delta(\theta')\delta(\varphi')}{\cos(\theta')\sin(\theta')} \kappa(\Omega) L(\Omega) \cos(\theta) \mathrm{d}\Omega. \tag{82}$$

Applied to a radiance distribution (L), $\mathfrak{B}'$ gives

$$[\mathfrak{B}' \circ L](.) = \pi^{-1} \int_{\Pi} w(\Omega \to .) L(\Omega) \cos(\Omega) \mathrm{d}\Omega. \tag{83}$$

Applied to a radiance distribution (L), $\mathfrak{v}$ and $\mathfrak{v}'$ give respectively

$$\begin{aligned} \mathfrak{v} \circ L &= \int_{\Pi} w(\Omega \to \Omega_o) L(\Omega) \cos(\Omega) \mathrm{d}\Omega, \\ \mathfrak{v}' \circ L &= \int_{\Pi} w'(\Omega \to \Omega_o) L(\Omega) \cos(\Omega) \mathrm{d}\Omega. \end{aligned} \tag{84}$$

A.2 Discretization

Next, concerning the implementation and in order to describe the upward and downward diffuses radiance hemispherical distribution, Verhoef (1998) proposes a discretization of hemispheres: zenithal and azimuthal angles into N segments. In this case, L_- and L_+ are replaced by sub-fluxes defined over the hemisphere segments forming together vectors called E_- and E_+, respectively. The operators of Eq. (72) are discretized accordingly, in particular, $\mathfrak{s}$, $\mathfrak{s}'$ become vectors called $\mathbf{s}$ and $\mathbf{s}'$, respectively, $\mathfrak{A}$, $\mathfrak{B}$ becomes square matrices called $\mathbf{A}$ and $\mathbf{B}$, respectively, and $\mathfrak{v}$ and $\mathfrak{v}'$ become vectors called $\mathbf{v}$ and $\mathbf{v}'$, respectively. Eqs. (72) (73) (74) (75) (76) become (Verhoef, 1998):

$$\frac{\mathrm{d}}{\mathrm{d}z}\begin{pmatrix} E_s \\ E_- \\ E_+ \\ E_o^+ \\ E_o^- \end{pmatrix} = \begin{pmatrix} k & 0 & 0 & 0 & 0 \\ -\mathbf{s}' & \mathbf{A} & -\mathbf{B} & 0 & 0 \\ \mathbf{s} & \mathbf{B} & -\mathbf{A} & 0 & 0 \\ w & \mathbf{v} & \mathbf{v}' & -K & 0 \\ -w' & -\mathbf{v}' & -\mathbf{v} & 0 & K \end{pmatrix}\begin{pmatrix} E_s \\ E_- \\ E_+ \\ E_o^+ \\ E_o^- \end{pmatrix}, \tag{85}$$

Note that, as in the continuous case [cf. Eq. (80)], $\mathbf{A}$ could be written as

$$\mathbf{A} = \boldsymbol{\kappa} - \mathbf{B}'. \tag{86}$$

with $\boldsymbol{\kappa}$ and $\mathbf{B}'$ the discrete scattering matrices corresponding to $\mathfrak{k}$ and $\mathfrak{B}'$, respectively.
The final solution linking the layer output fluxes to the input ones is (Verhoef, 1998)

$$\begin{pmatrix} E_s(L) \\ E_-(L) \\ E_+(t) \\ E_o^+(t) \\ E_o^-(L) \end{pmatrix} = \begin{pmatrix} \tau_{ss} & 0 & 0 & 0 & 0 \\ \boldsymbol{\tau}_{sd} & \mathrm{T} & \mathrm{R} & 0 & 0 \\ \boldsymbol{\rho}_{sd} & \mathrm{R} & \mathrm{T} & 0 & 0 \\ \rho_{so} & \boldsymbol{\rho}_{do}^T & \boldsymbol{\tau}_{do}^T & \tau_{oo} & 0 \\ \tau_{so} & \boldsymbol{\tau}_{do}^T & \boldsymbol{\rho}_{do}^T & 0 & \tau_{oo} \end{pmatrix}\begin{pmatrix} E_s(t) \\ E_-(t) \\ E_+(L) \\ E_o^+(t) \\ E_o^-(L) \end{pmatrix}, \tag{87}$$

where (L) and (t) refer to the bottom and top of the layer, respectively.
Now, let us consider the case when the source changes. This change includes both the direction and the way that the direct flux is scattered under the vegetation. Since the scattering properties depend only on the vegetation parameters and the source solid angle, the latter possibility of change does not have a physical meaning. However, it is needed in our case to define the scattering parameter when an effective vegetation density is considered. The variation has an impact over the scattering parameters of Eq. (85) as follows. The terms k, $\mathbf{s}'$, $\mathbf{s}$ and w change and the other matrix terms remain constant. The consequences over the boundary condition matrix concern elements that depend on the source, and are: τ_{ss}, $\boldsymbol{\tau}_{sd}$, $\boldsymbol{\rho}_{sd}$, ρ_{so} and τ_{so}. Thus, to allow their estimation, an explicit dependency of the boundary terms on the scattering ones has to be accomplished:

$$\{\tau_{ss} \Rightarrow \tau_{ss}(k), \boldsymbol{\tau}_{sd} \Rightarrow \boldsymbol{\tau}_{sd}(k,\mathbf{s}',\mathbf{s}), \boldsymbol{\rho}_{sd} \Rightarrow \boldsymbol{\rho}_{sd}(k,\mathbf{s}',\mathbf{s}), \rho_{so} \Rightarrow \rho_{so}(k,\mathbf{s}',\mathbf{s},w), \tau_{so} \Rightarrow \tau_{so}(k,\mathbf{s}',\mathbf{s},w')\}. \tag{88}$$

Moreover, in the discrete leaf case, the hot spot effect is taken into account in the computation of ρ_{so}, in this case it will be noted as ρ_{so}^{HS} (Verhoef, 1998).
To distinguish SAIL++ boundary matrix terms from our model terms, ++ will be added to SAIL++ terms as upperscript.

A.3 SAIL++ equation reformulation

In our study, we need to separate the upward diffuse fluxes created by the first collision with leaves of direct flux from the upward fluxes created by multiple collisions, the corresponding radiances are called L_+^1 and L_+^∞, respectively. Indeed, a specific processing for L_+^1 is proposed in this paper in order to take into account the hot spot effect as well as to conserve energy.
As defined, L_+^1 depends on E_s and can be extended when traveling under the vegetation. Compared to L_+ [cf. Eq. (74)], L_+^1 does not increases by L_- and L_+^1 itself scattering. Thus its variation is governed by [cf. Eq. (80)]

$$\frac{\mathrm{d}L_+^1(z,\Omega_+)}{\mathrm{d}z} = [\mathfrak{s} \circ E_s(z,\Omega_s)](\Omega_+) - [\mathfrak{k} \circ L_+^1(z)](\Omega_+). \tag{89}$$

Now, concerning L_+^∞, it does not depend any more on E_s. However it increases by L_+^1, L_- and L_+^∞ itself scattering and decreases, as usual, by extinction. It is given by

$$\frac{\mathrm{d}L_+^\infty(z,\Omega_+)}{\mathrm{d}z} = [\mathfrak{B}' \circ L_+^1(z)](\Omega_+) + [\mathfrak{B} \circ L_-(z)](\Omega_+) - [\mathfrak{A} \circ L_+^\infty(z)](\Omega_+), \tag{90}$$

According to this decomposition, the reformulation of SAIL++ equation set is as follows. Eq. (74) has to be replaced by Eqs. (89) and (90). In Eqs (73), (75) and (76), L^+ has to be replaced by $L_+^1 + L_+^\infty$. One obtains

$$\frac{\mathrm{d}L_-(z,\Omega_-)}{\mathrm{d}z} = -[\mathfrak{s}' \circ E_s(z,\Omega_s)](\Omega_-) + [\mathfrak{A} \circ L_-(z)](\Omega_-) - [\mathfrak{B} \circ L_+^1(z)](\Omega_-) - [\mathfrak{B} \circ L_+^\infty(z)](\Omega_-), \tag{91}$$

$$\frac{\mathrm{d}E_o^+(z,\Omega_o)}{\mathrm{d}z} = wE_s(z,\Omega_s) + [\mathfrak{v} \circ L_-(z)] + [\mathfrak{v}' \circ L_+^1(z)] + [\mathfrak{v}' \circ L_+^\infty(z)] - KE_o^+(z,\Omega_o), \tag{92}$$

$$\frac{\mathrm{d}E_o^-(z,\Omega_o)}{\mathrm{d}z} = -w'E_s(z,\Omega_s) - [\mathfrak{v}' \circ L_-(z)] - [\mathfrak{v} \circ L_+^1(z)] - [\mathfrak{v} \circ L_+^\infty(z)] + KE_o^-(z,\Omega_o). \tag{93}$$

The reformulated SAIL++ equation set is composed by Eqs. (72), (91), (89), (90) (92) and (93).

B. Vegetation local density

To define a realization of a vegetation distribution within the canopy in the discrete leaf case, Knyazikhin et al. (1998) propose the definition of an indicator function:

$$\chi(\vec{r}) = \begin{cases} 1, \text{ if } \vec{r} \in \text{ vegetation}, \\ 0, \text{ otherwise}, \end{cases} \tag{94}$$

where $\vec{r} = (x,y,z)$ is a point within the canopy. Then, they define a fine spatial mesh by dividing the layer into non-overlapping fine cells $(e(\vec{r}))$ with volume $V[e(\vec{r})]$. Thus, the foliage area volume density (FAVD) could be defined as follows:

$$u_L(\vec{r}) = \frac{1}{V[e(\vec{r})]} \int_{\vec{t} \in e(\vec{r})} \chi(\vec{t})\mathrm{d}\vec{t}. \tag{95}$$

By defining the average density of leaf area per unit volume, called d_L (depends only on leaf shape and orientation distribution), u_L is written simply as follows

$$u_L(\vec{r}) = d_L \chi(\vec{r}). \tag{96}$$

In a 1-D RT model, we always need an averaged value of u_L, called $\bar{u}_L$, rather than a unique realization. Assuming that we have a number, N_c, of canopy realizations, then

$$\bar{u}_L(\vec{r}) \approx \frac{\sum_{n=1}^{N_c} u_L^{(n)}(\vec{r})}{N_c}, \tag{97}$$

with $u_L^{(n)}$ the value of FAVD for the realization number n. Similarly, we can define the probability of finding foliage in $e(\vec{r})$ called P_χ as follows

$$P_\chi(\vec{r}) = \frac{\sum_{n=1}^{N_c} \chi^{(n)}(\vec{r})}{N_c}, \tag{98}$$

with $\chi^{(n)}$ the indicator function for the realization n. Finally, we obtain

$$\bar{u}_L(\vec{r}) = d_L P_\chi(\vec{r}). \tag{99}$$

C. Virtual flux decomposition validation

In this appendix, we will answer the following questions: why $\forall n \in \mathbb{N}$, L_1^n [cf. Eq. (17)] can be considered a radiance distribution and why the expression of $P_{\chi,n}$ [cf. Eq. (21)] is valid. The validity can be proved if we can show that the derived radiance hemispherical distributions L_- and L_+^∞, and radiances in observation direction E_o^+ and E_o^-, are correct. Since the proofs are similar, we will show only the validity of E_o^+ expression. As validation reference, we will adopt the AddingSD approach.
Recall that the upward elementary diffuse flux, $\mathrm{d}^3 E_+^1$, in an elementary solid angle $\mathrm{d}\Omega$, created by the first collision with the vegetation in an elementary volume at point N with thickness dt is given by [cf. Figure 1 and Eq. (14)]

$$\begin{aligned} \mathrm{d}^3 E_+^1(N \to M,\Omega) &= \mathrm{d}L_+^1(N \to M,\Omega)\cos(\theta)\mathrm{d}\Omega, \\ &= E_s(0)\exp[(k+K)(t-z)]\exp\left[\frac{\sqrt{kK}}{b}(1-\exp[-b(z-t)])\right] \\ &\quad \times \exp(kz)\pi^{-1}w(N,\Omega_s \to \Omega)dt\cos(\theta)\mathrm{d}\Omega. \end{aligned} \tag{100}$$

As defined in Section 2.1.3, the a posteriori extinction, K_{HS}, of a flux present on M collided only one time at N and initially coming from a source solid angle Ω_s is (cf. Figure 1)

$$\begin{aligned} K_{HS}(\Omega|\Omega_s,0,t-z) &= K + \lim_{u\to z}\frac{1}{b}\frac{\sqrt{kK}(\exp[b(t-u)]-\exp[b(t-z)])}{u-z}, \\ &= K - \sqrt{kK}\exp[-b(z-t)]. \end{aligned} \tag{101}$$

This decrease of extinction value means a decrease in the collision probability locally around M. Thus, in turn, means a decrease in the probability of finding foliage at M, P_χ (cf. Appendix B). Now, according to Eq. (99)

$$\left.\begin{aligned} K &= d_L P_\chi K_0 \\ K_{HS} &= d_L P_{\chi,HS}(\Omega|\Omega_s,0,t-z)K_0 \end{aligned}\right\} \Rightarrow P_{\chi,HS}(\Omega|\Omega_s,0,t-z) = \frac{K_{HS}}{K}P_\chi, \tag{102}$$

were K_0 is the normalized extinction parameter corresponding to K [cf. Eq. (77)], $P_{\chi,HS}(\Omega|\Omega_s,0,t-z)$ is the 'a posteriori' probability of finding vegetation at M. To be simpler, it will be noted $P_{\chi,HS}(\Omega|\Omega_s,t-z)$.
The angular differentiation of E_o^+ ($\mathrm{d}^3E_o^+(z,\Omega\rightarrow\Omega_o)$) that depends only on $\mathrm{d}^3E_+^1$ is

$$\begin{aligned}\frac{\mathrm{d}[\mathrm{d}^3E_o^+(t\rightarrow z,\Omega\rightarrow\Omega_o)]}{\mathrm{d}z} &= w'_{HS}(t\rightarrow z,\Omega\rightarrow\Omega_o)\mathrm{d}^3E_+^1(N\rightarrow M,\Omega),\\ &= w'_{HS}(\Omega|\Omega_s,t-z)L_+^1(t\rightarrow z,\Omega)\mathrm{d}t\cos(\theta)\mathrm{d}\Omega,\end{aligned} \tag{103}$$

where

$$w'_{HS}(\Omega|\Omega_s,t-z)=d_LP_{\chi,HS}(\Omega|\Omega_s,t-z)w'_0(\Omega\rightarrow\Omega_o). \tag{104}$$

Now,

$$\begin{aligned}L_+^1(z,\Omega)= &\; E_s(0)\exp(kz)\pi^{-1}w(\Omega_s\rightarrow\Omega)\\ &\times\int_{-H}^{z}\exp[(k+K)(t-z)]\exp\left[\frac{\sqrt{kK}}{b}(1-\exp[-b(z-t)])\right]\mathrm{d}t.\end{aligned} \tag{105}$$

Therefore,

$$\begin{aligned}\frac{\mathrm{d}[\mathrm{d}^2E_o^+(z,\Omega\rightarrow\Omega_o)]}{\mathrm{d}z} &= E_s(0)\exp(kz)\pi^{-1}w(\Omega_s\rightarrow\Omega)\cos(\theta)\mathrm{d}\Omega d_Lw'_0(\Omega\rightarrow\Omega_o)\\ &\quad\times\int_{-H}^{z}P_{\chi,HS}(\Omega|\Omega_s,t-z)\exp[(k+K)(t-z)]\\ &\quad\times\exp\left[\frac{\sqrt{kK}}{b}(1-\exp[-b(z-t)])\right]\mathrm{d}t.\end{aligned} \tag{106}$$

Now, it is straightforward to show that

$$\begin{aligned}&P_{\chi,HS}(\Omega|\Omega_s,t-z)\exp[(k+K)(t-z)]\exp\left[\frac{\sqrt{kK}}{b}(1-\exp[-b(z-t)])\right]\\ &=\sum_{n=0}^{+\infty}P_{\chi,n}A_n(-1)^n\exp[(k+K+nb)(t-z)].\end{aligned} \tag{107}$$

Then, Eq. (106) becomes

$$\begin{aligned}\frac{\mathrm{d}[\mathrm{d}^2E_o^+(z,\Omega\rightarrow\Omega_o)]}{\mathrm{d}z} &= E_s(0)\exp(kz)\pi^{-1}w(\Omega_s\rightarrow\Omega)\cos(\theta)\mathrm{d}\Omega d_Lw'_0(\Omega\rightarrow\Omega_o)\\ &\quad\times\int_{-H}^{z}\sum_{n=0}^{+\infty}P_{\chi,n}A_n(-1)^n\exp[(k+K+nb)(t-z)]\mathrm{d}t,\\ &= \sum_{n=0}^{+\infty}A_n(-1)^nE_s(0)\exp(kz)\pi^{-1}w(\Omega_s\rightarrow\Omega)\cos(\theta)\mathrm{d}\Omega\\ &\quad\times\int_{-H}^{z}w'_n(\Omega\rightarrow\Omega_o)\exp[(k+K+nb)(t-z)]\mathrm{d}t,\\ &= \sum_{n=0}^{+\infty}A_n(-1)^nw'_n(\Omega\rightarrow\Omega_o)L_+^{1,n}(z,\Omega)\cos(\theta)\mathrm{d}\Omega.\end{aligned} \tag{108}$$

Equations (30) and (108) are the same which implies the validity of our approach.

D. References

Bunnik, N. (1978). The multispectral reflectance of shortwave radiation of agricultural crops in relation with their morphological and optical properties, *Technical report*, Mededelingen Landbouwhogeschool, Wageningen, the Netherlands.

Campbell, G. S. (1990). Derivation of an angle density function for canopies with ellipsoidal leaf angle distribution, *Agricultural and Forest Meteorology* **49**: 173–176.

Chandrasekhar, S. (1950). *Radiative Transfer*, Dover, New-York.

Cooper, K., Smith, J. A. & Pitts, D. (1982). Reflectance of a vegetation canopy using the adding method, *Applied Optics* **21**(22): 4112–4118.

Gastellu-Etchegorry, J., Demarez, V., Pinel, V. & Zagolski, F. (1996). Modeling radiative transfer in heterogeneous 3-d vegetation canopies, *Rem. Sens. Env.* **58**: 131–156.

Gobron, N., Pinty, B., Verstraete, M. & Govaerts, Y. (1997). A semidiscrete model for the scattering of light by vegetation, *Journal of Geophysical Research* **102**: 9431–9446.

Govaerts, Y. & Verstraete, M. M. (1998). Raytran: A monte carlo ray tracing model to compute light scattering in three-dimensional heterogeneous media, *IEEE Transactions on Geoscience and Remote Sensing* **36**: 493–505.

Kallel, A. (2007). *Inversion d'images satellites 'haute résolution' visible/infrarouge pour le suivi de la couverture végétale des sols en hiver par modélisation du transfert radiatif, fusion de donnes et classification*, PhD thesis, Orsay University, France.

Kallel, A., Le Hégarat-Mascle, S., Ottlé, C. & Hubert-Moy, L. (2007). Determination of vegetation cover fraction by inversion of a four-parameter model based on isoline parametrization, *Rem. Sens. Env.* **111**(4): 553–566.

Kallel, A., Verhoef, W., Le Hégarat-Mascle, S., Ottlé, C. & Hubert-Moy, L. (2008). Canopy bidirectional reflectance calculation based on adding method and sail formalism: Addings/addingsd, *Rem. Sens. Env.* **112**(9): 3639–3655.

Knyazikhin, Y., Kranigk, J., Myneni, R. B., Panfyorov, O. & Gravenhorst, G. (1998). Influence of small-scale structure on radiative transfer and photosynthesis in vegetation canopies, *Journal of Geophysical Research* **103**(D6): 6133–6144.

Kuusk, A. (1985). The hot spot effect of a uniform vegetative cover, *Sovietic Jornal of Remote Sensing* **3**(4): 645–658.

Kuusk, A., Kuusk, J. & Lang, M. (2008). A dataset for the validation of reflectance models, *The 4S Symposium - Small Satellites Systems and Services*, Rhodes, Greece, p. 10.

Kuusk, A. & Nilson, T. (2000). A directional multispectral forest reflectance model, *Rem. Sens. Env.* **72**(2): 244–252.

Lewis, P. (1999). Three-dimensional plant modelling for remote sensing simulation studies using the botanical plant modelling system, *Agronomie-Agriculture and Environment* **19**: 185–210.

North, P. (1996). Three-dimensional forest light interaction model using a monte carlo method, *IEEE Transactions on Geoscience and Remote Sensing* **34**(946–956).

Pinty, B., Gobron, N., Widlowski, J., Gerstl, S., Verstraete, M., Antunes, M., Bacour, C., Gascon, F., Gastellu, J., Goel, N., Jacquemoud, S., North, P., Qin, W. & Richard, T. (2001). The RAdiation transfer Model Intercomparison (RAMI) exercise, *Journal of Geophysical Research* **106**: 11937–11956.

Pinty, B., Widlowski, J., Taberner, M., Gobron, N., Verstraete, M., Disney, M., Gascon, F., Gastellu, J., Jiang, L., Kuusk, A., Lewis, P., Li, X., Ni-Meister, W., Nilson, T., North, P., Qin, W., Su, L., Tang, R., Thompson, R., Verhoef, W., Wang, H., Wang, J., Yan, G.

& Zang, H. (2004). The RAdiation transfer Model Intercomparison (RAMI) exercise: Results from the second phase, *Journal of Geophysical Research* **109**.

Qin, W. & Sig, A. (2000). 3-d scene modeling of semi-desert vegetation cover and its radiation regime, *Rem. Sens. Env.* **74**: 145–162.

Suits, G. H. (1972). The calculation of the directional reflectance of a vegetative canopy, *Rem. Sens. Env.* **2**: 117–125.

Thompson, R. & Goel, N. S. (1998). Two models for rapidly calculating bidirectional reflectance: Photon spread (ps) model and statistical photon spread (sps) model, *Remote Sensing Reviews* **16**: 157–207.

Van de Hulst, H. C. (1980). *Multiple Light Scattering: Tables, Formulas, and Applications*, Academic press, Inc., New York.

Verhoef, W. (1984). Light scattering by leaf layers with application to canopy reflectance modelling : the sail model, *Rem. Sens. Env.* **16**: 125–141.

Verhoef, W. (1985). Earth observation modeling based on layer scattering matrices, *Rem. Sens. Env.* **17**: 165–178.

Verhoef, W. (1998). *Theory of Radiative Transfer Models Applied to Optical Remote Sensing of Vegetation Canopies*, PhD thesis, Agricultural University, Wageningen, The Netherlands.

Remote Sensing of Aerosol Over Vegetation Cover Based on Pixel Level Multi-Wavelength Polarized Measurements

Xinli Hu*[abc], Xingfa Gu[ac] and Tao Yu[ac]
[a]*State Key Laboratory of Remote Sensing Science, Jointly Sponsored by the Institute of Remote Sensing Applications, Chinese Academy of Sciences, Beijing 100101, China;*
[b]*Graduate University of Chinese Academy of Sciences, Beijing 100049, China;*
[c]*The Center for National Space-borne Demonstration, Beijing 100101,China*

Abstract

Often the aerosol contribution is small compared to the surface covered vegetation. while, atmospheric scattering is much more polarized than the surface reflection. In essence, the polarized light is much more sensitive to atmospheric scattering than to reflection by vegetative cover surface. Using polarized information could solve the inverse problem of separating the surface and atmospheric scattering contributions. This paper presents retrieval of aerosols properties from multi-wavelength polarized measurements. The results suggest that it is feasible and possibility for discriminating the aerosol contribution from the surface in the aerosol retrieval procedure using multidirectional and multi-wavelength polarization measurements.
Keywords: Aerosol, remote sensing, polarized measurements, short wave infrared

1. Introduction

Atmospheric aerosol forcing is one of the greatest uncertainties in our understanding of the climate system. To address this issue, many scientists are using Earth observations from satellites because the information provided is both timely and global in coverage [2], [4]. Aerosol properties over land have mainly been retrieved using passive optical satellite techniques, but it is well known that this is a very complex task [1]. Often the aerosol contribution is small compared to the surface scattering, particularly over bright surfaces [5]. On the other hand, atmospheric scattering is much more polarized than ground surface reflection [3]. This paper presents a set of spectral and directional signature of the polarized

---

*Xinli Hu (1978-), Male, in 2005 graduated from Northeast Normal University Geographic Information System, obtained his master's degree. Now, working for a doctorate at the Institute of Remote Sensing Applications, Chinese Academy of Sciences, mainly quantitative remote sensing, virtual simulation.Huxl688@hotmail.com

reflectance acquired over various vegetative cover. We found that the polarization characteristics of the surface concerned with the physical and chemical properties, wavelength and the geometric structure factors. Moreover, we also found that under the same observation geometric conditions, the Change of polarization characteristics caused by the surface geometric structure could be effectively removed by computing the ratio between the short wave infrared bands (SWIR) polarized reflectance with those in the visible channels, especially over crop canopies surface. For this crop canopies studied, our results suggest that using this kind of the correlation between the SWIR polarized reflectance with those in the visible can precisely eliminate the effect of surface polarized characteristic which caused by the vegetative surface geometric structure. The algorithm of computing the ratio of polarization bands have been applied to satellite polarization datasets to solve the inverse problem of separating the surface and atmospheric scattering contributions over land surface covered vegetation. The results suggest that compared to using a typically based on theoretical modeling to represent complex ground surface, the method does not require the ground polarized reflectance and minimizes the effect of land surface. This makes it possible to accurately discriminating the aerosol contribution from the ground surface in the retrieval procedure.

2. Theory and backgrand

Polarization (Brit. polarisation) is a property of waves that describes the orientation of their oscillations. The polarization is described by specifying the direction of the wave's electric field. According to the Maxwell equations, the direction of the magnetic field is uniquely determined for a specific electric field distribution and polarization. The simplest manifestation of polarization to visualize is that of a plane wave, which is a good approximation of most light waves. For plane waves the transverse condition requires that the electric and magnetic field be perpendicular to the direction of propagation and to each other. Conventionally, when considering polarization, the electric field vector is described and the magnetic field is ignored since it is perpendicular to the electric field and proportional to it. The electric field vector of a plane wave may be arbitrarily divided into two perpendicular components labeled x (0^0) and y (90^0) (with z indicating the direction of travel). The two components have exactly the same frequency. However, these components have two other defining characteristics that can differ. First, the two components may not have the same amplitude. Second, the two components may not have the same phase. That is they may not reach their maxima and minima at the same time.

Although direct, unscattered sunlight is unpolarized, sunlight reflected by the Earth's atmosphere is generally polarized because of scattering by atmospheric gaseous molecules and aerosol particles. Linearly polarized light can be described by the Stokes parameters (The Stokes parameters are a set of values that describe the polarization state of electromagnetic radiation (including visible light). They were defined by George Gabriel Stokes in 1852) I, Q, and U, which are defined, relative to any reference plane, as follows:

$$I=I0^\circ+ I90^\circ \tag{1}$$

$$Q=I0^\circ-I90^\circ \tag{2}$$

$$U = I45° - 135° \quad (3)$$

where I is the total intensity and Q and U fully represent the linear polarization. In Eqs.(1)-(3) the angles denote the direction of the transmission axis of a linear polarizer relative to the reference plane. The degree of linear polarization P is given by

$$P = \frac{\sqrt{Q^2 + U^2}}{I} \quad (4)$$

and the direction of polarization x relative to the reference plane is given by

$$\tan 2x = \frac{U}{Q} \quad (5)$$

For the unique definition of x, see Figure 1.

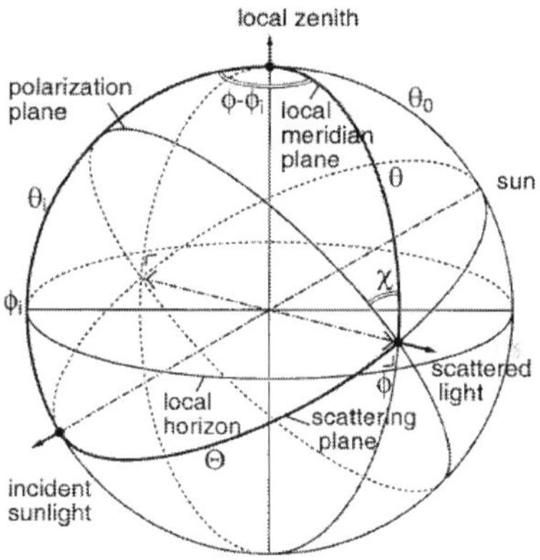

Fig. 1. Geometry of scattering by an atmospheric volume element. The volume element is located in the origin

In Figure 1 the local zenith and the incident and scattered light rays define three points on the unit circle. Applying the sine rule to this spherical triangle (thicker curves in the figure) yields.

$$\cos x = \frac{\sin\theta_i \sin(\phi - \phi_i)}{\sin\Theta} \frac{\sin(\frac{\pi}{2} + x)}{\sin\theta} = \frac{\sin(\phi - \phi_i)}{\sin\Theta} \quad (6)$$

Therefore polarization angle x, i.e., the angle between the polarization plane and the local meridian plane, is given by

$$\cos x = \frac{\sin\theta_i \sin(\phi - \phi_i)}{\sin\Theta} \tag{7}$$

The zenith and azimuth angles of the incident sunlight are (θ_i, ϕ_i), and the zenith and azimuth angles of the scattered light ray (the observer) are (θ, ϕ). The solar zenith angle is $\theta_0 = \pi - \theta_i$. With these definitions, scattering angle Θ is given by

$$\cos\Theta = \cos\theta\cos\theta_i + \sin\theta\sin\theta_i\cos(\phi - \phi_i)\ ,\ 0 \le \Theta \le \pi \tag{8}$$

3. Aerosol Polarization

From Mie calculation of the light scattered by spherical particles with dimensions representative of terrestrial aerosols, one can guess that polarization should be very informative about the particle size distribution and refractive index. Inversely, because polarization is very sensitive to the particle properties, this information is nearly untractable without a priori knowledge of the particle shape (Mishchenko and Travis, 1994). Over the past decades, considerable effort has been devoted to the study of aerosol polarization properties. One uses appropriate radiative transfer calculations to evaluate the contribution of aerosol polarization scattering. The aerosol's size distribution and refractive index are derived simultaneously from their scattering properties.

The simulations are performed by a successive order of scattering (SOS) code. We assume a plane-parallel atmosphere on top of a Lambertian ground surface with uniform reflectance $\rho = 0.3$ and a bi-direction reflectance with BPDF model, a typical and bi-direction reflectance value of ground reflectance at the near-infrared wavelength considered. The aerosols are mixed uniformly with the molecules. The code accounts for multiple scattering by molecules and aerosols and reflection on the surface. Polarization ellipticity is neglected. The results are expressed in terms of polarized radiance Lp, defined by

$$Lp = \sqrt{Q^2 + U^2} \tag{9}$$

3.1 Relationship between aerosol polarization phase function and particle physical properties

Aerosol polarization phase function is known to be highly sensitive to aerosol optical properties, especially aerosol absorption properties, as was shown by Vermeulen et al [6] and Li et al [7]. Polarization phase function provided important information for aerosol scattering properties. Figure 2(a) [Li et al]shows the calculated polarization phase function in the principal plane as a function of the scattering angle. The calculations correspond to those for aerosol with four types of refractive index and the size distribution given by the bimodal log-normal model for sensitivity of polarization phase function to the aerosol real (scattering) and imaginary (absorption) part of refractive index. It can be seen from

Figure 2(a) that the aerosol refractive index (including real and imaginary part) is highly sensitive to polarized phase function. Typically, we consider that the difference among polarized phase function curves of various aerosol refractive index at the range of scattering angle 30^0--90^0 is quite significant, showing a characteristic of the sensitivity of aerosol polarized phase function to refractive index. Moreover, the maximum value at the scattering angle from 30^0 to 90^0 is more accessible in the principal-plane geometry.

For the size distribution model of aerosol particle, Figure 2(b) is the curve of polarized phase function of three size distribution models with the same index of refractive. It can be seen from Figure 2(b) that the aerosol size distribution models is also highly sensitive to polarized phase function[7]. The aerosol size distribution models can significantly affect the polarization function. That is to say, the polarization phase function of aerosol can be used to be important information to retrieve the size distribution model of aerosol.

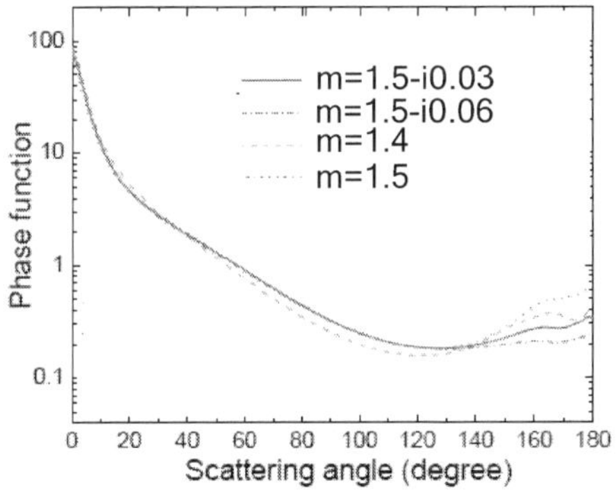

Fig. 2(a)

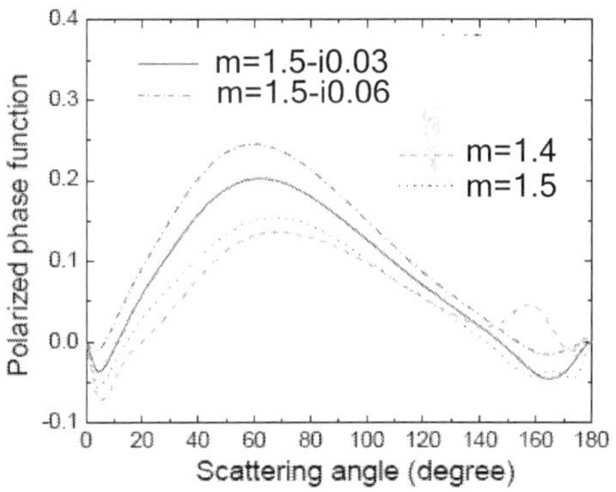

Fig. 2(b)

3.2 Polarization radiance response to aerosol optical thickness and wave lengths

Aerosol polarization radiance is sensitive to aerosol optical thickness. For remote sensing of aerosol, polarization radiance is nearly additive with respect to the contributions of molecules, aerosols. Figure 2(c) shows the calculated aerosol polarization radiance in the principal plane as a function of the observation zenith angle. The curves of the aerosol polarized radiance are calculated at 865nm, for different aerosol optical thickness with the size distribution given by the bimodal log-normal model for the sensitivity of aerosol polarization radiance to the aerosol optical thickness. It can be seen from Figure 2(c) that the aerosol polarization radiance is highly sensitive to aerosol optical thickness. Aerosol optical thickness can be derived from aerosol polarization radiance measurements. Aerosol polarization measurements can be used to retrieve the aerosol optical properties.

For the spectral wavelength, Figure 2(d) shows typical results for the sensitivity of aerosol polarized radiance to the spectral wavelength. The different curves correspond to different aerosol polarized radiance at 865nm, 670nm and 1640nm. It can be seen from Figure 2(d) and 3(d) that polarization will allow to retrieve aerosol key parameters concerning spectral wavelength.

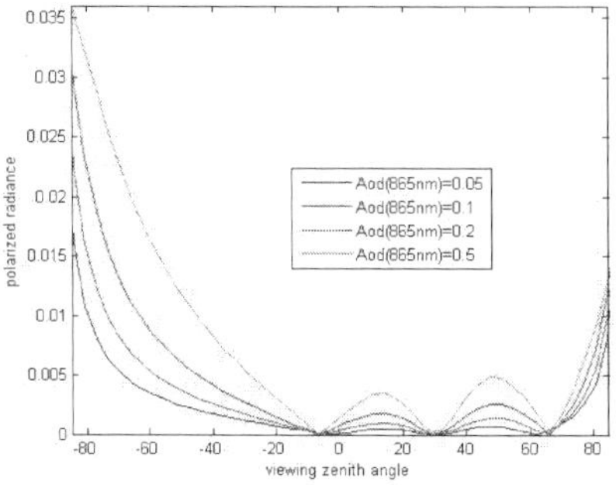

Fig. 2(c)

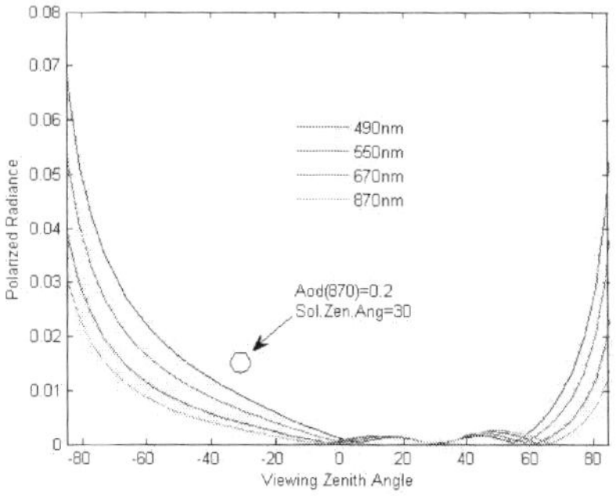

Fig. 2(d)

4. Vegetation Polarization model

In the remote sensing of aerosol over land surface, a parameterization of the surface polarized reflectance is needed for the characterization of atmospheric aerosol over land surface. Because the aerosols properties are efficient at polarizing scattered light. Whereas the surface reflectance is little polarized, that is the reason why the polarization measurements can be used to estimate the atmospheric aerosol properties over land surface [2]. Although small, the surface contribution to the top of the atmosphere (TOA) polarized reflectance cannot be neglected and some parameterization is required. In addition, the parameter of the surface is used as a boundary condition for solving vector radiative transfer (VRT) in both direct and inverse problems. In general, the bidirectional polarization distribution functions (BPDF) is used to estimate the atmospheric contribution to the TOA signal. The function will be used as a boundary condition for the estimate of atmospheric aerosol from polarization remote sensing measurements over land surface.

In most cases, measurements of linear polarization of solar radiation reflected by a plant canopy just provide a simple and relatively cheap way to obtain the characterization of a plant canopy polarized reflectance. Although it demonstrates the relationships between polarized light scattering properties and plant canopies properties, more research is needed if the complexity and diversity inherent in plant canopies is to be modeled, especially more practical BPDF model as a boundary condition for the estimate of atmospheric aerosol.

For remote sensing of aerosol over land surface including polarization information, the vector radiative transfer equation accounting for radiation polarization provides the power simulation of a satellite signal in the solar spectrum in a mixed molecular-aerosol atmosphere and surface polarized reflectance. In order to present the characterization of the TOA polarized reflectance of vegetated surface, some simulation accounting for radiation polarization in atmosphere and surface were made. In what follows, we used the method of successive orders of scattering (SOS) approximations to compute photons scattered one, two, three times, and etc. Rondeaux's , Breon's and Nadal's BPDF models were used to calculate the contribution of the land surface covered plant canopies polarized reflectance as a boundary condition to solve vector radiative transfer equation. It is noticed that:

4.1 The TOA polarized reflectance of vegetation cover depends on zenith angle of sunlight.

Upward polarization radiation at the top of the atmosphere was computed by the successive orders of scattering (SOS) approximations method for wavelengths (λ) of 443 μm . Polarization radiation at the TOA varies according to the angle of incidence. As shows in Figure 3(a) and (b).

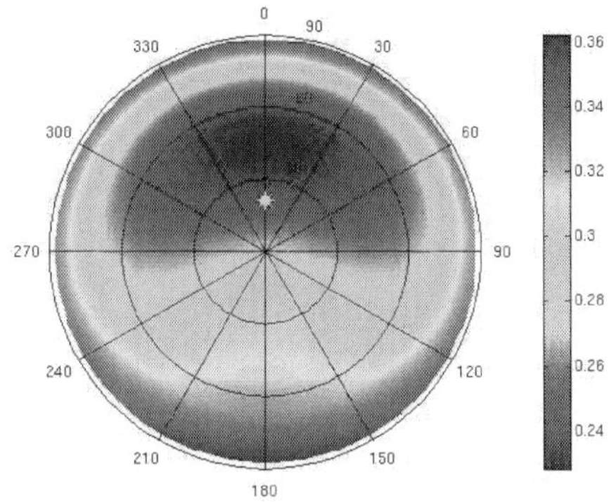

Fig. 3(a)

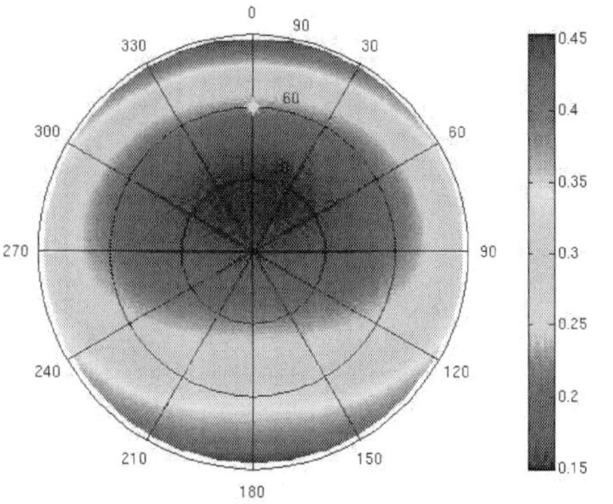

Fig. 3(b)

4.2 Surface BPDF with different land cover types or model

Characterization of the polarizing properties of land surfaces raises probably a more complicated problem than for the atmosphere, on account of the large diversity of ground targets. Concerning the underlying polarizing mechanism, it is usually admitted that land surfaces are partly composed of elementary specular reflectors (water facets, leaves, small mineral surfaces) which, according to Fresnel's law, reflect partially polarized light when illuminated by the direct sunbeam. There is convincing evidence that it is correct in the important case of vegetation cover [Vanderbilt and Grant, 1985; Vanderbilt et al., 1985; Rondeaux and herman, 1991]. By assuming this hypothesis and restricting to singly reflected light, we can anticipate that the main parameters governing the land surface bidirectional polarization distribution function (BPDF), apart from the Fresnel coefficients for reflection, should be the relative surface occupied by specular reflectors, the distribution function of the orientation of these reflectors, and the shadowing effects resulting from the medium structure [7]. Plant canopies structure is difficult to model with single BPDF.

As example of land surface canopy BPDF predicted within this context, we consider the TOA polarized radiance contribution of the model for vegetative cover depends on canopy structure, cellular pigments and refractive indices of vegetation, as the Figure 3(c) and (d) shown. It can be seen in Figure 3(c) that the polarized radiance distribution in 2π space is controlled by directions of both incidence and reflection, and by the main parameters governing the vegetative structures. Comparison of Figure 3(c) and Figure 3(d) shows that according to difference of the refractive indices, the reflective distribution of the polarized radiance varies correspondingly.

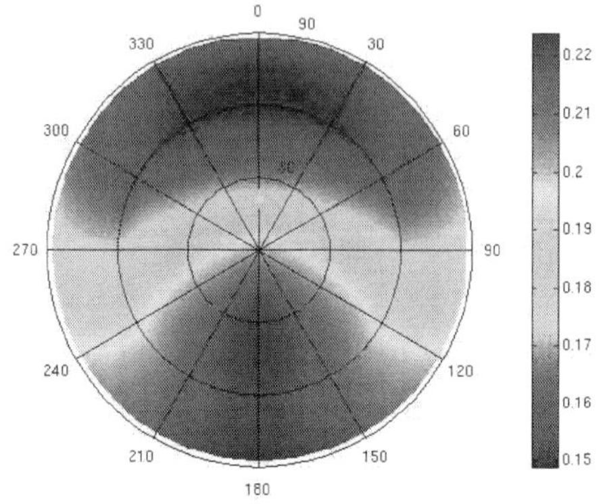

Fig. 3(c)

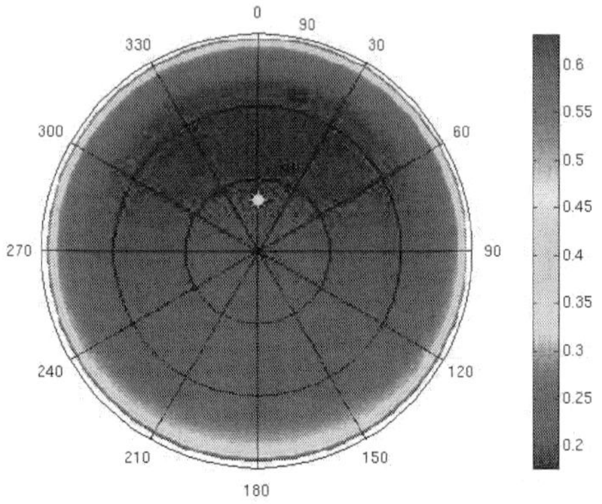

Fig. 3(d)

4.3 Retrieval of TOA contribution of aerosol and land surface polarization

The TOA measured polarized radiance is the sum of 3 contributions: aerosol scattering, Rayleigh scattering, and the reflection of sun light by the land surface, attenuated by the atmospheric transmission on the down-welling and upwelling paths. In order to find out the influence of aerosol and land surface polarization on the TOA polarized contribution, we choose different aerosol model and aerosol optical thickness at a certain land surface BPDF model condition as study parameters.

In this study, the contribution of land surface was calculated by BPDF derived from ground-based measurements for vegetative cover [Rondeaux and herman, 1991], for the atmospheric aerosol, an externally mixed model of these aerosol components is assumed [15]. The size distribution for each aerosol model is expressed by the log-normal function,

$$\frac{dn(r)}{d\ln r} = \frac{1}{\sqrt{2\pi}\ln\sigma}\exp\left(-\frac{(\ln r - \ln r_m)^2}{2\ln^2\sigma}\right) \tag{10}$$

Where rm is the median radius and ln r is the standard deviation. The rm and r values are 0.3 μm and 2.51 μm for the OC model [5], the refractive indices at $\lambda = 443\mu m$ is 1.38±i8.01 for the OC model, and 1.53±i0.005 and 1.52±i0.012 for the WS model. The scattering matrices are computed by the Mie scattering theory for radii ranging from 0.001 to 10.0 μm assuming the shape of aerosol particles to be spherical. We can see from the experiment result that the TOA polarized radiance in 2π space is obvious difference, varying according to the aerosol optical thickness Figure 3(e) and 3(f). Comparison of Figure 3(g) and 3(h) also shows that this difference in aerosol model implies influence on polarized radiance distribution in 2π space. Clearly, different assumptions about the aerosol model have large difference in the TOA polarized radiance.

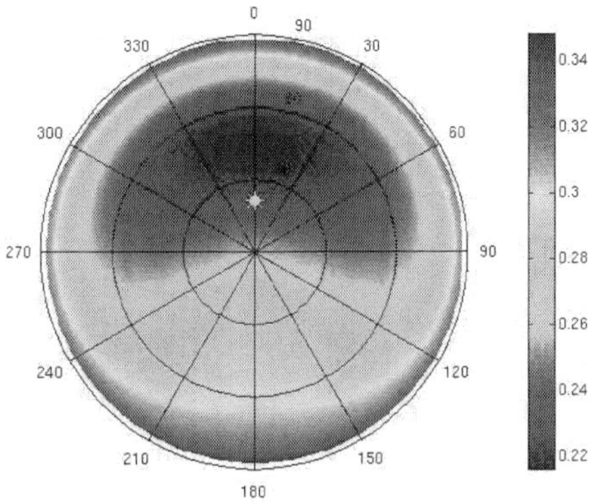

Fig. 3(e) aerosol optical depth is 0.2

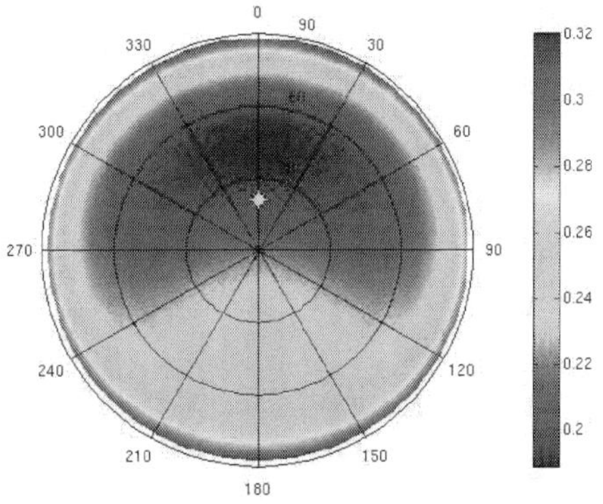

Fig. 3(f) aerosol optical depth is 0.5

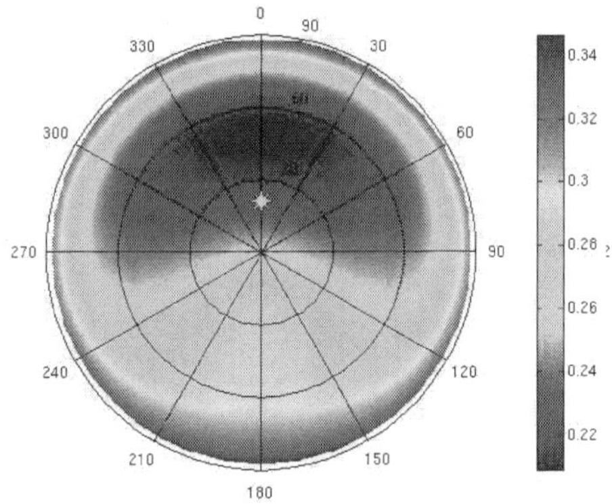

Fig. 3(g) Aerosol model is Jung model

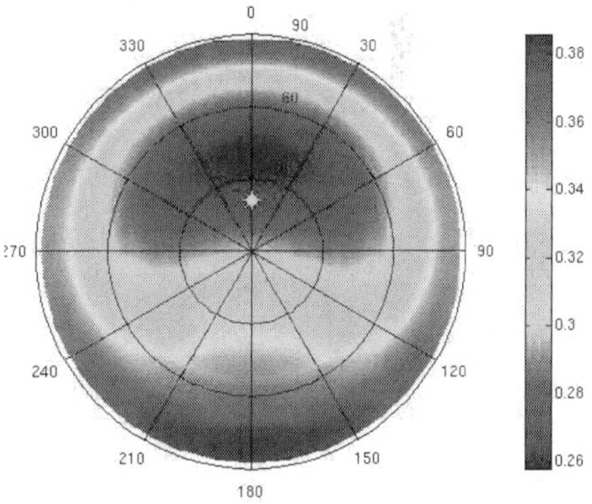

Fig. 3(h) aerosol model is WMO

4. Based on short-wave infrared band polarized model

Solar light reflected by natural surfaces is partly polarized. The degree of polarization, and the polarization direction, may yield some information about the surface such as its roughness, its water content, or the leaf inclination distribution. It is believed that polarized light is generated at the surface by specular reflection on the leaf surfaces. This hypothesis has been used to elaborate analytical models for the polarized reflectance of vegetation. Because of this fact and because the refractive index of natural targets (e.g. leaf of vegetation) varies little within the spectral domain of interest (visible and near IR), the surface polarized reflectance is spectrally neutral, in contrast with the total reflectance. Based on this polarization information and the requirement of the surface polarized reflectance, we can choose to study space-borne polarized reflectance with multi-wavelengths and multi-direction measurements.

The observations of the earth from space that have included polarization measurements are those in an exploratory project aboard the space Shuttle (Coulson et at. ,1986). By the nature of the problem, however, solar radiation directed to space at the level of the Shuttle and other spacecraft contains a significant component due to scattering by the atmosphere, meanwhile that due to surface reflection. For atmospheric characterization and discrimination, however, such surface reflection contamination of the radiation field should be minimized or corrected for by use of radiative transfer models applicable to the conditions of observation. For maximum information content, of course, both intensity and state of polarization of the scattering by the atmosphere should be included.

Light would consist of components $E_{\perp}$ and $E_{=}$, normal and parallel, respectively, to the principal plane. Fresnel's laws of reflection give the reflected electric intensity components.

$$E'_{\perp} = -\frac{\sin(\alpha-\beta)}{\sin(\alpha+\beta)} E_{\perp} \tag{11}$$

$$E'_{=} = \frac{tg(\alpha-\beta)}{tg(\alpha+\beta)} E_{=} \tag{12}$$

Thus from the definition of the Fresnel degree of polarization, we have

$$P = \frac{E'^2_{\perp} - E'^2_{=}}{E'^2_{\perp} + E'^2_{=}} = \frac{E^2_{\perp}\dfrac{\sin^2(\alpha-\beta)}{\sin^2(\alpha+\beta)} - E^2_{=}\dfrac{tg^2(\alpha-\beta)}{tg^2(\alpha+\beta)}}{E^2_{\perp}\dfrac{\sin^2(\alpha-\beta)}{\sin^2(\alpha+\beta)} + E^2_{=}\dfrac{tg^2(\alpha-\beta)}{tg^2(\alpha+\beta)}} \tag{13}$$

Since Fresnel's laws of refraction, in which case Eq. (11) and Eq. (12) reduces to

$$\frac{E'_{\perp}}{E_{\perp}} = -\frac{\sqrt{N^2 - \sin^2\alpha} - \cos\alpha}{\sqrt{N^2 - \sin^2\alpha} + \cos\alpha} \tag{14}$$

$$\frac{E'_{=}}{E_{=}} = \frac{N^2\cos\alpha - \sqrt{N^2 - \sin^2\alpha}}{N^2\cos\alpha + \sqrt{N^2 - \sin^2\alpha}} \tag{15}$$

Obviously, for unpolarized light $E_{\perp}^2 = E_{=}^2$, for convenience, we summarize the relations Eq.(13) and Eq. (14) as follows:

$$P = \frac{2\cos\alpha\sqrt{1 - \dfrac{\sin^2\alpha}{N^2}}\sin\alpha\dfrac{\sin\alpha}{N}}{\cos^2\alpha\dfrac{N^2 - \sin^2\alpha}{N^2} + \sin^2\alpha\dfrac{\sin^2\alpha}{N^2}} = \frac{2\sin\alpha tg\alpha\sqrt{N^2 - \sin^2\alpha}}{N^2 - \sin^2\alpha + \sin^2\alpha tg^2\alpha} \tag{16}$$

Here N is the index of refraction of the medium, and α is the angle of incidence or reflection. The index of refraction N is related to the wavelength.

This shows that the degree of polarization is related to the wavelength and the angle of incidence or reflection Figure 4(a). Furthermore, under the same observation geometric conditions, this important relationship also shows that the degree of polarization of the SWIR (short wave infrared band) is related to that of the visible rang Figure 4(b).

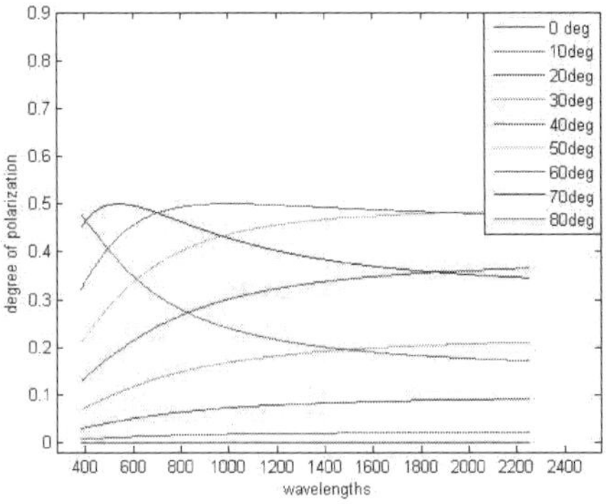

Fig. 4(a)

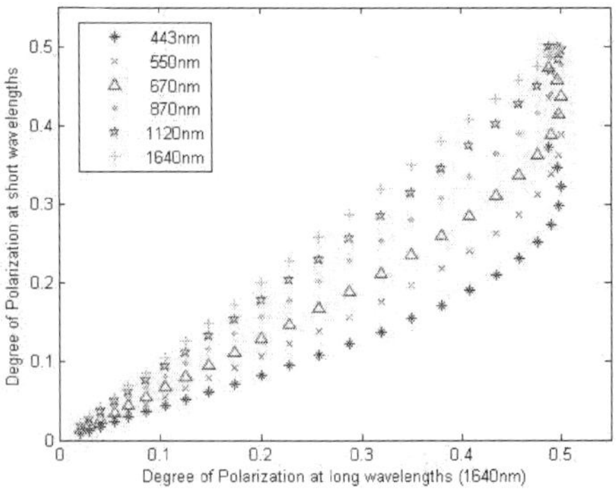

Fig. 4(b)

Figure 4(a) shows the relationship between degree of polarization and wavelengths and the angle of incidence or reflection and (b) between degree of polarization at long wavelengths 1640nm and that at short wavelengths

In Figure 4(b), we found that the SWIR is similar to the visible channels by polarized. That is, the polarized reflectance in SWIR could be used similarly to quantify that in the visible wavelength. This fact would find important applications in solving the inverse problem of separating the surface and atmospheric scattering contributions.

With these and atmospheric conditions, we find, after some algebraic manipulation that

$$\frac{L_{\lambda swir}}{L_{\lambda vi}} = \frac{R_p(\omega,\lambda_{swir}) * e^{-\tau_1(\lambda_{swir})/u_s}}{R_p(\omega,\lambda_{vi}) * e^{-\tau_1(\lambda_{vi})/u_s}} = \frac{R_p(\omega,\lambda_{swir})}{R_p(\omega,\lambda_{vi})} * e^{-(\tau_1-\tau_2)/u_s} \tag{17}$$

Where $R_p(\omega,\lambda)$ is given by

$$R_p(\omega,\lambda) = \frac{1}{2}\left[\frac{\sqrt{N^2-\sin^2\omega}-\cos\omega}{\sqrt{N^2-\sin^2\omega}+\cos\omega}\right]^2 \cdot \left[\frac{N^2\cos\omega-\sqrt{N^2-\sin^2\omega}}{N^2\cos\omega+\sqrt{N^2-\sin^2\omega}}\right]^2 \tag{18}$$

Where $L_{\lambda swir}$ and $L_{\lambda vi}$ are the polarized reflected radiance at long wavelengths and that at short wavelengths, respectively, ω is the scattering angle, and τ is atmospheric optical thickness.

The principle of the algorithm can be seen in Eq. (17). The relationship of the degree of polarization between two wavelengths (the visible rang and short wave infrared band) from the radiative transfer calculation is shown as a function of the aerosol optical thickness at the visible rang and the short wave infrared band.

5. Based on pixel-level multi-angle remote sensing of aerosol

The first space-based polarization measurements were undertaken by ADEOS/POLDER. The POLDER has supplied the observed data not only in the multi-wavelength bands but also at the multi-viewing angles. These directional measurements include significant information of atmospheric aerosols. This work is a feasibility study of multi-directional data for the retrieval of aerosols characteristics. The basic algorithm for aerosol retrieval is based on light scattering simulations of polarization field, where the heterogeneous aerosol model according to Maxwell-Garnett mixing rule is considered. It is shown that polarization data observed at multi-angles is a powerful tool to retrieve aerosol characteristics.
The information provided polarization space-borne sensor permit the development of a new approach to retrieving the aerosol loading at a global scale. The main contribution to the TOA polarized radiance at short wavelengths is due to the aerosols and molecules of the atmosphere, while the contribution of the surface is generally smaller than that of the aerosols. The contribution of atmospheric molecules, although significant at short wavelengths, is nearly invariant and can be easily modeled. That of the surface is more variable but the Eq. (17) and Eq. (18) show that it can be modeled with the polarized reflectance in SWIR, since the contribution of atmospheric aerosol at long wavelengths is generally small and always possible to be negligible. The contribution of the surface at long wavelengths could be used similarly to quantify that at short wavelength. Thus, the aerosol contribution to the polarized radiance can in principle be extracted from the measurement with computing the ratio between the SWIR ground polarized reflectance and those in the visible channels.

The measured polarized radiance L_{pol} is modeled as

$$L_{pol} = L_{aer} + L_{mol} + L_{surf} \tag{19}$$

that is, the sum of 3 contributions : L_{aer}, generated by aerosol single scattering, L_{mol}, by Rayleigh scattering, and L_{surf}, due to the reflection of sun light by the surface, attenuated by the atmospheric transmission on the down-welling and upwelling paths. These terms are expressed as:

$$L_{aer}(\lambda,\theta_s,\theta_v,\phi) = \frac{\tau_a(\lambda)}{4\cos\theta_v}\frac{E_s}{\pi}Q_a(\lambda,\omega,n) \tag{20}$$

$$L_{mol}(\lambda,\theta_s,\theta_v,\phi) = \frac{\tau_m(\lambda)}{4\cos\theta_v}\frac{E_s}{\pi}Q_m(\lambda,\omega) \tag{21}$$

$$L_{surf}(\lambda,\theta_s,\theta_v,\phi) = \cos\theta_s\frac{E_s}{\pi}R_p(\omega,\lambda)*\exp(\tau_m(\lambda)(\frac{1}{\cos\theta_v}+\frac{1}{\cos\theta_s})) \tag{22}$$

Where $\tau_a(\lambda)$ and $\tau_m(\lambda)$ are the optical thickness of the aerosols and of the molecules, respectively. E_s is the TOA solar irradiance. $Q_m(\lambda, \omega)$ and $Q_a(\lambda, \omega, n)$ are pre-calculated functions, which depend on the geometric angles θ_s, θ_v, ϕ only through the scattering angle ω . By performing some algebraic manipulation from Eq. (17)-(22), it is seen that the contribution of the surface at short wavelengths could be quantified with that at long wavelengths.

Inversions were performed with this important relationship. The TOA Polarized reflectance measurements were screened for cloud contamination and corrected for gas absorption. Based on lookup tables composed of optical contributions from mono-modal lognormal aerosol size distributions with fixed standard deviations, but with several values of the modal radius and refractive index, we made use of one week of space-borne POLDER acquisition on from November 7 to 12 , 2007 Beijing China, (latitude 39°58′37″, longitude 116°22′51″). The retrieval method described above for AOD from POLDER yields the AOD composite images of Figure 5a.

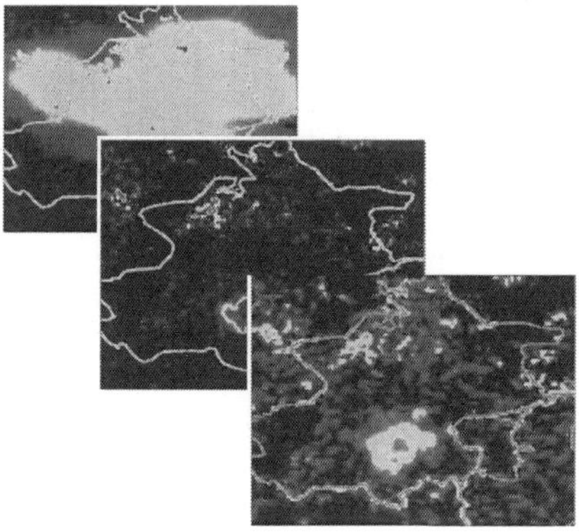

Fig. 6a the composite images of aerosol optical thickness

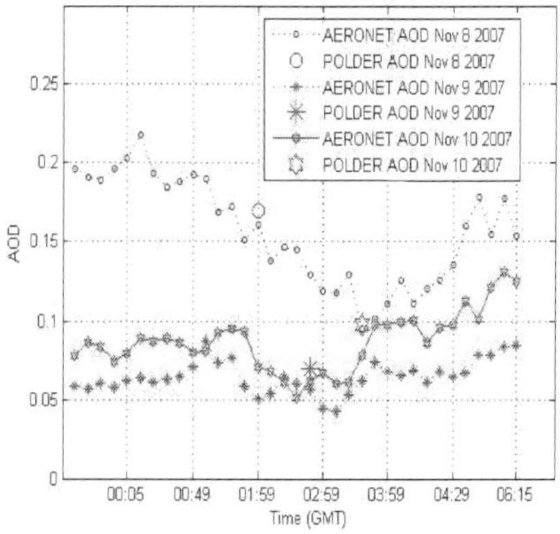

Fig. 6b compared AOD derived from POLDER and AERONET

In order to analyze the accuracy of aerosol inversion, we interpolated the AERONET AOD corresponding to time of the satellite overpass. We used the data provided by Beijing AERONET stations to analyze the accuracy of aerosol inversion from POLDER. Figure 5b shows the results. The validation shown in Figure 5b compares the AOD at 865nm derived from POLDER and AERONET instruments. As the result shows (Figure 5b), the retrieval method for the AOD from POLDER yields a nearly closer values compared with that from AERONET.

6. Summary

In this paper, the accuracy of AOD retrieved from the POLDER multi-wavelengths-based inversion scheme for the sample studied over Beijing remain relatively closer compared to

ground-based sun-photometer measurements. The comparisons between the AERONET AOD and the POLDER-derived AOD using the relations between the polarized reflectance of the surface at long wavelengths and that at short wavelengths show an agreement in most cases. The agreement is much better than when using the physical functions which were derived for bare soils and vegetation. Some effect caused by the surface geometric structure can precisely eliminate for dense vegetation cover or for bare soils at least.
The results suggest that the algorithm, the contribution of the surface at short wavelengths could be quantified with that at long wavelengths in pixel level, can be used as an alternative method in the aerosol retrieval procedure from Multi-wavelength polarization space-borne sensors.
The POLDER results also provide convincing evidence that remote sensing of the terrestrial aerosols over land surfaces by way of polarization measurements is feasible and possibility for discriminating the aerosol contribution from the surface and show the potential of measurements of polarized light scattered by aerosols to retrieve optical depth.

Acknowledgment

The paper is supported by Project supported by the Chinese Defence Advance Research Program of Science and Technology, China (DPC, KJSX0601)

7. References

[1] A.A.Kokhanovsky, F.-M. Breon, A.Cacciari, et al, "Aerosol Remote Sensing over Land: A Comparison of Satellite Retrievals using Different Algorithms and Instruments.", Atmospheric Research. 85, 372-394, 2007.

[2] F.Vachon, A. Royer, M. Aube, B.Toubbe, et al, "Remote Sensing of aerosols over North American Land surfaces from POLDER and MODIS measurements", Atmospheric Environment Vol. 38, 3501-3515, 2004.

[3] K. Arai, Y. Iisasa, X. Liang, "Aerosol Parameter estimation with changing Observation angle of ground based Polarization Radiometer", Advances in Space Research. 39, 28-31, 2007.

[4] Florence Nadal and Francois-Marie Breon, "Parameterization of surface Polarized Reflectance Derived from POLDER Space-borne measurements", IEEE Transactions On Geoscience and Remote Sensing. Vol. 37.NO. 3. MAY 1999.

[5] Von Hoyningen-Huene, W. Freitag, M. Burrows, J.B., et al, "Retrieval of Aerosol Optical Thickness over Land Surface from top-of-atmosphere Radiance", J. Geophys. Res. 108, 4260.doi:10, 1029/2001JD002018, 2003.

[6] Deuze, J.L, Breon, F.M, Devaux, C., et al, "Remote Sensing of Aerosols over Land Surface from POLDER-ADEOS-1 Polarized Measurements", J. Geophys. Res. 106, 4913-4926, 2001.

[7] Li et al, "Retrieval of aerosol optical and physical properties from ground-based spectral, multi-angular, and polarized sun-photometer measurements", remote sensing of environment , 101 (2006) 519-533.

18

Methods and performances for multi-pass SAR Interferometry

Stefano Tebaldini and Andrea Monti Guarnieri
Dipartimento di Elettronica e Informazione
Politecnico di Milano
Piazza Leonardo da Vinci, 32
20133 MILANO

1. Introduction

Thanks to the several space missions accomplished since ERS-1, the scientific community has been provided with a huge amount of data suitable for interferometric processing. The innovation was the availability of multiple compatible images of the same areas. Such images, achieved by looking from slightly different point of view different orbits, and/or by different frequencies, and/or at different times, has largely extended the capabilities of InSAR with respect to the traditional dual image case. The advantage granted by the possibility to form multiple interferograms, instead than just one, is two folded. On the one hand, the estimation of the parameters of interest, be them related to the DEM or the terrain deformations, is driven by a larger data set, resulting in more accurate estimates. On the other hand, new parameters may be added to the set of the unknowns, allowing to study complex phenomena, such as the temporal evolution of the atmospheric and deformation fields. A major issue with multi-image InSAR is that targets are, in general, affected by temporal and spatial decorrelation phenomena, which hinders the exploitation of large spatial and/or temporal baselines. For this reason, most of literature about multi-image InSAR has focused mainly on *targets that stay coherent in all the acquisitions*, which has resulted in a substantial lack of a systemic approach to deal with decorrelating targets in the field of InSAR.
The aim of this chapter is to propose a general approach to exploit all the available information, that is the stack of interferometric SAR images, and that formally accounts for the impact of target decorrelation. This approach is based on the optimal estimate of the data in a statistical sense. The basic idea is to split the estimation process into two steps. In the first step, a maximum likelihood (ML) estimator is used that jointly exploits all the $N \times (N-1)/2$ interferograms available with N acquisitions, in order to yield the best estimates of the $N-1$ phases that correspond to the optical path differences between the target and the sensors. Target decorrelation is accounted for by properly weighting each interferogram in dependence on the target statistics. The estimated phases will be referred to as Linked Phases, to remind that these terms are the result of the joint processing of all the $N(N-1)/2$ interferograms. Once the first estimation step has yielded the estimates of the interferometric phases, the second step is required to separate the contributions of the APSs and the decorrelation noise from the parameters of interest, such as the Line of Sight Deformation Field (LDF) and the topography.

The same two-step approach can be followed to derive the performances of the Multi-Baseline Interferometry, in the frame of the Hybrid Cramér Rao Bound (HCRB) that we will discuss in this same chapter.

2. A brief review of multi-image InSAR techniques

2.1 Permanent Scatterers Interferometry (PSI)

This approach, developed by *Ferretti et al.*(1), represents the first attempt to give a formal framework to the problem of multi-baseline InSAR. The analysis is based on the selection of a number of highly coherent, temporally stable, point-like targets within the imaged scene, which may be identified by analyzing the amplitude time series extracted from the whole set of images in correspondence with every pixel (2), (3), (4). Such targets, named Permanent Scatterers (PS), are typically represented by man made objects, but also isolated trees or stable rocks may serve as PSs. However, the highest density of PSs is expected to be found within urban areas. For every selected PS the time series of the phase differences of every image with respect to a reference one is extracted. Since the selected targets are by definition the ones which remain coherent in all of the images, it can be assumed that no decorrelation phenomena occur. Therefore, the phase difference time series may be effectively represented by a linear model plus noise. At this point, an effective separation of the various contributions, such as topography, displacement rate, and APSs, may be carried out by exploiting the time-space statistical properties of each. In this sense, this approach is similar to Wiener filtering, and could be in principle solved by such technique. However, because of the high computational burden and the non linearity due to the 2π ambiguity, *Ferretti et al.* proposed instead an iterative algorithm, involving 2D frequency estimation, phase unwrapping and linear filtering (5). The main limitation within this approach lies in the sparsity of the grid for the selected PS. To overcome this limitation, a second step is performed, consisting in resampling the APSs estimates on the uniform image grid, remove these terms, and look for a more dense set of PSs basing on phase stability, rather than amplitude. This process, however, is likely to fail in areas where the initial PS selection, based on amplitude stability, does not suffice to cover the whole imaged scene, as it may be the case of non urban areas, especially for data set suffering from amplitude calibration problems. A first solution to this problem has been proposed by *Hooper et al.* (6), who defined an iterative point selection algorithm basing directly on a phase stability criterion. The selected point are called Persistent Scatterers. This method has been shown to yield a more dense point grid on rock areas than the amplitude based algorithms exploited in PSI.

2.2 Techniques based on interferogram selection

Several approaches have been presented in literature to perform SAR interferometric analysis over scenes where the PS assumption may not be retained. A number of these works share the idea to minimize the effect of target decorrelation by forming the interferograms from properly selected pairs, rather than with respect to a fixed reference image, as done in PS processing. Despite the good results achieved in the applications, however, there's no clear and formal assessment of the criteria which should drive the selection of the image pairs to be used. As a result, the processing is heuristically based on the exploitation of a set of interferograms taken with the shortest temporal and/or spatial baselines possible (7), (8), (6).

2.3 The Small Baseline Subsets (SBAS) approach

A more sophisticated approach is the one by *Berardino et al.*, exploiting the concept of Small Baseline Subsets (SBAS) (9), (10). This approach may be somehow considered as the complement of the PS approach. While the latter looks for targets which remain coherent throughout the whole data set, the SBAS algorithm tries to extract information basing on every single interferogram available. The algorithm accounts for spatial decorrelation phenomena by partitioning the data set into a number of subsets, each of which is constituted by images acquired from orbits close to each other. In this way, interferograms corresponding to large baselines are discarded. After unwrapping the phases of the interferograms within the subsets, the estimation of the physical quantities of interest, such as the topographic profile and the deformation field, is carried out through singular value decomposition. The choice of this inversion technique accounts for the rank deficiency caused by partitioning the data set into subsets, resulting in the solution being chosen on the basis of a minimum norm criterion. Further processing, similar to that indicated in (5), carries out the removal of the atmospheric artifacts.

2.4 Maximum Likelihood Estimation Techniques

The application of Maximum Likelihood Estimation techniques for InSAR processing has been considered by *Fornaro et al* (11), *De Zan* (12), *Rocca* (13), and in two works by *Tebaldini and Monti Guarnieri* (14), of which this chapter represents an extension. The rationale of ML techniques, as applied to InSAR, is to exploit target statistics, represented by the ensemble of the coherences of every available interferogram, to design a statistically optimal estimator for the parameters of interest. An advantage granted by these techniques is that the criteria which determine the role of each interferogram in the estimation process are directly derived from the coherences, through a rigorous mathematical approach. Furthermore, by virtue of the properties of the ML estimator (MLE), the estimates of the parameters of interest are asymptotically (we.e. large signal to noise ratio, large data space) unbiased an minimum variance. On the other hand, a common drawback of these techniques is the need for a reliable information about target statistics, required to drive the estimation algorithm. The main differences among the works by *De Zan*, *Rocca*, and the one to be depicted in this chapter are relative to the initial parametrization of the data statistics. The ML approach proposed by *De Zan* consists in estimating residual topography and LOS subsidence rate directly from the data. Conversely, in the work by *Rocca* (13), similarly to the approach within this chapter, the estimation process is split into two steps, in that first $N(N-1)/2$ interferograms are formed out of N acquisitions, and then the second order statistics of the interferograms are exploited to derive the optimal linear estimator of the parameters of interest, under the small phase approximation. After the Extended Invariance Principle (EXIP), it follows that the condition under which the splitting of the MLE into two steps does not entail any loss of information about the original parametrization of the problem, $\boldsymbol{\theta}$, is that the covariance of the estimate errors committed in the first step actually approaches the CRB. Therefore, the estimation of topography and Line Of Sight (LOS) subsidence rate directly from the data proposed by *De Zan* in (12), is intrinsically the most robust, since the estimation of the whole structure of the model is performed in a single step. This approach, however, would result in an overwhelming computational burden if applied to a large set of parameters. On the contrary, in the approach followed by *Rocca*, (13), the first step may be interpreted as a totally unstructured estimation of the model, since each of the $N(N-1)/2$ phases of all the available interferograms is estimated separately. It follows that the computational burden is kept very low, but the performance of the one step ML estimator may be approached only under the condition that the $N(N-1)/2$

phases are estimated with sufficient accuracy, as it happens by exploiting a large estimation window and/or at high SNR (Signal To Noise Ratio). Finally, the two step estimator to be depicted in this chapter may be placed in between the two solutions here exposed, the first step being devoted to carrying out a joint estimation of $N-1$ phases from the data. This solution corresponds to the estimation of a weak structured model, based on the hypothesis of phase triangularity herewith discussed.

3. Model for the multi-pass observations of a single target

Let us consider the multibaseline geometry in Fig. 1: this geometry is fairly conventional, and the reader is referred to like (15–17) for a general view of SAR interferometry, or (18) for a tutorial. The same target P is observed by a set on N sensors in different parallel tracks or, identically, by N repeated acquisitions of the same sensor. Each observation is focused getting an high resolution image of the whole scene. Two complex focused images can be combined to compute the interferogram, that their Hermitian. The interferogram phase, shown in Fig. 1 on the right, is proportional to the resultant of the travel phase difference between the two acquisitions and the difference between the Atmospheric Phase Screen (APS) of the two acquisitions:

$$\begin{aligned} \varphi_{nm} &= \varphi_n - \varphi_m \\ &= \left(\frac{4\pi}{\lambda}R_n(P) + \alpha(n)\right) - \left(\frac{4\pi}{\lambda}R_m(P) + \alpha(m)\right), \end{aligned} \tag{1}$$

$R_n(P)$ and $R_m(P)$ being respectively the slant range of the n-th and m-th antennas to the target point P, and $\alpha(n)$, $\alpha(m)$ the phase errors due to the propagation in the atmosphere in the two acquisitions(19). In turn, the slant range can be thought of a fixed contribution, due to topography, and a time-varying LOS displacement. As an example, for a linear deformation:

$$R_n(P) = R_{n0} + v(P) \cdot t_n$$

$v(P)$ being the Linear Deformation Rate and t_n the time of the n-th acquisition.
The interferogram phases keeps then information on both the geometry of the system, that depends on the topography, hence the DEM, and the a possible Line Of Sight displacement of the target in the time between the two acquisitions. In the following, we assume that all the focused images are coregistered on the same range, azimuth reference of one image, that we will define as the master image, so that the same target contributes in the same pixels of all the N images in the stack (18).

3.1 Single target model

We assume that each pixel in the SAR images is described by a distributed target, i.e., the contributions of many independent scatterers in the resolution cell. The result is a realization of a stochastic process, whose pdf conditioned on the interferometric phases may be regarded as being a zero-mean, multivariate circular normal distribution (15). Therefore, the ensemble of the second order moments represents a sufficient statistics to infer information from the data. With reference to a particular location in the slant range - azimuth plane, the expression of the second order moment for the $nm-th$ interferometric pair may be expressed, under the assumption of phase triangularity, as:

$$E\left[y_n y_m^*\right] = \gamma_{nm} \exp\left(j\left(\varphi_n - \varphi_m\right)\right) \tag{2}$$

where:

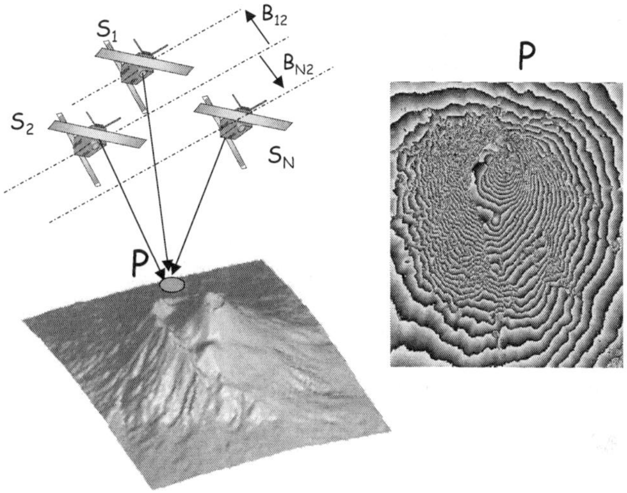

Fig. 1. Interferometric SAR geometry: N sensors, at same azimuth position are shown on the left. On the right, the interferogram's phases obtained by combining two images are shown.

- y_n represents a pixel in the $n-th$ SLC SAR image at the considered slant range - azimuth location;
- γ_{nm} is the coherence of the $nm-th$ interferometric pair; $\gamma_{nn}=1$ for every n;
- φ_n is the interferometric phase for the $n-th$ acquisition.

Note that the images are supposed to be normalized such that $E\left[|y_n|^2\right]=1\ \forall\ n$.
Following (1), the interferometric phases will be expressed in vectors as:

$$\boldsymbol{\varphi}=\boldsymbol{\psi}\left(\boldsymbol{\theta}\right)+\boldsymbol{\alpha} \tag{3}$$

where:

- $\boldsymbol{\varphi}=\left[\begin{array}{ccc}\varphi_0 & \dots & \varphi_{N-1}\end{array}\right]^T$ is the vector of the interferometric phases, with respect to an arbitrary reference;
- $\boldsymbol{\theta}$ is the vector of the unknown parameters which describe the LDF and residual topography to be estimated;
- $\boldsymbol{\psi}\left(\boldsymbol{\theta}\right)=\left[\begin{array}{ccc}\psi_0\left(\boldsymbol{\theta}\right) & \dots & \psi_{N-1}\left(\boldsymbol{\theta}\right)\end{array}\right]^T$ is a vector of known functions of $\boldsymbol{\theta}$;
- $\boldsymbol{\alpha}=\left[\begin{array}{ccc}\alpha_0 & \dots & \alpha_{N-1}\end{array}\right]^T$ represent the atmospheric fields, or APS, affecting the N acquisitions.

The APS may be modeled as a stochastic process, highly correlated over space and uncorrelated from one acquisition to the other, at least under the assumption that SAR images are taken with a repeat interval longer than one day (19; 20). Furthermore, we will here force the hypothesis that the ensemble of the APSs may be modeled as a normal, zero mean stochastic process with variance σ_α^2:

$$\begin{aligned} \boldsymbol{\alpha} &\sim N\left(\mathbf{0}, \mathbf{R}_\alpha\right) \\ \mathbf{R}_\alpha &= \sigma_\alpha^2 \mathbf{I}_N \end{aligned} \tag{4}$$

where $\mathbf{I}_N$ is the $N \times N$ identity matrix. The reason for this assumption is that it will allow to reach a simple analytical form to lower bound the estimate accuracy. Otherwise, the estimate accuracy would depend not only on the APS variance, but also on the shape of its distribution. As a first approximation, however, this assumption may be retained (21). Note that the characterization of the APS as a zero mean process is clearly unphysical, since the propagation through the atmosphere should result in an additional delay of the received echoes, therefore giving rise to a bias. Such bias, however, is automatically canceled by evaluating the phase differences, $\varphi_n - \varphi_m$, in such a way that model (4) can be retained in the framework of InSAR analyses. As a further remark, note that the assumption that the APSs are uncorrelated from an acquisition to another is not strictly necessary for the theoretical developments to follow.

The set of the coherences, γ_{nm}, of each interferometric pair accounts for decorrelation sources like superficial, volumetric, temporal, etc, is the following product form, suggested in (22):

$$\gamma_{nm} = \gamma_0 \gamma_{nm}^{(sup)} \gamma_{nm}^{(vol)} \gamma_{nm}^{(temp)} \tag{5}$$

Model (2) may be expressed in a conveniently compact form as:

$$\mathbf{R} \stackrel{def}{=} E\left[\mathbf{y}\mathbf{y}^H\right] = \boldsymbol{\phi}\boldsymbol{\Gamma}\boldsymbol{\phi}^H \tag{6}$$

where:

- $\mathbf{R}$ is the data covariance matrix.
- $\mathbf{y} = \begin{bmatrix} y_0 & \cdots & \cdots & y_{N-1} \end{bmatrix}^T$ is the stack of the SLC images at a given location, (r, x), in the slant range, azimuth plane;
- $\boldsymbol{\Gamma}$ is an $N \times N$ symmetric matrix whose elements are given by the interferometric coherences: $\{\boldsymbol{\Gamma}\}_{nm} = \gamma_{nm}$;
- $\boldsymbol{\phi}$ is an $N \times N$ diagonal matrix whose elements are given by the interferometric phases:

$$\boldsymbol{\phi} = \operatorname{diag}\left\{ \begin{matrix} \exp\left(j\varphi_0\right) & \cdots & \cdots & \exp\left(j\varphi_{N-1}\right) \end{matrix} \right\}$$

Throughout this chapter we will assume that the estimation of the parameters of interest is performed by exploiting **a conveniently small estimation window**, Ω. The pdf of the data sample within the estimation window will be expressed as:

$$p\left(\mathbf{y}_\Omega | \boldsymbol{\varphi}\right) = \text{const} \cdot \prod_{l=1}^{L} \exp\left(-\mathbf{y}^H\left(r_l, x_l\right) \boldsymbol{\phi}\boldsymbol{\Gamma}^{-1}\boldsymbol{\phi}^H \mathbf{y}\left(r_l, x_l\right)\right) \tag{7}$$

where (r_l, x_l) denote the location in the slant range, azimuth plane and L is the number of pixels within the estimation window.
Expression (7) is conditioned to two hypothesis: the stationarity and the incorrelation of the pixels in the window. The hypothesis of stationarity requires that the unknowns, $\boldsymbol{\theta}$, the APSs, $\boldsymbol{\alpha}$, and the coherences, γ_{nm}, are constant inside Ω. Therefore, in practical applications this hypothesis may be retained provided that Ω is small. The hypothesis that no correlation exists among neighboring pixels requires that either one of these two conditions is met:

- all the pixels within the estimation window are referred to the same slant range location;
- the normal baselines are small, in such a way as not to give rise to correlation among neighboring pixels.

Note that the latter represent the most suitable condition for estimating the LDF, which makes it sensible to retain its validity, especially in light of the capabilities which will be provided by the next generation of spaceborne SARs (13). As a first approximation, however, the effect of a non null normal baseline set can be included by letting the correlation coefficients γ_{nm} account for the contribution of spatial decorrelation.

4. Performances estimation in Multi Baseline Interferometry

This section is dedicated to multi-baseline SAR Interferometry (InSAR). We will associate InSAR with the hypothesis that a single target, either point-like or distributed, is present within the SAR slant range, azimuth resolution cell.
The aim here is to derive a lower bounds of the performance achievable by spaceborne InSAR as for the estimation of the LOS deformation field (LDF) from acquisitions over scenes characterized by a distributed scattering mechanism, such as forests, agricultural fields, soil or rock surfaces, and even ice shelves. Mainly two factors affect the accuracy of InSAR estimates. One arises from decorrelation phenomena among SAR images, due to the spatial structure and temporal behavior of the targets. The other is given by the presence of uncompensated propagation disturbances. As long as spaceborne applications are considered, it is reasonable to assume that uncompensated propagation disturbances are mainly caused by atmospheric fluctuations, and hence in this chapter propagation disturbances will be identified with the Atmospheric Phase Screens (APSs). The ensemble of target decorrelation and APS results in a non Gaussian distribution of the data, which complicates the a-priori assessment of the LDF estimator performance for any given scenario. In the existing literature the computation of InSAR lower bounds is commonly approached as the problem of estimating a set of deterministic parameters in presence of decorrelation noise (23), (24), whereas the role of the APS is in general neglected. Within this chapter it will be shown that, under the hypothesis that a single, distributed, target is present within the SAR resolution cell, the roles of target decorrelation and APSs may be jointly treated by exploiting the Hybrid Cramér-Rao Bound (HCRB), where the unknowns are both deterministic parameters and stochastic variables. This approach results in a compact formulation of the problem, from which it is possible to achieve some closed-form formulas that constitute a useful tool for system design and tuning (25).

4.1 The Hybrid Cramér-Rao bound for InSAR

The data formulation described in the previous section is suited for exploiting the Hybrid Cramér-Rao Bound (HCRB) for lower bounding the accuracy on the estimate of $\boldsymbol{\theta}$. The HCRB (26), (27), (28) applies in the case where some of the unknowns are deterministic and others are random; it unifies the deterministic and Bayesian CRB in such a way as to simultaneously bound the covariance matrix of the unbiased estimates of the deterministic parameters and the mean square errors on the estimates of the random variables (26), (27).
Let $\widehat{\boldsymbol{\theta}}$ be an unbiased estimator of the deterministic parameters θ, and denote $\widehat{\boldsymbol{\alpha}}$ an estimator of the random variables $\boldsymbol{\alpha}$. The HCRB assures that, for every estimator,

$$E_{\mathbf{y},\boldsymbol{\alpha}}\left[\begin{array}{cc} \left(\widehat{\boldsymbol{\theta}}-\boldsymbol{\theta}\right)\left(\widehat{\boldsymbol{\theta}}-\boldsymbol{\theta}\right)^T & \left(\widehat{\boldsymbol{\theta}}-\boldsymbol{\theta}\right)\left(\widehat{\boldsymbol{\alpha}}-\boldsymbol{\alpha}\right)^T \\ \left(\widehat{\boldsymbol{\alpha}}-\boldsymbol{\alpha}\right)\left(\widehat{\boldsymbol{\theta}}-\boldsymbol{\theta}\right)^T & \left(\widehat{\boldsymbol{\alpha}}-\boldsymbol{\alpha}\right)\left(\widehat{\boldsymbol{\alpha}}-\boldsymbol{\alpha}\right)^T \end{array}\right] \geq \mathbf{J}^{-1} \tag{8}$$

where $E_{\mathbf{y},\boldsymbol{\alpha}}[\ldots]$ denotes expectation with respect to the joint pdf of the data and the APS, $p(\mathbf{y},\boldsymbol{\alpha}|\boldsymbol{\theta})$, and the inequality means that the difference between the left and the right sides of (8) is a nonnegative definite matrix. The matrix $\mathbf{J}$ is called the Hybrid Information Matrix (HIM). It may be obtained as the sum of the standard Fisher Information Matrix (FIM), $\mathbf{F}$, averaged with respect to α, and the prior information matrix $\mathbf{I}_{\alpha}$.

$$\mathbf{J} = E_{\alpha}[\mathbf{F}] + \mathbf{I}_{\alpha} \tag{9}$$

Define $\Delta_{\mathbf{x}}^{\mathbf{y}}$ as a matrix of the second order partial derivatives with respect to two multi-dimensional variables $(\mathbf{x},\mathbf{y})$:

$$\left\{\Delta_{\mathbf{x}}^{\mathbf{y}}\right\}_{nm} = \frac{\partial^2}{\partial x_n \partial y_m} \tag{10}$$

Then, the matrices $\mathbf{F}$ and $\mathbf{I}_{\alpha}$ are given by:

$$\mathbf{F} = -E_{\mathbf{y}|\alpha}\left\{\begin{bmatrix} \Delta_{\boldsymbol{\theta}}^{\boldsymbol{\theta}} \log p(\mathbf{y}|\boldsymbol{\theta},\boldsymbol{\alpha}) & \Delta_{\boldsymbol{\theta}}^{\boldsymbol{\alpha}} \log p(\mathbf{y}|\boldsymbol{\theta},\boldsymbol{\alpha}) \\ \Delta_{\boldsymbol{\alpha}}^{\boldsymbol{\theta}} \log p(\mathbf{y}|\boldsymbol{\theta},\boldsymbol{\alpha}) & \Delta_{\boldsymbol{\alpha}}^{\boldsymbol{\alpha}} \log p(\mathbf{y}|\boldsymbol{\theta},\boldsymbol{\alpha}) \end{bmatrix}\right\} \tag{11}$$

$$\mathbf{I}_{\alpha} = -E_{\alpha}\left\{\begin{bmatrix} 0 & 0 \\ 0 & \Delta_{\boldsymbol{\alpha}}^{\boldsymbol{\alpha}} \log p(\boldsymbol{\alpha}) \end{bmatrix}\right\} \tag{12}$$

where $E_{\mathbf{y}|\alpha}[\ldots]$ denotes expectation with respect to $p(\mathbf{y}|\boldsymbol{\alpha},\boldsymbol{\theta})$, and $E_{\alpha}[\ldots]$ denotes expectation with respect to $p(\boldsymbol{\alpha})$.

4.1.1 Fisher Information Matrix

Under the assumption that the distribution of the data conditioned on the interferometric is normal it is possible to express the FIM in (11) in a relatively simple form, by exploiting the result:

$$E_{\mathbf{x}|\boldsymbol{\tau}}\left[\frac{\partial^2(\log(p(\mathbf{x}|\boldsymbol{\tau})))}{\partial\tau_n\partial\tau_m}\right] = trace\left\{\frac{\partial\mathbf{R}}{\partial\tau_n}\frac{\partial\mathbf{R}^{-1}}{\partial\tau_m}\right\} \tag{13}$$

that holds for any process distributed as $\mathbf{x} \sim N(\mathbf{0},\mathbf{R}(\boldsymbol{\tau}))$, where $\boldsymbol{\tau}$ is an arbitrary set of coefficients that parameterize the covariance matrix of $\mathbf{x}$.
From (11) and (13), a straightforward application of the chain rule leads to the following expression for the FIM:

$$\mathbf{F} = \begin{bmatrix} \boldsymbol{\Theta}^T\mathbf{X}\boldsymbol{\Theta} & \boldsymbol{\Theta}^T\mathbf{X} \\ \mathbf{X}\boldsymbol{\Theta} & \mathbf{X} \end{bmatrix} \tag{14}$$

where $\boldsymbol{\Theta}$ is the matrix of the first order partial derivatives of the set of functions $\{\psi_n(\boldsymbol{\theta})\}_{n=0}^{N-1}$ with respect to $\boldsymbol{\theta}$,

$$\{\boldsymbol{\Theta}\}_{nm} = \frac{\partial\psi_n(\boldsymbol{\theta})}{\partial\theta_m}, \tag{15}$$

and $\mathbf{X}$ is an $N \times N$ matrix representing the FIM associated to the estimates of the interferometric phases, φ_n. After (13), $\mathbf{X}$ is obtained as:

$$\{\mathbf{X}\}_{nm} = -trace\left\{\frac{\partial\mathbf{R}}{\partial\varphi_n}\frac{\partial\mathbf{R}^{-1}}{\partial\varphi_m}\right\} \tag{16}$$

A straightforward, yet rather long and tedious, evaluation of (16) leads to express $\mathbf{X}$ as:

$$\mathbf{X} = 2L \cdot \left(\boldsymbol{\Gamma} \circ \boldsymbol{\Gamma}^{-1} - \mathbf{I}_N\right) \tag{17}$$

where Γ is the coherence matrix in (6), $\circ$ indicates the Hadamard (i.e. entry-wise) product between two matrices and L is the number of (independent) looks inside Ω. The role of $\mathbf{X}$ is to account for the loss of information about the interferometric phases due to target decorrelation. A remarkable property of $\mathbf{X}$ is that its rank is strictly lower than the number of images which constitute the data-set. This fact must not surprise, since the interferometric phases affect the data covariance matrix only through their differences, $\varphi_n - \varphi_m$. Therefore, at least one interferometric phase may be defined arbitrarily. Consider now the case where the data-set may be partitioned into two statistically independent subsets, namely:

$$\mathbf{y} = \left[\begin{array}{cc} \mathbf{y}_1^H & \mathbf{y}_2^H \end{array} \right]^H \tag{18}$$
$$\left\{ E\left[\mathbf{y}_1 \mathbf{y}_2^H \right] \right\}_{nm} = 0 \quad \forall\, n, m$$

where $\mathbf{y}_1$ is the stack of $N_1 < N$ SAR images and $\mathbf{y}_2$ is the stack of the remaining $N_2 = N - N_1$ SAR images. By virtue of the statistical independence between the two subsets, no information is available about the differences between any of the phases of the first and the second subsets. This means that the estimation of the interferometric phases may be carried out separately for each subsets, which implies that at most $N_1 - 1 + N_2 - 1 = N - 2$ phases can be retrieved. Extending this argument to the case of N_S independent subsets, it turns out that at most $N - N_S$ phases can be retrieved, corresponding to $rank\,(\mathbf{X}) \leq N - N_S$. Conversely, if the data cannot be partitioned into two subsets, then it means that there are at least $N - 1$ interferometric pairs with a non null coherence. As a consequence, at least $N - 1$ phase differences can be measured. Since the data cannot be partitioned, each of the interferometric phases, φ_n, has to appear in at least one of such $N - 1$ phase differences. Therefore, the ensemble of the $N - 1$ phase difference forms a system of $N - 1$ independent equations, resulting in the possibility to retrieve exactly $N - 1$ interferometric phases. Applying this argument to each subset, one gets that:

$$rank\,(\mathbf{X}) = N - N_S \tag{19}$$

where N is the number of images and N_S is the maximum number of subsets in which the data can be partitioned.

4.1.2 Prior Information Matrix

Under the hypothesis of normally distributed APSs, the computation of the lower right block of the matrix $\mathbf{I}_\alpha$ proceeds directly from (12). In this case, it is readily found that:

$$\mathbf{I}_\alpha = \left[\begin{array}{cc} 0 & 0 \\ 0 & \mathbf{R}_\alpha^{-1} \end{array} \right] \tag{20}$$

where $\mathbf{R}_\alpha$ is the covariance matrix of the APSs.

4.1.3 HCRB

From (17), $\mathbf{X}$ does not depend on $\boldsymbol{\alpha}$, and thus $E_\alpha\left[\mathbf{F}\right] = \mathbf{F}$. Therefore, the hybrid information matrix may be written as

$$\mathbf{J} = \left[\begin{array}{cc} \Theta^T \mathbf{X} \Theta & \Theta^T \mathbf{X} \\ \mathbf{X} \Theta & \mathbf{X} + \sigma_\alpha^{-2} \mathbf{I}_N \end{array} \right]. \tag{21}$$

Now, computing the inverse of $\mathbf{J}$ and extracting the upper left block, the HCRB for the estimate of $\boldsymbol{\theta}$ results:

$$E_{\mathbf{y},\boldsymbol{\alpha}}\left[\left(\widehat{\theta}-\theta\right)\left(\widehat{\theta}-\theta\right)^{T}\right] \geq \left(\boldsymbol{\Theta}^{T}\mathbf{X}\boldsymbol{\Theta}-\boldsymbol{\Theta}^{T}\mathbf{X}\left(\mathbf{X}+\mathbf{R}_{\alpha}^{-1}\right)^{-1}\mathbf{X}\boldsymbol{\Theta}\right)^{-1} \tag{22}$$

where it may be shown that the a necessary and sufficient condition for the bound to exist is that the matrix $\boldsymbol{\Theta}^{T}\mathbf{X}\boldsymbol{\Theta}$ is full rank.

4.2 A physical interpretation

In order to provide physically meaningful insights about the mechanisms which rule the estimate accuracy in SAR interferometry, we find it convenient to recast (22) in the following form:

$$E_{\mathbf{y},\boldsymbol{\alpha}}\left[\left(\widehat{\theta}-\theta\right)\left(\widehat{\theta}-\theta\right)^{T}\right] \geq \lim_{\varepsilon\to 0}\left(\boldsymbol{\Theta}^{T}\left(\mathbf{X}_{\varepsilon}^{-1}+\mathbf{R}_{\alpha}\right)^{-1}\boldsymbol{\Theta}\right)^{-1} \tag{23}$$

where $\mathbf{X}_{\varepsilon}=\mathbf{X}+\varepsilon\mathbf{I}_{N}$ is a perturbation of $\mathbf{X}$. After some matrix manipulations and under the condition that $\boldsymbol{\Theta}^{T}\mathbf{X}\boldsymbol{\Theta}$ is full rank, it may be proved that (22) and (23) provide the same, finite, bound as $\varepsilon\to 0$. By definition, the matrix $\mathbf{X}$ represents the information that is available about the interferometric phases in dependence of target decorrelation. For this reason, it is natural to associate the inverse of its perturbation, $\mathbf{X}_{\varepsilon}$, to the covariance matrix of the phase noise due to target decorrelation. Notice that, although such matrix diverges as $\varepsilon\to 0$, if the condition that $\boldsymbol{\Theta}^{T}\mathbf{X}\boldsymbol{\Theta}$ is full rank is fulfilled the bound in (23) exists finite, and thus the interpretation of $\mathbf{X}_{\varepsilon}^{-1}$ as a covariance matrix can be retained. After model (3), the presence of the APSs results in a further phase noise, independent of target decorrelation, whose covariance matrix is given by $\mathbf{R}_{\alpha}$. Hence, the kernel $\left(\mathbf{X}_{\varepsilon}^{-1}+\mathbf{R}_{\alpha}\right)$ in (23) represents the covariance matrix of the total phase noise, due to both target decorrelation and the APSs. It follows that the term $\left(\mathbf{X}_{\varepsilon}^{-1}+\mathbf{R}_{\alpha}\right)^{-1}$ represent the global information about the interferometric phases, and hence $\boldsymbol{\Theta}^{T}\left(\mathbf{X}_{\varepsilon}^{-1}+\mathbf{R}_{\alpha}\right)^{-1}\boldsymbol{\Theta}$ represents the total information about the parameters of interest $\boldsymbol{\theta}$. The practical implications of this interpretation will be discussed in the next chapter.

4.3 On the validity of the HCRB for InSAR applications

To give an intuitive idea of the conditions that must be met for the HCRB to represent a realistic bound, it is again useful to consider separately the roles of $\mathbf{X}_{\varepsilon}^{-1}$ and $\mathbf{R}_{\alpha}$ in (23). Assume that the APS can be neglected with respect to the contribution coming from target decorrelation. In this case, the bound for the variance of the LDF predicted by (23) defaults to the standard CRB, since it ignores the statistical component represented by the APSs. In formula:

$$E_{\mathbf{y}}\left[\left(\widehat{\theta}-\theta\right)\left(\widehat{\theta}-\theta\right)^{T}\right] \geq \left(\boldsymbol{\Theta}^{T}\mathbf{X}\boldsymbol{\Theta}\right)^{-1} \tag{24}$$

This bound is quite closely approached by maximum-likelihood estimators at sufficiently large signal-to-noise ratios, or when the number of available data is sufficiently large. In the

framework of InSAR, this means that either the correlation coefficients, the number of images, or the estimation window must be large.
On the other hand, when the APS noise dominates with respect to the source decorrelation, the term $\widetilde{\mathbf{X}}^{-1}$ can be ignored, and from (4), the HCRB for $\boldsymbol{\theta}$ is proportional to the variance of the APS:

$$E_{\mathbf{y},\boldsymbol{\alpha}}\left[\left(\widehat{\boldsymbol{\theta}}-\boldsymbol{\theta}\right)\left(\widehat{\boldsymbol{\theta}}-\boldsymbol{\theta}\right)^T\right] \geq \left(\widetilde{\boldsymbol{\Theta}}^T\widetilde{\mathbf{R}}_\alpha^{-1}\widetilde{\boldsymbol{\Theta}}\right)^{-1} \tag{25}$$

Attention must be paid prior to retaining this result. The role of the variance of the APSs in the estimation process may be described through a threshold like behavior. The fact that this aspect is not handled by the HCRB is an intrinsic limitation of the method, whereas other bounds have been proposed in literature that account properly for threshold effects (27). Practical conditions for retaining the validity of the results will be provided in the following.

4.4 Closed form solutions for InSAR

This last section provides a simple and practical example of an application of the proposed bound (23). For simplicity's sake, the LDF is modeled as a LOS subsidence with constant velocity, v, even though an extension of this analysis to a larger set of unknowns is immediate. In this case the set of the unknowns reduces to a single scalar, v, and the phase functions $\psi_n(v)$ are given by:

$$\psi_n(v) = \frac{4\pi}{\lambda} n\delta t \cdot v \tag{26}$$

δt being the time interval between two nearby acquisitions and λ the wavelength.
To properly characterize the source statistics, we assume that γ_{nm} is determined by an exponential temporal decorrelation (22), plus a thermal noise, uncorrelated from one acquisition to the other:

$$\gamma_{nm} = \gamma_0\rho^{|n-m|\delta t} + (1-\gamma_0)\,\delta_{n-m} \tag{27}$$

where ρ is a parameter describing temporal decorrelation and γ_0 is related to the signal-to-noise ratio via

$$\gamma_0 = \frac{SNR}{1+SNR}.$$

Finally, let N denote the number of available images and L the number of independent looks within the estimation window.
From the decorrelation model (27), the HCRB for the estimate of the subsidence velocity is easily computed through (17) and (23); see the left panel of Fig. (2). However, to achieve deeper insight on how temporal decorrelation, thermal noise, and APSs impact the accuracy of the estimate, it is convenient to analyze these phenomena separately.

4.4.1 Temporal decorrelation

Neglecting the contribution of thermal noise, the decorrelation model (27) reduces to $\gamma_{nm} = \rho^{|n-m|\delta t}$, for which the matrix $\boldsymbol{\Gamma}^{-1}$ is tridiagonal and computable in a closed form. Neglecting also the contribution of the APSs, from (17) and (23), and after some matrix manipulations it is possible to obtain:

$$\sigma_v^2 = \left(\frac{\lambda}{4\pi\delta t}\right)^2 \frac{1-\rho^{2\delta t}}{2L\rho^{2\delta t}} \frac{1}{N-1} \overset{def}{=} \frac{\sigma_{temp}^2}{N-1} \tag{28}$$

which shows that, in absence of thermal noise and APS, temporal decorrelation acts in such a way as to pose the problem of the estimate of velocity as the estimate of the mean of a normal white process with variance σ_{temp}^2.

4.4.2 Thermal noise and APS

The thermal noise and APS corresponds to the case of **the long-term coherent target**, the same assumption of the Persistent Scatterer (PS), however here we generalize to the case when the scatterer is made by N pixels. By neglecting the temporal decorrelation is neglected, the model (27) becomes $\gamma_{nm} = \gamma_0 + (1-\gamma_0)\,\delta_{n-m}$, for which the matrix $\mathbf{\Gamma}^{-1}$ is again achievable in a closed form. The computation of $\mathbf{X}$, through (17), leads to:

$$\mathbf{X} = \frac{2L\gamma_0^2}{(1-\gamma_0)\,(1+(N-1)\,\gamma_0)}\left(N \cdot \mathbf{I}_N - \mathbf{1}_N\mathbf{1}_N^T\right) \tag{29}$$

where $\mathbf{1}_N$ denotes the vector in $\mathbb{R}^N$ for which all elements are equal to 1. It is easy to see that the null space of $\mathbf{X}$ is the one dimensional subspace spanned by $\mathbf{1}_N$, therefore $rank\,(\mathbf{X}) = N-1$. It is then convenient to define an $N-1 \times N$ phase transformation matrix $\mathbf{D}$ such that the transformed phases are given by the differences of the interferometric phases with respect to a common reference, in formula:

$$\begin{aligned} \widetilde{\boldsymbol{\varphi}} &= \mathbf{D}\boldsymbol{\varphi} \quad \text{such that} \\ \widetilde{\varphi}_n &= \varphi_n - \varphi_0 \quad \forall\; n = 1,2...N-1 \end{aligned} \tag{30}$$

The covariance matrix for the transformed phases, $\widetilde{\boldsymbol{\varphi}}$, may be readily obtained after (22), yielding:

$$\begin{aligned} \widetilde{\mathbf{X}}^{-1} &= \frac{1-\gamma_0}{2L\gamma_0^2}\frac{1+(N-1)\,\gamma_0}{N}\left(\mathbf{I}_{N-1} + \mathbf{1}_{N-1}\mathbf{1}_{N-1}^T\right) \\ &\stackrel{def}{=} \sigma_{noise}^2\left(\mathbf{I}_{N-1} + \mathbf{1}_{N-1}\mathbf{1}_{N-1}^T\right) \end{aligned} \tag{31}$$

As for the covariance of the transformed APSs, $\widetilde{\boldsymbol{\alpha}}$, by letting $\mathbf{R}_\alpha = \sigma_\alpha^2\mathbf{I}_N$ it is immediate to see that:

$$E\left[\widetilde{\boldsymbol{\alpha}}\widetilde{\boldsymbol{\alpha}}^T\right] = \widetilde{\mathbf{R}}_\alpha = \mathbf{D}\mathbf{R}_\alpha\mathbf{D}^T = \sigma_\alpha^2\left(\mathbf{I}_{N-1} + \mathbf{1}_{N-1}\mathbf{1}_{N-1}^T\right) \tag{32}$$

Therefore $\widetilde{\mathbf{R}}_\alpha$ has exactly the same structure as (31), from which it follows the important result that, besides a scale factor, thermal noise and APSs affect the estimate accuracy in the same way.

Finally, from (23), the accuracy on the estimate of the subsidence rate is given by:

$$\sigma_v^2 = \left(\frac{\lambda}{4\pi\delta t}\right)^2 \frac{12}{N^3 - N}\left(\sigma_\alpha^2 + \sigma_{noise}^2\right) \tag{33}$$

which is analogous to the CRB for a linear regression in presence of a normal white process with variance $\sigma_\alpha^2 + \sigma_{noise}^2$.

4.4.3 Temporal decorrelation plus thermal noise and APSs

At this point, the behavior of the HCRB curve may be qualitatively explained as a mixture of (33) and (28). When the number of images is large, the $1/(N-1)$ mechanism is dominant; thus, the HCRB may be assumed to be given by (28). On the other hand, when N is small, the dominant contribution is given by thermal noise and APSs, and thus we expect the curve to

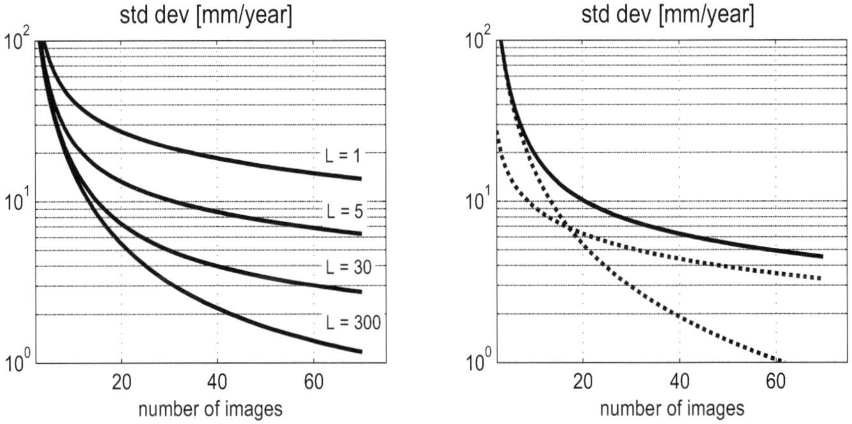

Fig. 2. HCRB for the standard deviation of the estimate of the subsidence velocity. The parameters of this scenario are: $\lambda = 56\ mm$, $\gamma_0 = 0.7$, $\rho = 0.975$, $\sigma_\alpha = 1$ rad, and $\delta t = 12$ days. The value of ρ was determined according to the work of (13). Left: standard deviation curves obtained corresponding to a number of pixels L equal to 1, 5, 30, and 300. Right: comparison between the HCRB (continuous line) and the approximations derived in (28) and (35) (dashed lines) for $L = 10$ pixels.

be proportional to $1/\left(N^3 - N\right)$. To find the proportionality factor it suffices to compute the variance for $N = 2$ images. In this case, $\mathbf{X}$ becomes a scalar, and the result is immediate:

$$\sigma_v^2(2) = \left(\frac{\lambda}{4\pi\delta t}\right)^2 \left(2\sigma_\alpha^2 + \frac{1 - \gamma_0^2\rho^{2\delta t}}{2\gamma_0^2\rho^{2\delta t} L}\right) \tag{34}$$

and thus

$$\sigma_v^2(N) = \sigma_v^2(2)\,\frac{6}{N^3 - N} \quad for\ N\ small \tag{35}$$

The standard deviation of the LDF is plotted in Fig. 2 (left) as a function of the number of images and for different number of pixels, L. The GMES-Sentinel-1 case was assumed, with a repeat pass interval of 12 days, a temporal decorrelation constant of 40 days, a coherence $\gamma_0 = 0.7$ and an APS standard deviation of 1 rad. The two asymptotic behaviors of the LDF standard deviation, corresponding to the PS, $\sigma_v \propto N^{-3/2}$ and to the pure thermal decorrelation, $N^{-1/2}$, are shown on the right.

The value of N where the HCRB curve changes its behavior is determined by intersecting (28) and (35), yielding

$$\widetilde{N} \simeq \sqrt{6\frac{\sigma_v^2(2)}{\sigma_{temp}^2}} \tag{36}$$

This value corresponds to the intersection of the dashed lines in Fig. 2.

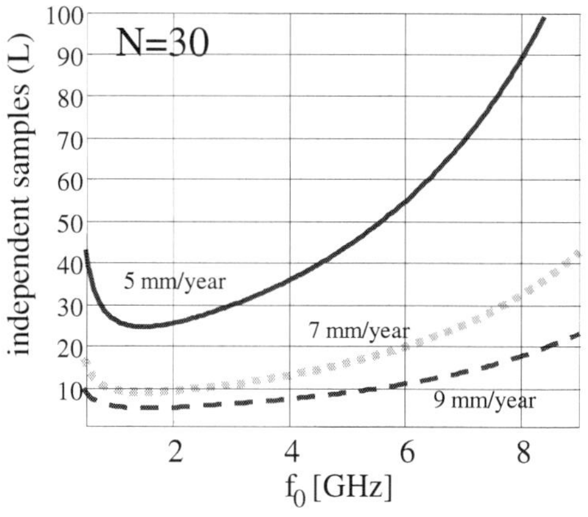

Fig. 3. Number of independent samples to be exploited for each target to get a standard deviation of the estimate of the subsidence velocity of 5-7-9 mm/year. Frequencies from L to X band have been exploited.

As a further example, the HCRB allowed us to compute the performances at different frequencies. The number of independent samples to be used to get $\sigma_v = 5,7$ and 9 mm/year is plotted in Fig. (3).In computing the HCRB, the temporal decorrelation constant has been updated with the square of the wavelength according to the Markov model in (13), and the APS phase standard deviation has been updated inversely to the wavelength, the APS delay being frequency-independent. As a result, the performances drops at the lower frequencies (the L band), due to the scarce sensitivity of phase to displacements, hence the poor SNR. Likewise, there is a drop at the high frequencies due to both the temporal and the APS noises. However, the behavior is flat in the frequencies between S and C band.

4.4.4 Single baseline interferometry

In case of single baseline interferometry, N=2 and there is no way to distinguish between temporal decorrelation and long term stability. Moreover the phase to be estimated is now a scalar. Expression (29) leads to the well known CRB (15):

$$\sigma_{\phi}^2 = \frac{1-\gamma^2}{2L\gamma^2}$$

4.5 Conclusions

In this chapter a bound for the parametric estimation of the LDF through InSAR has been discussed. This bound was derived by formulating the problem in such a way as to be handled by the HCRB. This methodology allows for a unified treatment of source decorrelation (target changes, thermal noise, volumetric effect, etc.) and APS under a consistent statistical approach. By introducing some reasonable assumptions, we could obtain some closed form solutions of practical use in InSAR applications. These solutions provide a quick performance assessment of an InSAR system as a function of its configuration (wavelength, resolution, SNR), the intrinsic scene decorrelation, and the APS variance. Although some limitations may arise at higher wavelengths, due to phase wrapping, the result may still be useful for the design and tuning of the overall system.

5. Phase Linking

The scope of this section is to introduce an algorithm to estimate the set of the interferometric phases, φ_n, comprehensive of the APS contribution. As discussed in previous chapter, assuming such model is equivalent to retaining phase triangularity, namely $\varphi_{nm} = \varphi_n - \varphi_m$. In other words, we are forcing the problem to be structured in such a way as to explain the phases of the data covariance matrix simply through $N-1$ real numbers, instead than $N(N-1)/2$. For this reason, the estimated phases will be referred to as Linked Phases, meaning that these terms are the result of the joint processing of all the $N(N-1)/2$ interferograms. Accordingly, the algorithm to be described in this section will be referred to as Phase Linking (PL).
An overview of the algorithm is given in the block diagram of Fig. 4. The algorithm is made of two steps, the first is the phase linking, where the set of N linked phases are optimally estimated by exploiting the $N(N-1)/2$ interferograms. These phases corresponds to the optical path, hence at a second step, the APS, the DEM (the target heights) and the deformation parameters are retrieved.

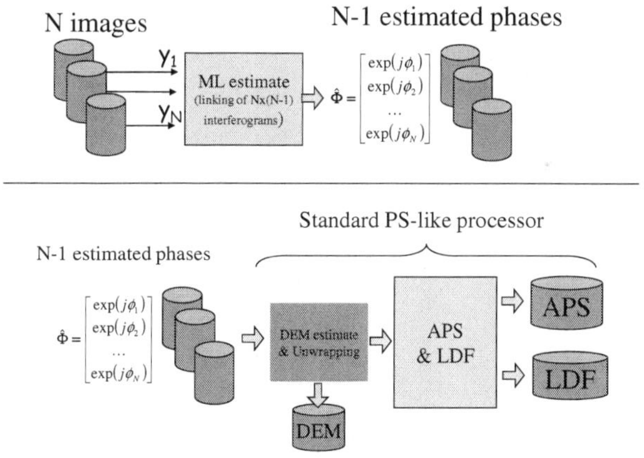

Fig. 4. Block diagram of the two step algorithm for estimating topography and subsidences.

Before going into details, it is important to note that phase triangularity is automatically satisfied if the data covariance matrix is estimated through a single sample of the data, since $\angle(y_n y_m^*) = \angle(y_n) - \angle(y_m)$. It follows that a necessary condition for the PL algorithm to be effective is that a suitable estimation window is exploited.

Since the interferometric phases affect the data covariance matrix only through their differences, one phase (say, $n = 0$) will be conventionally used as the reference, in such a way as to estimate the $N-1$ phase differences with respect to such reference. Notice that this is equivalent to estimating N phases under the constraint that $\varphi_0 = 0$. Therefore, not to add any further notation, in the following the $N-1$ phase differences will be denoted through $\{\varphi_n\}_1^{N-1}$. From (7), the log-likelihood function (times -1) is proportional to:

$$\begin{aligned} f(\varphi_1,..\varphi_{N-1}) &\propto \sum_{l=1}^{L} \mathbf{y}^H(r_l,x_l)\,\phi\boldsymbol{\Gamma}^{-1}\phi^H\mathbf{y}(r_l,x_l) \\ &\propto trace\left(\phi\boldsymbol{\Gamma}^{-1}\phi^H\widehat{\mathbf{R}}\right) \end{aligned} \tag{37}$$

where $\widehat{\mathbf{R}}$ is the sample estimate of $\mathbf{R}$ or, in other words, it is the matrix of all the available interferograms averaged over Ω. Rewriting (37), it turns out that the log-likelihood function may be posed as the following form:

$$f(\varphi_1,..\varphi_{N-1}) \propto \boldsymbol{\xi}^H\left(\boldsymbol{\Gamma}^{-1}\circ\widehat{\mathbf{R}}\right)\boldsymbol{\xi} \tag{38}$$

where $\boldsymbol{\xi}^H = \begin{bmatrix} 1 & \exp(j\varphi_1) & ... & \exp(j\varphi_{N-1}) \end{bmatrix}$. Hence, the ML estimation of the phases $\{\varphi_n\}_1^{N-1}$ is equivalent to the minimization of the quadratic form of the matrix $\boldsymbol{\Gamma}^{-1}\circ\widehat{\mathbf{R}}$ under the constraint that $\boldsymbol{\xi}$ is a vector of complex exponentials. Unfortunately, we could not find any closed form solution to this problem, and thus we resorted to an iterative minimization with respect to each phase, which can be done quite efficiently in closed form:

$$\widehat{\varphi}_p^{(k)} = \angle\left\{\sum_{n\neq p}^{N}\left\{\boldsymbol{\Gamma}^{-1}\right\}_{np}\left\{\widehat{\mathbf{R}}\right\}_{np}\exp\left(j\widehat{\varphi}_n^{(k-1)}\right)\right\} \tag{39}$$

where k is the iteration step. The starting point of the iteration was assumed as the phase of the vector minimizing the quadratic form in (38) under the constraint $\xi_0 = 1$.
Figures (5 - 7) show the behavior of the variance of the estimates of the $N-1$ phases $\{\varphi_n\}_1^{N-1}$ achieved by running Monte-Carlo simulations with three different scenarios, represented by the matrices $\boldsymbol{\Gamma}$. In order to prove the effectiveness of the PL algorithm, we considered two phase estimators commonly used in literature. The trivial solution, consisting in evaluating the phase of the corresponding L-pixel averaged interferograms formed with respect to the first ($n=0$) image, namely

$$\widehat{\varphi}_n = \angle\left(\left\{\widehat{\mathbf{R}}\right\}_{0n}\right) \tag{40}$$

is named PS-like. The estimator referred to as AR(1) is obtained by evaluating the phases of the interferograms formed by consecutive acquisitions (i.e. n and $n-1$) and integrating the result. In formula:

$$\widehat{\zeta}_n = \angle\left(\left\{\widehat{\mathbf{R}}\right\}_{n,n-1}\right); \quad \widehat{\varphi}_n = \sum_{k=1}^{n}\widehat{\zeta}_n \tag{41}$$

The name AR(1) was chosen for this phase estimator because it yields the global minimizer of (38) in the case where the sources decorrelate as an AR(1) process, namely $\gamma_{nm} = \rho^{|n-m|}$, where $\rho \in (0,1)$. This statement may be easily proved by noticing that if $\{\boldsymbol{\Gamma}\}_{nm} = \rho^{|n-m|}$, then $\boldsymbol{\Gamma}^{-1}$ is tridiagonal, and thus $\widehat{\zeta}_n$, in (41), represents the optimal estimator of the phase difference $\varphi_n - \varphi_{n-1}$. In literature this solution has been applied to compensate for temporal decorrelation in (7), (8), (6), even though in all of these works such choice was made after heuristical considerations. Finally, the CRB for the phase estimates has been computed by

zeroing the variance of the APSs. In all the simulations it has been exploited an estimation window as large as 5 independent samples.
In Fig. (5) it has been assumed a coherence matrix determined by exponential decorrelation. As stated above, in this case the AR(1) estimator yields the global minimizers of (38), and so does the PL algorithm, which defaults to this simple solution. The PS-like estimator, instead, yields significantly worse estimates, due to the progressive loss of coherence induced by the exponential decorrelation. In Fig. (6) it is considered the case of a constant decorrelation throughout all of the interferograms. The result provided by the AR(1) estimator is clearly unacceptable, due to the propagation of the errors caused by the integration step. Conversely, both the PS-like and the PL estimators produce a stationary phase noise, which is consistent with the kind of decorrelation used for this simulation. Furthermore, it is interesting to note that the Linked Phases are less dispersed, proving the effectiveness of the algorithm also in this simple scenario. Finally, a complex scenario is simulated in Fig. (7) by randomly choosing the coherence matrix, under the sole constraints that $\{\mathbf{\Gamma}\}_{nm} > 0 \; \forall \; n, m$ and that $\mathbf{\Gamma}$ is positive definite. As expected, none of the AR(1) and the PS-like estimators is able to handle this scenario properly, either due to error propagation and coherence losses. In this case, only through the joint processing of all the interferograms it is possible to retrieve reliable phase estimates.

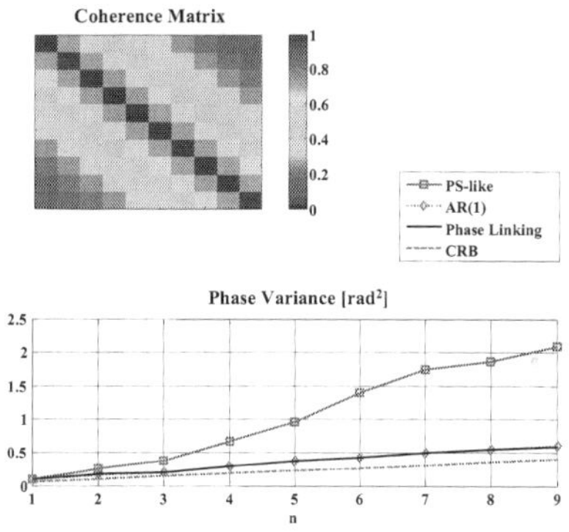

Fig. 5. Variance of the phase estimates. Coherence model: $\{\mathbf{\Gamma}\}_{nm} = \rho^{|n-m|}; \rho = 0.8$.

5.1 Phase unwrapping

As stated above, the splitting of the MLE into two steps is advantageous provided that the two resulting sub-problems are actually easier to solve than the original problem. Despite we could not find a closed form solution to the PL problem, it must be highlighted that the algorithm does not require the exploration of the parameter space, thus granting an interesting computational advantage over the one step MLE, especially in the case of a complex initial parametrization. Instead, difficulties may arise when dealing with the estimation of the original parameters from the linked phases, since the PL algorithm does not solve for the 2π ambiguity. As a consequence, a Phase Unwrapping (PU) step is required prior to the moving to the estimation of the parameters of interest. However, the discussion of a PU technique is out of the scope of this chapter, we just observe that, once a set of liked phases phases $\widehat{\varphi}_n$has

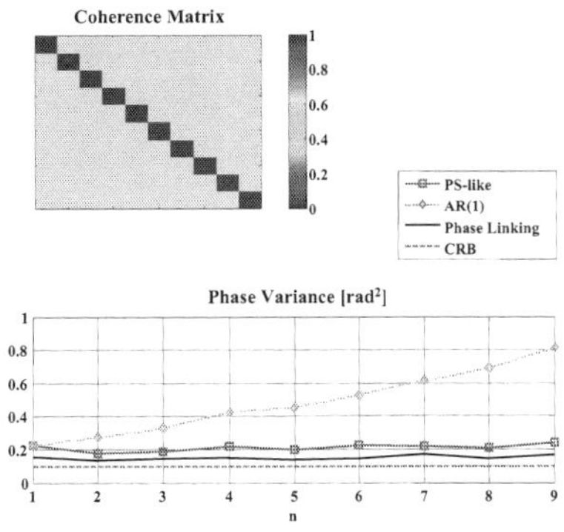

Fig. 6. Variance of the phase estimates. Coherence model: $\{\mathbf{\Gamma}\}_{nm} = \gamma_0 + (1-\gamma_0)\,\delta_{n-m}$; $\gamma_0 = 0.6$.

been estimated, we just approach PU as in conventional PS processing, that is quite simple and well tested (1), (5).

6. Parameter estimation

Once the 2π ambiguity has been solved, the linked phases may be expressed in a simple fashion by modifying the phase model in (3) in such a way as to include the estimate error committed in the first step. In formula:

$$\widehat{\boldsymbol{\varphi}} = \boldsymbol{\psi}(\boldsymbol{\theta}) + \boldsymbol{\alpha} + \boldsymbol{v} \tag{42}$$

where $\boldsymbol{v}$ represents the estimate error committed by the PL algorithm or, in other words, the phase noise due to target decorrelation. After the properties of the MLE, v is asymptotically distributed as a zero-mean multivariate normal process, with the same covariance matrix as the one predicted by the CRB (30). In the case of InSAR, the term "asymptotically" is to be understood to mean that either the estimation window is large or there is a sufficient number of high coherence interferometric pairs. If these conditions are met, then it sensible to model the pdf of v as:

$$\boldsymbol{v} \sim N\left(\mathbf{0}, \lim_{\varepsilon \to 0} (\mathbf{X} + \varepsilon \mathbf{I}_N)^{-1}\right) \tag{43}$$

where the covariance matrix of $\boldsymbol{v}$ has been determined after (23), by zeroing the contribution of the APSs. Notice that the limit operation could be easily removed by considering a proper transformation of the linked phases in (42), as discussed in section 4.2. Nevertheless, we regard that dealing with non transformed phases provides a more natural exposition of how parameter estimation is performed, and thus we will retain the phase model in (42).

After the discussion in the previous chapter, the APS may be modeled as a zero-mean stochastic process, highly correlated over space, uncorrelated from one acquisition to the other and, as a first approximation, normally distributed. This leads to expressing the pdf of the linked phases in as

$$\widehat{\boldsymbol{\varphi}} \sim N\left(\boldsymbol{\psi}(\boldsymbol{\theta}), \lim_{\varepsilon \to 0} (\mathbf{W}_\varepsilon)\right)$$

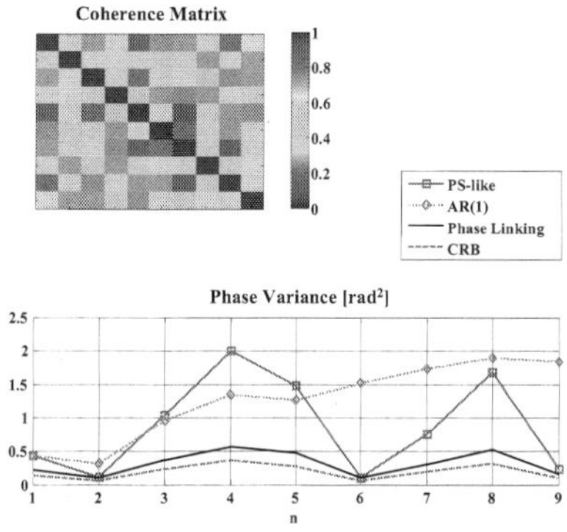

Fig. 7. Variance of the phase estimates. Coherence model: random.

where $\mathbf{W}_\varepsilon$ is the covariance matrix of the total phase noise,

$$\mathbf{W}_\varepsilon = (\mathbf{X} + \varepsilon \mathbf{I}_N)^{-1} + \sigma_\alpha^2 \mathbf{I}_N, \tag{44}$$

and σ_α^2 is the variance of the APS.
In order to provide a closed form solution for the estimation of $\boldsymbol{\theta}$ from the linked phase, $\widehat{\boldsymbol{\varphi}}$, we will focus on the case where the relation between the terms $\boldsymbol{\psi}(\boldsymbol{\theta})$ and $\boldsymbol{\theta}$ is linear, namely $\boldsymbol{\psi}(\boldsymbol{\theta}) = \boldsymbol{\Theta}\boldsymbol{\theta}$. This passage does not involve any loss of generality, as long as that $\boldsymbol{\theta}$ is interpreted as the set of weights which represent $\boldsymbol{\psi}(\boldsymbol{\theta})$ in some basis (such as a polynomial basis). At this point, the MLE of $\boldsymbol{\theta}$ from $\widehat{\boldsymbol{\varphi}}$ may be easily derived by minimizing with respect to $\boldsymbol{\theta}$ the quadratic form:

$$(\widehat{\boldsymbol{\varphi}} - \boldsymbol{\Theta}\boldsymbol{\theta})^T \mathbf{W}_\varepsilon^{-1} (\widehat{\boldsymbol{\varphi}} - \boldsymbol{\Theta}\boldsymbol{\theta}), \tag{45}$$

which yields the linear estimator

$$\widehat{\boldsymbol{\theta}} = \mathbf{Q}\widehat{\boldsymbol{\varphi}}, \tag{46}$$

where

$$\mathbf{Q} = \lim_{\varepsilon \to 0} \left(\boldsymbol{\Theta}^T \mathbf{W}_\varepsilon^{-1} \boldsymbol{\Theta} \right)^{-1} \boldsymbol{\Theta}^T \mathbf{W}_\varepsilon^{-1} \tag{47}$$

Therefore, the MLE of $\boldsymbol{\theta}$ from $\widehat{\boldsymbol{\varphi}}$ is implemented through a weighted L2 norm fit of the model $\boldsymbol{\psi}(\boldsymbol{\theta}) = \boldsymbol{\Theta}\boldsymbol{\theta}$, and $\mathbf{W}_\varepsilon^{-1}$ may be interpreted as the set of weights which allows to fit the model accounting for target decorrelation and the APSs. It can be shown that the condition that $\boldsymbol{\Theta}^T \mathbf{X} \boldsymbol{\Theta}$ is full rank is sufficient to ensure the finiteness of the matrix $\mathbf{Q}$.
By plugging (47) into (46) it turns out that $\widehat{\boldsymbol{\theta}}$ is an unbiased estimator of $\boldsymbol{\theta}$ and that the covariance matrix of the estimates is given by:

$$E\left[\left(\widehat{\boldsymbol{\theta}} - \boldsymbol{\theta}\right) \left(\widehat{\boldsymbol{\theta}} - \boldsymbol{\theta}\right)^T \right] = \mathbf{Q}\mathbf{W}_\varepsilon \mathbf{Q}^T \tag{48}$$

$$= \lim_{\varepsilon \to 0} \left(\boldsymbol{\Theta}^T \left((\mathbf{X} + \varepsilon \mathbf{I})^{-1} + \sigma_\alpha^2 \mathbf{I} \right)^{-1} \boldsymbol{\Theta} \right)^{-1}$$

which is the same as (23). The equivalence between (23) and (48) shows that the two step procedure herein described is asymptotically consistent with the HCRB, and thus it may be regarded as an optimal solution at sufficiently large signal-to-noise ratios, or when the data space is large.
It is important to note that the peculiarity of the phase model (42), on which parameter estimation has been based, is constituted by the inclusion of phase noise due to target decorrelation, represented by $\boldsymbol{v}$.In the case where this term is dominated by the APS noise, model (42) would tends to default to the standard model exploited in PS processing. Accordingly, in this case the weighted fit carried out by (47) substantially provides the same results as an unweighted fit. In the framework of InSAR, this is the case where the LDF is to be investigated over distances larger than the spatial correlation length of the APS. Therefore, the usage of a proper weighting matrix $\mathbf{W}_{\varepsilon}^{-1}$ is expected to prove its effectiveness in cases where not only the average displacement of an area is under analysis, but also the local strains.

7. Conditions for the validity of the HCRB for InSAR applications

The equivalence between (23) and (48) provides an alternative methodology to compute the lower bounds for InSAR performance, through which it is possible to achieve further insights on the mechanisms that rule the InSAR estimate accuracy. In particular, (48) has been derived under two hypotheses:

1. the accuracy of the linked phases is close to the CRB;
2. the linked phases can be correctly unwrapped.

As previously discussed, the condition for the validity of hypothesis 1) is that either the estimation window is large or there is a sufficient number of high coherence interferometric pairs. Approximately, this hypothesis may be considered valid provided that the CRB standard deviation of each of the linked phases is much lower than π. Provided that hypothesis 1) is satisfied, a correct phase unwrapping can be performed provided that both the displacement field and the APSs are sufficiently smooth functions of the slant range, azimuth coordinates (15), (31). Accordingly, as far as InSAR applications are concerned, the results predicted by the HCRB in are meaningful as long as phase unwrapping is not a concern.

8. An experiment on real data

This section is reports an example of application of the two step MLE so far developed. The data-set available is given by 18 SAR images acquired by ENVISAT[1] over a $4.5 \times 4\ Km^2$ (slant range, azimuth) area near Las Vegas, US. The scene is characterized by elevations up 600 meters and strong lay-over areas. The normal and temporal baseline spans are about 1400 meters and 912 days, respectively. The scene is supposed to exhibit a high temporal stability. Therefore, both temporal decorrelation and the LDF are expected to be negligible. However, many image pairs are affected by a severe baseline decorrelation. Fig. (8) shows the interferometric coherence for three image pairs, computed after removing the topographical contributions to the phase. The first and the third panels (high normal baseline) are characterized by very low coherence values throughout the whole scene, but for areas in backslope, corresponding to the bottom right portion of each panel. These panels fully confirm the hypothesis that the scene

[1] The SAR sensor aboard ENVISAT operates in C-Band ($\lambda = 5.6\ cm$) with a resolution of about $9 \times 6\ m^2$ (slant range - azimuth) in the Image mode.

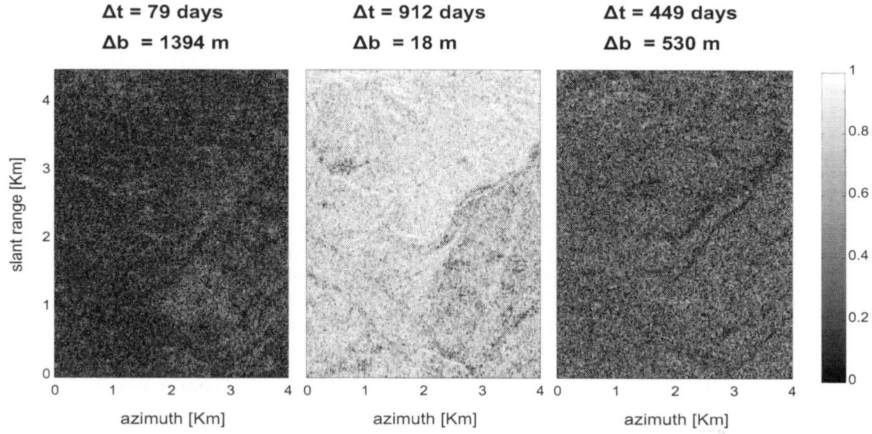

Fig. 8. Scene coherence computed for three image pairs. The coherences have been computed by exploiting a 3×9 pixel window. The topographical contributions to phase have been compensated for by exploiting the estimated DEM.

is to be characterized as being constituted by distributed targets, affected by spatial decorrelation. On the other side, the high coherence values in the middle panel (low normal baseline, high temporal baseline) confirms the hypothesis of a high temporal stability. The aim of this section is to show the effectiveness of the two step MLE previously depicted by performing a pixel by pixel estimation of the local topography and the LDF, accounting for the target decorrelation affecting the data. There are two reasons why the choice of such a data-set is suited to this goal:

- an a priori information about target statistics, represented by the matrix $\mathbf{\Gamma}$, is easily available by using an SRTM DEM;
- the absence of a relevant LDF in the imaged scene represents the best condition to assess the accuracy.

8.1 Phase Linking and topography estimation

Prior to running the PL algorithm, each SAR image have been demodulated by the interferometric phase due to topographic contributions, computed by exploiting the SRTM DEM. In order to avoid problems due to spectral aliasing, each image have been oversampled by a factor 2 in both the slant range and the azimuth directions. Then the sample covariance matrix has been computed by averaging all the interferograms over the estimation window, namely:

$$\left\{\widehat{\mathbf{R}}\right\}_{nm} = \mathbf{y}_n^H \mathbf{y}_m \tag{49}$$

where $\mathbf{y}_n$ is a vector corresponding to the pixels of the $n-th$ image within the estimation window. The size of the estimation window has been fixed in 3×9 pixels (slant range, azimuth), corresponding to about 5 independent samples and an imaged area as large as 12×20 m^2 in the slant range, azimuth plane.

The PL algorithm has been implemented as shown by equations (38), (39), where the matrix $\mathbf{\Gamma}$ has been computed at every slant range, azimuth location as a linear combination between the sample estimate within the estimation window and the a priori information provided by the SRTM DEM. Then, all the interferograms have been normalized in amplitude, flattened by the linked phases, and added up, in such a way as to define an index to assess the phase stability at each slant range, azimuth location. In formula:

$$\Upsilon = \sum_{nm} \frac{\mathbf{y}_n^H \mathbf{y}_m}{\|\mathbf{y}_n\| \, \|\mathbf{y}_m\|} \exp\left(j\left(\widehat{\varphi}_m - \widehat{\varphi}_n\right)\right) \tag{50}$$

The precise topography has been estimated by plugging the phase stability index defined in (50) and the linked phases, $\widehat{\varphi}_n$, into a standard PS processors. More explicitly, the phase stability index has been used as a figure of merit for sampling the phase estimates on a sparse grid of reliable points, to be used for APS estimation and removal. After removal of the APS, the residual topography has been estimated on the full grid by means of a Fourier Transform (1), (5), namely:

$$\widehat{q} = \arg\max_{q} \left\{ \left| \sum_n \exp\left(j\left(\widehat{\varphi}_n - k_z(n)\, q\right)\right) \right| \right\} \tag{51}$$

where $\widehat{q}$ is the topographic error with respect to the SRTM DEM and $k_z(n)$ is the height to phase conversion factor for the $n - th$ image.

The resulting elevation map shows a remarkable improvement in the planimetric and altimetric resolution, see Fig. (9). In order to test the DEM accuracy, the interferograms for three different image pairs have been formed and compensated for the precise DEM and the APS, as shown in Fig. (10, top row). Notice that the interferograms decorrelate as the baseline increases, but for the areas in backslope. In these areas, it is possible to appreciate that the phases are rather good, showing no relevant residual fringes.

The effectiveness of the Phase Linking algorithm in compensating for spatial decorrelation phenomena is visible in Fig. (10, bottom row), where the three panels represent the phases of the same three interferograms as in the top row obtained by computing the (wrapped) differences among the LPs: $\widehat{\varphi}_{nm} = \widehat{\varphi}_n - \widehat{\varphi}_m$. It may be noticed that the estimated phases exhibit the same fringe patterns as the original interferogram phases, but the phase noise is significantly reduced, whatever the slope.

This is remarked in Fig. 11, where the histogram of the residual phases of the 1394 m interferogram (continuous line) is compared to the histogram of the estimated phases of the same interferogram (dashed line). The width of the central peak may be assessed in about 1 rad, corresponding to a standard deviation of the elevation of about 1 m.

Finally, Fig. 12 reports the error with respect to the SRTM DEM as estimated by the approach depicted above (left) and by a conventional PS analysis (right). More precisely, the result in the right panel has been achieved by substituting the linked phases with the interferogram phases in (51). Note that APS estimation and removal has been based in both cases on the linked phases, in such a way as to eliminate the problem of the PS candidate selection in the PS algorithm. The reason for the discrepancy in the results provided by the Phase Linking and the PS algorithms is that the data is affected by a severe spatial decorrelation, causing the Permanent Scatterer model to break down for a large portion of pixels.

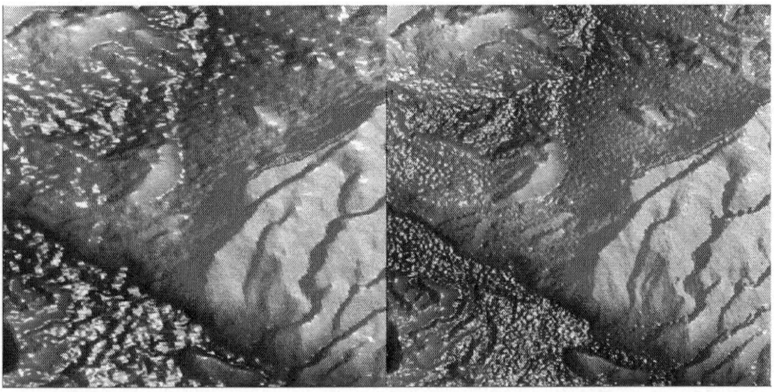

Fig. 9. Absolute height map in slant range - azimuth coordinates. Left: elevation map provided by the SRTM DEM. Right: estimated elevation map

8.2 LDF estimation

A first analysis of the residual fringes (see Fig. 10, middle panels) shows that, as expected, no relevant displacement occurred during the temporal span of 912 days under analysis. This result confirms that the residual phases may be mostly attributed to decorrelation noise and to the residual APSs. Thereafter, all the $N-1$ estimated residual phases have been unwrapped, in order to estimate the LDF as depicted in section 6. For sake of simplicity, we assumed a linear subsidence model for each pixel, that is

$$\Theta = \frac{4\pi}{\lambda} \left[\begin{array}{cccc} \Delta t_1 & \Delta t_2 & \cdots & \Delta t_N \end{array} \right]^T \tag{52}$$

being λ the wavelength and Δt_n the acquisition time of the $n-th$ image with respect to the reference image. The weights of the estimator (47) have been derived from the estimates of $\mathbf{\Gamma}$, according to (44). As pointed up in section 6, the weighted estimator (47) is expected to prove its effectiveness over a standard fit (in this case, a linear fitting) in the estimation of local scale displacements, for which the major source of phase noise is due to target decorrelation. To this aim, the estimated phases have been selectively high-pass filtered along the slant range, azimuth plane, in such a way as to remove most of the APS contributions and deal only with local deformations.

Figure (13) shows the histograms of the estimated LOS velocities obtained by the weighted estimator (47) and the standard linear fitting. As expected, the scene does not show any relevant subsidence and the weighted estimator achieves a lower dispersion of the estimates than the standard linear fitting. The standard deviation of the estimates of the LOS velocity produced by the weighted estimator (47) may be quantified in about 0.5 $mm/year$, whereas the HCRB standard deviation for the estimate of the LOS velocity is 0.36 $mm/year$, basing on the average scene coherence.

The reliability of the LOS velocity estimates has been assessed by computing the mean square error between the phase history and the fitted model at every slant range, azimuth location, see Fig. (14). It is worth noting that among the points exhibiting high reliability, few also exhibit a velocity value significantly higher that the estimate dispersion.

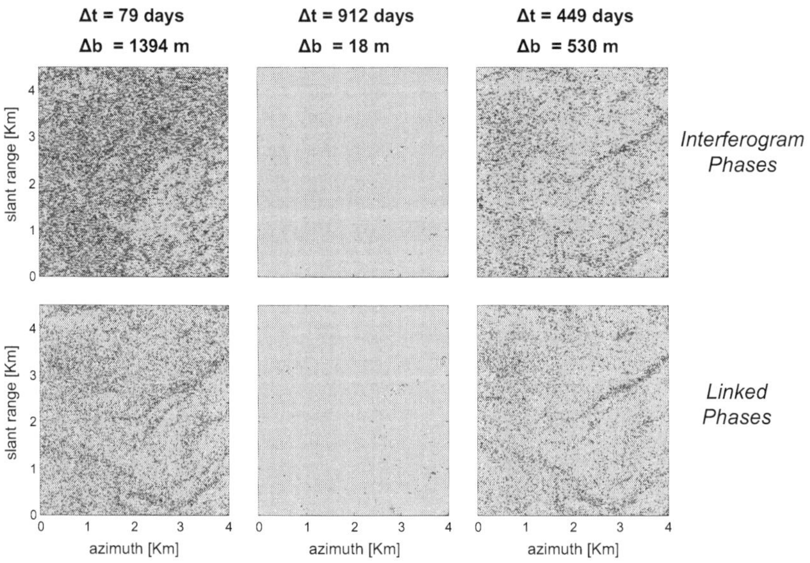

Fig. 10. Top row: wrapped phases of three interferograms after subtracting the estimated topographical and APS contributions. Each panel has been filtered, in order yield the same spatial resolution as the estimated interferometric phases (3 $\times$ 9 pixel). Bottom row: wrapped phases of the same three interferograms obtained as the differences of the corresponding LPs, after subtracting the estimated topographical and APS contributions.

9. Conclusions

This section has provided an analysis of the problems that may arise when performing interferometric analysis over scenes characterized by decorrelating scatterers. This analysis has been performed mainly from a statistical point of view, in order to design algorithms yielding the lowest variance of the estimates. The PL algorithm has been proposed as a MLE of the (wrapped) interferometric phases directly from the focused SAR images, capable of com-

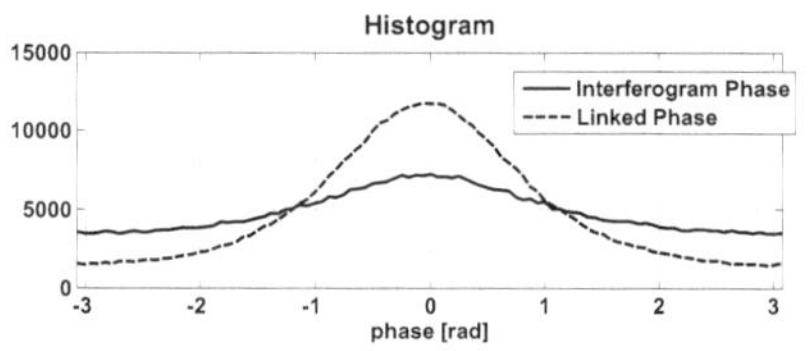

Fig. 11. Histograms of the phase residuals shown in the top and bottom left panels of Fig. 10, corresponding to a normal baseline of 1394 m.

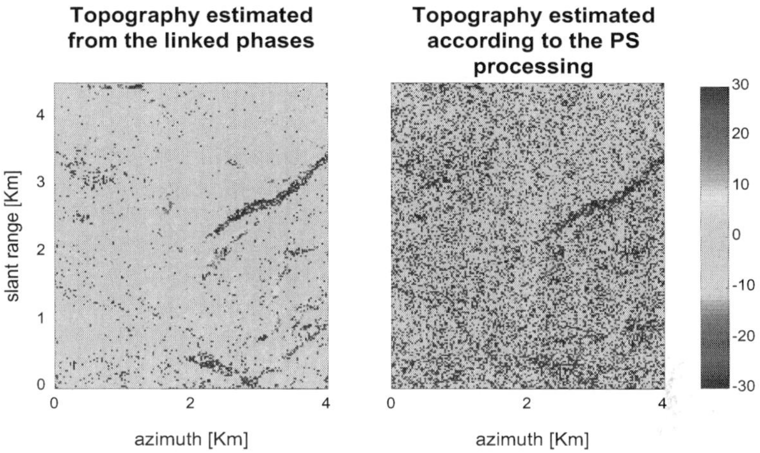

Fig. 12. Left: topography estimated from the linked phases. Right; topography estimated according to the PS processing. The color scale ranges from −30 to 30 meters.

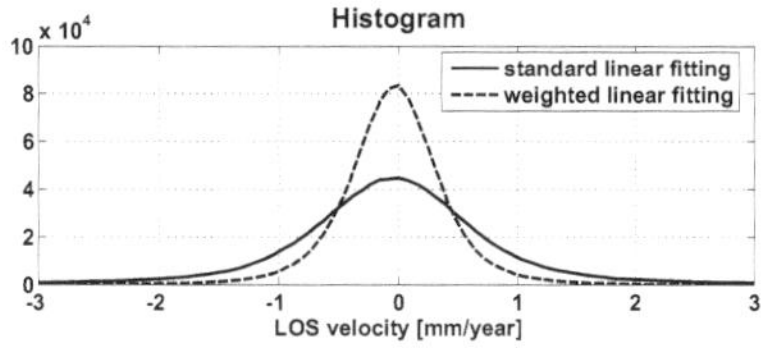

Fig. 13. Histograms of the estimates of the LOS velocity obtained by a standard linear fitting and the weighted estimator (47).

pensating the loss of information due to target decorrelation by combining all the available interferograms. This technique has been proven to be very effective in the case where the target statistics are at least approximately known, getting close to the CRB even for highly decorrelated sources. Basing on the asymptotic properties of the statistics of the phase estimates, a second MLE has been proposed to optimally fit an arbitrary LDF model from the unwrapped estimated phases, taking into account both the phase noise due target decorrelation and the presence of the APSs. The estimates have been to shown to be asymptotically unbiased and minimum variance.

The concepts presented in this chapter have been experimentally tested on an 18 image dataset spanning a temporal interval of about 30 months and a total normal baseline of about 1400 m. As a result, a DEM of the scene has been produced with $12 \times 20\ m^2$ spatial resolution and an elevation dispersion of about 1 m. The dispersion of the LOS subsidence velocity estimate has been assessed to be about 0.5 $mm/year$.

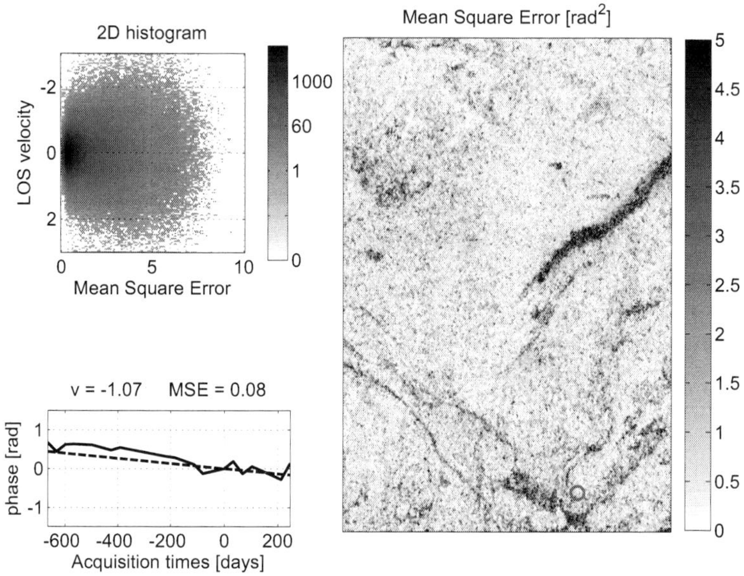

Fig. 14. Right: map of the Mean Square Errors. Top left: 2D histogram of LOS velocities estimated through weighted linear fitting and Mean Square Errors. Bottom left: phase history of a selected point (continuous line) and the correspondent fitted LDF model (dashed line). The location of this point is indicated by a red circle in the right panel.

One critical issue of this approach, common to any ML estimation technique, is the need for a reliable estimate of the scene coherence for every interferometric pair, required to drive the algorithms. In the case where target decorrelation is mainly determined by the target spatial distribution, it has been shown that a viable solution is to exploit the availability of a DEM in order to provide an initial estimate of the coherences. The case where temporal decorrelation is dominant is clearly more critical, due to the intrinsic difficulty in foreseeing the temporal behavior of the targets. Solving this problem requires the exploitation of either a very large estimation window or, which would be better, of a proper physical modeling of temporal decorrelation, accounting for Brownian Motion, seasonality effects, and other phenomena.

10. References

[1] A. Ferretti, C. Prati, and F. Rocca, "Permanent scatterers in SAR interferometry," in *International Geoscience and Remote Sensing Symposium, Hamburg, Germany, 28 June–2 July 1999*, 1999, pp. 1–3.

[2] ——, "Permanent scatterers in SAR interferometry," *IEEE Transactions on Geoscience and Remote Sensing*, vol. 39, no. 1, pp. 8–20, Jan. 2001.

[3] N. Adam, B. M. Kampes, M. Eineder, J. Worawattanamateekul, and M. Kircher, "The development of a scientific permanent scatterer system," in *ISPRS Workshop High Resolution Mapping from Space, Hannover, Germany, 2003*, 2003, p. 6 pp.

[4] C. Werner, U. Wegmuller, T. Strozzi, and A. Wiesmann, "Interferometric point target analysis for deformation mapping," in *International Geoscience and Remote Sensing Symposium, Toulouse, France, 21–25 July 2003*, 2003, pp. 3 pages, cdrom.

[5] A. Ferretti, C. Prati, and F. Rocca, "Nonlinear subsidence rate estimation using permanent scatterers in differential SAR interferometry," *IEEE Transactions on Geoscience and Remote Sensing*, vol. 38, no. 5, pp. 2202–2212, Sep. 2000.

[6] A. Hooper, H. Zebker, P. Segall, and B. Kampes, "A new method for measuring deformation on volcanoes and other non-urban areas using InSAR persistent scatterers," *Geophysical Research Letters*, vol. 31, pp. L23 611, doi:10.1029/2004GL021 737, Dec. 2004.

[7] R. Hanssen, D. Moisseev, and S. Businger, "Resolving the acquisition ambiguity for atmospheric monitoring in multi-pass radar interferometry," in *International Geoscience and Remote Sensing Symposium, Toulouse, France, 21–25 July 2003*, 2003, pp. cdrom, 4 pages.

[8] Y. Fialko, "Interseismic strain accumulation and the earthquake potential on the southern San Andreas fault system," *Nature*, vol. 441, pp. 968–971, Jun. 2006.

[9] P. Berardino, G. Fornaro, R. Lanari, and E. Sansosti, "A new algorithm for surface deformation monitoring based on small baseline differential SAR interferograms," *IEEE Transactions on Geoscience and Remote Sensing*, vol. 40, no. 11, pp. 2375–2383, 2002.

[10] P. Berardino, F. Casu, G. Fornaro, R. Lanari, M. Manunta, M. Manzo, and E. Sansosti, "A quantitative analysis of the SBAS algorithm performance," *International Geoscience and Remote Sensing Symposium, Anchorage, Alaska, 20–24 September 2004*, pp. 3321–3324, 2004.

[11] G. Fornaro, A. Monti Guarnieri, A. Pauciullo, and F. De-Zan, "Maximum liklehood multi-baseline sar interferometry," *Radar, Sonar and Navigation, IEE Proceedings -*, vol. 153, no. 3, pp. 279–288, June 2006.

[12] A. Ferretti, F. Novali, D. Z. F, C. Prati, and F. Rocca, "Moving from ps to slowly decorrelating targets: a prospective view,," in *European Conference on Synthetic Aperture Radar, Friedrichshafen, Germany, 2–5 June 2008*, 2008, pp. 1–4.

[13] F. Rocca, "Modeling interferogram stacks," *Geoscience and Remote Sensing, IEEE Transactions on*, vol. 45, no. 10, pp. 3289–3299, Oct. 2007.

[14] A. Monti Guarnieri and S. Tebaldini, "On the exploitation of target statistics for sar interferometry applications," *Geoscience and Remote Sensing, IEEE Transactions on*, vol. 46, no. 11, pp. 3436–3443, Nov. 2008.

[15] R. Bamler and P. Hartl, "Synthetic aperture radar interferometry," *Inverse Problems*, vol. 14, pp. R1–R54, 1998.

[16] G. Franceschetti and G. Fornaro, "Synthetic aperture radar interferometry," in *Synthetic Aperture Radar processing*, G. Franceschetti and R. Lanari, Eds. CRC Press, 1999, ch. 4, pp. 167–223.

[17] P. Rosen, S. Hensley, I. R. Joughin, F. K. Li, S. Madsen, E. Rodríguez, and R. Goldstein, "Synthetic aperture radar interferometry," *Proceedings of the IEEE*, vol. 88, no. 3, pp. 333–382, Mar. 2000.

[18] A. Ferretti, A. Monti Guarnieri, C. Prati, F. Rocca, and D. Massonnet, *InSAR Principles: Guidelines for SAR Interferometry Processing and Interpretation*, esa tm-19 feb 2007 ed. ESA, 2007.

[19] R. F. Hanssen, *Radar Interferometry: Data Interpretation and Error Analysis.* Dordrecht: Kluwer Academic Publishers, 2001.

[20] ——, *Radar Interferometry: Data Interpretation and Error Analysis*, 2nd ed. Heidelberg: Springer Verlag, 2005, in preparation.

[21] J. Muñoz Sabater, R. Hanssen, B. M. Kampes, A. Fusco, and N. Adam, "Physical analysis of atmospheric delay signal observed in stacked radar interferometric data," in *International Geoscience and Remote Sensing Symposium, Toulouse, France, 21–25 July 2003*, 2003, pp. cdrom, 4 pages.

[22] H. A. Zebker and J. Villasenor, "Decorrelation in interferometric radar echoes," *IEEE Transactions on Geoscience and Remote Sensing*, vol. 30, no. 5, pp. 950–959, Sep. 1992.

[23] V. Pascazio and G. Schirinzi, "Multifrequency insar height reconstruction through maximum likelihood estimation of local planes parameters," *Image Processing, IEEE Transactions on*, vol. 11, no. 12, pp. 1478–1489, Dec 2002.

[24] F. Gini, F. Lombardini, and M. Montanari, "Layover solution in multibaseline sar interferometry," *Aerospace and Electronic Systems, IEEE Transactions on*, vol. 38, no. 4, pp. 1344–1356, Oct 2002.

[25] S. Monti Guarnieri, A; Tebaldini, "Hybrid cramÉrŰrao bounds for crustal displacement field estimators in sar interferometry," *Signal Processing Letters, IEEE*, vol. 14, no. 12, pp. 1012–1015, Dec. 2007.

[26] Y. Rockah and P. Schultheiss, "Array shape calibration using sources in unknown locations–part ii: Near-field sources and estimator implementation," *Acoustics, Speech and Signal Processing, IEEE Transactions on*, vol. 35, no. 6, pp. 724–735, Jun 1987.

[27] I. Reuven and H. Messer, "A barankin-type lower bound on the estimation error of a hybrid parameter vector," *IEEE Transactions on Information Theory*, vol. 43, no. 3, pp. 1084–1093, May 1997.

[28] H. L. Van Trees, *Optimum array processing*, W. Interscience, Ed. New York: John Wiley & Sons, 2002.

[29] F. Rocca, "Synthetic aperture radar: A new application for wave equation techniques," *Stanford Exploration Project Report*, vol. SEP-56, pp. 167–189, 1987.

[30] A. Papoulis, *Probability, Random variables, and stochastic processes*, ser. McGraw-Hill series in Electrical Engineering. New York: McGraw-Hill, 1991.

[31] D. C. Ghiglia and M. D. Pritt, *Two-dimensional phase unwrapping: theory, algorithms, and software.* New York: John Wiley & Sons, Inc, 1998.

Integration of high-resolution, Active and Passive Remote Sensing in support to Tsunami Preparedness and Contingency Planning

Fabrizio Ferrucci
Università della Calabria Italy

1. Introduction

Known from time immemorial to the inhabitants of the Pacific region, tsunamis became worldwide known with the great Indian Ocean disaster of December 26, 2004, and its toll of about 234'000 deaths, 14'000 missing and over 2,000,000 displaced persons. Beyond triggering the international help in managing the immediate post-event, and sustaining eventual rehabilitation of about 10'000 km^2 of hit coastal areas, the disaster scenario was intensively focused on by spaceborne remote sensing. The latter, was the only fast and appropriate mean of collecting updated information in as much as 14 hit countries, stretching from Indonesia to South Africa across the Indian Ocean.

Short-term, institutional satellite observation response was mostly centered on the International Charter on Space and Major Disasters, a joint endeavor of 17 public and private satellite owners worldwide (including the three founding agencies: ESA-European Space Agency, CNES-Centre National d'Etudes Spatiales, and CCRS-Canadian Center for Remote Sensing) that provided emergency spaceborne imaging and rapid mapping support (www.disasterscharter.org/web/charter/activations).

In disaster response, remote sensing information needs are usually restrained to damage assessment, thus have limited duration. This implies that information must be timely and timely useable, and be provided with high-to-very high spatial resolution.

Conversely, high temporal resolution - useful in repeated damage assessment across moderate or long lasting events, as for example storm sequences, earthquake swarms and volcanic unrests - is generally unnecessary in the tsunami case, where damage presents large amplitude but is assessed once and for all after the main wavetrain has struck.

A much wider community of institutional and private users of remote sensing information, in form of special cartography products, and much longer lasting benefits are experienced if information is used for tsunami flooding risk mapping, impact scenario building and the inherent contingency planning.

Benefits are intimately connected to the characteristics of tsunamis that occur seldom, propagate at top speeds close to 200 m/s on deep ocean floors, and can hit in a few hours areas distant thousands of kilometers from the source. On account of these parameters, tsunami impact mitigation cannot simply rely upon response.

In 2004, once the earthquake originating the tsunami was felt, it would have been possible to give a 2-hour advance impact notice in distant countries as India, Sri Lanka and Maldives. This did not happen, because a monitoring-and-alert system as the current PTWC-Pacific Tsunami Warning Center managed by NOAA-National Ocean and Atmosphere Administration (www.prh.noaa.gov/ptwc/) did not exist yet in the Indian Ocean. However, since slowest velocities of tsunami waves are much larger than humans can run for escaping them, in lack of efficient emergency plans to enact immediately, it is clear that the alert system alone would not have solved the problem.

We can conclude that the risk can be mitigated acting principally on early warnings and preparedness. The latter is by far the leading issue, as preparedness measures can be effective even without early warning, whereas early warning is useless without accompanying measures.

Here, we discuss how a multi-technique, integrated remote sensing approach provides the essential information to satisfy prevention and response needs in a tsunami prone area, located in the heart of the theater of the great 2004 Indian Ocean tsunami.

2. Tsunamis and Storm Surges

Tsunamis are liquid gravitational waves that are triggered by sudden displacement of water bodies by co-seismic seafloor dislocation or underwater landslide mass push/pull. The speed (celerity) of tsunami waves is

$$V=\sqrt{\frac{g\lambda}{2\pi}\tanh\left(2\pi\frac{d}{\lambda}\right)} \quad (1)$$

with g the gravity acceleration, d the thickness of the water layer in meters and λ the wavelength. If the argument of the hyperbolic tangent is large with $d > \lambda/2$, equation (1) reduces to

$$V_{max} \approx \sqrt{\frac{g\lambda}{2\pi}} \quad (2)$$

whereas in shallow waters and $d < \lambda/20$, equation (1) becomes

$$V_{min} \approx \sqrt{gd} \quad (3)$$

On account of the steadily large ratio between wavelength and thickness of the water layer, the shallow water approximation of equation (3) applies generally.

The main parameter that discriminates tsunamis from swell, is wavelength: wind generated waves present near-constant wavelengths up to a few hundred meters, and periods between seconds and tens of seconds.

Conversely, a tsunami wave as in equation (2) travelling in a 4000m thick ocean water layer, locally reaches 200m/s with periods of 100-120 minutes (or wavelenghts of several hundred kilometers) and unnoticeable amplitude with respect to wavelength. When approaching the

shore ('shoaling') with velocity dropping below 20 m/s, wavelengths shorten to kilometers, and wave amplitudes increase (run-up) before penetrating coastal areas.
Outstanding wave heights are obtained as a combination of steep seafloor topographic gradient, and a short distance from the source. The worst documented such case occurred in the near field of a M_W=8.0 earthquake in 1946 at Unimak Island, Alaska, where the Scotch Cap lighthouse was flushed away by a 35-meter high wave.
Reportedly, wave heights for the great Indian Ocean tsunami of 26th December 2004, may have exceeded 15 m along northern Sumatra coasts (Geist et al., 2007). In Sri Lanka, about 2000 km away from the epicentre of the M_W=9.2±0.1 earthquake, largest wave heights may have exceeded 10 m in the East, whereas at least 5000 lives were taken by wavetrains not higher than 4 m, in the South and the Southwest of the island.

YEAR	DAMAGE AREA (SOURCE AREA)	SOURCE TYPE	CASUALTIES (approx.)
2004	Eastern and Central Indian Ocean (Sumatra)	Earthquake	240000
1991	Bangladesh, Chittagong (category-5 tropical cyclone)	Storm surge	138000
1970	Bangladesh (Bhola category-4 tropical cyclone)	Storm surge	500000
1908	southern Italy, Messina and Reggio Calabria	Earthquake	100000
1896	Honshu (off-Sanriku, Japan)	Earthquake	27000
1883	Indonesia, Sunda strait (Krakatau)	Volcanic eruption	35000
1868	South America Pacific coasts (Peru-Chile, Arica)	Earthquake	70000
1771	Japan, Ryukyu Islands	Earthquake	13000
1755	Portugal, Lisbon (Alentejo fault and Carrincho bank)	Earthquake	60000
1741	Japan, Oshima and Hokkaido (controversial amplitude)	Volcano landslide	2000-15000

Table 1. Top-10 deadly seawater floodings worldwide in the last three Centuries, in inverse temporal order. Most frequent tsunami triggers relate to earthquakes, either directly (co-seismic displacement) or indirectly (submarine landslides; Tinti *et al.*, 2005): in terms of ground floor dislocation alone, earthquake Magnitudes M_w<7 are not believed to trigger tsunamis. In tropical areas of strong cyclogenetic activity as the Bay of Bengal and the Gulf of Mexico, the combination of strong tropical storms and low topographic gradient of coastal areas, may lead to massive inland penetration of sea waters called 'storm surge'.

With little modifications, the above concepts may consistently apply to storm driven water surges, or 'storm surges', a threat provided with much higher repeat frequency (yearly) than tsunamis. Storm surges, typically associated to tropical cyclones, are a near-permanent elevation of the sealevel for the duration of the event, arising from the combination of extreme atmospheric pressure drop and push of the associated strong winds. Storm surges are common in tropical areas worldwide. Storm surges were responsible of the largest, flood related, mass casualty ever scored (in Bangladesh, Bengal Bay, 1970; ca. 500'000, see Table 1). In economic terms, the costliest tropical storm surge was that associated to hurricane Katrina, August 2005, with over 100 Billion USD of direct and indirect losses.

3. Rationale

As stated earlier, operational effectiveness in tsunami impact mitigation requires taking major preparedness measures to allow exposed populations moving fast to the closest safe area nearby. This solution may allow avoiding blanket evacuation of tsunami jeopardized areas, that may imply permanent activity banning in large, critical portions of the territory, especially if the topografic gradient is very low (as in Sri Lanka and Bangladesh, e.g.) and small increase of water levels lead to deep inland flooding.

In terms of preparedness, this means that escape way solutions must be addressed well in advance. Considering that unnoticeably elevated areas close to the shoreline can be good, and sometimes unexpected escape places to single out, map and include in emergency plans, protection against tsunamis and timeliness of response require the advance drawing of quantitative impact scenarios.

Emergency cartography must be frequently updated to mirror the modifications with time in location and value of vulnerable elements (inhabitants, buildings and infrastructures). This calls for the use of fast, synoptic and high-to-very high resolution mapping technologies: a need that can be satisfied by airborne and spaceborne remote sensing only. These concepts drove the design and the carrying out in 2006 - upon request of the Government of Sri Lanka to the Italian Government - of a thorough field investigation aimed to ease, provide with quantitative grounds and speed-up the national emergency planning in tsunami-prone areas. The request addressed the need of drawing a realistic set of flooding scenarios for most of the coastal areas of the island, with special emphasis on settlements and infrastructures in the reach of a model tsunami or a model storm surge. The basic criteria of investigation were broadly inspired by the format of early risk assessment and scenario simulation in the reference cases of Northwest USA (Mendocino and Humboldt in northern California, Tacoma in Washington, e.g.).

This portion of the Pacific coast is subjected to frequent tsunami impact from local seismic sources in the unresting, undersea Mendocino fault zone (Oppenheimer et al., 1993), and is focused on by the US National Tsunami Hazard Mitigation Program (Lander et al., 1993; Eisner et al., 2001; Priest et al., 2001; Venturato et al., 2007).

Downstream to US NTHMP, US Geological Survey provided dissemination of impact maps, portraying different scenarios based on possible tsunami impacts heights, and listing the number of people that would be affected by tsunamis of 5m, 10m, and 15m height respectively, with elevation data based on the SRTM (Shuttle Radar Topography Mission) Digital Elevation Model. The latter, is available worldwide. It displays planimetric resolution of 90 meters and absolute vertical accuracy of 9.6m (mission specifications). In the case of Sri Lanka, these parameters were considered not sufficient for reaching the required level of horizontal and vertical resolution compatible with a terrain heterogeneous at all scales, densely vegetated, provided with scattered manufacts eventually hidden or partly covered by tropical vegetation, and displaying negligible topographic gradients as low as 1-2% over much of the coastal zones of interest.

The drawing of quantitative flooding scenarios required collecting the information needed for completing the following steps, at the suitable scale:

i) model tsunami (at sea, before impact): requires detailed 3D knowledge of the seabed, aimed to model and forecast, spot by spot, the wavetrain pattern, the energy distribution and the run-up before impact. On account of the expected wavelengths to deal with, the ideal working scale for accurate modelling was considered to lie between 1/10000 and 1/20000 within at least 10 km from the shoreline. In lack of such information, and on account of unfavorable time and cost implications of an ad-hoc campaign, it was decided to rely upon the existing, loose seafloor cartographies by NOAA and British Admiralty, and the few wave heights observed in December 2004 (Liu et al., 2005).

ii) Model flooding (on land, after impact): requires very high-resolution 3-D terrain model, to simulate the hydraulic behavior of flooded zones at scales of 1/5000 or better, and to draw the limits of the impact zone, the expected severity of the areal impacts and, if appropriate, the energy absorption on impacted manufacts. In brief, the risk model and the scenarios, to permit emergency deciders to plan evacuation and safety measures, and urban planners to adopt structural measures finalized to ease citizens' escape in case of alert. According to urban planners, this target requires ground resolutions in the order of 1 m, and elevation precisions in the order of 0.2+0.3 m to be achieved uniformly over large areas.

Since the 2004 tsunami losses concentrated in ocean-bound strips of variable width, up to observed maxima of as much as 8 km in the East of the island (Batticaloa), the width of coastal areas to map and model was fixed at 3 km in average.

This pointed to an expected 1800 km^2 to map in 3D, in very short times (maximum one month), and with the resolutions/precisions as above: such target - clearly out of reach for standard topography missions - could be achieved only with use of State-of-the-Art active and passive remote sensing techniques.

It was chosen to combine airborne LiDAR and Hyperspectral - for top 3D resolution and simultaneous confidence qualification of elevation data - and spaceborne RaDAR (Prati et al., 1994) with multispectral mapping (Hirn & Ferrucci, 2005, 2006), aimed to extend Digital Elevation Model building and thematic mapping, to the whole of the areas requested by the Sri Lankan Government via the Disaster Management Center in Colombo. As a good balance between high resolution needs and feasibility issues, operational costs and security issues, the inter-Government agreement converged on mapping in 3D and at high-to-very high resolution, a portion of the coastal areas hosting at least two-thirds of damage and casualties observed in 2004.

Overall, the island had suffered 34'000 casualties and has experienced - for various reasons - over 1'100'000 displaced persons, ca. 500'000 of of which directly related to the tsunami destruction. The percentage of tsunami affected coastal populations ranged from 35% in the northern coastal districts of Kilinochi, to 80% in the eastern districts of Mullaitivu and 78% in Ampara, whereas the southern districts of Galle, Matara, and Hambantota displayed about 20% impact, albeit with scattered pockets of severe damage. The location map and the survey plan are shown in Figure 1.

4. The HyperDEM campaign

Following establishment of the inter-Government agreement five months after the 2004 tsunami, the operational project "'HyperDEM - The precise Digital Elevation Model of the coastal areas of Sri Lanka", was launched early in September 2005.

The work was completed in summer 2006 after acquisition of an overall data volume of 2.7 TeraBytes. Upon completion of the work, the End Users - the Disaster Management Center and the Ministry of Disaster Management and Humanitarian Affairs - were provided with ca. 2'500 km^2 of Digital Elevation Models of the coastal areas (location maps in Figure 1)

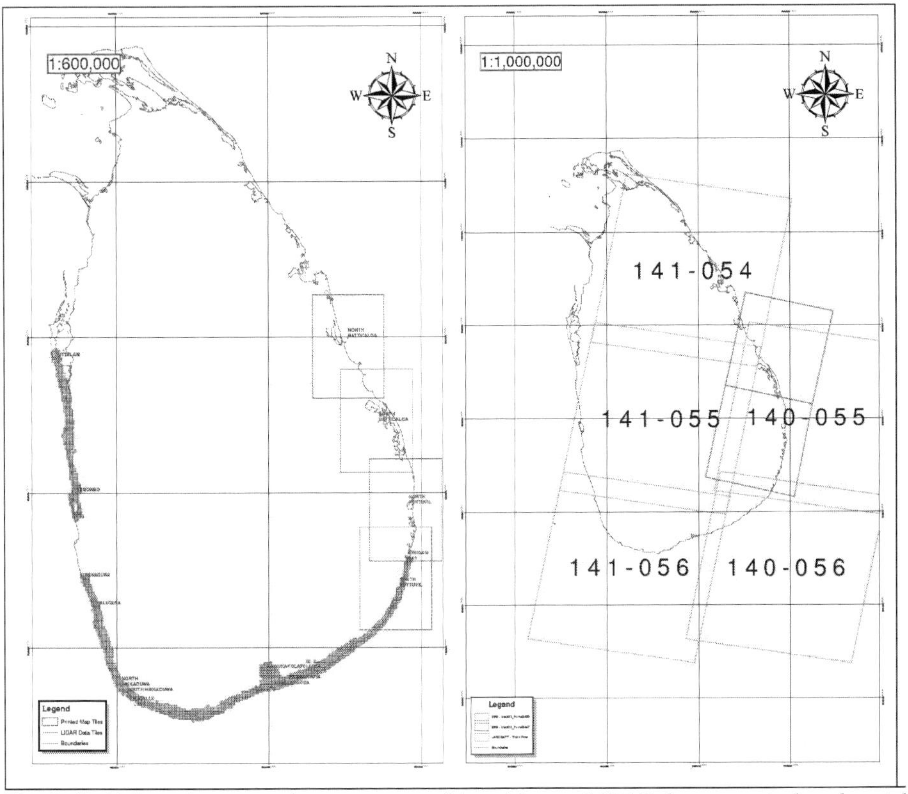

Fig. 1. (Left) Location map of areas surveyed by airborne LiDAR, hyperspectral and aerial photo (red squares) and spaceborne RaDAR (blue open squares). In the former, both Digital Elevation and Digital Surface Models were obtained at 1m resolution; in the latter, only DSM, at 30m resolution. (Right) Location of Landsat-7/ETM+ (green) and ERS-1 /ERS-2/ENVISAT (violet) satellite frames used in HyperDEM. ASTER and QuickBird imagery was also used for satisfying interpretation needs eventually arisen during processing of the 2.7 TeraByte dataset.

4.1 Airborne campaign

The airborne campaign and the related technical activities, were set up and carried out by the Istituto Nazionale di Oceanografia e Geofisica Sperimentale-OGS of Trieste, Italy. The survey, planned for integrated operation and combined acquisition of active and passive instruments at once, was designed on target ground resolutions of 1 m^2 for LiDAR (Figure 2), and 4 m^2 for hyperspectral (Figures 3, 4).

Fig. 2. Example of 3D rendering of combined LiDAR (1m planimetric resolution, 0.3 m precision in elevation on steady reflectors) and digital camera aerial scenery (resolution of 0.2 m). Picture taken over the artificial lake of Angunakolapelessa, north of Hambantota, south Sri Lanka.

After a long waiting because of a long lasting Autumn Monsoon, the survey was finally carried out in about one month after move-in of instruments to Colombo, early on February, 2006.

About 1'780 km^2 were LiDAR mapped airborne, at the planimetric resolution of 1 meter and the elevation precision of 0.3 metres (Figure 1, left), with the following payloads installed on the airborne platform, a De Havilland DHC-3 single-propeller "Otter" operated by the Sri Lankan private operator Air Taxi :

- a LiDAR system Optech ALTM 3033. The instrument consisted of a Near Infrared (Λ=1064 nm) Laser beam with pulse repetition rate of 33KHz. A scanning mirror directs the Laser optical pulses across the flight path, providing coverage to either sides of the flight direction. The forward motion of the aircraft provides coverage in the direction of flight.

ALTM 3033 incorporates a GPS receiver and an Inertial Measurement Unit (IMU), that acquires flight attitude data at the frequency of 200 Hz.

- A hyperspectral radiometer AISA Eagle 1K by the Finnish firm SPECIM. It is a pushbroom scanner made up of a V-NIR hyperspectral sensor, a GPS/INS Applanix sensor, and a laptop implemented data acquisition unit.

AISA Eagle 1K operates at wavelengths between 400-970 nm; it is able to record up to 244 bands (with spectral sampling of 2.3 nm/pixel) and 1024 spatial pixels. The system is flexible enough to allow acquiring data in almost every band combination, simultaneously acting on the number of bands and the bandwidth by use of a computer assisted procedure. We operated the system with 42-channel configuration, aimed at improving the signal-noise ratio in individual spectral bands.

- A semi-metric digital camera ROLLEI 6008 db45, with digital back Phase-One, model H2O. The camera presented a pixel spacing of 9 micrometers, in a scene composed of 4080 x 5440 pixel with 48-bit dynamics. Acquisition is assisted by a camera compensation system to adjust the roll and pitch variations due to aircraft position and flight attitude.

The decision to operate simultaneously the semi-metric digital camera with typical footprint in the order of 0.2 m (when operated at the same flight level useful in obtaining the nominal LiDAR resolution of 1 m) for assisting in the interpretation of ambiguous elevation features in the very-high resolution LiDAR and hyperspectral datasets.

In cartography applications, indeed, LiDAR raw elevation data are systematically purged of false or misleading information as those due to lateral backscattering, multiple scattering, returns from strongly reflecting physical surfaces, and so forth (Baltsavias, 1999; Kraus & Pfeifer, 1998).

Such information-cleaning process is performed through a classification process that allows assigning physical meaning to scatterers provided with variable signal/noise ratios. First pulses are typically associated to strongly reflecting objects, like trees, wires, roofs and bridges, whereas later (and weaker) pulses are attributed to returns from "ground" (Kraus & Pfeifer, 1998).

As stated earlier, the average inland extension of prospected area is about 3 km, with an isolated maximum of over 10 km in the sensitive area of the artificial basin and the dam of Angunakolapelessa (Figures 1-left and 2), immediate north of Hambantota in the south. Airborne LiDAR, orthophotos and hyperspectral data were acquired from February 11th and 21st, in two legs, separated by a four-day interval (17th to 20th February) devoted to process acquired data, assess the dataset completeness and plan eventual recoveries. The flight zone (Figure 1, left) spanned between Puttalam, in the West, and Pottuvil, in the Southeast. For security reasons, authorized flight plans did not include the capital, Colombo, nor some specific damaged coastal zones in the East (Trincomalee, Batticaloa, Ampara). Instead, eastern areas (Figure 1) were covered by spaceborne RaDAR, and qualified by high resolution spaceborne multispectral observation (Figure 1, right). Flight heights ranged between 900-2700 metres, as a function of the desired ground resolution, the morphology and land-cover of surveyed areas, and the meteorological conditions.

Flight paths were computed in real time by DGPS (differential kinematic GPS), using data simultaneously acquired by one GPS receiver onboard the aircraft and two, twin-frequency geodetic GPS receivers Ashtech (mod. Z-Extreme) at the fixed rate of one measurement per second. Twin-frequency GPS receivers were operated only on the benchmarks of an ad-hoc geodetic frame created by OGS, starting from a re-calculated benchmark of the Sri Lanka Survey Department, at the Katunayake International airport, north of Colombo.

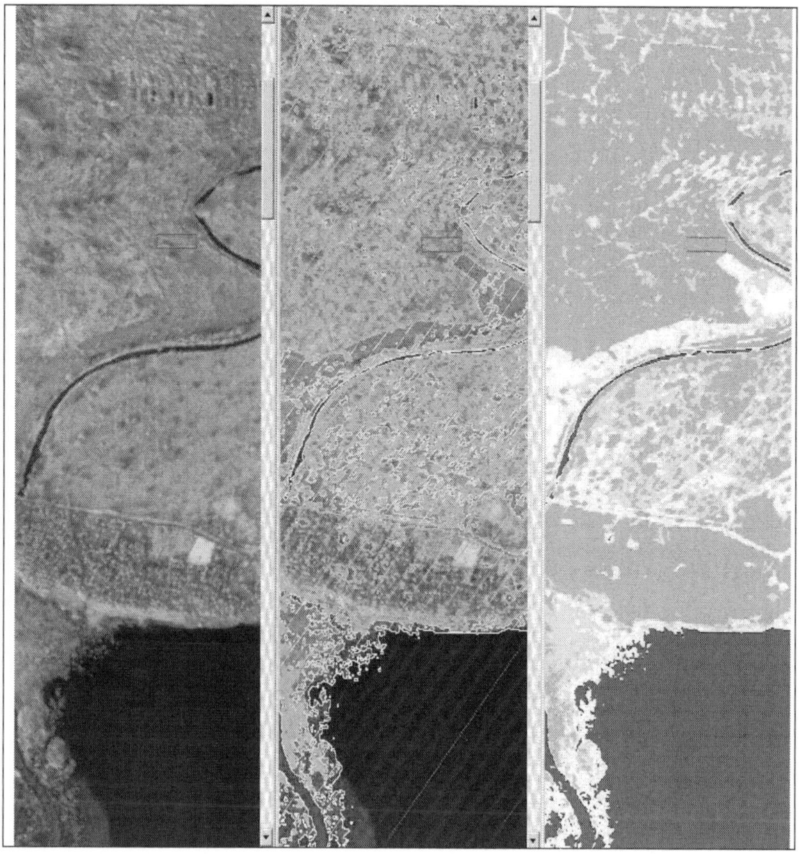

Fig. 3. Automated identification and contouring of 4x4 m^2 pixels unprovided with vegetation, done on AISA hyperspectral V-NIR data by use of a patented method, mutuated by burn scar analysis (Ferrucci & Hirn, 2005). Processing was conducted on raw data (left), aimed to prepare and carry out future operations in real-time. In contoured pixels (center), LiDAR elevation measurement are expected to be precise within the error estimate (±0.15m averaged over buildings and bare soils). Unlike vegetation, bare rocks, soils and buildings are the essential constituents of DSMs (see Figure 5) for flooding and tsunami impact simulation. The Level-2 classification (right) was used for pixel-by-pixel elevation quality assessment (Figure 4).

All benchmarks of the new geodetic frame were calculated and located on ellipsoids WGS84 and Everest 1830 in the Transverse Mercator projection. Upon completion of the campaign, the Sri Lanka Survey Dept. was provided with the monographs of newly established benchmarks.

The best estimate aircraft trajectory (SBET), made up of fixes spaced 0.15 cm in average, presented rms residual errors < 0.3m, that are compatible with the required precision in elevation. Range data were geo-referenced by use of spatial and orientation parameters; basic products are vectors of points, including the information on position, GPS time and backscattered LiDAR amplitude. All products were delivered in UTM-44N projection, WGS84 datum.

Fig. 4. Sample output of the automated identification and contouring process of buildings and vegetation, done on 56-channel AISA hyperspectral VNIR data by use of a patented method, mutuated by burn scar analysis (Ferrucci & Hirn, 2005). In these pixels, LiDAR elevation measurement are expected to be precise within the error estimate (±0.15m averaged over buildings and bare soils).

Finally, bare pixels (without vegetation) were weighted 1, vegetated pixels weighted 0, and vegetated pixels for which two LiDAR returns are available (an early reflection from the top of canopy, and a late reflection from the underlying ground) were marked 0.5. This procedure allowed creating automatically (i) a mask including all points whose elevation is fully reliable within the nominal error range (Figure 4), and (ii) a three-dimensional, Level-2 land-cover of subsets weighted 0.5 and 1.

The information was completed by carrying out same bare soil classification on multispectral, very high-resolution, pre-/post-tsunami QuickBird data. In spite of the comparable pixel footprint, however, the 4-band Visible/Near-Infrared spectral content of QuickBird provided much poorer information than the airborne 56-band Hyperspectral airborne radiometer.

LiDAR data were also corrected by use of a geodic model derived from the EGM96 model. In particular, Digital Elevation Models obtained by airborne LiDAR, were associated to co-registered airborne Hyperspectral data that underwent unsupervised, Level-2 classification for automatically discriminating bare soil from vegetation.

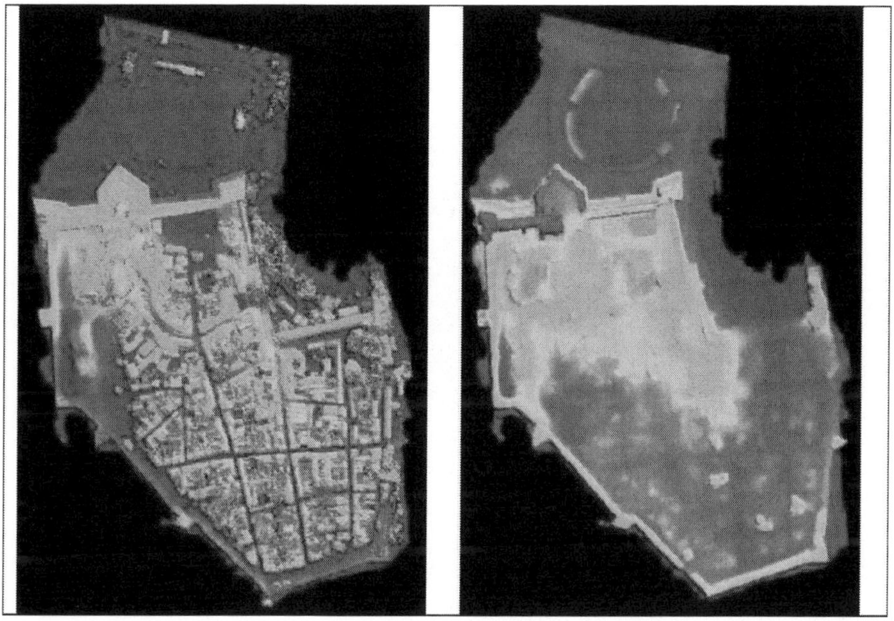

Fig. 5. LiDAR-derived Digital Surface Model (DSM, left) and Digital Ground Model (DGM, right). In the DGM, thick walls are emphasized by removal of most of buildings and vegetation. Because of such removal, DGMs are suited to standard cartography, but they are not to tsunami or storm surge flood modelling since they do not contain anymore relevant obstacles and vulnerable structures. The example relates to the 17th Century Dutch fort in Galle, southern Sri Lanka.

4.2 Spaceborne campaign

The spaceborne campaign was conducted synergetically by the Department of Electronics and Information of the Politecnico di Milano, that manufactured products in Synthetic Aperture Radar interferometry with the proprietary procedure PS-InSAR™ (Prati et al., 1994; Ferretti et al., 1999, 2001), and the University of Calabria, that manufactured multispectral and cartography products exploiting the proprietary procedure MyME2 (Hirn & Ferrucci, 2005; Ferrucci & Hirn, 2005).

The overall process relied upon same strategy as in the air campaign, with elevation data founded upon interferometric Synthetic Aperture RaDAR techniques, and pixel qualification carried out on Infra-Red multispectral satellite scenery.

Pixel qualification was based on the automated discrimination of bare soils, buildings and infrastructures from vegetation. These classes return highest confidence weight to RaDAR measured elevation values in the same pixel, whereas dense canopy returns lower or zero values. Overall, the space dataset was composed of 67 images, both RaDAR and multispectral, with resolutions ranging from metric (QuickBird) to decametric (ASTER, Landsat-7, ERS-1, ERS-2, Envisat). To fit the requirements of HyperDEM, repeat-pass interferometry was carried out to provide for two different products: Permanent Scatterers

(PS-InSAR™) data and DEM using ERS-1/ERS-2 'Tandem' pair combinations. PS-InSAR is a trademark of Politecnico di Milano.

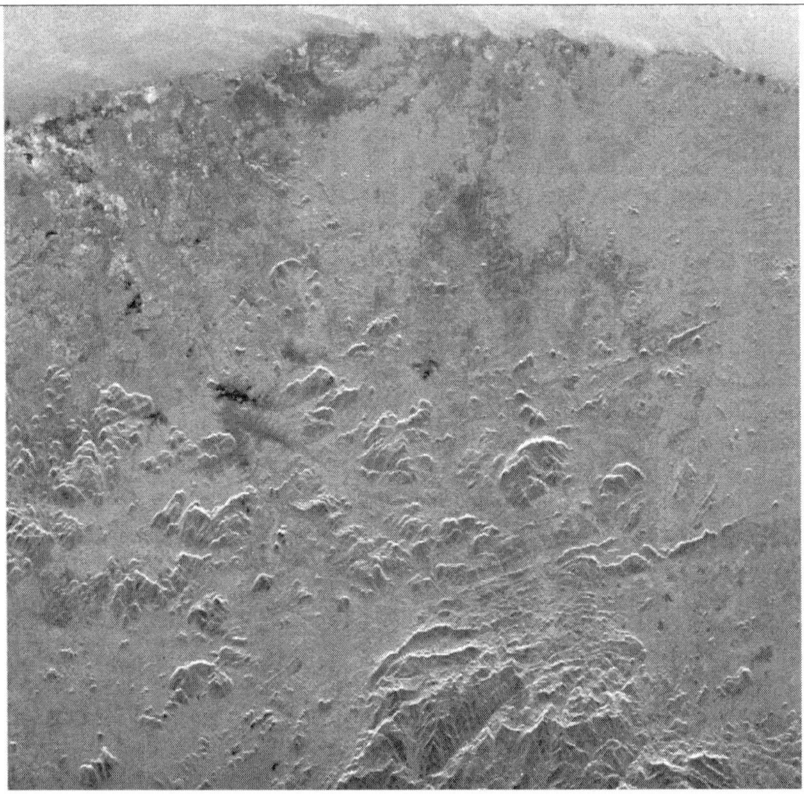

Fig. 6. View of the eastern coast (North is to the left) from radar satellites ERS (Track 33 - Frame 3465) of the European Space Agency. Because of the overall limited dataset, and the characteristics of the eastern coast areas (alternating rain forest, extremely flat terrains and frequent water bodies), the technical choice switched from PS-InSAR™ to Multi-Baseline InSAR technique.

The available SAR dataset was composed of 23 scenes along ERS/Envisat track 33 (Descending orbit), frames 3447 and 3465, acquired since 1992 at uneven rates. Accounting for the pixel dimensions (20m x 20m), and the lack of penetration of C-band RaDAR radiation across canopy, the elevation model mirrors the envelope of the Earth surface, including vegetation. In particular, the characteristics of the land cover make this area very sensitive to temporal decorrelation, that is, the loss of coherence between two successive images due to a large time interval elapsed between acquisitions. For this reasons, basic

RaDAR analysis was recentred on past ERS-1/ERS-2 tandem pairs, with 1-day intervals between over passes and 35-day repeat times.

Fig. 7. SAR Multi-image, reflectivity map of a portion of the eastern coast of Sri Lanka (North is to the left). Dark shading indicates dense vegetation cover, whereas black areas correspond to internal water bodies. Radar scattering is de-organized by foliage, and water bodies favor forward scattering instead of backscattering towards the Radar platform. Both features in combination give rise to incoherent behaviour within multi-temporal sequences.

The useable SAR dataset, composed of 42 scenes, was theoretically sufficient for carrying out thorough PS-InSAR analysis. Conversely, the characteristics of land cover - in combination with the characteristics specified above of Synthetic Aperture RaDARs onboard the ESA spacecrafts - did not reveal suitable for thorough, Permanent Scatterer analysis (Figures 6, 7).

After elimination of tandem pairs with baseline larger than 1km, only 5 pairs for frame 3465, and only 3 for frame 3447 were left. This forced moving from the PS-InSAR™ technique to the Multi-Interferogram approach (Prati et al., 1994; Ferretti et al., 1999): which is less precise, but less sensitive to the quantity of data (Figure 8).

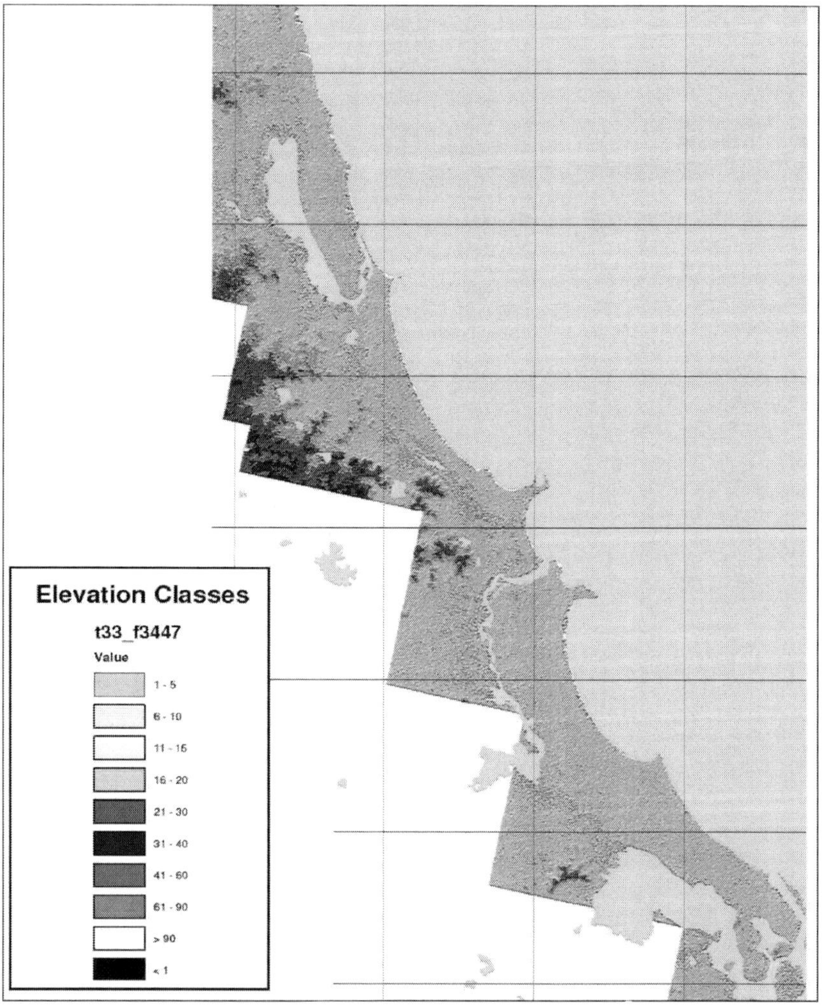

Fig. 8. Multi-interferogram DSM of eastern Sri Lanka: particular of the area of Batticaloa. Legend of elevation classes witnesses of a very flat topography that, with rich vegetation cover and spatial frequence of water bodies, leads to limited success of PS interferometry

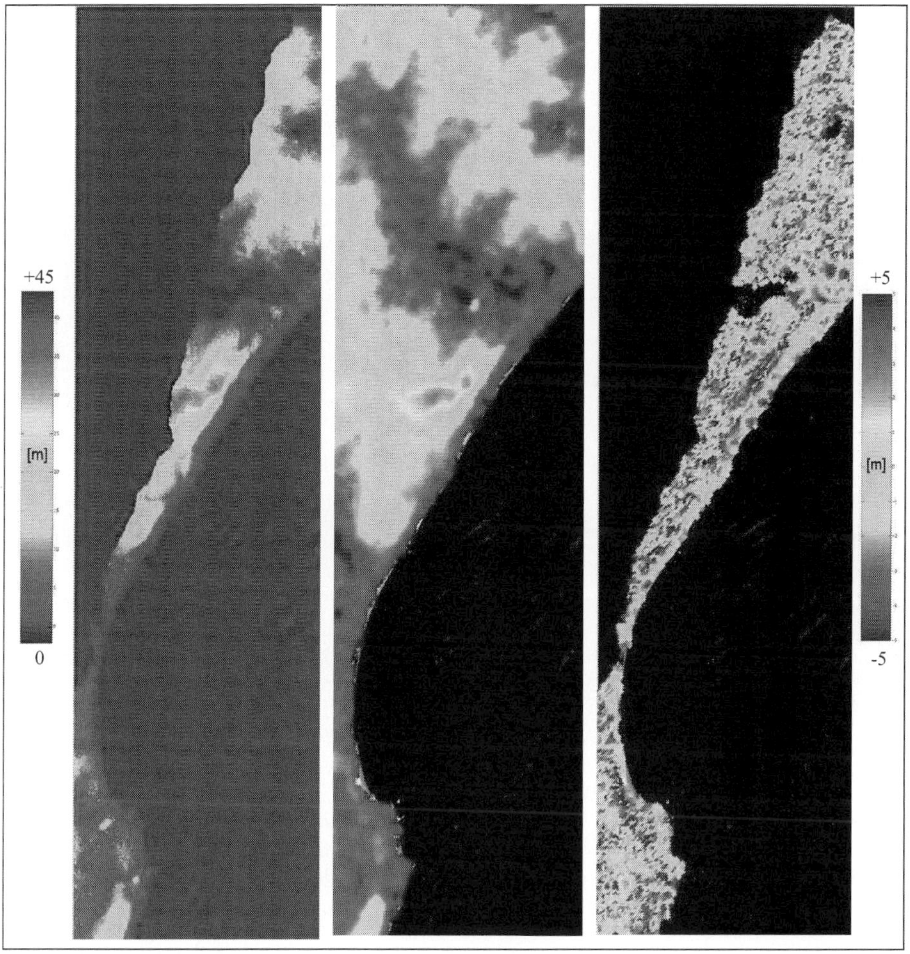

Fig. 9. DSM obtained by LiDAR (left), by Multi-Interferogram InSAR (centre) and map of the LiDAR-InSAR elevation differences (right). Colorbars shown for elevation (left) and errors (right). Minimum standard deviation of 2.56 m is a good estimate of InSAR topography accuracy in the whole area.

As for merge of the very-high resolution LiDAR, and moderate-to-high resolution other spatial data, it is worth recalling that raw data coming from the LiDAR airborne acquisition are in dual form, "first pulse" and "last pulse". First-pulses, allow mapping the reflecting envelope surface and give rise to DSM; whereas the last-pulses subset, the DGM or "ground", is constituted of rays bouncing back from the ground after crossing void spaces in the canopy.

The generic definition of "DEM" (Digital Elevation Model), applies to elevation of terrain referred to bare-Earth without vegetation and/or buildings (Figure 5-right). In order to deal with LiDAR and InSAR data at once, conversely, we had to split models into DGM ("ground") and DSM ("surface"). Definitely, DGM refers only to LiDAR, whereas DSM (Figures 5-left and 8) - that envelopes the whole of reflecting structures above the Earth's surface - refers also to RaDAR. Consequently, a comparison between different results obtained in the East - by space and RaDAR - and in the West and South - by air and LiDAR - can be performed only on Digital Surface Models.

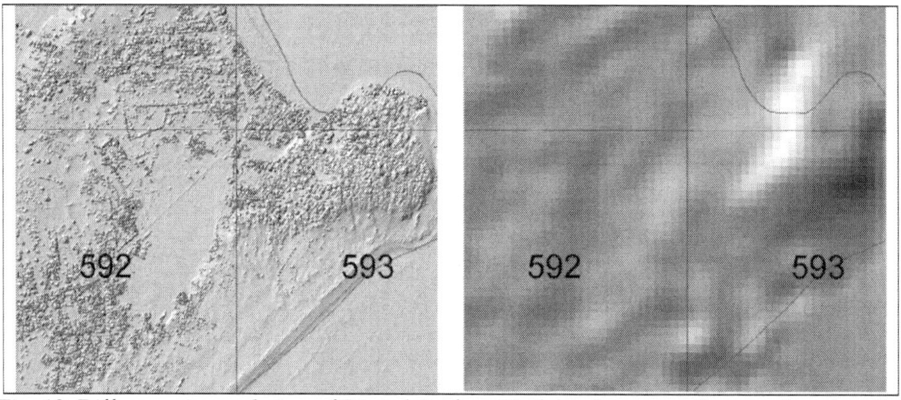

Fig. 10. Difference in resolution of Digital Surface Models of Arugam Bay, southeastern Sri Lanka, obtained by LiDAR (left) and Synthetic Aperture Radar (right). The shaded relief images show the first-pulse, 1-meter LiDAR shaded relief DSM (left) and its equivalent from the 20-meter PS-InSAR™ interferometric DEM. The shoreline of the official cartography 1/50'000 is also shown (gray line).

Comparison, done over the LiDAR-InSAR, DSM overlap area of Arugam Bay in the southeast (see Figures 9 and 10) leads to a satisfactory, least standard deviation value of 2.56 meters.

It is worth noting, however, that such comparison is carried out on products displaying a two-order of magnitude difference in ground resolution, that is, 1 square meter per pixel for LiDAR, vs. 400 square meters for InSAR.

5. Results and products

Raw data leading to the construction of precise 3-D models of the explored coastal areas of Sri Lanka, came from spaceborne Synthetic Aperture RaDAR (SAR) and airborne LiDAR surveying.

Both types of data data were post-processed, to remove errors and fill by resampling and interpolation voids arising from acquisition, and to transform clouds of points in (X, Y, Z) in a grid of X-Y evenly spaced points endowed with the inherent Z elevation fields. Information that can be extracted from LiDAR is dual: "first pulse" and "last pulse".

First pulses relate to Laser beam reflections from the external envelope of objects (canopy, roofs, electric wires, etc.), whereas last detectable pulses in a Laser beam reflection sequence can be associated to the last reflector, that is, bare Earth. Such dual LiDAR (Laser Scanner)

datasets, give rise to two, 3-D cartographic products (Figure 5): Digital Ground Model (DGM) and Digital Surface Model (DSM) , to be used alone or in combination.
DGM represents the bare-Earth elevation cleaned of vegetation and manufacts, whereas DSM represent the elevation of LIDAR first pulses, including manufacts. DGM is suitable for mapping the water penetration in vegetated areas, provided with smooth topography and little or nil 3-D manufacts.
DSM is indicated for detailed inundation mapping in urban areas since it contains 3-D footprints of manufacts, that are of utmost relevance in risk assessment if such vulnerable elements are in the reach of tsunami or storm surge generated flooding.
Conversely, spaceborne RaDAR data allow creating one product - DSM - Due to much longer wavelength ($\sim 6*10^{-2}$ metres for Radar against $\sim 1*10^{-6}$ metres for LiDAR) and pixel size (~ 500 m^2 for Radar against ~ 1 m^2 for LiDAR). However, spaceborne DSM obtained by PS-InSAR™ RaDAR interferometry, are accurate enough to approximate realistically the terrain in areas with sparse or nil vegetation.
Post-processed LiDAR products keep a horizontal resolution of 1 meter, displaying an accuracy in elevation (pseudo-vertical) better than 30 cm.
The interferometric RaDAR product presents a horizontal resolution of 20 meters, an average vertical precision of ±3 meters, with a resolution in elevation better than 1cm in multi-temporal, differential mode only.
Accounting for the huge data volume, the process of map generation required developing an automated procedure to process the dataset, preserve the surface information, and minimize time consumption. The nearest neighbour interpolation method was used to generate DGMs and DSMs from raw data.
This kind of interpolation method has the property of not extrapolating above or below actual data values coming from input. This has appeared essential, because of (a) the very close spacing of input data points, and (b) the fact that other potential methods (e.g. polynomial functions or kriging) may substantially modify the representation of some terrain attributes like buildings or tree canopy.
LiDAR models were arranged in tiles of 1000 x 1000 x 1 metre (excepting those along the shoreline), for as much as 4600 billion grid points measured in elevation (Baltsavias, 1999; Kraus & Pfeifer, 1998; Axelsson, 2000).
Spaceborne RaDAR Digital Surface Models were arranged in two frames of 1811x3497 and 894x3202 (columns x rows) respectively, with 20-metre spacing of points, allowing for total 2.4 million grid points measured in elevation.

6. Simulations

According to the inter-Government agreement referred to above, the Disaster Management Centre in Colombo was provided also with a few inundation examples (Figures 11 and 12), aimed to demonstrate the procedures for tsunami and/or storm surge scenario building - whose responsibility and exploitation rights stay with the national Authority.

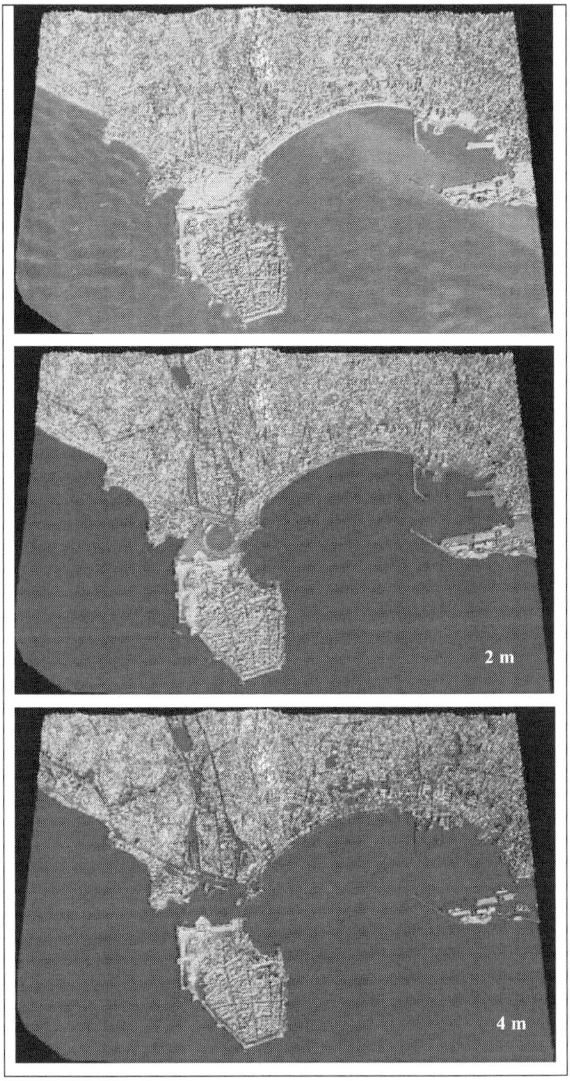

Fig. 11. Shaded relief, Digital Surface Model of the town of Galle, in the Southwest, with static demonstration of flooding simulation by a 2-meter (center) and a 4-meter (bottom) model surge. The synthetic urban flooding scenario shown here, is satisfactorily consistent with field evidence observed at Galle in the aftermath of the event of December 26, 2004. The full wave cut the 17th Century Dutch Fort (bottom) off from mainland, but did not hit the internal streets. This simulation demonstrates that areas allowing safe escape from a 4-meter tsunami, storm surge or tidal wave, can be found and better reached close to the shoreline instead than inland.

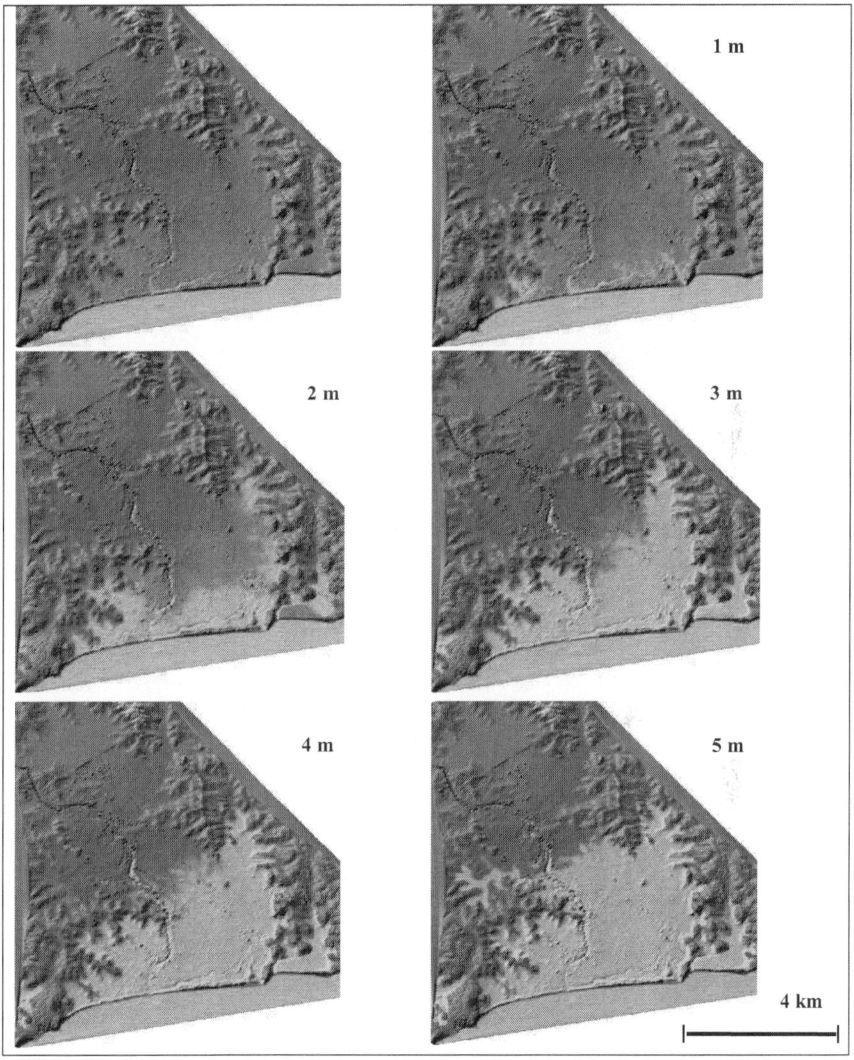

Fig. 12. Shaded relief, Digital Ground Model in the area of Hambantota, southeastern coast, with static flooding simulation by a model surge corresponding to a ca. 5-meter tsunami wave estimated by eye-witnesses and later field investigations. The combination of very low topographic gradient, and presence of lagoons, rivers and ponds close to the shoreline, led to significant inland impact of the incoming wavetrain (over 4 km).

In lack of specific works in scientific literature - that is more focused on the propagation at sea and the impact effects on the sea-shore interface - the demonstrations were carried out following a simplified static approach, consisting in the consecutive piling-up of 1-meter thick layers. Since the procedure does not account for the significant, dynamic component of tsunami wavetrains, it is probably better suited to storm surges or tidal waves. The sites chosen for demontration are the urban area of Galle - in the southwest - and the area of Hambantota, in the southeast.

Galle (Figure 11) and the surrounding coast were the site of major damage and casualty (more than 4'000), engendered by a relatively small wave height reported in the order of 4 meters. In Hambantota, (Figure 12) the impacting tsunami wave was reportedly steeper, higher (5-6 meters, according to eye-witnesses) and penetrating the coast by a few kilometers. Such significant water ingression is explained by the combination of little topographic gradient over ca. 4-5 kilometers, and presence of large lagoons (at the foot of Hambantota itself), ponds, and rivers. As in Galle with a 4-meter wave, in Hambantota the simulation with a 5-meter wave satisfactorily fits the observed extent of flooding.

7. Conclusions

In the broad aftermath of the Sri Lanka tsunami disaster, the stack of synoptic procedures and remote sensing techniques chosen for satisfying the urgent needs of the User, presented the undebated advantage of : (a) allowing to start the work immediately, (b) without relying upon ground logistics until the onset of the air campaign, (c) minimizing the duration of the work on spot, (d) covering fast - and at an otherwise unreacheable resolution - large portions of a difficult-to-penetrate territory, (e) keeping the work sustainable and, overall, (f) allowing to carry out the work. This combination of airborne and spaceborne techniques was, and is ready-to-use worldwide, and the techniques for flooding simulation and scenario building, can be chosen at whatever level of complexity - choosing preferably robustness.

It is also worth noting further that the new generation of metric resolution, X-band Radar satellite constellations (as TerraSAR-X and Cosmo-SkyMED), may allow creating LiDAR-like products avoiding the air work on spot. Conversely, much is missing on the standpoint of Infra-Red observation, that has currently become poor and poorly resoluted in terms of SWIR spectral resolution (necessary for vegetation and bare soils applications). Whatever the choice of the platform, however, the technique combination holds valid and robust for further applications. In conclusion, the HyperDEM products were handed over by the Ambassador of Italy in Sri Lanka to the Minister for Disaster Management and Humanitarian Affairs, on 7th December 2006, in Colombo, Sri Lanka.

Acknowledgements

The project HyperDEM on tsunami preparedness in Sri Lanka was funded by the Italian Ministry of Foreign Affairs, Directorate General for Cooperation, and promoted by the Sri Lanka Government, Ministry of Disaster Management and Humanitarian Affairs, and Ministry of Science and Technology. Its carrying out overseas involved more than 20 scientists, engineers and technicians, guided by Franco Coren, Giuliano Savio, Barbara Hirn, Michela Vellico, Paolo Sterzai and Gianluca Calabretta, under the coordination of the author. Field effort and post-processing of a ca. 3-TeraByte data gather benefited of the support of three technological partners, owning and exploiting 7 international patents involved in the whole process (HeliOGS Consortium in airborne acquisition and LiDAR; IES Consulting - Intelligence for Environment and Security, in hyperspectral and multispectral data processing; TRE-Telerilevamento Europa in RaDAR interferometry and PS-InSAR™).

Pre/post-tsunami, very high-resolution multispectral QuickBird scenes were granted by the Pacific Disaster Center of Maui, Hawaii. The whole work was supported by the Sri Lankan Air Force, the Disaster Management Centre, Sri Lankan Airlines and the pilots of Air Taxi, and the Italian Embassy in Colombo. Special acknowledgements are due to Amb. Salvatore Zotta and Maj. Gen. Gamini Hettiarachchi, who allowed solving any criticality incurred when designing, launching and carrying out HyperDEM.

8. References

Axelsson, P. (2000), DEM Generation from Laser Scanner data Using Adaptive TIN Models, International Archives of Photogrammetry and Remote Sensing, Vol. XXXIII, B4. Amsterdam.

Baltsavias, E.P. (1999), Airborne laser scanning: basic relations and formulas. ISPRS Jour. Photogramm. Rem. Sens., 54, 199-214.

Colesanti C., Ferretti A., Ferrucci F., Prati C., and F. Rocca (2000). Monitoring Known Seismic Faults Using the Permanent Scatterers (PS) Technique, Procs. IEEE International Geoscience and Remote Sensing Symposium - IGARSS 2000 (Honolulu, USA, 24-28 July), 5, 2221-2223.

Eisner R.K., Borrero J.C., and C.E. Synolakis (2001). Inundation maps for the State of California. Procs. ITS-International Tsunami Symposium 2001 and NTHMP Review Session (Seattle, USA, 7-10 August); R-4, 67-81.

Ferretti, A., C. Prati, and F. Rocca (1999), Multibaseline InSAR DEM Reconstruction: the Wavelet Approach, IEEE Trans. Geosci. Rem. Sens., 37, 2; 705-715.

Ferretti, A., C. Prati, and F. Rocca (2001), Permanent scatterers in SAR interferometry, IEEE Trans. Geosci. Rem. Sens., 39, 1; 1528-1530.

Ferrucci, F., and B. Hirn (2005), An automated method for detecting and mapping, particularly for areas without vegetation; International Patent, PCT WO2005/005926A1, published: January 26, 2005.

Geist, E.L., V.V. Titov, D. Arcas, F.F. Pollitz, and S.L. Bilek (2007): Implications of the 26 December 2004 Sumatra-Andaman Earthquake on tsunami forecast and assessment models for great subduction-zone earthquakes. Bull. Seism. Soc. of Am., 97, 1A; 249-270.

Gregg, C.E.; Houghton, B.F.; Paton, D; Johnston, D.M., Swanson, D., and B. A. Yanagi,(2007). Tsunami Warnings: Understanding in Hawaii. Nat. Hazards 40:71-87.

Hammack, J. L. (1973), A Note on Tsunamis: Their Generation and Propagation in an Ocean of Uniform Depth, J. Fluid Mech., 60, 769-799.

Hirn, B., and F. Ferrucci (2005). MYME2: a Multi-Payload Integrated Procedure for the Automated, High-Resolution Remote Sensing of Burn Scars. Procs. IEEE International Geoscience and Remote Sensing Symposium - IGARSS 2005, (Seoul, South Korea, 25-29 July) paper no. 20238; 5721-5724.

Kraus, K., and N. Pfeifer (1998), Determination of Terrain Models in wooded areas with Airborne laser Scanner data, ISPRS Jour. Photogramm. Rem. Sens., 53, 4: 193-203.

Lander, J.F., P.A. Lockridge, and M.J. Kozuch (1993). West coast of the United States 1806-1992. NGDC Key to Geophysical Record Documentation 29, NOAA, 242 pp.

Leblond, P. H. and Mysak, L. A. (1978) Waves in the Ocean (Elsevier Scientific Publishing Company.

Liu P.L.F., Lynett P., Fernando H., Jaffe B.E., Fritz H., Higman B., Morton R., Goff J., and C Synolakis (2005). Observations by the International Tsunami Survey Team in Sri Lanka. Science; 308, 5728, p. 1595. DOI: 10.1126/science.1110730

Monti Guarnieri, and A. and Ferretti (2000). Visibility of Permanent Scatters by ScanSAR. Procs. EUSAR 2000 (Munich, Germany, May 23-25), 725-728.

Monti Guarnieri, and Y-L. Desnos (1999), Optimizing performances of the ENVISAT ASAR ScanSAR modes. Procs. IEEE International Geoscience and Remote Sensing Symposium - IGARSS 1999 (Hamburg, Germany, June 28-July 2), 1758-1760.

Oppenheimer, D., Beroza, G., Carver, G., Dengler, L., Eaton, L., Gee, L., González, F., Jayko, A., Li, W. H., Lisowski, M., Magee, M., Marshall, G., Murray, M., McPherson, R., Romanowicz, B., Satake, K., Simpson, R., Somerville, P., Stein, R. and Valentine, D. (1993), The Cape Mendocino, California, Earthquakes of April 1992: Subduction at the Triple Junction, Science, 261, 433-438.

Prati, C., F. Rocca, A. Monti Guarnieri (1994), Topographic capabilities of SAR exemplified with ERS-1. Geo Information System, 7, 1; 17-23.

Priest G.R., Baptista A.M., Myers E.P.III, and R.A. Kampahaus (2001). Tsunami hazard assessment in Oregon. Procs. ITS-International Tsunami Symposium 2001 and NTHMP Review Session (Seattle, USA, 7-10 August) ; R-3, 55-65.

Satake, K., Bourgeois, J., Abe, K., Abe, K., Tsuji, Y., Imamura, F., Iio, Y., Katao, H., Noguera, E. and Estrada, F. (1993), Tsunami Field Survey of the 1992 Nicaragua Earthquake, Eos, Trans., Am. Geophys. Union, pp. 74, 145 and 156-157.

Sibuet, J-C., Rangin, C., Le Pichon, X., Singh, S., Cattaneo, A., Graindorge, D., Klingelhoefer, F., Lin, J-Y., Malod, J., Maury, T., Schneider, J-L., Sultan, N., Umber, M., Yamaguchi, H., and the "Sumatra aftershocks" team (2007). 26th December 2004 great Sumatra-Andaman earthquake: Co-seismic and Post-seismic motions in northern Sumatra; Earth Plan. Sci. Lett., 263, 1-2, 88-103.

Tinti, S., A. Armigliato, A. Manucci, G. Pagnoni, and F. Zaniboni (2005), Landslides and tsunamis of 30th December 2002 at Stromboli, Italy: numerical simulations, Boll. Geofis. Teor. Appl., 46, 153-168.

Tinti, S., A. Armigliato, A. Manucci, G. Pagnoni, F. Zaniboni, A.C. Yalginer, and Y. Altinok (2006), The generating mechanism of the August 17, 1999 Izmit Bay (Turkey) tsunami: Regional (tectonic) and local (mass instabilities) causes; Marine Geol., 225, 311-330.

Titov, V.V., and C.E. Synolakis (1998). Numerical modeling of tidal wave runup. J. Waterw. Port Coast. Ocean Eng., 124, 4; 157-171.

Venturato, A.J., D. Arcas, V.V. Titov, H.O. Mofjeld, C.C. Chamberlin, and F.I. Gonzalez (2007): Tacoma, Washington, tsunami hazard mapping project: Modeling tsunami inundation from Tacoma and Seattle fault earthquakes. NOAA Tech. Memo. OAR PMEL-132, 23pp.

Weiss R., Wünnemann K., and H. Bahlburg (2006). Numerical modelling of generation, propagation and run-up of tsunamis caused by oceanic impacts: model strategy and technical solutions. Geophys. Jour. Int.; 167, 1; 77-88.

Whitmore, P.M. (1993), Expected Tsunami Amplitudes and Currents along the North American Coast for Cascadia Subduction Zone Earthquakes, Nat. Hazards, 8, 59-73.

Wiegel, R. L. (1976). Tsunamis. In: Seismic Risk and Engineering Decisions; C. Lomnitz and E. Rosenblueth, eds.; Elsevier Scientific Publishing Co., Amsterdam (NL) 225-286.

20

3D Measurement of Speed and Direction of Turbulent Air Movement

Shirokov Igor and Gimpilevich Yuri
Sevastopol National Technical University
Ukraine

1. Introduction

Measurement of air streams movement, particularly speed and direction, always has been a subject of steadfast scientific investigations in all areas of human life and activity. It is especially important to supervise moving of turbulent air when the researches on microwave propagation are carried out. Only when we have full representation in behaviour of the turbulent air and synchronous measured parameters of an electromagnetic wave it is possible to determine the laws of influence of turbulent air moving on parameters of an electromagnetic field (Shirokov et al., 2003). On the other hand it is possible to solve reverse task – to control meteorological environment with direct measurements of propagated microwave parameters (Shirokov, 2007).

Investigations in a field of turbulent air movement are not limited by the meteorological one or by the researches in microwave propagation. Local measurements of air movements are especially useful in industry where the bodies of various mechanisms design. In a last case the great attention is paid to aero-dynamic characteristics of mechanisms bodies, taking into consideration possible mechanisms move in different gases or liquids.

Widely used in meteorological supervision mechanical anemometers and instruments for measurement of a wind speed and direction are essentially unsuitable when the investigations of microwave propagation are carried out. Owing to its inertia, these devices allow to get only integrated values of measured magnitudes (Kremlevsky, 1989). At the same time, there is certain interest to supervise the air turbulence which some times can change the value during carrying out of measurements with mechanical devices.

The dynamic range and accuracy of mechanical devices are low. Measurements can be implemented only in a plane, at the best case.

In the mentioned above industry applications the mechanical instruments for supervising the turbulent air movement are quite unsuitable.

Other ways of measurements (radar, optical) are unsuitable for local measurements, as they demand the extended distances (Nakatani et al., 1980)

In this paper the acoustic method of measurement of speed and direction of turbulent air movement is discussed (Bobrovnikov, 1985) and (Waller, 1980). The working algorithm and the block diagram of a measuring instrument are described. The spectrum analysis of signals and their contribution to the general error of described measuring system is discussed.

2. Approach to a Problem

For a possibility of measurement of a direction and speed of a turbulent air movement in three-dimensional space, are necessary, at least, three independent measuring channels located upon orthogonal coordinates. Thus each of them will measure scalar value of a projection of moving air speed. Accordingly, the direction of moving and value of speed of a stream can be obtained, due to the processing of signals simultaneously in all channels of measuring equipment.

The principle of operating of a similar measuring instrument is described in (Shirokov et al. 2006) and (Shirokov et al., 2007).

The measuring instrument consists of two modules: the sensor unit, which contains of ultrasonic transmitter transducer TXT and three ultrasonic receiver transducers RXTi and the processing block which carries out the handling of signals from the sensor unit. It will consist of three identical mutually perpendicular measuring channels realizing measurement of components V_X, V_Y and V_Z of air stream speed vector V, as shown in Figure 1.

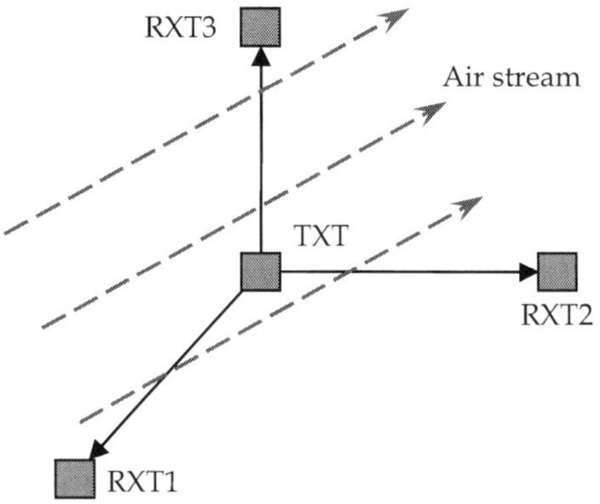

Fig. 1. Transducers separation of measuring device

The measured values of components V_i pass to the processing block which carries out the calculating of speed of an air stream, and also value of corresponding corners.

The major requirement to the sensor unit: it must insert the minimal distortions to the structure of an air stream, speed and direction of which is measured. For maintenance of performance of this requirement sensors should have minimal aperture; radiating and receiving elements must have whenever possible small dimensions.

Let's consider a principle of operation of one of the measuring instrument channels. The processing block forms a harmonious signal of a kind:

$$s_T(t) = A_0 \cos(\omega_0 t + \varphi_0). \tag{1}$$

This signal is radiated by an ultrasonic radiator in a direction of this channel receiver. When the component of the wind directed along an axis of ultrasonic signal propagation of the considered channel is absent the signal on an output of the receive ultrasonic converter will be:

$$s_R(t) = A_0 \cdot K(t) \cdot \cos(\omega_0 t + \varphi_0 + \Delta\varphi). \tag{2}$$

The amplitude factor $K(t)$ we will not take into consideration because the only argument of equation (2) is of interest for our measurements. We can eliminate the influence of $K(t)$ by the deep limiting of received signal. Further we will assume this factor is equal to K.
The phase progression $\Delta\varphi$ of a signal $s_{TR}(t)$ at its propagation from transmitting to receiving transducers will be determined as:

$$\Delta\varphi = \frac{2\pi \cdot f_0}{c} \cdot l, \tag{3}$$

where f_0 is the frequency of a signal; c is the speed of a sound in the environment (air); d is the distance between the transmitting and the receiving ultrasonic transducers.

When the component of the wind directed along an axis of propagation of an ultrasonic signal of the considered channel is present, the signal on an output of the receiving converter of the considered channel will be:

$$s_R(t) = A_0 K \cos(\omega_0 t + \varphi_0 + \Delta\varphi + \varphi_W), \tag{4}$$

where φW is the value of the component caused by the moving of air, as environment carrier of sound.
Additional phase shift φ_W will be determined as:

$$\varphi_W = \pm \frac{2\pi \cdot f_0 \cdot l \cdot v_\omega}{(c \pm v_\omega) \cdot c}, \tag{5}$$

where v_ω is the value of the component of the wind directed along an axis of propagation of an ultrasonic signal of the considered channel.
Value φ_W can be both positive and negative, as the component of speed of wind can be directed as along, as contrary in relation to a direction of propagation of an ultrasonic signal.
If the speed of the moving of air is negligible, comparing with the speed of sound, this formula can be rewritten:

$$\varphi_W \approx \pm \frac{2\pi \cdot f_0 \cdot l \cdot v_\omega}{c^2}. \tag{6}$$

When we carry out the analysis of (6) we can find the resolution of phase measurements will be the higher the distance l will be the longer. So, for frequency of ultrasonic 40 kHz and for measurement of moving air speed in 0,01 m/c with phase resolution in 1°, we must set distance l equal to 1 m. For the meteorological measurements we have taken into account

that real wind speed can exceeds 30 m/s. When speed of moving air will reach this value the additional difference of phases will reach the value about 4000°. In (Shirokov et al. 2006) there was presented an algorithm of processing such values of phase difference, where the number of phase cycles was counted. This approach to the problem will be discussed later.
This approach assumes the measuring of not only phase difference between two signals, which itself possible only at orthodox measurements of phase difference, when the frequencies of signals are strictly equal and phase difference can change from 0 up to 360°, but also it assumes the measurements of cumulative phase of signal, where the number of phase cycles is counted. In this case we will measure the difference of total phases of two signals. Taking into account such approach, there is an opportunity to carry out the phase measurements, when the frequency of one of two signals changes in some range. There is nothing non ordinary in this approach, if we will remember that eigenfrequency of any oscillations is the derivation of phase of ones:

$$\omega(t) = \frac{d\Psi(t)}{dt} = \omega_0 + \frac{d\varphi(t)}{dt}. \qquad (7)$$

If the phase progression of ultrasonic signal increases or decreases continuously for a certain time interval the frequency of received signal will change adequately at that interval. The solving of task with this manner assumes the assignment of the certain requirements on stability of frequency and phase of all signals.
The frequency stability of mentioned above signals determines the accuracy of measurements. Because there is no problem to realize all of signals with frequency stability at several parts per million (ppm), and taking into account that real measured data are of interest in 3-4 decimal digits, we can claim: there is no error determined with frequency stability. The only thing we must do is to use the crystal clock.
All of mentioned reasoning will be valid if the length of acoustic link not exceed 3000 acoustic wave length with frequency stability we have assumed. In other words the changing of acoustic wave phase progression kd ($k = \frac{2\pi \cdot f}{c}$ is the acoustic wavelength constant, d is the link length) because of frequency instability must not exceed 1°. Taking into consideration the length of acoustic wave is near 8 mm, the maximum length of acoustic link will be 25 m for the error of phase measurements in 1°. Really, for local air turbulence movement measurements we assume the link length to be less than 1 m. So in this case the error of phase measurements will be less than 0.04° for frequency instability in 1 ppm we had assumed.
For the improving of the resolution of measurements of low-level moving air speeds we must increase the resolution of phase measurements up to 0.1° or even better. For the frequency of ultrasonic oscillations in 40 kHz it seems some problematic to implement the measuring process, because the clock frequency must be equal to 144 MHz or even more in this case. In (Shirokov et al. 2006) it was proposed to transform this frequency with traditional heterodyne manner up to 4 kHz. For the increasing of resolution of measurements in (Shirokov et al., 2007) it's proposed to transform the initial frequency up to 400 Hz. It is suggested to form the frequency of heterodyne signal shifted on 1% with respect to frequency of acoustic wave signal (result frequency of heterodyne signal will be

40.4 kHz or 39.6 kHz), so that the frequency of mixer's output signal will be 400 Hz. Therefore, the reference signal frequency must be equal to 400 Hz too.
With discussed measurement approach, the phase difference between all of mentioned above signals must be strictly constant. In other words all of these signals must be derived from single oscillator.

3. Some Aspects of Realization of Homodyne Frequency Converter

Because we are tending to carry out the phase measurements, the heterodyne signal must be obtained from initial signal with homodyne method (Gimpilevich & Shirokov, 2006). Such approach can be realized with using of phase shifter. The changing of phase of any signal on 2π over the period of the control signal T is tantamount to the frequency shift of the initial signal on the value $\Omega=2\pi/T$, according to the well known expression (7). The initial phase of frequency transformed signal will be the same as initial phase of origin signal plus initial phase of control signal. This fact lets us to carry out the phase measurements without any phase errors caused by the using of different oscillators with different derivation of frequencies.
The practical realization of phase shifters, which realises the linear rule of phase changing, is a complex problem. In (Jaffe & Mackey, 1965). and (Shirokov et al., 1989) it was shown, that for investigations of phase characteristics of objects, the discrete phase shifters with number of steps higher than 2 can be used. Discrete phase shifters have very stable repetition parameters, and there is the possibility of realization of any rule of phase changing. The basic question, which appears on design of this device is how much of steps must be in phase shifter (Shirokov & Polivkin, 2004).
If discrete phase shifter is used in homodyne measuring system, the higher harmonicas of main frequency (1) will appear on mixer output. Let's carry out the spectrum analysis and estimate the harmonic factor of this signal by using of different number of steps of phase shifter. We will define the level of first harmonic of signal, which approximates the sinusoid oscillation by the 3, 4, 5, 8 and 16 steps.
As it's well known, any periodic signal $s(t)$ can be written as:

$$s(t) = \frac{a_0}{2} + \sum_{n=1}^{\infty} A_n \cos(n\omega_1 t + \Theta_n), \tag{8}$$

where $a_0/2$ is the constant component of signal, n is the number of harmonica of signal, A_n is the amplitude of harmonicas of signal, ω_1 is the frequency of the first harmonica of signal.
In general case, the amplitudes of harmonicas are defined by Fourier transformation of signal. Let's write this transformation for odd function as:

$$A_n = |b_n| = \frac{4}{T} \cdot \int_0^{T/2} s(t) \cdot \sin(n\omega_1 t) dt. \tag{9}$$

It is significant, that in our case $s(t)$ is the stepping approximation of sinusoid function. By the increasing of number of steps, the approximation step function will be approach to the harmonic sinusoid function.

The approximate signals for m=3, 4, 5, 8 and 16 of steps of approximation is shown in Figure 2. The calculation of levels of step we can define by:

$$K_{i,m} = \sin\left[\frac{2 \cdot \pi}{m} \cdot (i-1)\right], \tag{10}$$

where $K_{i,m}$ is the ith sample of signal, that represents the step of approximation of sinusoid oscillation, $i \in [1...m]$.

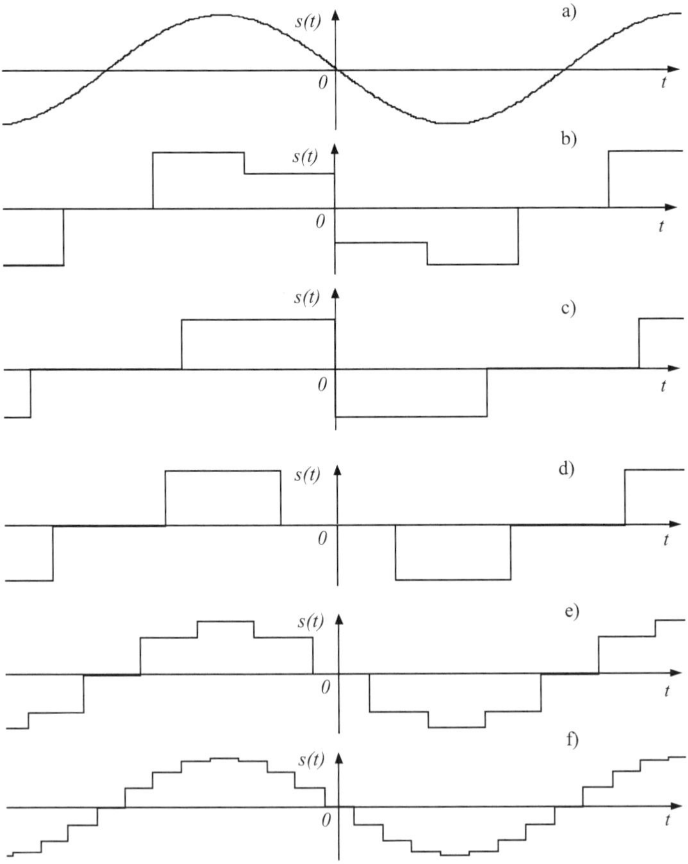

Fig. 2. Stepping signals approximating the sinusoidal function (a) at the different number of steps: 5 (b), 3 (c), 4 (d), 8 (e), 16 (f)

The stepping signal can be represented as:

$$s(t)=\begin{cases} K_{1,m},\ by\ \dfrac{-T}{2}\le t<\dfrac{-T}{2}+\dfrac{T}{2\cdot m} \\ K_{i,m},\ by\ \begin{cases}\dfrac{-T}{2}+\dfrac{T}{2\cdot m}+\dfrac{T\cdot(i-2)}{m}\le t \\ t<\dfrac{-T}{2}+\dfrac{T}{2\cdot m}+\dfrac{T\cdot(i-1)}{m}\end{cases},\ i\in[2....m] \\ 0\ ,\ by\ \dfrac{T}{2}-\dfrac{T}{2\cdot m}\le t<\dfrac{T}{2}\end{cases} \quad (11)$$

or, after transformation:

$$s(t)=\begin{cases} K_{1,m},\ by\ \dfrac{-T}{2}\le t<-\dfrac{T(m-1)}{2m}; \\ K_{i,m},\ by\begin{cases}-\dfrac{T}{2}\cdot\left(\dfrac{m-2\cdot i+3}{m}\right)\le t \\ t<\dfrac{-T}{2}\cdot\left(\dfrac{m-2\cdot i+1}{m}\right)\end{cases},\ i\in[2....m] \\ 0\ ,\ by\ \dfrac{T(m-1)}{2m}\cdot\le t<\dfrac{T}{2}.\end{cases} \quad (12)$$

Equation (12) represents the step signal, which approximates the sinusoid at different number of steps m.

For substitution $s(t)$ in (9) it is enough to assign it on a part of period $t\in\left(0...{}^{T}\!/_{2}\right]$. For this transformation:

$$s(t)=\begin{cases} K_{i+\frac{m+1}{2},m},\ by\ \dfrac{T\cdot(i-1)}{m}\le t<\dfrac{T\cdot i}{m},\ i\in[1...\dfrac{m-1}{2}]; \\ 0,\ by\ \dfrac{T(m-1)}{2m}\cdot\le t<\dfrac{T}{2}.\end{cases} \quad (13)$$

$$s(t)=\begin{cases} K_{\frac{m}{2}+1,m}\ \ by\ 0\le t<\dfrac{T}{2m}; \\ K_{\frac{m}{2}+i,m}\ by\ \dfrac{T\cdot(2\cdot i-3)}{2\cdot m}\le t<\dfrac{T\cdot(2\cdot i-1)}{2\cdot m},i\in[2...\dfrac{m}{2}]; \\ 0,\ by\ \dfrac{T(m-1)}{2\cdot m}\cdot\le t<\dfrac{T}{2}.\end{cases} \quad (14)$$

Equation (13) describes the sampling signal at odd number of steps, (14) – at even number of steps.

Let's put (13) and (14) into (11) and define the amplitudes of spectrum components of signal:

$$A_n = \frac{4}{n\omega_1 T}\left| K_{1+\frac{m+1}{2},m} \cdot \cos(n\omega_1 t)\Big|_0^{\frac{T}{m}} + K_{2+\frac{m+1}{2},m} \cdot \cos(n\omega_1 t)\Big|_{\frac{T}{m}}^{\frac{T\cdot 2}{m}} + ... + K_{m,m} \cdot \sin(n\omega_1 t)\Big|_{T\cdot(m-3)/2\cdot m}^{T\cdot(m-1)/2\cdot m} \right., \quad (15)$$

$$A_n = \frac{4}{n\omega_1 T}\left| K_{\frac{m}{2}+2,m} \cdot \cos(n\omega_1 t)\Big|_{\frac{T}{2\cdot m}}^{\frac{3\cdot T}{2\cdot m}} + K_{\frac{m}{2}+3,m} \cdot \cos(n\omega_1 t)\Big|_{\frac{3\cdot T}{2\cdot m}}^{\frac{5\cdot T}{2\cdot m}} + ... + K_{m,m} \cdot \sin(n\omega_1 t)\Big|_{T\cdot(m-3)/2\cdot m}^{T\cdot(m-1)/2\cdot m} \right.. \quad (16)$$

Expressions (15) and (16) allow us to determine the amplitudes of spectrum components of signal at odd and even number of steps. Results of calculations are summarized in Table 1.

Number of steps	Level of harmonicas							
	1	2	3	4	5	6	7	8
3	0,82	0,41	0	0,21	0,16	0	0,12	0,1
4	0,90	0	0,30	0	0,18	0	0,13	0
5	0,93	0	0	0,23	0	0,16	0	0
8	0,98	0	0	0	0	0	0,14	0
16	0,99	0	0	0	0	0	0	0

Table 1. Level of harmonicas of approximating signal

From table 1, we can see that if number of steps is 4 or higher, the level of first harmonic is more than 90 percent from theoretically possible. At increasing of number of steps from 3 to 4 the growth of level of first harmonica reaches 8 percent. The increasing of number of steps to 5 results in the growth of level in 3 percent and additional 5 percent at the increasing of number of steps from 5 to 8. When the number of steps increases from 8 to 16, the growth of level reaches 1 percent only.

The number of steps, obviously, must satisfy to the binary law. Such approach simplifies the controlling unit and one lets to reduce the number of phase shifter cells. The cells must be weighted according the binary law in this case. Consequently, more optimal is the use of 4 or 8 of steps of phase shifter for using in homodyne measuring systems. If critical condition is the simplicity of control unit at normal quality, it's recommended to use 4-step phase shifter. If critical condition is the quality of signal, it's recommended to use 8-step phase shifter. The application of 16 and more steps of phase shifter complicates the control unit, but it not gives considerable advantages and it is unjustified.

From table 1 one more law is traced. Besides the basic harmonica, the nearest harmonious component with an essential level, has a serial number m - 1, where m is the number of steps. This fact allows us to determine unequivocally requirements to filtering parts of measuring equipment. And with the increasing of number of steps, the filter cut-off frequency increases adequately in relation to the frequency of the basic harmonica.

As mentioned above, the number of steps must satisfy to the binary law. The ultrasonic frequency f_0 in 40 kHz is relative low frequency from the point of view of operating of modern integrated circuits and discrete semiconductors. Thus, there are no any technical restrictions to increase the number of steps of phase shifter. Obviously it's recommended to use the reasonably maximal number of steps. Those steps would be 8, what corresponds to using of 3 cells of phase shifter in 180°, 90° and 45° respectively. The step of phase shift will be 45°. The ultrasonic signal phase shift sequence must be 0°, 45°, 90°, 135° etc. or 0°, 315°, 270°, 225° etc. The changing of phase of ultrasonic signal on 2π over the period of the control signal with lowest frequency F in 400 Hz (for 180° phase shifter cell) is tantamount to the frequency shift of the initial signal frequency f_0 on the value $F = 400$ Hz. So, the first law of phase changing results in forming of transformed signal with frequency $f_0 - F = 39.6$ kHz, the second law – $f_0 + F = 40.4$ kHz.

4. Technical Solutions

The main problem of measuring device design is the implementation of phase shifter. There is no need to implement the phase shifter separately, but we can form all of needed signals by means of unit, the block-diagram of which is shown in Figure 3.

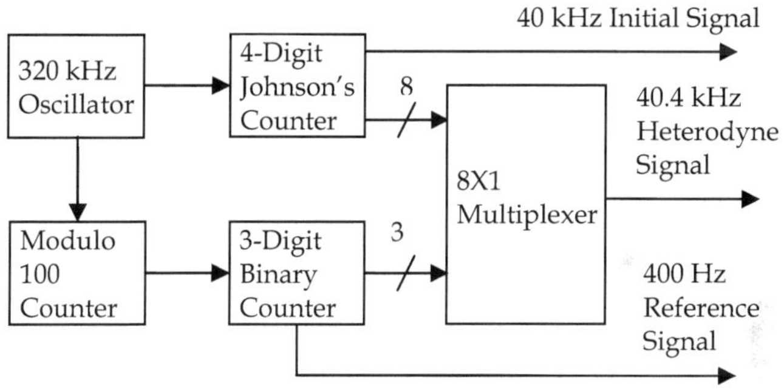

Fig. 3. Block-diagram of signal forming unit

All of signals are synchronized with the single 320 kHz Oscillator. The oscillator realization is not on principle. The use of the crystal inexpensive 8 MHz oscillator with modulo 25 counter is the best solution of the problem.

The 4-Digit Johnson's Counter forms multiphase clock. The frequency of each clock is 40 kHz, number of clocks is 8 and phase difference between neighbour sequences is 45°. This multiphase clock or outputs of Johnson's Counter are connected with 8 inputs of Multiplexer. One of these clocks represents 40 kHz Initial Signal, which feeds ultrasonic transmitting transducer.

Transmitting and receiving air ultrasonic transducers for these frequencies are well supported, for example electronic parts EC4010-EC4018, Sencera Co. Ltd.

The Modulo 100 Counter in conjunction with 3-Digit Binary Counter form 400 Hz Reference Signal and three meanders with 1.6 kHz, 800 Hz and 400 Hz frequencies. These meanders

control the address inputs of Multiplexer. The meander with 400 Hz frequency controls the highest address input, meander with 1.6 kHz frequency controls the lowest address input.
The 8X1 Multiplexer commutes multiphase clock in single output with certain periodicity and certain law. So, the commutation period is determined with lowest control frequency in 400 Hz. The signal phase sequence must be 0°, 45°, 90°, 135° etc. or 0°, 315°, 270°, 225° etc. The changing of signal phase over the period T of the controlling signal with lowest frequency F=400 Hz by 2π is tantamount to the frequency shift of the initial signal by the frequency $\Omega=2\pi/T$. In this case the initial phase of the control signal is transferred into initial signal argument as well as frequency shift. Thus, the first law of phase changing results in forming on the multiplexer output of 39.6 kHz heterodyne signal, the second law results in forming of 40.4 kHz one. These laws of commutation are determined with the rule of operation of 3-Digit Binary Counter. The first law is obtained when this counter operate as the summing one. The second law is obtained when this counter operate as subtracting one.
On the output of Multiplexer the complicated-form signal is formed. Primarily this signal is digital-level signal with the frequency of pulses repetition in 40 kHz and periodical phase hops in 45°.
The main harmonica of Multiplexer output signal (heterodyne signal) will be:

$$s_{HET}(t) = A_0 \cos\left[(\omega_0 \pm \Omega)t + \varphi_0 + \varphi_0/8 + 2\pi i/m\right], \tag{17}$$

where $i \in [1...m]$ is the phase uncertainty. This uncertainty takes place in reference signal:

$$s_{REF}(t) = A_0 \cos(\Omega t + \varphi_0/8 + 2\pi i/m) \tag{18}$$

and one is eliminated at the phase measurements.
So, the initial, heterodyne and reference signals of device for measurements of turbulent air movement are formed with high frequency stability and strictly constant phase difference.
The block-diagram of one of receiving channel units is shown in Figure 4.

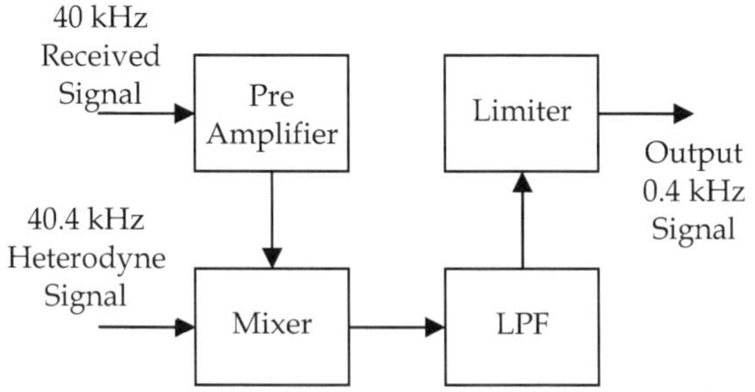

Fig. 4. Receiving channel block-diagram

The signal from each receiving transducer is amplified and mixed with 40.4 (39.6) kHz heterodyne signal.
The mixer output signal will be:

$$s_R(t) = A_0 K \cos\left(\Omega t + \Delta\varphi + \varphi_W + \varphi_0/8 + 2\pi i/m\right). \tag{19}$$

The low pass filter picks up difference signal on the output of each mixer.
Initial phase of these signals is determined by acoustical length of corresponding link and by the corresponding component of air movement.
After limiting operation there are three signals, the initial phases of which adequately represent three orthogonal components of air movement vector. Each of signals is compared in phase with the reference signal, described by expression (18). The comparison is carried out by means of microcontroller on program manner and in a result we will obtain three codes, each of which is proportional to corresponding value of $\Delta\varphi_i$ and φ_{Wi}:

$$CODE_i = R\left(\Delta\varphi + \varphi_W\right), \tag{20}$$

where R is the rating coefficient of phase measurements.
The value $\Delta\varphi$ is constant and there is a possibility to eliminate it during the calibration procedure.
In turn, the next term of sum of phase difference is proportional to corresponding component of turbulent air movement vector, which is described by expressions (5) or (6).
Corresponding component of turbulent air movement vector we can write down as:

$$v_i = v_{AIR} \cdot \cos\left(\alpha_i\right), \tag{21}$$

where α_i is the angle which is formed with air movement vector and one of orthogonal vector respectively.
Thus, when we carry out the measurements of phase difference of two low-frequency signals with phase resolution in 0.036° (clock frequency of microcontrollers is equal to 4 MHz), we can reach the resolution of measurements of weak moving of air up to 0.0003 m/c.
Certainly, the amplitude and phase of acoustic wave, which is propagated through air turbulence, change own amounts with relation to turbulence composition. The turbulence composition depends on meteorological parameters (temperature, pressure) and on the presenting in atmosphere of various gases, dust and other capacity distributed turbulences. All of them must be taken into account during measurements.
Certainly, the phase characteristics of all of parts of equipment must be taken into consideration. But these characteristics are constant and can be eliminated by calibration procedure.
The physical lengths of acoustical links are constant. But acoustical length depends on medium quality and must be taken into account in conjunction with measurements of air temperature, pressure, humidity etc. Certainly, the acoustical wave propagation constant, which depends on all off mentioned above factors, determines value $\Delta\varphi$ directly. So, taking into account the initial phase of all of these signals, we can consider the changes of medium characteristics and carry out the measuring of air movement with high accuracy.
We can use two different approaches for the solving of this problem.

The first of them assumes the measurements of air main parameters, such as temperature, pressure, humidity and gas composition. Such approach requires the presence of calibration line and assumes the implementing of calibration procedures. This approach involves in complicating of measuring process.

The second approach is the creation of additional measurement channel or reference channel, where there is no any air movement, but the air has the same parameters as the turbulent air. For example the separate semi-closed chamber can be used inside of which the transmitting-receiving pair of transducer is placed. By the fixing of all distances of measuring channels and reference channel we can eliminate the destabilizing factors influence, by the subtracting of result of reference channel measurement from the useful channels measurement results. This approach involves in complicating of measuring equipment.

Both approaches can be realized by means of separate calculating device.

5. Measurements of Phase Difference and Calculating of Phase Cycles

In this paper we assume do not measure the phase difference of test signals with standard measuring devices, but we assume to combine the calculating and measuring of this parameter. The algorithms of calculating of phase difference and total phase are different for different values of measured magnitudes.

The resolution of measurement of phase difference will be depended on resolution of measuring device as well as calculating one. There are no reasonable limitations of increasing of resolution of measurement procedure. For 0.4 kHz test signal and reference one the time clock 4 MHz will be more than enough. So, the phase difference measurement resolution will be 0.036°. There are no difficulties to increase the frequency of time clock up to 40 MHz and more with increasing of corresponding phase difference measurement resolution. The modern microcontrollers with RISC-architecture let us to do that.

There are no any limitations of increasing of resolution of calculating methods at principle. In any case, the resolution of calculating methods with high-order magnitude will be realized easily.

Further, it will be very important to distribute correct the roles between measuring and calculating procedures and to assign corresponding microcontroller for each one. There will be reasonable to assign for each of channel of receiving the separate microcontroller, which will be measure and pre-calculate the required magnitude for each channel apart. The fourth microcontroller will collect all of measured data from measuring microcontrollers. This calculating microcontroller will control by the measuring microcontrollers and will carry out only calculating procedure and will represent the required data.

According technical solution we have assumed we can not measure the phase difference of useful and reference signals directly for obtaining information concerning of large scale speed of turbulent air movement, because the phase difference will change in wide range and exceeds the value 360° many times. Furthermore, owing to use of combining measurement and calculating methods, we have an opportunity to accumulate the history of changing of phase difference and obtain the real value of any reasonable phase difference up to 4000° and more (air movement speed up to 30 m/c) at any time without any reasonable delay. So, we can obtain the phase difference data every 2.5 ms (400 Hz useful and reference signals) with high resolution and obtain, thus, the air movement vector data every 5 ms.

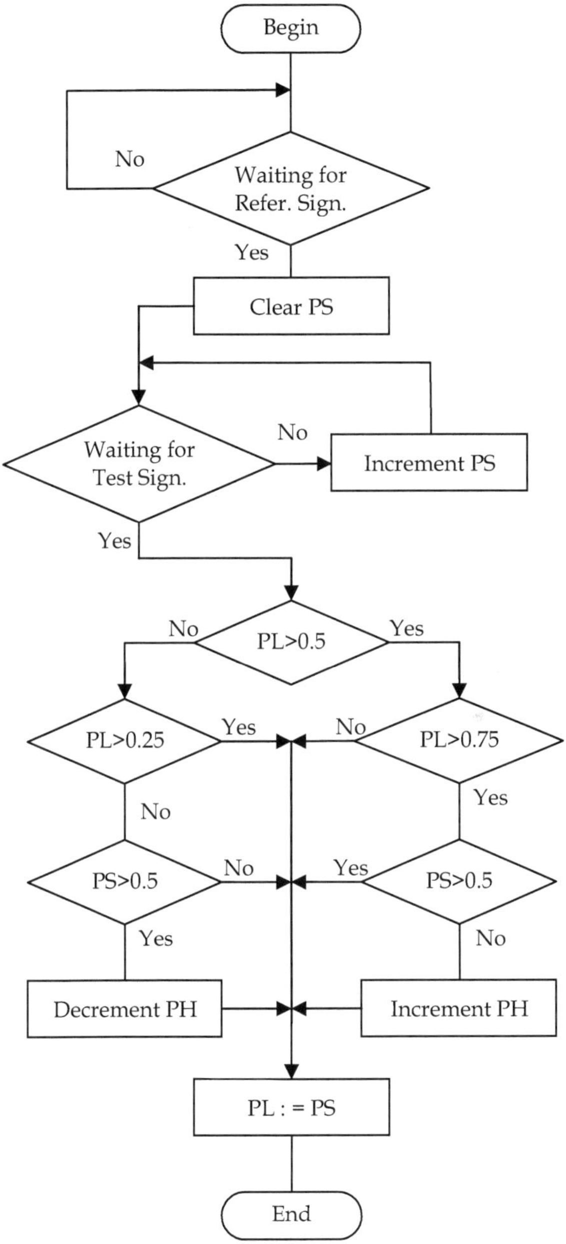

Fig. 5. The algorithm of calculating of total phase

The only thing we must to do is not to measure only phase difference, but obtain total or cumulative phase of test signal with respect to reference one. The algorithm of calculating of total phase of signal is presented on Figure 5.

Here, symbols PS, PL, PH there mean: the register of current phase difference measurement, the low register of total phase and the high register of total phase respectively. The abbreviation Sig. means "Signal". The numbers 0.25, 0.5 and 0.75 mean the filling of corresponding register. So, the register PS contains the current data of phase measurements. The register PL contains the phase difference data too. These data can vary from 0 up to 360° too. But this register holds the previous measured data. In other words by the each measurement the data into this register are reloaded. Certainly, the digit capacity of both registers must be equal.

If the register is the 8-binary-digit one, the filling 1.0 (hexadecimal number FFh) corresponds to phase difference 360°.

The resolution of phase measurements will be restricted by the digit capacity of counters PS and PL, and this resolution will be 1.4° for previous case.

The register PH contains the data of number of phase cycles. The concatenation of register PL and register PH represents the data of total phase of signal. By the analyzing of the contents of pair of these registers, we can obtain the air movement data every 0.5 ms (the calculating time is negligible).

Certainly, there are some restrictions on measurement procedure with the mentioned above algorithm. So, the changing of phase difference from one to another measurement procedure must not exceed 90°. In other words the obtained data will be valid if the gradient of air speed not exceeds 0.7 m/s for 2.5 ms time interval, according the formulae (6). These restrictions are determined with verification of 25% filling of registers we have assumed in this algorithm. For the measurement of larger air movement speed gradient there is need to use another algorithm or measuring, based on the reducing of measuring interval.

6. Simulation and Spectral Measurement

There were carried out the simulation of frequency transformations in discussed signal forming unit.

The controlled phase shifter simulates the operating of 4-Digit Johnson's Counter, Multiplexer and 3-Digit Binary Counter.

The controlling signal of controlled phase shifter results in changing of phase of initial ultrasonic oscillation by 2π over the period T of this controlling signal. For simulation this period T in 2.5 ms was chosen. The resulting frequency shift Fn will be 400 Hz. The number of steps of controlled phase shifter was chosen equal to 8 for simulation.

The simulation was carried out in environment MathCAD. For simulation there were taken the initial ultrasonic oscillations which are described by the equation (1) where the initial phase of these oscillations was equal to 0 and the amplitude factor was equal to 1.

The law of phase changing of ultrasonic oscillations is described by the following equation:

$$\Phi(t) = 180\{0.5\mathrm{sign}[-\sin(2\pi F_n t)] + 0.5\} + \\ + 90\{0.5\mathrm{sign}[-\sin(4\pi F_n t)] + 0.5\} + \\ + 45\{0.5\mathrm{sign}[-\sin(8\pi F_n t)] + 0.5\}. \quad (22)$$

This law of phase changing of ultrasonic oscillations is shown in Figure 6.

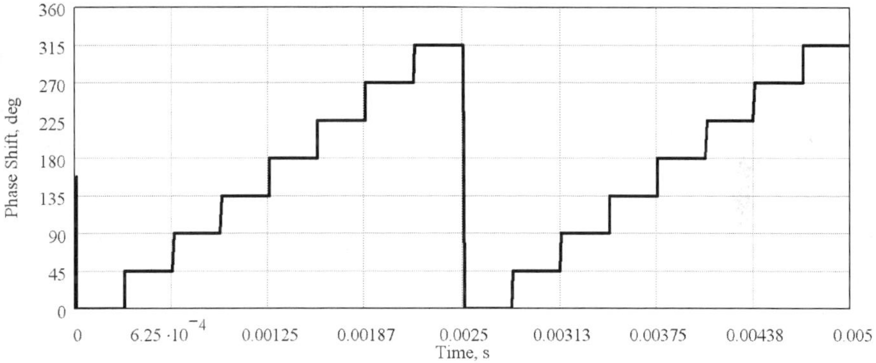

Fig. 6. The law of initial phase changing of ultrasonic oscillations

Signal on the output of controlled phase shifter will be:

$$u_{\mathrm{CPHS}}(t) = \sin\left[2\pi f_0 t + \Phi(t)\frac{\pi}{180}\right]. \quad (23)$$

Here the initial phase of controlling signal was accepted equal to 0.
The fragment of periodical signal on the output of phase shifter is shown in Figure 7.

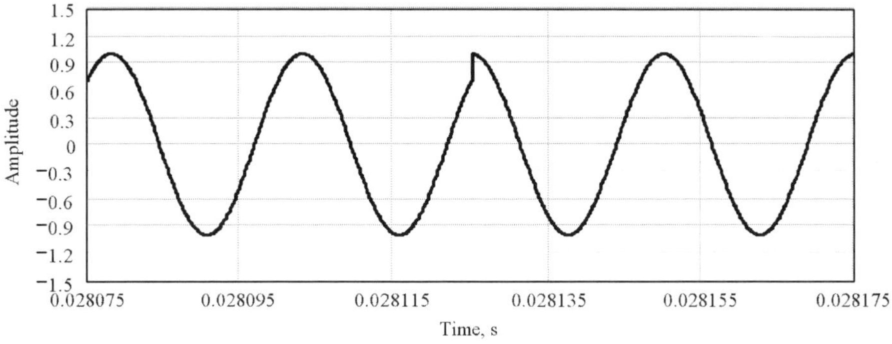

Fig. 7. The hop of phase of ultrasonic oscillations

The calculated spectrum of signal (22) is shown in Figure 8 and measured spectrum is shown in Figure 9.

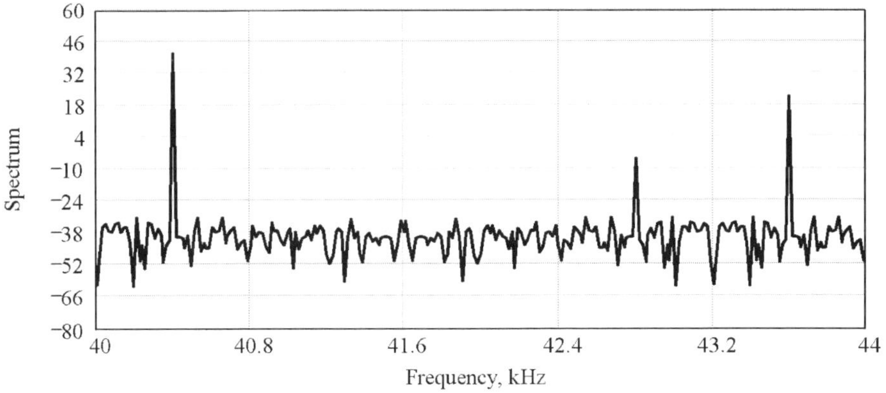

Fig. 8. The spectrum of ultrasonic signal on the output of controlled phase shifter

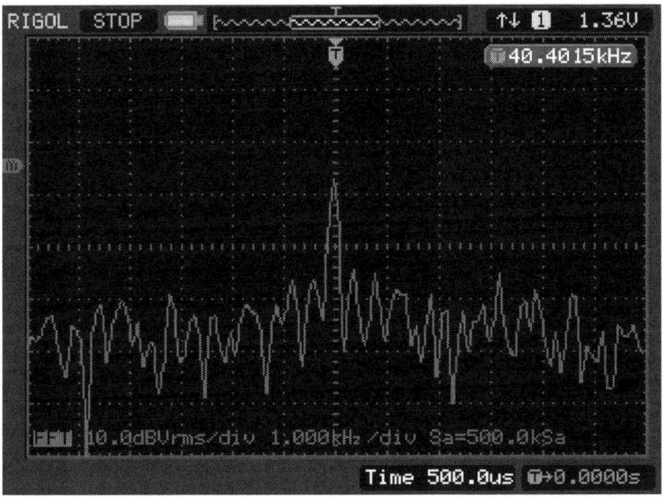

Fig. 9. The spectrum of digital-level signal on the output of multiplexer

As we can see from the Figure 8 the ultrasonic oscillations obtain the frequency shift in 400 Hz and the frequency of main harmonica of transformed ultrasonic oscillations on the output of phase shifter is equal to 40.4 kHz. The order of nearest harmonica with essential level is equal to 7, as it was pointed out in previous section (see Table 1). The frequency of this harmonica is equal to 42.8 kHz.

All of mentioned above simulations were carried out for a case of sinusoidal signal of initial ultrasonic oscillations and for sinusoidal signal on the output of phase shifter with the phase hops, as shown in Figure 7.

Really, the discussed ultrasonic signal has digital-level nature as the digital multiplexer and digital counters are used in our case.

The spectrum of output multiplexer signal was measured with the digital oscilloscope RIGOL DS1052D. This spectrum of digital-level signal on the output of multiplexer is shown In Figure 9.

We must understand the mentioned oscilloscope is not spectrum analyzer. The shown spectrum is a result of Fast Fourier Transformation of sequence under the test. This is the calculated value, caused with some restrictions and assumptions. Often we can watch amazing pictures on screens of similar devices. These pictures can radically overthrow the established views on a problem. Often we can watch on the screen so called "sub-harmonicas" of signals. But in our case, on the output of the multiplexer the signal is present and we watch the well known spectrum, which is well agreed with theoretical knowledge.

7. Conclusion

The considered manner of equipment design for 3D measurements of speed and direction of an air stream allows constructing on its basis modern technological measuring instruments which can find application in the industry, meteorological researches, etc. Using a data file of such measuring instruments, it is possible to receive a picture of spatial moving of air in real time. The absence of mobile parts in a considered measuring instrument excludes its mechanical deterioration that favourably distinguishes it from existing analogues with mechanical converters.

Thus, placing of three orthogonal acoustical links with single transmitter and three receivers it can get an accurate account about local 3D air turbulence with high resolution and without any inertia.

Certainly, the amplitude and phase of acoustic wave, which is propagated through air turbulence, change own amounts with relation to turbulence composition. The turbulence composition depends on meteorological parameters (temperature, pressure and so on) and on the presenting in atmosphere of various gases, dust and other capacity distributed turbulences. All of them must be taken into account by the measurements.

Therefore, in this paper it was shown, that there is a good opportunity to solve the problem of ecological monitoring with mentioned above method and (or) to carry out scientific investigations on microwave propagation. Furthermore, it can be find another application in industry, such as aerodynamics (motor-car- , aircraft- construction) and others.

Discussed device for supervising the turbulent air movement consists of not expensive equipment, which is ended by the microcontroller. Such device can be stand-alone one as well as a part of more complicated equipment. Several local turbulent air measuring instruments we can joint into a distributed digital system of measurement as each device has anyone digital interface according to the accepted definition. Such system let us supervise the large scale turbulences of air and predict such natural disasters as tornado and so on.

8. References

Jaffe, J. & Mackey, R. (1965). Microwave Frequency Translator, *IEEE Trans. MTT*, Vol. 13, Issue 3, May, 1965, pp. 371-378, ISSN: 0018-9480.

Kremlevsky, P (1989). *Flowmeters and quantity counters*, Mashinostroenie, 701 p., ISBN 5-217-00412-6, Leningrad.

Bobrovnikov, G; Novozilov, B & Cfhfafyjd, V. (1985). *Contactless flowmeters*, Mashinostroenie, 128 p., Moscow.

Waller, J. (1980). Guidelines for applying Doppler acoustic flowmeters. – *Instrument Technol.*, , 27, Vol. 10, pp. 55–57.

Nakatani N., Nishikawa T. & Iamada T. (1980). LDV optical system with multifrequency shifting for simultaneous measurement of flow velocities at several points, *Journal of Physics. E: Scientific Instruments*, Vol. 13, #. 2, pp. 172–173.

Shirokov, I.; Shaban, S.; Polivkin, S. & Sinitsyn D. (2003). Theoretical Modeling and Experimental Investigations of Amplitude and Phase Progression Fluctuations on Microwave Line-of-Sight Links in Relation with Natural Medium Conditions, *IEEE Proceedings of International Geoscience and Remote Sensing Symposium, IGARSS'03*, Vol. VII, pp. 4177-4179, ISBN: 0-7803-7929-2, Toulouse, France, July 21-25, 2003.

Shirokov, I. (2007). Measurements of the Radius of Atmosphere Surface Layer Pollution near the Plant with Microwave", *Proceedings of Urban Remote Sensing Joint Event*, pp. 1-5, ISBN: 1-4244-0712-5, Paris, France, April 11-12, 2007.

Shirokov, I.; Shabalina, O. & Palgov, F. (2006) The 3D Measurement of Speed and Direction of Turbulent Air Movement, *IEEE Proceedings of International Geoscience and Remote Sensing Symposium, IGARSS'06*, Vol. VII, pp. 3470-3473, ISBN: 0-7803-9510-7, Denver, CO, USA, July 31 -August 4, 2006.

Shirokov, I.; Polivkin, S.; Korobitsyn, A. & Dyurba, V. (2007). The device for 3D measurement of speed and direction of turbulent air movement *IEEE Proceedings of International Geoscience and Remote Sensing Symposium, IGARSS'007*, pp. 635-638, ISBN: 978-1-4244-1211-2, Barcelona, Spain, July 23-28, 2007.

Gimpilevich, Yu. & Shirokov, I. (2006). Generalized Mathematical Model of Homodyne Frequency Transfer Method with Periodical Changing of Phase Shift of Testing Signal. *Radiotekhnika: All-Ukr. Sci. Interdep. Mag.*, 2006, Vol. 145, pp. 185–189, ISSN 0485-8972.

Shirokov, I. et al. (1989). Amplitude and Phase Difference Measurement Device, *Author Cert. USSR*, SU 1486942 A1, publ. 15 June 1989, Bull. 22, MPC G01R 19/04, 25/00.

Shirokov, I. & Polivkin, S. (2004). The selection of the number of phase shifter discrete in tasks of investigations of channel characteristics, carried out with homodyne methods. *Radiotekhnika: All-Ukr. Sci. Interdep. Mag.*, 2004, Vol. 137, pp 36–43, ISSN 0485-8972.

21

Observing marine pollution with Synthetic Aperture Radar

Paolo Trivero and Walter Biamino
Università del Piemonte Orientale "Amedeo Avogadro", Alessandria
Italy

1. Introduction

Marine pollution is a matter of public concern because of its strong influence on various human activities such as fisheries and tourism, as well as for consequences on health. In this context, particular attention is being paid to pollution phenomena on the sea surface, where even a small amount of substance can spread over a large area in the form of a thin film.
A great aid in the effort of monitoring sea surface pollution comes from remote sensing techniques. Satellite - borne instruments are able to monitor wide areas and to detect the presence of surface slicks; optical instrument can do this by evaluating the change in spectral components of visible and infrared radiation, but they are unable to work during the night or in bad weather (clouds) conditions. For this reason, active microwave instruments play a key role in sea surface observation because electromagnetic waves freely propagate in atmosphere and in clouds.
The aim of this chapter is to explain the usefulness of the Synthetic Aperture Radar (SAR) as a tool for sea surface monitoring, especially to detect pollution. This happens because a number of pollutant substances produce huge areas of surface film which reduce water surface roughness and therefore they can be detected by the Normalized Radar Cross-Section (NRCS) on SAR images where they appear as dark areas.
Theoretical basis and practical applications will be described by reviewing literature, in order to give a comprehensive view about fundamental concept and the latest advances.
Theoretical and experimental studies, carried out over the last decades, demonstrate that the presence of a monomolecular film is able to modify the spectra of short sea waves. The damping ratio, e.g. the ratio between the spectra with clean and slick covered water, shows a maximum in the frequency domain, strongly dependent on slick composition and thickness.
Sea surface roughness is due to the short waves (wavelength up to a few tenths of centimetres) appearing on sea surface due to external forcing such as wind. The dynamics of those short waves (wavelength, velocity, etc.) is driven by the physical characteristics of sea water such as density and surface tension. The presence of a surface film modifies the surface tension and therefore causes a noticeable damping of centimetric waves: the slick covered area appears "flatter" than the surrounding sea.

Water surface strongly reflects microwaves; water vapour is transparent instead. SAR is a powerful instrument to detect the presence of surface active pollutants, able to operate regardless of sunlight and weather conditions. The SAR sends microwaves towards the earth and collects the echoes from many radar pulses, processing them into a single radar image, allowing high spatial resolutions; radio pulses are sent with high incidence angles and therefore scattered by sea surface roughness. Radio wavelengths currently used by SAR are Bragg resonant with centimetric water waves: different scattered signals are summed with constructive interference and therefore easily detected.

Marine ecosystems are threatened by various pollution phenomena with possible consequences for vegetal and animal forms of life. Some pollutants appear as thin films on sea surface, spreading over large areas: this is the case of insoluble surfactant substances such as hydrocarbons, coming from pipelines or tank leakage, as well as illegal discharges in open seas or natural seeps. Other pollutants, whilst being water soluble, may produce macroscopic effects on the surface: a typical example is given by organic substances from sewage and land runoff, carried by rivers and then dissolved in sea; chemical modifications in seawater composition can cause algae to bloom, which in turn produces mucilage on the surface.

Surface films are able to modify water dynamics, inhibiting gas exchanges and strongly modifying the formation of short waves. This is the key point for understanding how SAR can be used for remote sensing of marine pollution episodes.

Satellite - borne SARs have been used since 1978 for sea surface monitoring, as well as for mapping applications; there are today various different satellites carrying SAR instruments with different technical characteristics.

The state-of-the-art of SAR instruments and data analysis procedures will be presented, with a special focus on algorithm for automatic features extraction from SAR images. The limits of those technologies will be also evidenced; front-end technologies and future planned advances will be pointed out.

A number of operational services are currently managed and maintained by public and private bodies. A review will be carried out, in order to give a comprehensive view on practical issues and advantages.

2. Water surface slicks

Water surface slicks have several terrestrial and marine sources. Most of them are constituted by hydrophobic material naturally yielded, for instance as surfactant exuding from phytoplankton, composed mainly by homo- and hetero-polysaccharides, found at sea surface during phytoplankton blossoming (Zutic et al. 1981). Other natural sources come from land, such as the products of vegetables degradation carried by rivers to sea, and can have man-made origins such as industrial and oil plants or agricultural activities; furthermore, high concentrations of surfactants are found in urban waste water (Liss et al. 1997). Both soluble and insoluble surface-active substances are present at air-sea interfaces. The chemical nature and surface concentrations of these materials are influenced by environmental factors, such as distance from shore, local bio ecology, influx of man-made effluents from ships and meteorological conditions. Wave motion tends to select and accumulate organic materials in relation at their surface activity. With age, the films become progressively more water insoluble. Aged films and slicks generally involve multilayered

structure and weak cohesion under wind action, manifesting a tendency to break up to into macroscopic discontinuities. These films, concentrated at air-sea interface, cover large oceanic surfaces. Even when their concentration is low, they can show important effects, such as alterations in the structure of surface waves, foam formation, modification of gas exchange at interface and changes in the behaviour of backscattering of electromagnetic waves at sea surface. Natural surfactants reduce gas transfer and short waves amplitude (Goldmann and Dennet 1983, Bock et al. 1999) and in general films at sea surface can influence energy dissipation of capillary waves (Lucassen-Reynders and Lucassen, 1969; Huhnerfuss et al. 1987) and gas exchange rates (Frew et al. 1990).
In the more soluble adsorption films the relaxation process is essentially of a diffusional nature. The intermolecular forces between the adsorbed film molecules resist complete displacement from the surface by wind and wave dynamics and are of the same order as that of the solvent, since surface-active molecules are completely hydrated. In the more water-insoluble spreading films, however, when the surface concentration is high, interaction forces among hydrophobic chains are strong, and may even reach two-dimensional micellar conditions. Here the relaxation phenomenon involves structural rearrangement. Consequently, one should expect ripple-damping effects, which are greater for insoluble films than for films with greater seawater solubility.
The damping of short ocean surface waves by surfactant films is a well investigated phenomenon (Lucassen-Reynders and Lucassen, 1969; Huhnerfuss and Garrett, 1981; Lucassen, 1982; Huhnerfuss, 1986; Ermakov et al., 1986; Alpers and Huhnerfuss, 1988; 1989; Wu, 1989; Wei and Wu, 1992; Frysinger et al., 1992; Onstott and Rufenach, 1992; Huhnerfuss et al., 1994; 1996).
The theory of rheology of air-water interfaces predicts a maximum in the frequency response of the ratio of the damping coefficient of short-gravity waves for water covered by an organic surface film to the coefficient for a pure water surface (Cini and Lombardini 1978). The theoretical analysis, based upon the Navier-Stokes equation and developed for the case of small ripples on an interface covered by a surface-active substance, has been extended by with a formalism which includes both soluble and insoluble monomolecular films for the two coexisting modal solutions: the Laplace or transversal mode and the Marangoni or longitudinal mode (Lombardini et al. 1982, Fiscella et al. 1985a).
According to Lombardini et al. (1989), the analytical form which describes the ratio between real parts of the complex radian frequencies on pure water to that for water covered by slick (damping ratio) can be given by the semi – empirical formula:

$$y(f)_s = \frac{1 \pm 2\tau + 2\tau^2 - X + Y(X + \tau)}{1 \pm 2\tau + 2\tau^2 - 2X + 2X^2} \tag{1}$$

Where:

$$\tau = \sqrt{\frac{\omega_D}{2\omega}}, \quad X = \frac{\varepsilon_0 k^2}{\rho\sqrt{2\nu\omega^3}}, \quad Y = \frac{\varepsilon_0 k}{4\nu\rho\omega}, \quad \varepsilon_0 = \frac{d\sigma}{d(\ln\Gamma)}$$

are adimensional quantities and:

$$f = \frac{\sqrt{(\sigma/\rho)k^3 + gk}}{2\pi}$$

the dispersion law, σ surface tension, ρ water density, g acceleration of gravity, k wave number, ν kinematics viscosity; the constant characteristic parameters of the film are: elasticity modulus ε_0, surface concentration Γ, and characteristic frequency ω_D, which, for soluble films, depends upon the diffusional relaxation, and for insoluble films, depends upon structural relaxation between intermolecular forces. In (1) a plus sign refers to soluble films, while a minus sign indicates insoluble films.

Spectral measurements carried out both in tanks and in many oceanic sites on slicked waters clearly show this damping effect. The ratios between spectra measured in pure water and in water covered by film have a maximum in the 2-10 Hz range (Cini et al. 1983). From observed ratios and theory it is possible to deduce rheological parameters, such as the relaxation characteristic frequency and the visco-elastic modulus, as well as the film concentration or fragmentation (Fiscella et al. 1995).

Soluble (adsorption) films have been thoroughly investigated (Lucassen-Reynders and Lucassen, 1969). Typical values of the diffusional characteristic frequency ω_D observed in saturated conditions are in the order of 10^{-2} rad/s, or smaller (Loglio et al. 1986). Hence, in good approximation, soluble films are characterized by setting $\omega_D = 0$. With this condition one can verify that in soluble films the Marangoni waves are too highly attenuated to be of practical interest. The study of insoluble (spreading) films on sea surface (Lucassen 1982, Cini et al. 1983) have indicated the possibility of obtaining qualitative data on polluting films by analysing the short gravity portion of the wave spectrum of a breezy sea.

2.1 Experimental evidences

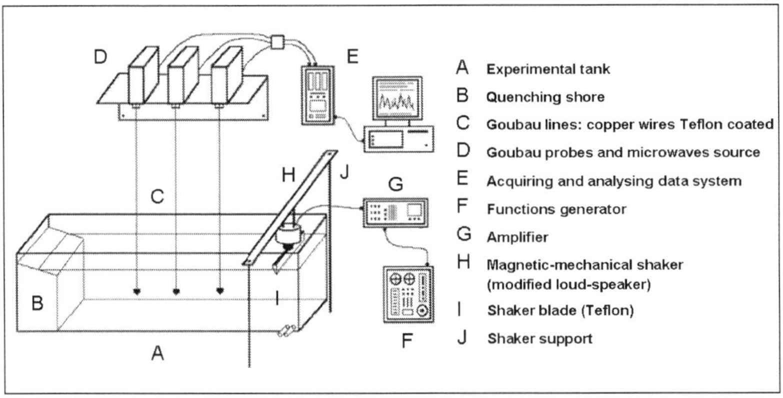

Fig. 1. Experimental setup for wave damping measurements

By means of a microwave probe (Fiscella et al. 1982), short gravity and capillary wave domain of the sea spectra have been investigated in a variety of field situations. Viscoelastic characteristics of insoluble films prepared in laboratory from pure surfactants (e.g. palmitic acid methyl ester, hexadecyl trimethyl ammonium bromide) have been then studied by tests including spectral measurements performed in a wind tunnel, and attenuation measurements of several monochromatic mechanically generated waves in the maximum damping ratio range (Fiscella et al. 1985b). Comparisons between the observed data and

theory have produced relaxation characteristic frequencies ω_D in the range 7.5 to 11 rad/s, and elasticity modulus ε_0 in the range $5.0 \cdot 10^{-3}$ to $2.2\ 10^{-2}$ N/m. Such values produce a damping ratio for the Marangoni mode revealing that the insoluble films sustain both wave modes.
The results of measurements obtained in laboratory using oleyl alcohol as surfactant are presented below; the surfactant organic compound, in fact, have been already used in past experiments as a good representative of hydrophobic surface substances (Trivero et al. 2001).
Oleyl alcohol was used as surfactant substance to study damping effect by meanS of a laboratory tempered glass tank (dim. 298 x 27.3 x 29 cm; 235 litres volume) and an interferometric microwave wave gauge which measures wave heights on an absolute, self-calibrating scale with high accuracy; this apparatus has been already used in sea surface measurements (Fiscella et al. 1982). The basic element of this probe is a Teflon coated wire. The lower end of this wire is held vertically straight and dipped in water, while the other end is fed by a microwave source. The microwave energy travels downwards, confined to a close proximity of the coated wire (Goubau line). The contact with the water acts as a short circuit, giving origin to a reflected wave. In condition of good matching of the microwave system the field in the transmission line has a standing wave pattern, which is uniquely determined by the location of the water contact with the coated wire.

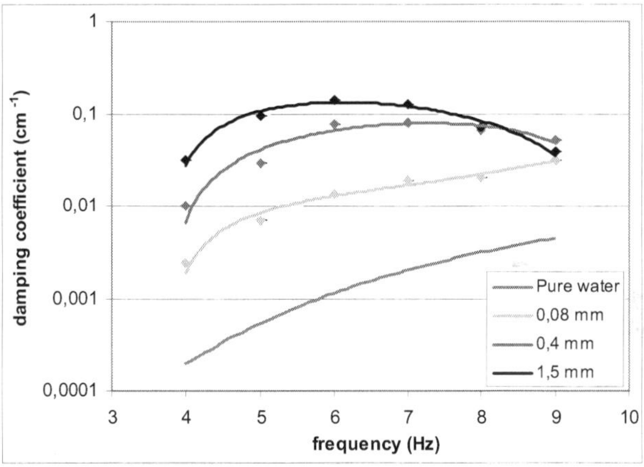

Fig. 2. Damping coefficient vs. frequency for different film thickness

Power spectra are obtained by data segmentation, Hanning windowing, FFT operation and subsequent power spectra meaning. In a Goubau line with a copper wire radius of 0.6 mm and coating thickness of 0.3 mm, the radius of the area in which 50% of the propagated power is concentrated is: $\rho_o = 2.4$ mm. This area includes the meniscus (for clean water) and implies a Voltage Standing Wave Ratio ≥ 2. In this case the liquid wavelength $4\rho_o$, i.e., 26 Hz, may be considered the upper frequency limit of the probe. The measurement of z can thus be accomplished with accuracy of the order of few micrometers. In laboratory and clean water conditions the time series of the sea water elevation are affected by instrumental errors of few micrometers and frequency spectra can be obtained without distortion up to 20

Hz. The results obtained are in accord to theory of rheology and confirm even in laboratory the damping wave effect showed by surfactant substances at sea surface.
The experimental apparatus, shown in figure 1, consists of a three Goubau line coated wire system. The wires are positioned along wave direction at proper distance in order to measure spatial damping of the waves mechanically generated. The same apparatus can be used at the sea to obtain information about directional wave spectra.
Damping measurements versus frequency were performed adding a known oleyl alcohol quantity in order to obtain a fixed growing film thickness. Figure 2 shows the damping coefficients for pure water and different film thicknesses.

3. SAR observation of marine surface

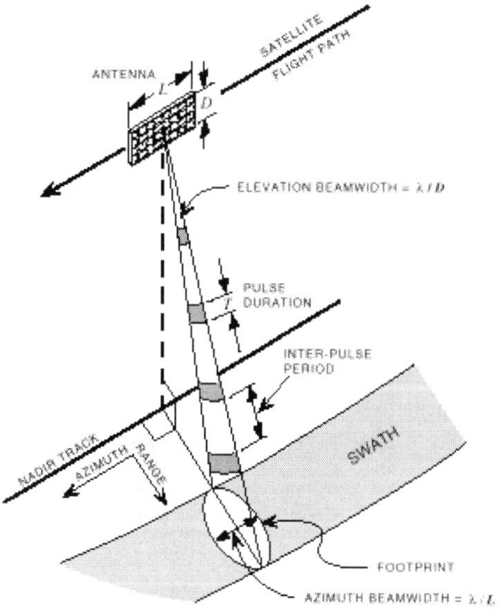

Fig. 3. SAR acquisition geometry

The SAR is basically a conventional radar instrument, carried by a mobile system such as aircraft or satellite. The main principle of SAR is the antenna synthesis: when moving, the target is observed from different angles and the backscattered signals are put together. Observation is lateral rather than perpendicular to the earth's surface (figure 3).
The electromagnetic waves, used by SAR, are in the microwave region. Wavelengths are in the range from 0.1 to 100 cm and it is divided in different bands with a standard nomenclature. Table 1 summarizes the most utilised frequencies and their characteristics.
The basic mechanism involved is the normalised radar cross-section which, for incidence angles higher than 20°, is proportional to the spectral energy density of the sea waves having wavelength Λ that obey the Bragg resonance condition:

$$\Lambda = \frac{\lambda}{2\sin\vartheta} \tag{2}$$

where λ is the radar wavelength and θ the incidence angle of radar beam; electromagnetic waves, backscattered from every water wave front, sum in phase producing a well detectable echo (figure 4). For low incidence angles the backscatter is due to specular reflection. The sea waves, which are Bragg resonant with microwaves employed by the SAR systems, fall in the short gravity wave region where is found a maximum in the ratio between spectra measured in pure water and in water covered by film.

Band designator	*Frequencies (GHz)*	*Wavelength in Free Space (cm)*
L band	1 to 2	30.0 to 15.0
S band	2 to 4	15.0 to 7.5
C band	4 to 8	7.5 to 3.8
X band	8 to 12	3.8 to 2.5
Ku band	12 to 18	2.5 to 1.ò7
K band	18 to 27	1.7 to 1.1
Ka band	27 to 40	1.1 to 0.75
V band	40 to 75	0.75 to 0.40
W band	75 to 110	0.40 to 0.27

Table 1. radar bands

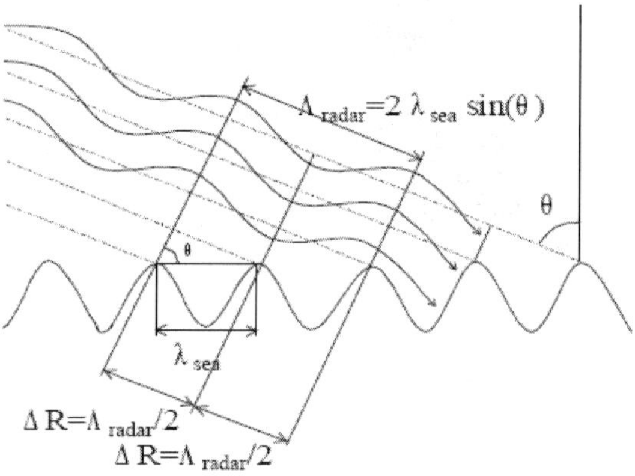

Fig. 4. Bragg condition between water waves and radio waves

Remote sensing radars are usually designed to transmit either vertically polarised or horizontally polarised radiation. Likewise, the radar can receive either vertically or horizontally polarised radiation, or sometimes both. Polarisation planes are designated by the letters H for Horizontal and V for Vertical. When the polarisation of received radiation is the same as the transmitted radiation, the image is said to be like-polarised When the polarisation of received radiation is the opposite of the transmitted radiation, the image is said to be cross-polarised .Cross polarisation requires multiple-scattering by the target and therefore results in weaker backscatter than like-polarisation. Cross-polarised signals are sometimes too weak to produce a good image, although the use of multiple polarization can help in revealing sea surface characteristics (Brekke and Solberg 2005) .

3.1 Wind vector extraction

Sea surface roughness is directly related to external forces such as wind and currents; for this reason, data from SAR observation can be used to compute sea surface parameters.
When dealing with surface slick observation, the knowledge of wind vector plays a key role for at least two reasons. First, surface slicks can be detected if wind intensity is within a range of 2 m/s and 20 m/s; a weaker wind cannot produce roughness and therefore the sea area appears as flat whether a slick is present or not. Conversely, a very strong wind can produce roughness even on a slick covered area. Secondly, wind forcing and wind - induced currents are directly responsible for slick evolution (drift and weathering); therefore the wind vector is fundamental in estimating the fate of a surface slick.
Wind data are can be get from in-situ measurements (e.g. buoys), meteorological models or satellite - borne scatterometers (with poor spatial resolution). In last years, studies have been carried out in order to find suitable methods to extract wind field from the SAR image itself.
A popular approach is to use a geophysical model function (GMF), e.g. a nonlinear function who describes the NRCS as a function of, wind speed (normalized to 10 m height), wind direction, incidence angle and azimuth angle with respect to wind direction); radar frequency is also taken into account. Using numerical techniques to invert such a function, wind speed can be obtained from SAR image (assuming to have the wind direction). For satellite - borne SAR operating at C-band and vertical (VV) polarization in transmission and reception, several empirical GMFs have been developed and validated; the most popular are:

- CMOD4 (Stoffolen and Anderson, 1997), utilising ECMWF (the European Center for Medium-Range Weather Forecast) weather model results as calibration data;
- CMOD_IFR2 (Quilfen et al., 1998) developed at Ifremer-France and calibrated against in situ measurements, (buoys data and ECMWF model results);
- CMOD5 (Hersbach, 2003), an upgraded version of CMOD4.

These functions are applied on SAR data from ERS-1, ERS-2 and ENVISAT satellites. Other similar approaches have been defined for L-band data or from C-band horizontal polarization data.
However, to measure wind speeds from SAR images using such kind of methods, it is necessary to have the wind direction. A first approach is to assume a fixed direction for a whole SAR image, for example interpolating the wind direction from scatterometer data or atmospheric models.
In last years, several attempt have been done to extract the wind direction field from the image itself, exploiting linear features aligned with the wind direction are visible.
A popular method is the Fast Fourier Transform (FFT) method (Gerling, 1986): the Fourier spectrum of the SAR image is computed and the main spectral energy is located perpendicular to the orientation of the wind streaks, giving a wind direction with a 180° directional ambiguity.
A recently developed wind direction estimation method is the local gradient (LG) method, which derives the orientation of the wind streak by evaluating the local gradient on different scales (Horstmann and Koch, 2005; Horstmann et al., 2002; Koch, 2004). Alternative

approaches such as wavelet analysis (Du et al., 2002; Fichaux and Rachin, 2002) and variance method (Wackerman et al., 2003) have also been applied.
A comprehensive review on wind estimation methods can be found in (Lin et al. 2008).

3.2 Surface slicks detection

It is well known that the surfactants are responsible for sea wave damping and reflectivity modulation over a broad range of frequencies from the visible to the microwave regions of the spectrum. Measurements of slick-induced damping of short-gravity ocean waves excited by the wind provide useful data for the investigation and characterization of ocean micro-layers on a thermodynamic basis. By means of a theoretical model one can infer the soluble or insoluble nature of the substance forming the film itself, the rheological parameters and the surface film fragmentation.
Among all monitoring ways, multi-frequency radar and SAR are powerful tools for the detection and characterization of substances forming sea surface film.
Wave damping, due to surface slick, modifies the backscattering of electromagnetic waves and hence it is possible to optically detect slicks (Scully-Power 1986) and measure their characteristics by means of radars from platform, airborne and satellite (Fiscella et al. 1985b, Espedal et al. 1996, Trivero et al. 2001). However, dark areas can be also due to atmospheric effects (Alpers, 1995; Melsheimer et al., 1998).

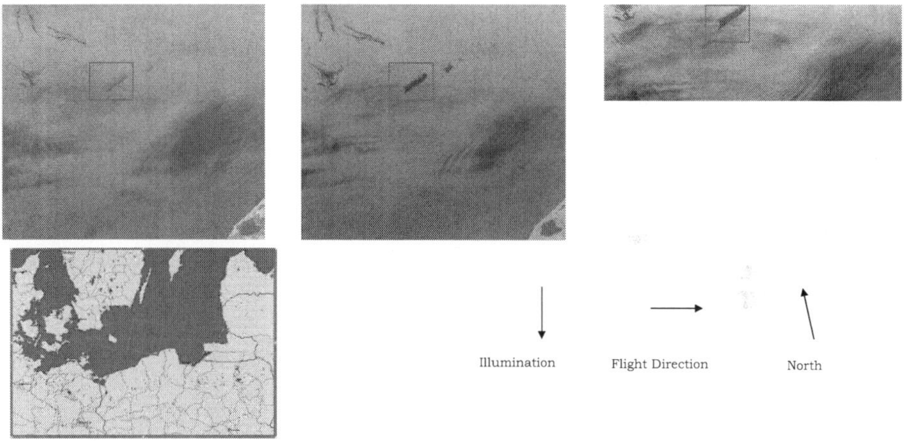

Fig. 5. Oil slick observed by L-band (left), C-band (center), X-band (right)

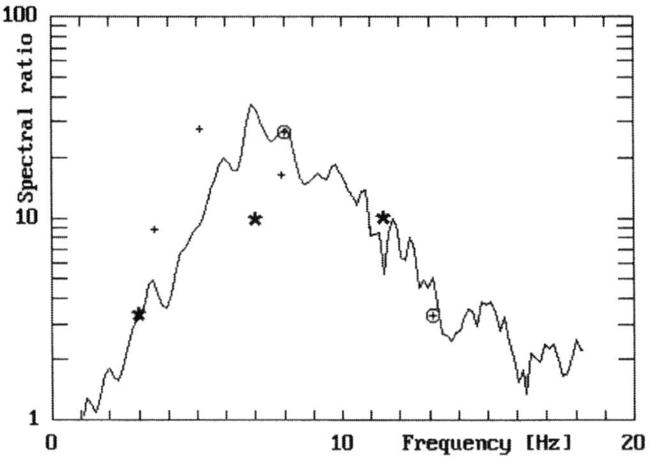

Fig. 6. Comparison among damping ratios obtained by radars and by the wave gauge.

The ability of multi-frequency SAR to characterize surface films was tested with data obtained during an experiment in October 1990 in the northern Adriatic Sea, when an airborne SAR flew over a research platform, on board of which, time series of radar backscatter as well as high frequency wave spectra were measured (Trivero et al. 2001). The obtained results were confirmed by analogous experiment performed in the North Sea during the first SIR-C/X-SAR mission in 1994 (Gade et al. 1998); in that mission, the NASA's space shuttle carried SAR instruments at three different bands simultaneously looking at the same areas. Figure 5 shows how the same surface slick is seen in a different way by three different frequencies;.

In June 1991 in the Gulf of Genoa (Italy) another similar experiment was performed observing slicked area after oil-tanker accident (Trivero et al. 1998).

In figure 6, the full line is the display of the ratio of the power spectral components from gauge data between pure and slick-covered water; the three (+) and the two (⊕) points are the plots of σ_0/σ_s from multi-frequency scatterometer data and from SAR-580 images, respectively. The three (*) points are the plots of σ_0/σ_s from SIR-C/X-SAR images.

The experimental results showed that multi-frequency SAR is an ideal instrument to monitor sea surface substances, since SAR data contain information about the spectral components affected by damping.

4. Available SAR data

The use of satellite - borne SARs for earth observation is a well tested and established technique. The history begun on the 28th June of 1978 when Seasat, the first Earth-orbiting satellite designed for remote sensing, was launched.

Seasat, managed by NASA's Jet Propulsion Laboratory (JPL), operated for 105 days until October 10, 1978, when a short circuit occurred in the satellite electrical system; during the mission, the onboard L-band SAR acquired approximately 42 hours of data. Despite this

unhappy ending, Seasat played a key role for earth observation by demonstrating the feasibility of global satellite monitoring of sea surface and by helping scientists in defining technical requirements (Evans et al. 2005).
In the following years a number of satellites were launched, carrying SAR instruments with very different characteristics. Here, we want to review the currently operating missions, with data easily available for scientific and/or commercial use; for these reasons, only currently civil operational missions are described, while military only missions have been omitted as well as sensors which have reached the end of their operational life.
For every sensor an overview is given about frequency, resolutions, revisitation frequency and data distribution policies; for detailed information, the satellite owner and contact details are reported. All the information is correct at the publishing date.

4.1 European Remote Sensing (ERS) satellites

Managed by European Space Agency (ESA), ERS-1 satellite was launched in 1991 and completed its operation in 2000, overlapping with the new ERS-2 launched in 1995 and still operating.
ERS-2 SAR works on C band at fixed VV polarisation. Its best spatial resolution is 12.5 m, but limited to 5km x 5km imagettes in "wave mode" acquisition. It can also operate in wave mode at 30m spatial resolution with a 100 km swath.
Full documentation is available on ESA earth observation website http://earth.esa.int/ers/ while data availability can be checked online via the EOLI catalogue (http://earth.esa.int/EOLi/EOLi.html), also allowing online ordering for ftp delivery. Different prices and data policies apply for scientific/non-commercial use rather than commercial exploitation.

4.2 Envisat satellite

Designed and built by ESA, launched in 2002 with the aim to be the ERS follower, Envisat carries an improved C-band SAR sensor as well as a number of other active and passive instruments.
Envisat SAR best spatial resolution is ~30m for a ~100 km swath; multiple polarization modes are available (VV, HH, VH, HV). A wide swath mode (~400 Km) with a 150m spatial resolution is available.
It has a dedicated section on the ESA website (http://envisat.esa.int/); also Envisat data are available on EOLI catalogue.

4.3 Radarsat satellites

Designed and built by the Canadian Space Agency, the Radarsat - 1 satellite was launched in 1995. It operates in C-band with fixed HH polarization; seven imaging modes are available with different swaths (from 50 to 500 km) and different spatial resolutions (from 8 m to 100 m)
Radarsat - 2, launched in December 2007, is an enhancement of the previous sensor; all polarization modes are now available and a new "ultra fine" acquisition mode (3 m pixel and 20 km swath) can be operated. Moreover, Radarsat - 2 is able to look on both right and left sides with a switch time of a few minutes, allowing more flexibility on selecting the target zone.

Radarsat images are distributed by MacDonald, Dettwiler and Associates Ltd. (MDA), a Canada based firm (http://gs.mdacorporation.com/products/sensor/index.asp).

4.4 Terrasar satellite

The TerraSAR-X Earth observation satellite is a joint venture being carried out under a public-private-partnership between the German Aerospace Center DLR and EADS Astrium GmbH.; TerraSAR-X was launched on June 15th, 2007 and has been in operational service since January 2008 ; it operates on X-band with single, dual and quad polarization. There are three operational imaging modes: SpotLight (1m resolution, 10 km swath); StripMap (3m resolution, 30 km swath) ; ScanSAR (18 m resolution, 100 km swath).
The exclusive commercial exploitation rights are held by the geo-information service provider Infoterra GmbH (http://www.infoterra.de/), while data access for scientific (non commercial) use is directly managed by DLR.

4.5 Advanced Land Observing Satellite (ALOS)

Designed and managed by the Japanese space agency JAXA, ALOS was launched in 2006 being operational in October. It carries, together with other instruments, an L-band SAR (named PALSAR) able to acquire in single polarisation (HH or VV) or dual polarisation (HH/HV or VV/VH) modes. The best available resolution is 10 m with 70 km swath; the widest swath is over 250km at 100m resolution.
More information can be found at JAXA (http://www.eorc.jaxa.jp/ALOS/en/index.htm) as well as ESA (http://earth.esa.int/ALOS/) websites; ESA is also responsible for ALOS data distribution in Europe, Africa and Middle East; ALOS images are available on the above mentioned EOLI catalogue.

4.6 Cosmo – SkyMed constallation

Since late 90s the Italian Space Agency ASI started to design a multi - purpose Earth Observation System devoted to providing products and services for military and civil use through an integrated approach (Dual Use System). The decision to build a constellation was driven by the need for the shortest revisit time with the aim to exploit data in critical applications such as risk management and environmental monitoring (Rum 2000).
The first three satellites of the COSMO-SkyMed constellation are already in orbit and operational; the launch of the fourth satellite is planned for 2010. The carried sensor is an X-band right and left looking SAR with a maximum spatial resolution of 1m (on a 10 km swath); the "huge" acquisition mode offers a 100 m resolution with a 200 km swath; all polarization modes are available.
Operational mode can be set up in three ways (routine, crisis and very urgent) allowing the system to respond to different needs in terms of required programming latency. A User Request (in the case of the first level of SAR standard products) can be satisfied in 72 hours for the system working in routine mode, 36 hours for the crisis mode and 18 hours for the very urgent mode. The management of a constellation rather than a single satellite, as well as the high number of degrees of freedom in acquisition management, pose new issues in operational management (Bianchessi and Righini, 2008).
The commercial distribution rights for Cosmo - SkyMed images have been recently granted to e-GEOS (http://www.e-geos.it/), a new entity owned 80% by the private firm Telespazio

and 20% by ASI. Technical documentation about Cosmo - SkyMed satellites and products can be found on e-GEOS website.

5. Pollution detection algorithms

Damped areas appear as dark spots on SAR images; however, different features are clearly identifiable because of their own geometric characteristics. For example, oil slicks usually have a linear shape with well defined contours, whilst natural surfactant appears in a different way (figure 7). An expert photo interpreter is able to distinguish between categories.

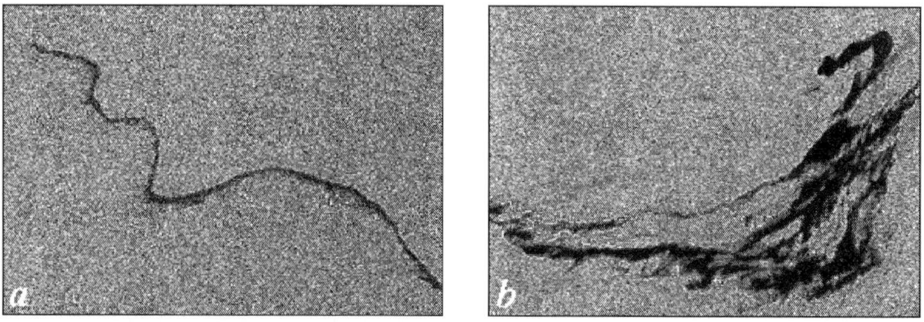

Fig. 7. Example of oil spill (a) and look-alike feature (b) in SAR images

Since SAR images were available, a number of attempts have been done in order to develop and test automatic procedures for oil spill detection, with the aim OF definING A new instrument for real time analysis of satellite images in order to prevent pollution. Here we want to present some exempla of different approaches based on various mathematical techniques.

5.1 Statistical techniques

Here we describe a procedure, named "Oil Spill Automatic Detector" (OSAD), able to distinguish oil spills from other similar sea surface features (look-alike) in SAR images using a statistical approach; during last years, the procedure has been updated (Fiscella et al. 2000, Nirchio et al. 2002, Nirchio et. al 2005a, Nirchio et al. 2005b). It considers both the radiometric and the geometric characteristics of the areas being tested. In order to minimize the operator intervention, it adopts automatic selection criteria to extract the potentially polluted areas from the images.

The related operational activities are carried out at the Matera Geodesy Space Centre, where the Italian Space Agency Processing and Archiving Facility (PAF) for the European Remote Sensing (ERS) satellite sensor data and the 'Telespazio' acquisition facility are located. A satellite ground station has been operative in Matera (Italy) since November 1999. It acquires data trasmitted down by the European satellites ERS 2 and Envisat. The facility is composed of an 8 m main dish, a down converter chain and a direct ingestion sub system. The acquisition area spans from the North Sea to the Red Sea including the entire Mediterranean basin.

The acquired data are first recorded and screened for evaluating the quality parameters, and then a browse image is generated. At this point data are available for processing and distribution to end-users. An operator inspects the browse image, covering an area about 4000 km long and 100 km wide, and selects those frames 100 km x 100 km which appear affected by oil spill. The next step foresees the production of the full resolution image. This can be analysed by OSAD that provides the probability that the suspected area is affected by an oil spill. The land, eventually present in the image, is masked to allow the identification of dark areas potentially interested by oil spills. A threshold is computed, its value depends on the average image intensity from which the image standard deviation has been subtracted. Those areas whose average backscattering is lower than the threshold are further screened for retaining those whose dimensions are larger than 0,3 km^2 and smaller than 10 km^2. The small regions are rejected because these eventual slicks are not significant; usually they disappear from images in a short time. The large areas are also rejected because they are probably due to lack of wind. At the end of the process, several analysis are performed on the remaining slicks candidates, the probability the area under test is affected by a spill is computed and a detection report is generated and sent to the responsible authority.
Before operational use, the system must be tuned; this is done by means of a training dataset, composed by images that have been classified by an expert photointerpreter as "oil spill" or "look-alike"; uncertain images are discarded. For every image family, geometric and radiometric characteristics are computed; for every characteristic, the data distribution is evaluated in order to find significant parameters for both oil spills and look-alikes.
In the operational use, every acquired image is first calibrated and the land is masked. On sea areas, the significant values are evaluated and then compared with the previously described statistical distributions, in order to define the probability to be an oil spill. This probability is given as a percentage ("score") where 0 is the lookalike and 100% in an oil spill.
The first tuning of OSAD system has been done on a set of SAR images corresponding to 153 cases of oil spill and 237 cases of look-alike detected during 1999 in the Mediterranean Sea using ERS-1/2 SAR Precision Image Product (PRI) which dimensions are 100 x 100 km. The method has demonstrated an a priori percentage of correct classification higher than 90%, it is easy to apply and able to determine the identification probability in an automated way (Nirchio et al. 2005b).
Another statistical approach was proposed by Solberg et al. (1999); the procedure consists of first detecting dark spots in the image, then computing a set of features for each dark spot, before the spot is classified as either an oil slick or a "look - alike". The classification rule is constructed by combining statistical modelling with a rule-based approach. Prior knowledge about the higher probability for the presence of oil slicks around ships and oil platforms is incorporated into the model. In addition, knowledge about the external conditions like wind level and slick surroundings are taken into account. The algorithm accuracy is 94% for the oil slicks and 99% for the look-alikes.

5.2 Neural network approach

An artificial neural network, usually called "neural network", is a mathematical model or computational model that tries to simulate the structure and/or functional aspects of biological neural networks. It consists of an interconnected group of artificial neurons and processes information using a connectionist approach to computation.

In more practical terms neural networks are non-linear statistical data modeling tools. They can be used to model complex relationships between inputs and outputs or to find patterns in data. It can be thought as an adaptive system that changes its structure based on external or internal information that flows through the network during the learning phase.
Using an approach similar to that described on previous paragraph, Calabresi et al. (1999) used neural networks in order to find specific values identifying oil slicks, chosen from a given set of parameters. Here the network input was a vector containing the values of a set of features previously calculated.
Usually the oil spill candidates (that is the dark areas on SAR image) are identified with their geometric and radiometric parameters, then a classification algorithm is applied. In Topouzelis et al. (2007) two different neural networks are used: one to detect dark spots on sea surface and another to classify the previously found areas as oil spills or look-alikes. The proposed method shows good results in detecting dark formations and discriminating oil spills from look-alikes as it detects with an overall accuracy of 94% the dark formations and discriminate correctly 89% of examined cases. For dark area detection the network unit is the pixel, while for classification is a vector made of ten parameters.
It is worth to note how a neural network approach has been used to classify samples of unknown crude oils and distilled fuels on the basis of the results from standard chemical analysis (Fonseca et al. 2006, Fernandez - Varela et al. 2008).

5.3 Other techniques

A number of different mathematic techniques can be found in literature for sea surface detection and characterisation of surface slicks.
As previously said, the first step is the determination of dark areas on SAR image, due to low backscattering levels. A first approach is described by Skøelv and Wahl (1993) for ERS-1 SAR images). A similar approach is described in Vachon et al. (1998) and Manore et al. (1998). Solberg et al. (1999, 2003) apply an adaptive algorithm where the threshold is dynamically set at k dB below the mean value estimated in a moving window.
Canny (1986) started the use of hysteresis thresholding, later applied by Kanaa et al. (2003)
An approach based on the Laplace of Gaussian (LoG) and Difference of Gaussian (DoG) operators is described in Change et al. (1996) and Chen et al. (1997).
Liu et al. (1997) and Wu and Liu (2003) proposed the use of wavelet analysis in ocean feature detection, including oil spills. Mercier et al. (2003) suggest a segmentation method based on detecting local variations of the wave spectra. Fuzzy logic was investigated as a since mid 90s (Barni et al. 1995)
Fractal mathematics is also used for classification purposes (Keller et al., 1989; Benelli and Garzelli 1999; Gade and Redondo 1999). The topic is still studied and new results presented (Marghany et al . 2009).
Even though we here focus on single frequency and single polarization SAR images, it is worth mentioning the possibility of a discrimination algorithm based on differences in multi-frequency and multi-polarization signatures. Gade et al. (1996) did some experiments to investigate whether spaceborne L-, C- and X-band multipolarization SARs are capable of discriminating between films of different chemical properties, and found that discrimination is only possible at low to moderate wind. Maio et al. (2001) propose such an algorithm for discrimination between oil spills and lookalikes.

5.4 SAR limitations

It must be observed that not always SAR is able to reveal oil spills; even when detection is made only the thicker part, typically covering only 10% of the whole oil spills area, is imaged (Sabins 1997).

The detection of oil slicks/spills in SAR images strongly depends on the wind speed at the sea surface. Under low wind speed, typically between 0 and 2-3 ms^{-1}, the sea surface looks dark on SAR images. In this case the wind-generated waves are not already developed and oil films look dark on a dark background: detection in this case is impossible. Wind speed between 3 and 6 ms^{-1} is ideal for oil slick detection, the sea surface roughness is developed and oil slicks appear as dark patches on a bright background. However, when wind speed reaches 10-12 ms^{-1}, detection is impossible again or obstructed due to the redistribution of oil spills/slicks by the surface waves and wind-induced mixing in the upper ocean layer (Scott 1986); in this case becomes determinant the compactness of film. As the result slick disappears from the sea surface and SAR imagery. The upper wind speed threshold for spill detection with SAR is suggested to be between 10 and 14 ms^{-1} (Gade and Ufermann 1998, Ivanov 2000). In the Mediterranean Sea it was demonstrated that oil spills can be detected from SAR images if the wind speed results between 2 ms^{-1} to 10 ms^{-1}; the SAR capability in detecting oil spills has given good results at open sea, while in the near-shore region the detection percentage drops quickly, because in these cases the effect of wind sheltering becomes determinant, that is, the wind screening effect caused by the local topography on the areas near-shore (Nirchio et al. 2005b).

When a slick is detected by SAR, it is not easy to know its evolution state and as a consequence its age. The behaviour of oil spill on the sea surface significantly depends on its important physical-chemical properties, such as viscosity, density, surface tension and elasticity. Moreover, crude oil is a complex mixture of different chemical components including heavy and light fractions. Typically, crude oil can be detected during its evolution in the sea in different phases of age: oil spill, oil film, emulsion (for the first time oil-water emulsion and then water-oil emulsion), blue shine and aggregates. During the lifetime of oil spill in the sea it will be exposed to a number of weathering processes, which dramatically influence physical-chemical properties: spreading, drift, evaporation, dispersion, emulsification, bacterial degradation and photo oxidation (Kotova et al. 1996). With time the physical-chemical properties of oil spills are changed due to effect of these processes. These processes play important role in oil spill detection using SAR sensors. But relative importance of each process is not still well understood. Direct observation seems to show that big slicks lifetime is greater than small ones; for example in the Prestige (Spain 2002) and Haven (Italy 1991) disasters, the great quantities of oil released into the sea stayed on for a long time, although their SAR detection was not been always possible. An estimate of the quantity of oil observed at sea is also crucial because oil thickness is difficult to measure especially the sea is rough. Moreover, water-in-oil emulsions and viscous oils like heavy crude and fuel oil can vary in thickness from millimetres to several centimetres.

6. Future advances

The State-Of-The-Art of sea surface SAR monitoring, as described by this paper, shows good results as well as the possibility of being applied for operational issues. However, some issues have yet to be resolved.

Revisit time is not yet optimal for real time applications; the preferred way to solve this issue is to combine observation with different satellites, by building constellations or by signing agreements between operators in order to exchange data.
Satellite availability is becoming wider with the launch, scheduled in 2010, of the fourth Cosmo-SkyMed satellite and the two SAOCOM (L-band SAR, managed by the Argentinean space agency CONAE).
While automatic oil spill detection shows good results, characterizing other pollutants is less straightforward; moreover, slick characterization (e.g. to define the chemical composition or slick thickness) is still an open issue. It is possible that the use of multi frequency and multi polarization data will lead to valuable results.

7. References

Alpers, W. (1995), Measurement of mesoscale oceanic and atmospheric phenomena by ERS-1 SAR. *URSI Radio Sci. Bull.* 275:14–22.

Alpers, W. & Hühnerfuss, H. (1988), Radar signatures of oil films floating on the sea and the Marangoni effect. *J. Geophys. Res.* 93:3642-3648.

Alpers, W. & Hühnerfuss, H. (1989), The damping of ocean waves by surface films: a new look at an old problem. *J. Geophys. Res.* 94:6251-6265.

Barni, M.; Betti, M. & Mecocci, A. (1995). A fuzzy approach to oil spill detection on SAR images. *Proc. IGARSS '95*, vol. 1 (pp. 157- 159).

Benelli, G. & Garzelli, A. (1999). Oil-spills detection in SAR images by fractal dimension estimation. *Proc. IGARSS'99*, vol. 1 (pp. 218- 220).

Bianchessi N. & Righini G. Planning and scheduling algorithms for the COSMO-SkyMed constellation. *Aerospace Science and Technology 12* (2008) 535–544

Bock, E. J.; Hara, T.; Frew, N. M. & McGillis, W. R. (1999) Relationship between air-sea gas transfer and short wind waves, *J. Geophys. Res.* 104(C11), 25,821–25,831.

Brekke, C. & Solberg, A. H. S. (2005) Oil spill detection by satellite remote sensing, *Remote Sensing of Environment* 95 1 -13

Calabresi, G.; Del Frate, F.; Lichtenegger, I.; Petrocchi, A. & Trivero P. (1999), Neural Networks for the oil spill detection using ERS-SAR data. *IGARSS '99 Workshop*, 28 June 1999 (Hamburg, Tammy Stein), 1, pp. 215-217, 1999.

Canny, J. (1986). A computational approach to edge detection. *IEEE Transactions on Pattern Analysis and Machine Intelligence, PAMI*, 8(6), 679- 698.

Change, L. Y.; Chen, K.; Chen, C.; & Chen, A. (1996). A multiplayer - multiresolution approach to detection of oil slicks using ERS - SAR image. *Proc. ACRS 1996*—17th Asian Conference of Remote Sensing, Sri Lanka.

Chen, C. F.; Chen, K. S.; Chang, L. Y.; & Chen, A. J. (1997). The use of satellite imagery for monitoring coastal environment in Taiwan. *Proc. IGARSS'97*, vol. 3. (pp. 1424-1426).

Cini, R. & Lombardini, P.P. (1978) Damping effect of monolayers on surface wave Motion in a liquid. *J. Colloid Interface Sci.*, 65, pp. 387-389.

Cini, R.; Lombardini, P.P. & Hünerfuss, H. (1983) Remote sensing of marine slicks utilizing their influence on wave spectra. *International Journal of Remote Sensing*, 4, 101-110.

Du, Y.; Vachon, P. W. & Wolf, J. (2002) Wind direction estimation from SAR images of the ocean using wavelet analysis. *Can J Remote Sens.* 28:498–509.

Ermakov, S. A.; Zujkova, E. M.; Panchenko, A. R.; Salashin, S. G.; Talipova, T. G. & Titov, V. I. (1986), Surface film effect on short wind waves. *Dyn. Atmos. Oceans* 10:31–50.

Espedal, H.; Johannessen, O. & Knulst, J (1996) Satellite detection of natural film on the ocean surface *Geophysical Research Letters*, 23, pp. 3151-3154.

Evans, D. L.; Alpers, W.; Cazenave, A.; Elachi, C.; Farr, T.; Glackin, D.; Holt, B.; Jones, L.; Liu, W. T.; McCandless, W.; Menard, Y.; Moore, R. & Njoku, E. (2005) Seasat—A 25-year legacy of success. *Remote Sensing of Environment* 94 384–404

Fernández-Varela, R.; Andrade, J.M.; Muniategui, S.; Prada, D. & Ramírez-Villalobos, F. (2008) Identification of fuel samples from the Prestige wreckage by pattern recognition methods. *Marine Pollution Bulletin* 56 335–347

Fichaux, N. & Rachin, T. (2002) Combined extraction of high spatial resolution wind speed and direction from SAR images: a new approach using wavelet transform. *Can J Remote Sens.*; 28:510–6.

Fiscella, B.; Giancaspro, A.; Nirchio, F.; Pavese, P. & Trivero, P. (2000) Oil spill detection using marine SAR images. *International Journal of Remote Sensing*, 21, 3561-3566.

Fiscella, B.; Gomez, F.; Pavese, P. & Trivero, P. (1995) Ocean surface films measured by interferential microwave probe *Nuovo Cimento C*, 18, 375-383.

Fiscella, B.; Lombardini, P.P. & Pavese P. (1982) Interferential microwave probe for measuring sea ripples. *Nuovo Cimento C*, 5, 247-255.

Fiscella, B.; Lombardini, P.P. & Trivero P. (1985a) Ripple Damping on Water Surface Covered by a Spreading Film: Theory and Experiment. *Il Nuovo Cimento*, vol.8C, no.5, pp. 491-500, 1985a.

Fiscella, B.; Lombardini, P.P.; Trivero, P.; Pavese, P. & Cini R. (1985b) Measurements of the Damping Effect of a Spreading on Wind-Excited Sea Ripples Using a Two Frequency Radar. *Nuovo Cimento*, C8, pp. 175-183,.

Fonseca, A. M.; Biscaya, J. L.; Aires-de-Sousa, J. & Loboa A. M. (2006) Geographical classification of crude oils by Kohonen self-organizing maps. *Analytica Chimica Acta* 556 374–382

Frew, N. M.; Goldman, J. C; Dennett, M. R. & Johnson A. S. (1990), Impact of phytoplankton-generated surfactants on air-sea gas exchange, *J. Geophys. Res.*, 95(C3), 3337–3352.

Frysinger, G. S.; Asher, W. E.; Korenowski, G. M.; Barger, W. R.; Klusty, M. A.; Frew, N. M. & Nelson R. K. (1992), Study of Ocean Slicks by Nonlinear Laser Processes 1. Second-Harmonic Generation, *J. Geophys. Res.*, 97(C4), 5253–5269.

Gade, M.; Alpers, W.; & Bao, M. (1996). Measurements of the radar backscattering over different oceanic surface films during the SIR-C/XSAR campaigns. *Proc. IGARSS'96* (pp. 860–862).

Gade, M.; Alpers, W.; Hühnerfuss, H.; Wismann, V. R. & Lange, P. A. (1998) On the Reduction of the Radar Backscatter by Oceanic Surface Films: Scatterometer Measurements and Their Theoretical Interpretation, *Remote Sens. Environ.* 66:52–70

Gade, M. & Redondo, J. (1999). Marine pollution in European coastal waters monitored by the ERS-2 SAR: a comprehensive statistical analysis. *OCEANS '99 MTS/IEEE Riding the Crest into the 21st century*, vol. 3 (pp. 1239- 1243).

Gade, M., & Ufermann, S. (1998) Using ERS-2 SAR images for routine observation of marine pollution in European coastal waters. *Proc. of IGARSS'98*, Seattle, USA, July 6-10,.

Gerling, T. W. (1986) Structure of the surface wind field from the Seasat SAR. *J Geophys Res.* 91:2308–20.

Goldman, J.C. & Dennet, M. R. (1983) Carbon dioxide exchange between air and sea water: no evidence for rate catalysis. *Science*, 220, 199-201.

Hersbach, H. (2003) CMOD5: an improved geophysical model function for ERS C-band scatterometry. *European Centre for Medium Range Weather Forecasting (ECMWF) Technical Memorandum*. 395:1–50.

Horstmann, J & Koch, W. (2005) Measurement of sea surface winds using synthetic aperture radars. *IEEE Trans Geosci Remote Sens*;30:508–15.

Horstmann, J; Koch, W; Lehner, S & Tonboe, R. (2002) Ocean winds from RADARSAT-1 ScanSAR. *Can J Remote Sens*. 28:524–33.

Hühnerfuss, H. (1986), The molecular structure of the system water/monomolecular surface film and its influence on water wave damping, *Habilitationsschrift, University of Hamburg*, Department of Chemistry, Hamburg, Germany, 245 pp.

Hühnerfuss, H.; Alpers, W.; Dannhauer, H.; Gade, M.; Lange, P. A.; Neumann, V. & Wismann, V. (1996), Natural and man-made sea slicks in the North Sea investigated by a helicopter-borne 5-frequency radar scatterometer. *Int. J. Remote Sens*. 17:1567–1582.

Hühnerfuss, H. and Garrett, W. D. (1981), Experimental seaslicks: their practical applications and utilization for basic studies of air-sea interactions. *J. Geophys. Res.* 86:439–447.

Hühnerfuss, H.; Gericke, A.; Alpers, W.; Theis, R.; Wismann, V. & Lange, P. A. (1994), Classification of sea slicks by multi- frequency radar techniques: new chemical insights and their geophysical implications. *J. Geophys. Res.* 99:9835–9845.

Hühnerfuss, H.; Walter, W.; Lange, P. A. & Alpers, W. (1987) Attenuation of wind waves by monomolecular sea slicks and the Marangoni effect. *J. Geophys. Res.*, 92 , 3961 - 3963.

Ivanov, A. (2000) Oil pollution of the sea on Kosmos -1870 and Almaz-1 radar imagery, *Earth Observation & Rem. Sensing*, 15(6), 949-966, 2000.

Kanaa, T.F.N.; Tonye, E.; Mercier, G.; Onana, V.P.; Ngono, J.M.; Frison, P.L.; Rudant, J.P. & Garello, R. (2003). Detection of oil slick signatures in SAR images by fusion of hysteresis thresholding responses. *Proc. IGARSS'03*, vol. 4 (pp. 2750– 2752).

Keller, J. M.; Chen, S. & Crownover, R. M. (1989). Texture description and segmentation through fractal geometry. *Computer Vision, Graphics, and Image Processing*, 45, 150–166.

Koch, W. (2004) Directional analysis of SAR images aiming at wind direction. *IEEE Trans Geosci Remote Sens*. 42:702–10.

Kotova, L.; Espedal, H.A. & Johannessen, O.M. (1998) Oil Spill Detection Using Spaceborne SAR: a Brief Review. *Proc. 27th Int. Symposium on Remote Sensing Environmental*, 8-12 June 1998, Tromso, Norway, 791-794.

Lin, H.; Xu, Q. & Zheng, Q. (2008)An overview on SAR measurements of sea surface wind, *Progress in Natural Science* 18 (2008) 913–919

Liu, A. K.; Peng, C. Y. & Chang, S. Y. (1997). Wavelet analysis of satellite images for coastal watch. *IEEE Journal of Oceanic Engineering*, 22(1), 9– 17.

Liss, P. S.; Watson, A. J.; Bock, E. J.; Jaehne, B.; Asher, W. E.; Frew N. M.; Hasse, L.; Korenowski, G. M.; Merlivat, L.; Phillips, L. F.; Schluessel, P. & Woolf. D. K. (1997) Physical processes in the microlayer and air-sea exchange of trace gases. In: *The sea surface and global change*, Cambridge University Press, New York, 1-33,.

Lombardini, P.P.; Fiscella, B.; Trivero, P.; Cappa, C. & Garrett, W. D. (1989) Modulation of the Spectra of Short Gravity Waves by Sea Surface Film: Slick Detection and Characterization with a Microwave Probe. *J. Atmosph. Ocean Techn.*, 6, pp. 882- 890.

Lombardini, P.P.; Piazzese, F. & Cini, R. (1982) The Marangoni Wave in Ripples on an Air-Water Interface Covered by Spreading Film. *Nuovo Cimento*, C5, pp. 256-263.

Loglio, G.; Tesei, U. & Cini, R. (1986) Viscoelastic dilatation processes of fluid/fluid interfaces: time-domain representation. *Colloid Polym. Sci.*, 264, 712.

Lucassen, J. (1982), Effect of surface-active material on the damping of gravity waves: a reappraisal. *J. Colloid Interface Sci.* 85:52–58.

Lucassen-Reynders, E. H. & Lucassen, J. (1969) Properties of Capillary Waves. *Advan. Colloid interface Sci.*, 2, pp. 347-395,.

Maio, A. D.; Ricci, G. & Tesauro, M. (2001). On CFAR detection of oil slicks on the ocean surface by multifrequency and/or multipolarization SAR. In Radar conference, 2001. *Proceedings of the 2001 IEEE* (pp. 351- 356).

Manore, M. J; Vachon, P. W.;, Bjerkelund, C.;, Edel, H. R. & Ramsay, B. (1998). Operational use of RADARSAT SAR in the coastal zone: The Canadian experience. *27th International Symposium on Remote Sensing of the Environment*, Tromsø, Norway, June 8–12 (pp. 115- 118).

Marghany, M.; Cracknell, A. P. & Hashim M. (2009) Modification of fractal algorithmfor oil spill detection from RADARSAT-1 SAR data. *International Journal of Applied Earth Observation and Geoinformation* 11 96–102.

Melsheimer, C.; Alpers, W &. Gade, M. (1998) Investigation of multifrequency - multipolarization radar signatures of rain cells over the ocean using SIR-C/X-SAR data, *J. Geophys. Res.*, 103(C9), 18,867–18,884.

Mercier, G.; Derrode, S.; Pieczynski, W.; Caillec, J. M. L. & Garello, R. (2003). Multiscale oil slick segmentation with Markov Chain Model. *Proc. IGARSS'03*, vol. 6 (pp. 3501–3503).

Nirchio, F.; Sorgente, N.; Giancaspro, A.; Pavese, P.; Ravera, R. & Trivero, P. (2002) A method to detect oil spill based on SAR images" Proc. OIL SPILL, Rhodes, Sept. 2002.

Nirchio, F.; Sorgente, N.; Giancaspro, A.; Biamino, W.; Parisato, E.; Ravera, R.; Trivero, P. & Pavese, P. (2005a) Oil spills automatic detection from SAR images. *Proc. of the 2004 Envisat & ERS Symposium*, Salzburg, Austria 6-10 September 2004, ESA SP-572.

Nirchio, F.; Sorgente, N.; Giancaspro, A.; Biamino, W.; Parisato, E.; Ravera, R. & Trivero, P. (2005b) Automatic detection of oil spills from SAR images. *International Journal of Remote Sensing*, Vol. 26, N. 6, 1157-1174

Onstott, R. & Rufenach, C. (1992), Shipboard active and passive microwave measurement of ocean surface slicks off the southern California coast. *J. Geophys. Res.* 97:5315-5323.

Quilfen, Y.; Chapron, B.; Elfouhaily, T.; Katsaros, K.; & Tournadre, J. (1998), Observation of tropical cyclones by high-resolution scatterometry, *J. Geophys. Res.*, 103(C4), 7767–7786.

Rum, G. (2000) The Interest of a Constellation: The COSMO-SkyMed Project. *Air & Space Europe*, 2 (4), 53-58

Sabins, F .F. (1997) Remote sensing - Principles and Interpretation. W.H. Freeman & Co, N.Y. Scott, J.C. 1986. *Surface films in oceanography. ONRL Workshop* Rep. C-11-86. Office of Nav. Res., London, 19-34.

Scott, J. C. (1986) Surface films in oceanography. *ONRL Workshop* Rep. C-11-86. Office of Nav. Res., London, 19-34,.

Scully-Power, P. (1986) Navy oceanographer shuttle observations, *Mission Report. Naval Underwater Center* 2 .2,5.3,.

Skøelv, Å. & Wahl, T. (1993). Oil spill detection using satellite based SAR, Phase 1B competition report. *Tech. rep., Norwegian Defence Research Establishment.*

Solberg, A. H. S.; Dokken, S. T. & Solberg, R. (2003). Automatic detection of oil spills in Envisat, Radarsat and ERS SAR images. *Proc. IGARSS'03*, vol. 4 (pp. 2747–2749).

Solberg, A. H. S.; Storvik, G.; Solberg, R. & Volden, E. (1999). Automatic detection of oil spills in ERS SAR images. *IEEE Transactions on Geoscience and Remote Sensing*, 37(4), 1916- 1924.

Stoffelen, A & Anderson, D. (1997) Scatterometer data interpretation: estimation and validation of the transfer function CMOD4. *J Geophys Res* 102:5767–80.

Topouzelis, K.; Karathanassi, V.; Pavlakis, P. & Rokos, D. (2007) Detection and discrimination between oil spills and look-alike phenomena through neural networks. *ISPRS Journal of Photogrammetry & Remote Sensing* 62, 264–270

Trivero, P.; Fiscella, B.; Gomez, F. & Pavese, P. (1998) SAR detection and characterization of sea surface slicks. *International Journal of Remote Sensing*, 19, 543-548,.

Trivero, P.; Fiscella, B. & Pavese, P. (2001) Sea surface slicks measured by SAR. *Il Nuovo Cimento*, 24 C, 1, pp. 99-111.

Vachon, P. W.; Thomas, S. J.; Cranton, J. A.; Bjerkelund, C.; Dobson, F. W. & Olsen, R. B. (1998). Monitoring the coastal zone with the RADARSAT satellite. *Oceanology International 98, UK,* March 10-13, 10 pages.

Wackerman, C. C.; Horstmann, J. & Koch, W. (2003) Operational estimation of coastal wind vectors from RADARSAT SAR imagery. In: *Geoscience and Remote Sensing Symposium, IGARSS '03.* 2:1270–2.

Wei, Y. and Wu, J. (1992), In situ measurements of surface tension, wave damping, and wind properties modified by natural films. *J. Geophys. Res.* 97:5307–5313.

Wu, J. (1989), Suppression of oceanic ripples by surfactant - spectral effects deduced from sun-glitter, wave-staff and microwave measurements. *J. Phys. Oceanogr.* 19:238–245.

Wu, S. Y. & Liu, A. K. (2003). Towards an automated ocean detection, extraction and classification scheme for SAR *International Journal of Remote Sensing*, 24(5), 935–951.

Zutic, V. B.; Cosovic, B.; Marcenko, E. & Bihari, N. (1981) Surfactant production by marine phytoplankton. *Mar. Chem.*, 10: 505-520.

22

Development of Flood Space Monitoring in Kazakhstan

O.P. Arkhipkin, L.F. Spivak and G.N. Sagatdinova
National Center of Space Research and Technology
Kazakhstan

1. Introduction

From the moment of appearance of the first systems of remote sensing of the Earth, allowing to carry out survey in a real time mode, one of the main directions of their use is the operative control of various emergency situations, first of all fires and high waters. At the same time the general principles and methods of space monitoring of high waters and flooding have been formulated (Barton & Bathols, 1989). Now this direction still intensively develops in the various countries.

In Kazakhstan such large flooding which are characteristic for the Siberian rivers and modern Europe were observed seldom. Therefore real conditions for development of system of space monitoring of flooding have appeared after installation at Space Research Institute (now it was a part of National Center of Space Research and Technologies) in 2001 reception station of the Russian firm SCANEX providing reception of Terra MODIS data (the resolution 250 m) in a real time mode. Since then this system actively developed in Kazakhstan (Spivak et al., 2004, 2005; Arkhipkin et al., 2007; Arkhipkin & Sagatdinova , 2008). It should provide the operative observation of republic territory, detection and mapping the real and potential centers of flooding of natural and technogenic character.

The brief description of technology of work of flood space monitoring system and the most important results is presented below.

2. Water system of Kazakhstan

Kazakhstan is located in the centre of the Eurasians continent and enters into ten greatest on the area of the countries of the world. Its area is 2,7 million square kilometers. In too time in such big territory there lives the small population (only 16 million people).

Despite the location and arid climate, Kazakhstan has considerable number of various water objects /figure 1/. Largest of them well-known all over the world, thanks to the environmental problems which make essential impact on considerable territories of the Eurasians continent. There are Caspian Sea and Aral Sea. We have also many lakes (more than 35 000), including such large as Balkhash, and many (more than 3 000) artificial water objects (water basins and ponds).

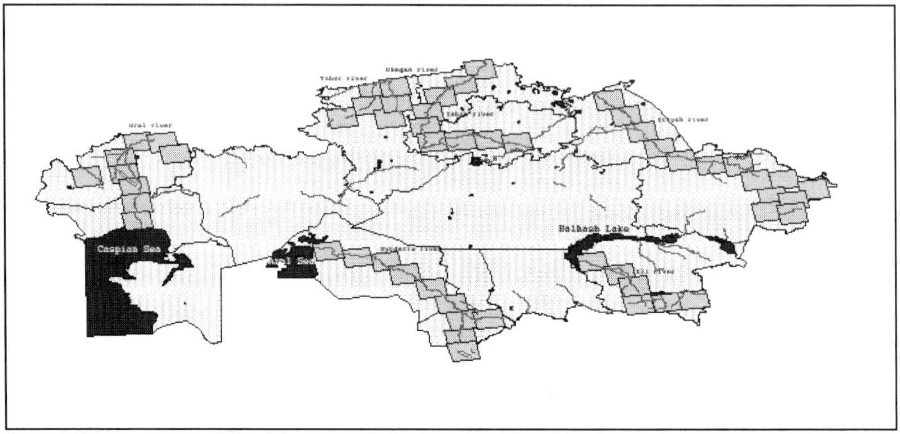

Fig. 1. Main regions of Kazakhstan with high risk of flooding by spring water and flood

In Kazakhstan there is a great number of the rivers. Mainly they have the average and small size. The largest rivers are Irtish, Ural, Tobol, Ishim, Syrdaria and Ili. During spring thawing snow the volume of flow of these rivers sharply increases, sometimes more than 1000 times. They can overflow banks and flood large territories. The maximal hoisting of water recorded in the river Ural was in 1942, when the water was pushed up on 10-11 m above the normal level.

Let's notice that all large river systems of Kazakhstan have transboundary character. Therefore their problems become the general for the several countries: Russia, China, Uzbekistan, the Kyrgyz Republic. In the past years we observed a very interesting situation in the region of the Syrdaria River. On the one hand, we observed big shortage of water in the Aral Sea, and on the other hand - during the winter-spring period in middle stream of the Syrdaria we observed its overabundance. Such situations arose because of inconsistency of questions of water use between Kazakhstan, Uzbekistan and Kyrgyzstan.

3. Main tasks of Space monitoring of a water objects of Kazakhstan

At the present time it is more efficient to control water objects on such big territory as Kazakhstan using space monitoring. Each class of water objects has its own tasks.

Environmental problems of the Caspian region are caused by rapid development of oil-extracting branch around Northern Caspian Sea and sea level rising. Therefore the main tasks of space monitoring of water surface of the Caspian Sea are:

- Operative detection and monitoring of migration of oil spills on the water surface during flooding of oil derrick during the period of high water on the Caspian Sea and movement of tankers on its water area of Caspian Sea;
- Monitoring of oil derrick in the Caspian shelf zone,
- Control of ice conditions in this region.

A number of organizations of Kazakhstan, including SRI take part in the solution of these problems. Base for the solution of these problems is RADARSAT (modes from Fine to ScanSAR Wide) and IRS data.

Environmental problems of the Aral region are caused by sharp and fast reduction of the water surface. As a result of it the surface of the dried up bottom of Aral Sea which is a source of the powerful dust storms extending on considerable distances increases also quickly. In this case the primary tasks of space monitoring are:
- Monitoring of a water surface of the Aral Sea,
- Detection and monitoring of dynamics of development of dusty storms,
-Detection and analysis of the dusty storms centers,
- Modeling of occurrence and dynamics of development of dusty storms.

Many organizations both in Kazakhstan and in other countries participate in the solution of these problems. SRI also takes part in it, including participation in the international grants (for example, CALTER).

The main task of Space monitoring of water objects is mapping of flooding zones during passage of freshet waters and flooding. Floods are a considerable menace for the part of the population of Kazakhstan living not only on the banks of large rivers, such as the Irtish, Ural,Tobol, Ishim, Syrdaria, but also on the banks of small rivers. In recent years the situation in the Kyzyl-Orda oblast has become especially aggravated.

At the previous and initial stage of high water passage we carried out mapping of destruction of the snow cover on the territory of the region, and recently we made mapping of destruction of an ice cover on large water reservoirs and lakes. In view of worsening of flood situation in the middle stream of the Syrdaria during the winter-spring period in the past years, the space monitoring of dynamics of filling of the Chardara reservoir was formulated as a special task. The development of flood situation in this region strongly depends on the Chardara reservoir. Space monitoring of dynamics of filling of the water surface of the Chardara reservoir is divided into two tasks. The first is satellite surveying of the current situation, and the second is comparison of this situation with the dynamics of the situation development in the past years.

4. Space monitoring of snow cover destruction

As it was mentioned above, the first stage of space monitoring of high waters is space monitoring of the snow cover destruction. Special programs are devised to automatize the process of process of mapping of snow, ice and water covers, which allows to carry out, in the automated mode, the procedure of transformation to a required projection, cutting of necessary territory, calculation of various indexes (NDSI, NDVI, VI) and classification of images in MODIS data. Simultaneously masks of a cloudy cover and maps of temperature of of the earth surface are formed. Masks of a cloudy cover specify territories on which the condition of an earth surface is not defined. The maps of temperatures combined with masks of a snow or ice cover, specify zones of active thawing on which the temperature is more $0^{o}C$ (melt snow).

By the results of space monitoring of destruction of a snow cover, the special maps indicative of the current condition of a snow cover /figure 2/, dynamics and calendar terms of destruction a snow cover are constructed /figure 3/. This information is very interesting for the forecast of a freshet situation. It allows to estimate terms of the beginning of thawing of a snow cover (early, normal, late), and also its rate (fast, normal, slow thawing).

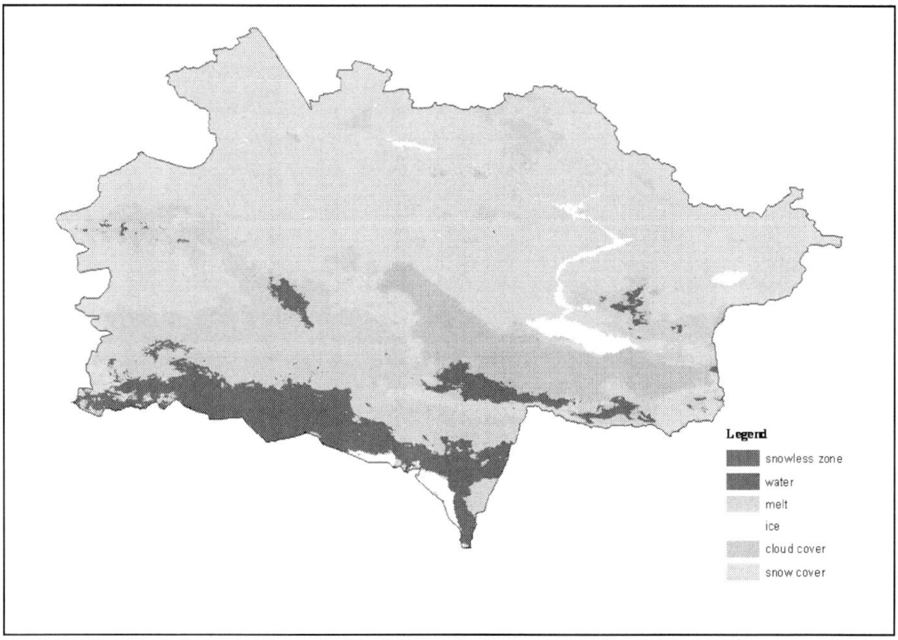

Fig. 2. Daily map of snow coverage (East-Kazakhstan oblast, March 22, 2009)

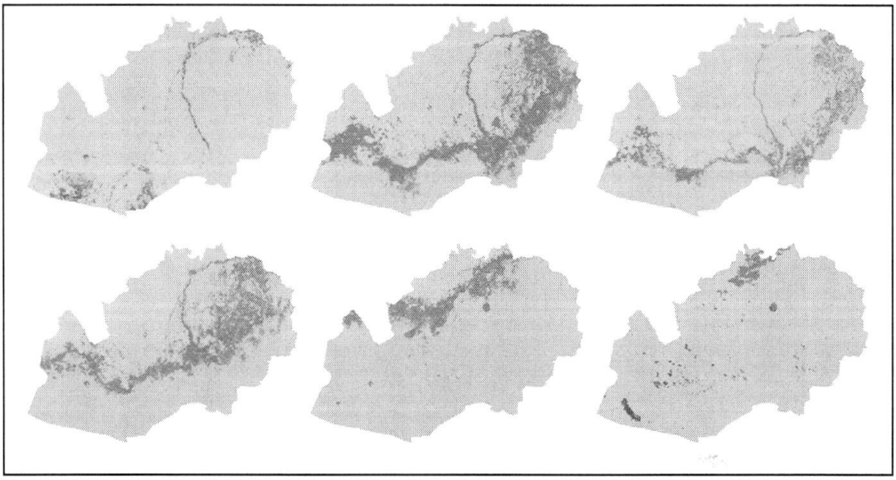

Fig. 3. Decade maps of snow melting (Western-Kazakhstan oblast)

Monitoring of destruction of the ice cover on large water objects (Balkhash, Alakol, Buhtarma, etc.), represents certain interest for regional emergency agencies. Figure 4 shows the dynamics of destruction of the ice cover on Lake Balkhash in March, 2007. The figure also shows zones of active thawing.

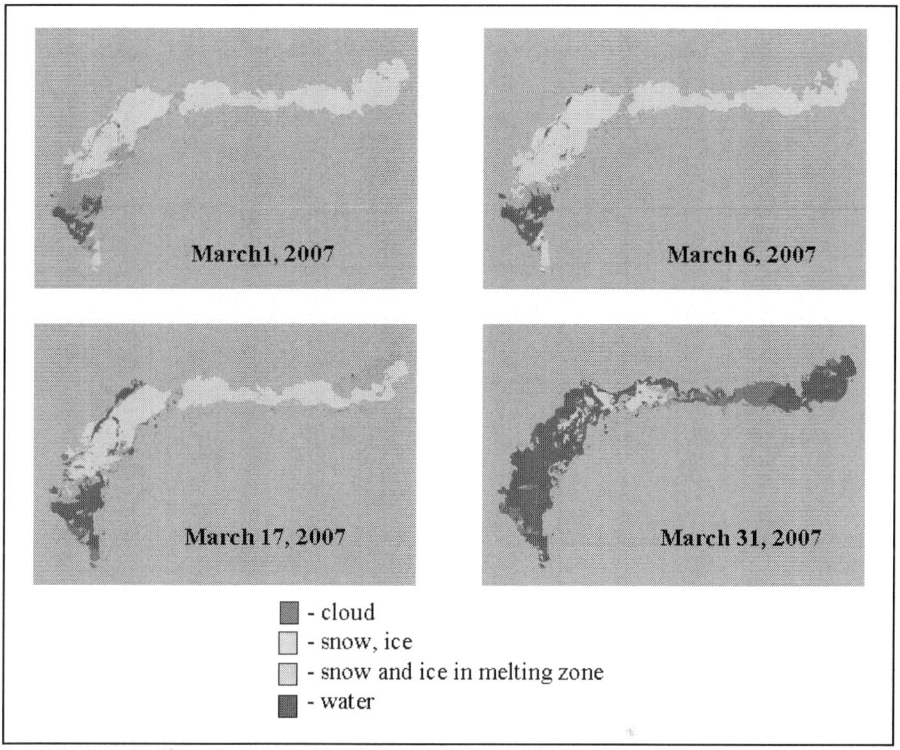

Fig. 4. Dynamic of ice destruction in Balkhash Lake in 2007

5. System of operative flood space monitoring

5.1 Space segment

Functional basis of work of flood space monitoring system in an operative mode are the reception stations of remote sensing data located in Astana and Almaty. They carry out regular receiving of data from NOAA, EOS Terra and Aqua, Indian satellites IRS and Canadian radar satellite RADARSAT-1. Zones of radio visibility of receiving stations cover Kazakhstan, a significant part of Russia and Asian region.

5.2 Technology of flood space monitoring

The technology of flood space monitoring /figure 5/ is based on daily EOS-AM Terra MODIS images of territories for which there is a high risk of flooding, including the parts located in the neighbor countries. The main task is operating mapping of flooding zones during floodwater passage. On the basis of these images after thematic processing, masks of flood areas are created. Zones of flooding are defined as a difference of water surfaces in normal conditions and during high water. To except cloudy cover and clouds shadows the cloud mask is used. To except wet soil the vegetation index NDVI is used. The operative situation is compared to maps for the previous day and the most dangerous territories with high dynamics of developing of increase water are identified.

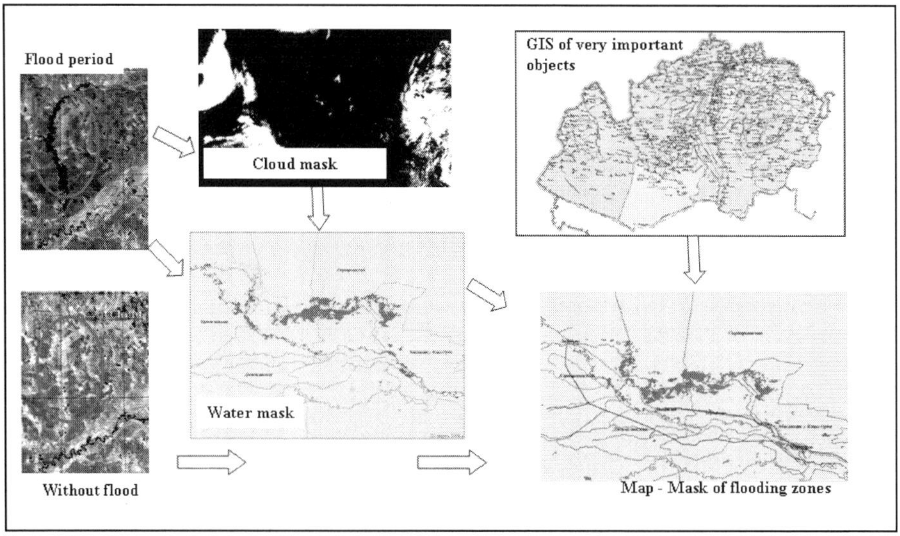

Fig. 5. Functional scheme of the GIS-technology of operative flood space monitoring

In order to estimate the potential danger of flooding special GIS is used. It contains the information about settlements and towns, road and railway networks, lines of the electric system, oil-and gas pipelines, forests, especially important objects, etc. Combining these layers and zones of flooding it is possible to define their location with respect to the nearest settlements and especially important objects, and distance up to them. Final maps of flooding zones /figure 6/ are sent by e-mail to regional emergency agencies.

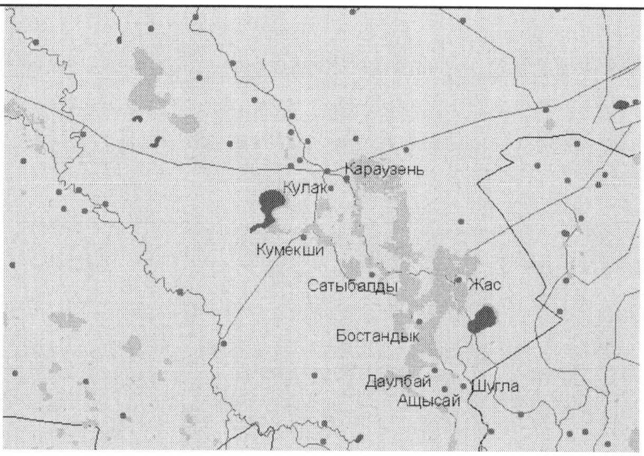

Fig. 6. Map of flood zones on the territory of Kaztalovsky raion of West-Kazakhstan oblast in April 8, 2004

5.3 Algorithm of allocation of flooding zones on space images

The problem of allocation of flooding zones on remote sensing data is solved using the algorithms of automatic classification in the programming environment ArcGIS-9.1 passes in three stages /figure 7/. At the first stage five basic classes of objects are allocated: a cloudy cover, a snow cover, a water surface covered with ice, a water surface and a terrestrial surface free from snow.

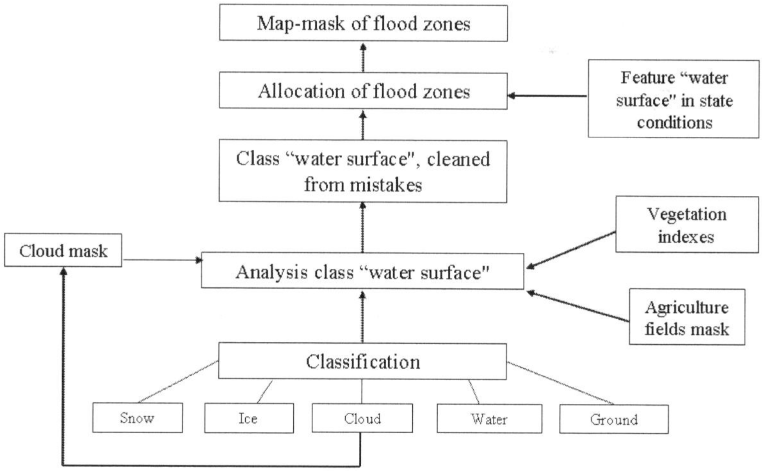

Fig. 7. Algorithm of allocation of flood zones on space images

The main interest is presented by a water surface. Therefore at the second stage the additional analysis of this class for the purpose of allocation of false objects is carried out. First of all, it is shades from clouds, lately ploughed fallows and wet soils. In order to except the cloud mask received at the first stage, and fact that shades on a configuration repeat clouds are used. In order to remove fallows which are false carried to water objects, masks of agricultural fields and analysis of their structure are used. Wet soils are excluded by means of analysis using vegetation indexes /figure 8/. Let's notice that sometimes sunlight dazzles from a water surface creating additional hindrances. Besides it is necessary to know the normal state of water objects in order to indicate deflections in their location.

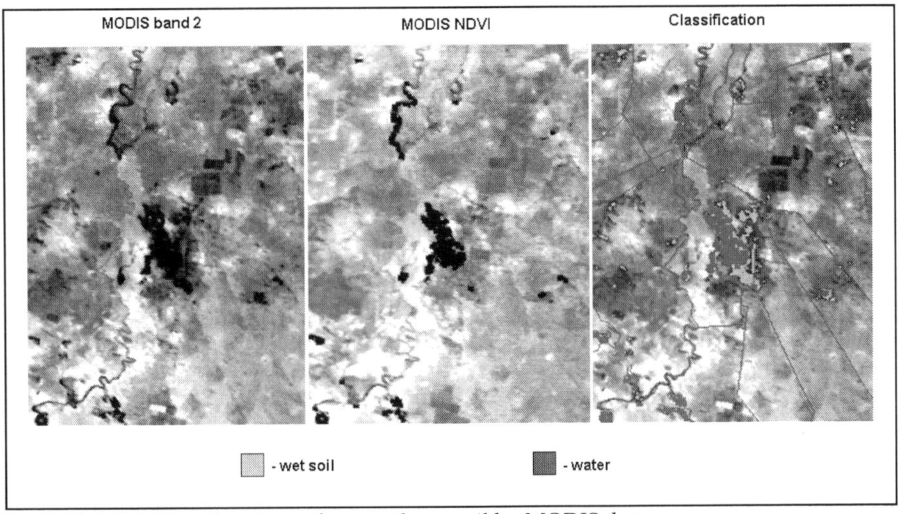

Fig. 8. Classification of water surfaces and wet soil by MODIS data

At the third stage operative maps of the flooded territories /figure 6/ at level of region and separate districts are formed using the mask of water surfaces in the normal conditions defined on the autumn space images.

6. Estimation of flood risk zones

First of all, it is necessary to note that concept «the estimation of risk of an emergency situation (fires, flooding, etc.) » has dual sense. On the one hand, it is a current estimation of a real condition of concrete territory during a concrete period of time which is formed by the characteristics of this territory (a relief, vegetation etc.) and its meteorological condition (temperature, amount of precipitation, their intensity, storm activity, etc.) during a concrete period of time. On the other hand, «the estimation of risk of an emergency situation» can be formed on the basis of a statistical estimation of results of long-term supervision of investigated territory, including the remote sensing data. Such approach allows to receive an integrated estimation at any time period in each concrete place. In this case it is not required to know the characteristic of the land environment. We follow the second method of estimation.

In process of accumulation of the information time series remote sensing data (seasonal and long-term) are formed. They enable to characterize the development of flood situations in time during a current season and to compare them with the previous seasons. Also on the basis of of analysis of all long-term series of remote sensing data we can estimate risks of flood situations on various territories. The territories are ranged by the degree of risk of flooding by freshet waters and floodwater. For this purpose available long-term series of remote sensing data are analyzed and the frequency of territory flooding is defined. The more often the territory flooded in this period, the higher is the risk of flooding.
The functional scheme of GIS-technology of zoning of investigated territory by the degree of flood risk using the long-term space monitoring data consists of three blocks corresponding to three stages of determination of resultant estimation in the programming environment ArcGIS-9.1 /figure 9/. The first is formed by the daily data about the areas which have suffered from flood (flooding zone), received in the process of operative monitoring. The second contains the annual data about the total areas which have suffered from flood which are formed of the data of the first block and represent total zones of flooding for each concrete year. The third block contains results of zoning of investigated territory by the degree of risk of flooding which are calculated from the analysis of the annual data.

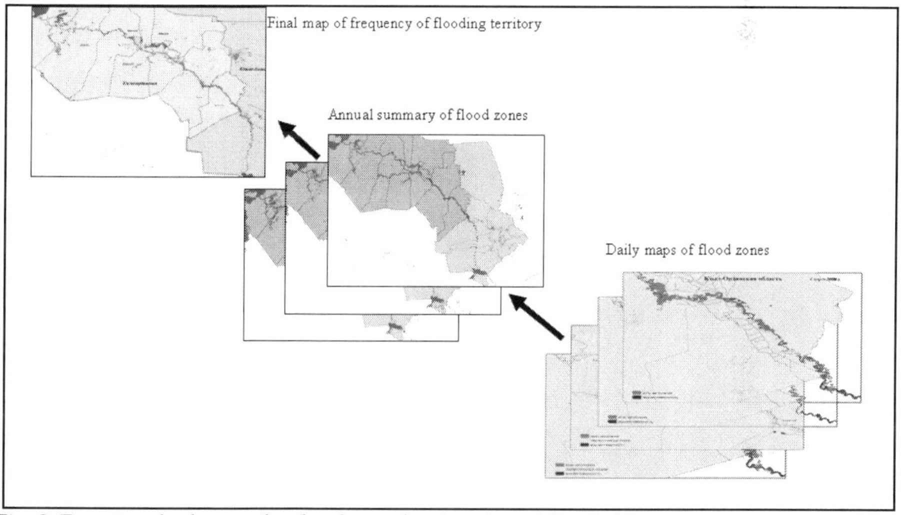

Fig. 9. Functional scheme of technology of estimation of flood risk zones

Such information is very useful for planning economic activities. Besides it is useful for planning protective actions against repeated emergency situations and for analyzing the efficiency of the measures accepted by local authorities for struggle against high waters. Not all zones of flooding present danger to life and the human economic activities. Some of them can be even useful. For example, places of gathering of freshet waters which are used by nature and man in the further.

7. Practical results of flood space monitoring

Flood space monitoring is exploited in West-Kazakhstan (2003-2009), Karaganda (2005-2009), East--Kazakhstan (2007-2009) oblasts and Syrdaria river region including a Chardara water basin (2003-2009) in real or near real time mode. Space monitoring of high waters is carried out in the spring, basically in March - April, and for a middle stream of Syrdarya during the winter-spring period. For all observable regions, except Syrdarya, in the recent years (2006-2009) weak high waters are observed. In the middle stream of Syrdarya there are critical situations every year. Also in 2009 there was a flooding of some settlements located around one of the lakes in East Kazakhstan which was caused by climatic features of the past years. Earlier these places were not flood. Some results of flood space monitoring are described described further.

7.1 Space flood monitoring of West Kazakhstan

In this region there are the Ural River and many small rivers running and not running into the Ural. Local agencies of emergency situations have defined the most important objects for space monitoring in a high water period /figure 10/. This group also included the objects located in territory of Russia which in many respects define passage of flood waters on the Ural River.

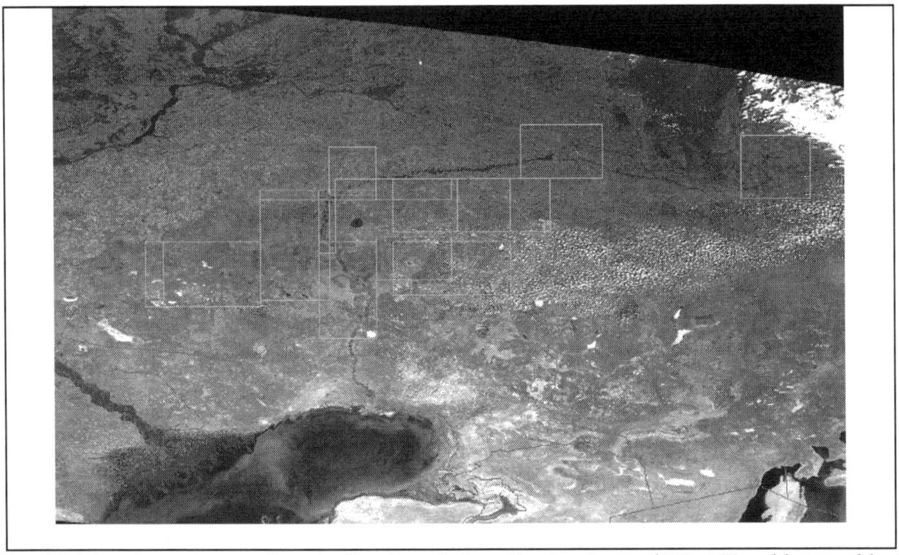

Fig. 10. Most important regions in flood period on the territory of West-Kazakhstan oblast and Russia

The results of long-term monitoring show that more or less intensive high water was observed in first three years. In 2003 the high water has begun in the second decade of April, reached its peak in the middle of April and practically vanished at the beginning of May. Thus the water level in the river Ural Mountains was below the average, and in a number of small rivers above the average (in some areas it was strong). In 2004 the water level was below the average practically everywhere. In 2005 the high water appeared early (at the beginning of April) and continued to increase till the middle of May, then it began to decrease, but at the beginning of June it did not vanish. Thus the basic water stream passed on the Ural River. Further intensity of high waters has considerably decreased.

It is necessary to note, that high overcast is a vital problem for carrying out of operative space monitoring of flooding, especially in the West Kazakhstan. So in 2005 during 45 days of high water development we managed to receive 16 space images of territories of West Kazakhstan, suitable for mapping of flooding zones. For separate areas this value was even less. For example, for the Uralsk area we got 9 images. In 2007 during the high water development it was not possible to receive any space images of this territory, suitable for mapping of flooding zones. In 2008 and 2009 the situation was better, only there was no high water.

In the West Kazakhstan often happens that at night the overcast is lower than during the daytime. Therefore the night images NOAA AVHRR and EOS-AM Terra MODIS in infra-red band are used for flooding monitoring /figure 11/. The water has higher thermal lag, therefore water surfaces at night are warmer than ground (soil). In the paper (Barton & Bathols, 1989) it is noticed that sometimes during the flood the thermal canals at night gave better land-flood discrimination than the visible data during the daytime. Certainly radical way of solving the a problem of overcast is monitoring of only areas with especially dangerous freshet situation of data RADARSAT, but it is limited due to their high cost and complexity of the operative order.

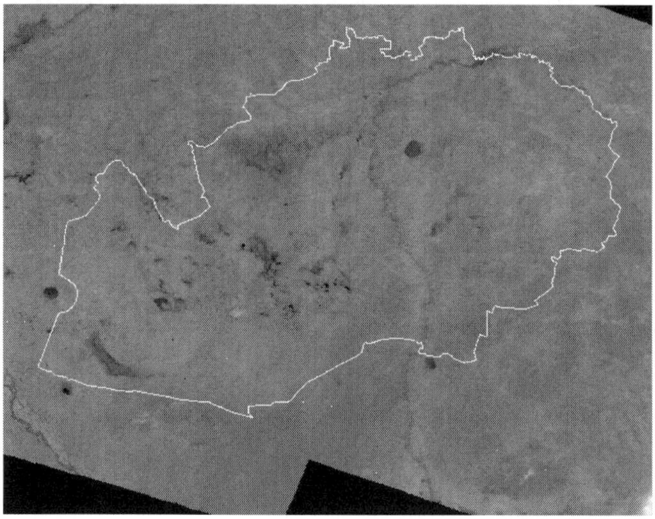

Fig. 11. Night image NOAA AVHRR (West-Kazakhstan, 28 April 2003)

7.2 Space Monitoring of Syrdaria Region

Last six years in this region the intense situation with flooding is observed. As is has been noticed above, the reason of it is not settled question of use of the river water between Kazakhstan, Uzbekistan and Kyrgyz Republic. In the Soviet period the Togtogul water basin in the Kyrgyz Republic, situated in riverheads, was used mainly for the agricultural purposes. In the winter it accumulated water, and in the summer water went down for watering of agricultural fields, mainly cotton. Now the Kyrgyz Republic actively uses this water basin for electric power generation, especially in winter. As a result, there is active discharge of water in winter. Surplus of water arrives at the Chardara water basin located downstream. It is located on border of Kazakhstan and Uzbekistan. For prevention of danger of overflow of a water basin during the spring high water the Kazakhstan authorities are forced to dump part of water downstream. It leads to flooding of territories in middle stream of Syrdaria river.

For the first time such situation arose during the winter of 2003-2004. During the winter of 2004-2005 a real danger of overflow of the Chardara reservoir was created. The government of Kazakhstan hardly managed to convince Uzbekistan to open dump of water in Arnasai hollow and stabilize the situation. In 2004-2005 the conditions turned out to be even more difficult. The winter of 2003-2004 was not snowy with very early spring. Precipitations were heavy in the winter of 2004-2005, they were much higher than the norm. Heavy snowfalls created snow stocks in the mountains. Besides, strong frosts held down a thick layer of ice on the shoaled river-bed of the Syrdaria. Late spring shifted terms of active high water (melting of snow). Therefore the most dangerous situation developed in the middle of March.

The situation on the Chardara water basin is one of the main characteristics of the freshet situation. Therefore from the end of 2003 the remote sensing control of filling of Chardara reservoir and development of high waters over this region in the real time mode has been carried out. Figure 12 shows dynamics of changes of the Chardara reservoir water surface in 2003-2008. As follows from figure 12 the situation was difficult almost every season. However, if during first two seasons the authorities were afraid threat of overflowing of the water basin, further the situation was under the control of regional authorities, except two critical situations in 2007 and 2008.

In 2007 the critical situation arose at the end of the first decade of February and led to flooding of the populated territories in area of Kzyl-Orda (figure 13). For find out of the reasons of this process we gave the regional authorities the information about the dynamics of filling of the Chardara reservoir during this period. From figure 13 shows that, firstly, at this time there was a discharge of water from the Chardara reservoir (left diagram), and, secondly, it was filled by two thirds of the maximal size, which was fixed during many years supervision (right diagram). Whether it was the reason of the critical situation, we could not prove. The special commission found out this reason.

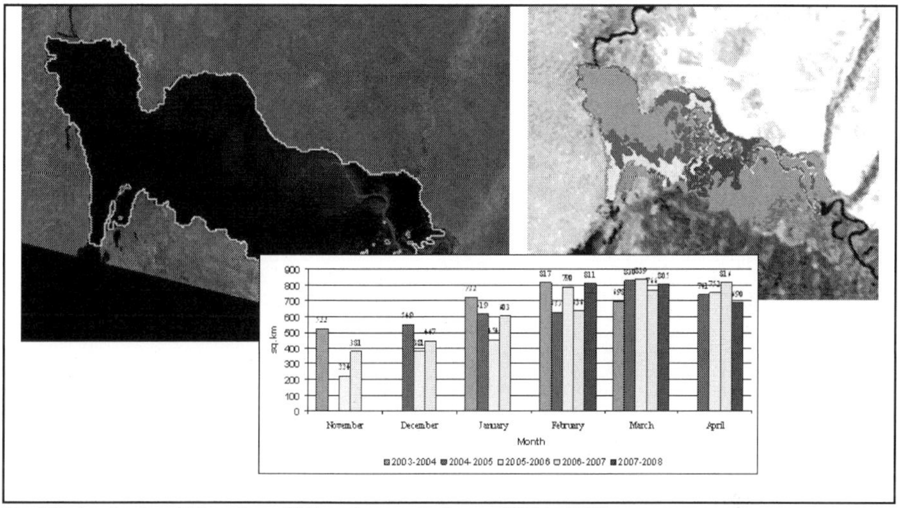

Fig. 12. Analysis of Chardara filling dynamic from 2003 to 2008

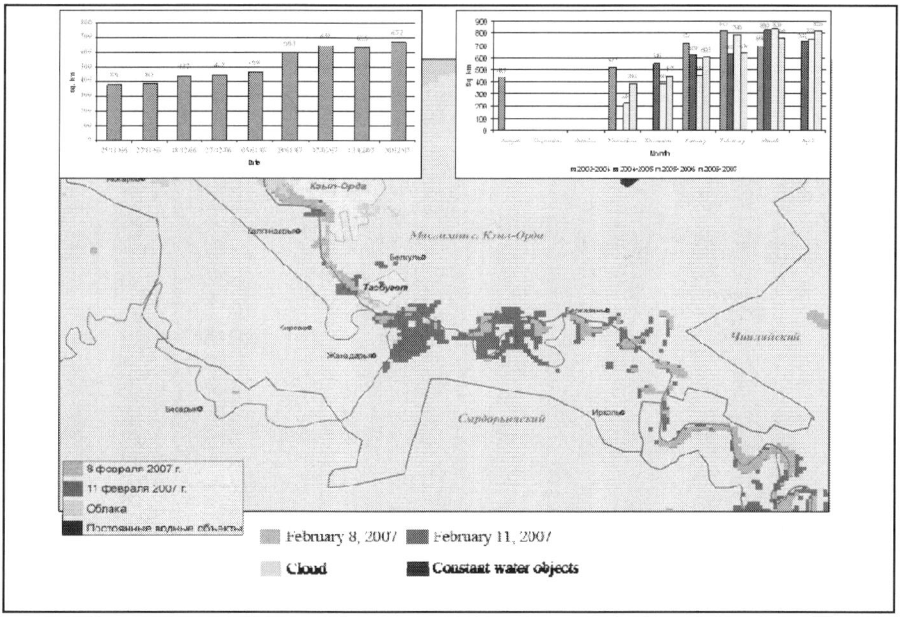

Fig. 13. The analysis of the reasons of break of a dam near Kzyl-Orda in first decade of February 2007.

Flooding in the third decade of February 2008 had much more serious consequences both in terms of the number of flood victims and destroyed houses. The reason of it was not the

Syrdaria. The reasons were storm rains abnormal for this period of the year and abnormal cold winter which turned the ground into an ice trench. The settlements which had never been flooded and were not ready to such situation suffered from flooding. Actually, high water on the Syrdaria passed without special problems (figure 14). We could not fix the critical situation during the peak period with the help of MODIS data because of the overcast, and used for these purposes RADARSAT data, which did not help either because of some objective and subjective reasons, in particular, because of impossibility of the operative order.

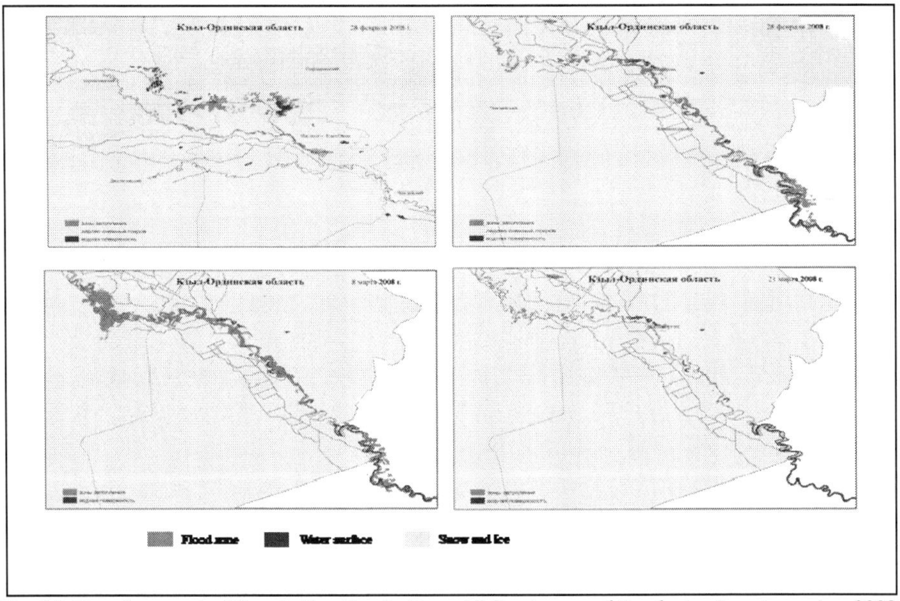

Fig. 14. Development of flood situation in the middle stream of Syrdaria river in spring 2008 (North part of Kzil-Orda oblast, 28 February; South part of Kzil-Orda oblast, 28 February; South part of Kzil-Orda oblast, 8 March; South part of Kzil-Orda oblast, 31 March)

Let's notice that same critical situation has arisen this year in East Kazakhstan (look above the text in paragraph 7 beginning).
In 2009 the high water has passed easy in the middle stream of Syrdaria river.

7.3 Zoning on degree of risk of flooding of territory of the West-Kazakhstan and a middle stream of Syrdaria River

The technology of estimation of flood risk zones described above has been used for zoning by the degree of risk of flooding of the territory West Kazakhstan (a six-year number of the remote sensing data is analyzed), and middle stream of the Syrdaria River (high waters for the last five winter-spring periods are considered).
Figure 15 maps of zones of risk of flooding on the territory of West-Kazakhstan oblast, as a whole and its separate fragments. For these years more or less intensive high water was

observed on the territory only two times, but the same information allows to carry out analysis of danger of high waters for various territories.

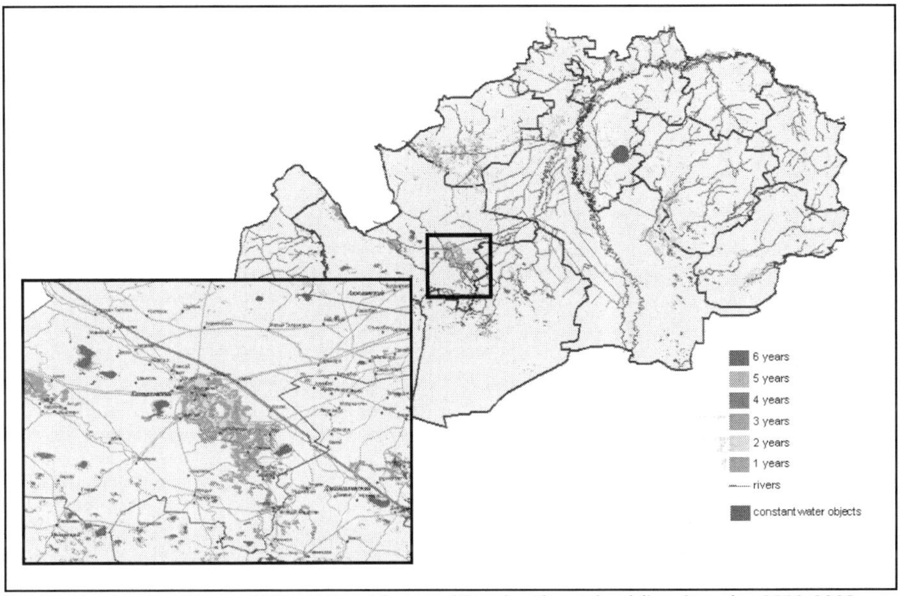

Fig. 15. Zoning territory of West-Kazakhstan oblast by the risk of flooding for 2003-2008

From the insert of figure 15 we can see that territories constantly flooded during 6 years are located far from settlements and roads. They do not represent danger and are regular places of gathering of flood waters as are located far from constant water objects. On the other hand the territories flooded only one or two times, are located in immediate proximity from settlements and roads. It is necessary to pay paramount attention to these territories in planning of protective actions.

More difficult situation was with high waters in the middle stream of Syrdaria River where practically every year there were critical situations. The results of zoning of the territory in the middle and the low stream of the Syrdaria River (Kazakhstan part of the river) by the degree of risk of flooding for the last five winter-spring periods are presented in Figure 16. In more details these results are presented in the insert of figure 16 for one of the main flooded territories in the middle stream of Syrdaria River. Figure 17 shows the same territory with spatial allocation of zones with high frequency of flooding (3-5 times for 5 seasons) is presented. Especially critical situation developed for settlement Dzhusaly where regularly flooded territories are located in immediate proximity from it and on the territory, where the railway is located.

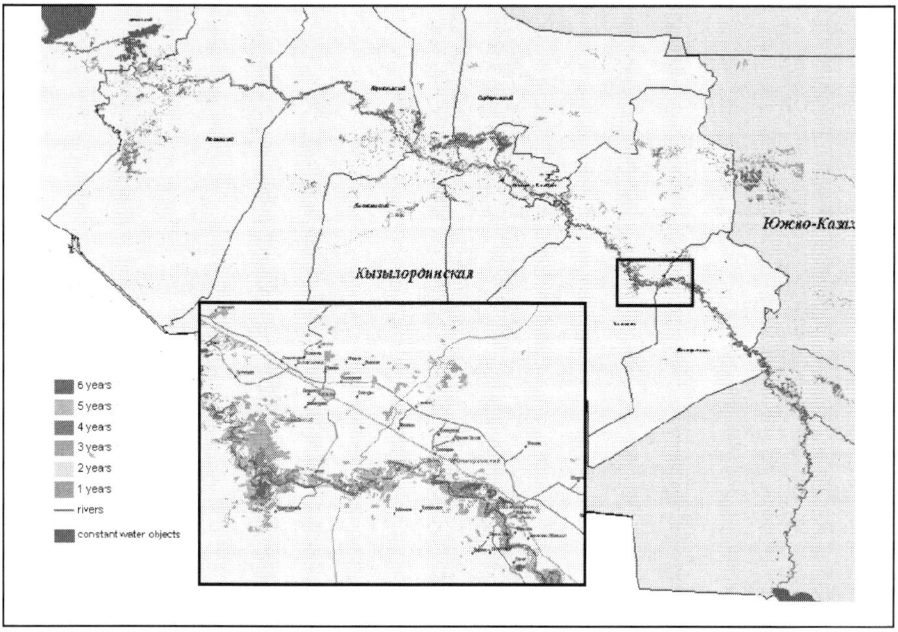

Fig. 16. Zoning territory in the middle part of Syrdaria river by the risk of flooding for winter-spring period of 2003-2008

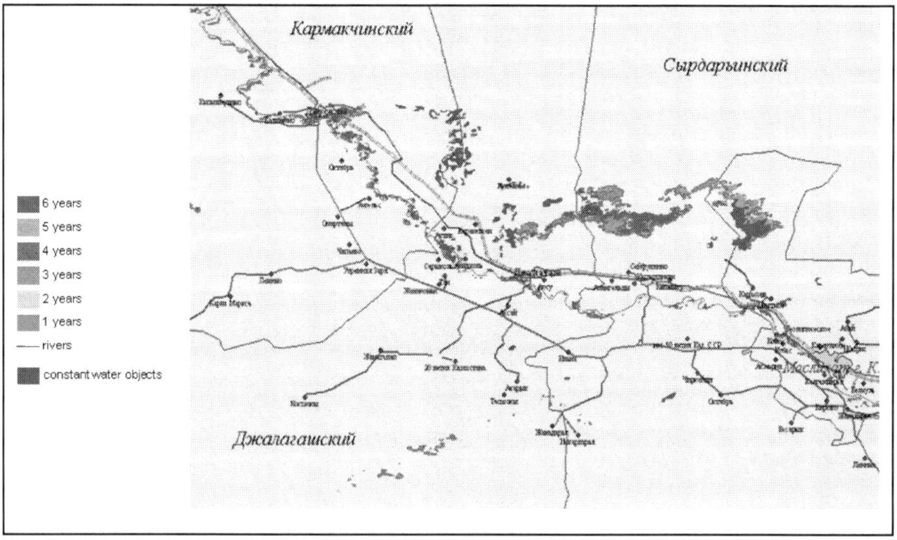

Fig. 17. Zoning territory in the middle part of Syrdaria river by the high risk of flooding for 2003-2008

7.4 Space monitoring of pollutions of a surface of a water basin

In the first decade of October 2008 on the water surface of the Shulbinsky water basin the green color pollutions, well identified in space images, were detected (figure 18). Pollution was observed on 5, 6 and 8 of October. At first, it was supposed that it is industrial pollution as drinking of the infected water led to cattle poisoning. As a result of analysis of space and ground-based data the agency of emergency situations came to a conclusion that the pollution was caused by vegetative objects (seaweed).

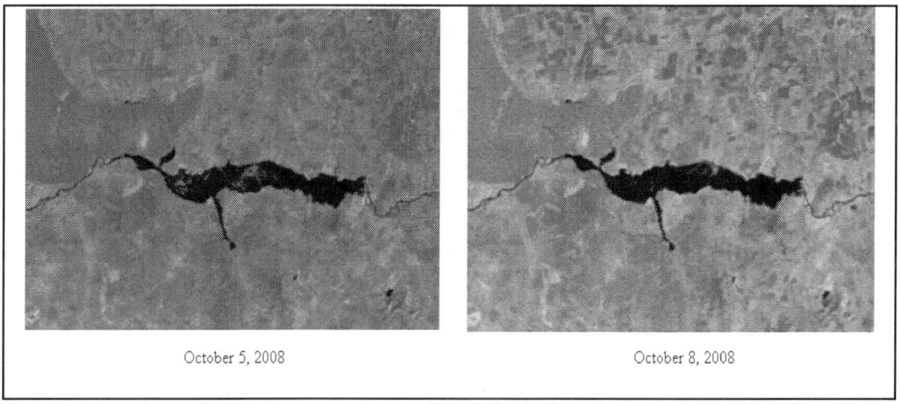

Fig. 18. Mapping of pollution on the surface of Shulninsky reservoir (East-Kazakhstan oblast)

8. Modeling

Recently emergency departments have shown interest to modeling of various extreme situations on water objects. It is modeling of zones of flooding as a result of flooding, high waters or break of dams; modeling of deformations of river-beds; and modeling of protective constructions. The solutions of these tasks are based on the software packages (BOR, RIVER, FLOOD) developed in Institute of Power Constructions (Russia) by Belikov V.V.

9. Conclusion

Application of the results of space monitoring of flooding in practical work of emergency agencies will enable to lower the damage due to early detection of freshet floods and realization of preventive measures decreasing the danger of flooding in zones of high risk of their occurrence, detected on the long-term remote sensing data.

Nowadays application of the results of space monitoring of passage of freshet waters and flooding has basically information character. One of mainstreams of development of flood space monitoring in Kazakhstan is search of optimum variants of use of high-resolution data resolution, including radar. In this case the efficiency of monitoring considerably increases. At present wide application of such data for operative space monitoring of high waters and flooding is limited by their high cost and complexities of the operative order.

The other direction of development of technologies of space monitoring of water objects in interests of regional emergency departments is use remote sensing data for modeling of potential extreme situations (flood, break of dams, washout of coast, etc.).
The results of territory zoning by the degree of risk of flooding can be useful for planning of economic activities: building of industrial enterprises, main gas pipelines, electric mains, etc. They can also be used for solving of some other problems. In particular, they are useful in planning of protective actions against repeated flooding, and in analyzing efficiency of measures taken by local authorities in struggle against them. Another possible application can be use of these data by insurance agencies.

10. References

Arkhipkin O.P., Spivak L.F., Sagatdinova G.N. Space monitoring of flood in Kazakhstan , *Proceedings of the IGARSS'2007, Spain, Barselona, 2007.*

Arkhipkin O.P., Sagatdinova G.N. Functioning of Fires and Flood Space Monitoring System in Kazakhstan, *Proceedings of the XXI Congress the International Society for Photogrammetry and Remote Sensing, Beijing, 3-11 July, 2008.* pp. 435-439.

Barton I.J.,Bathols J.M. Monitoring Floods with AVHRR // Rem. Sens. Environ. 1989. 30. 89-94.

Spivak L.F., Arkhipkin O.P., Pankratov V., Vitkovskaya I, Sagatdinova G. Space monitoring of floods in Kazakhstan. Mathematics and Computers in Simulation. 2004. 67. P. 365–370.

Spivak L.F., Arkhipkin O.P., Sagatdinova G.N. Development of Flood Monitoring Information System in Kazakhstan, *Proceedings of 31st International Symposium on Remote Sensing of Environment. Saint-Petersburg, 2005.*

The Role of DSD and Radio Wave Scattering in Rain Attenuation

Ondrej Fiser
Institute of Atmospheric Physics, Academy of Sciences of Czech Republic
Czech Republic

1. Introduction

The (rain) drop size distribution (DSD) plays a very important role in meteorology (determination of radar reflectivity and consequently rain rate, classification of precipitation, flood prediction), in microwave radio-communication (determination of rain attenuation) and also in many other applications like agriculture and insurance business.

1.1 Quick overview of literature

There is a large number of papers devoted to the DSD problems. The classical and very important study by Marshall-Palmer (1948) has to be mentioned. Almost every scientist is using this DSD model for average rain. Rain intensity is a parameter of this DSD.
The most important recent papers concerning the DSD problem are focused on following topics:
Carolin Richter (1995) declared the Gamma function to be an appropriate analytical approximation of DSD. She has studied the dependence of rain rate, median diameter and shape factor (μ-parameter in Gamma approximation) on synoptic events. Clear systematic trends in the drop spectra were found. It was possible to distinguish warm advection from the cold advection according to the numerical value of the shape factor μ. Carolin has also found that the rain intensity does not influence the shape of DSD as no relation between rain rate and shape factor could be found.
Albert Waldvogel (1974) discussed the intercept parameter N_o in the exponential approximation of DSD. Prof. Waldvogel has found that sudden variations of spectra can be recognised easily as N_o jump versus the time axis. An empirical model is proposed for the relation between the type of raindrop spectra and the convective activity of the precipitating mass.
Tokay A. and Short D. (1996) have observed dramatic change in the intercept parameter N_o in the Gamma approximation of DSD occurring during rainfall events with little change in rainfall rate. It can correspond to the transition from rain of convective origin to rain originating from the stratiform portion of tropical systems. The authors have presented an empirical stratiform-convective classification method based on N_o and rain rate scatterplot. It is worth noting that this study is related only to tropical rains.

J. Joss and E. Gori (1978) have defined an "integral" shape factor of DSD different from the "μ" shape factor, which is a parameter in the Gamma approximation. They discussed the role of sampling time on the shape factor founding that adding many instant distributions from different conditions leads to an exponential distribution such as proposed by Marshall and Palmer (1948).
J. Joss and A. Waldvogel (1968) have modified the Marshal-Palmer DSD for drizzle, continuous rain, shower and thunderstorm for conditions of Switzerland. This modification cannot be accurate because it is based on short term DSD measurement.
O.Fišer and M.Hagen (1998) have discussed the influence of integration time on resulting DSD. For larger integration time the increasing agreement between experimental DSD and its exponential distribution was shown. Also variations of parameters defining DSD (N_o ,λ and μ) with rain rate was illustrated and discussed. Two methods to determine parameters of analytical DSDs were shown and discussed (linear regression, method of moments).
O.Fišer, D.Řezáčová,P.Pešice, Z.Sokol and O.Školoud (1998) realised an attempt to estimate the basic rain type from existing rain rate records using the rain event duration, rain amount, average and standard deviation of rain rate as predictors.
O. Fišer (2002b) discussed the role of particular rain drop size on resulting specific rain attenuation im micro and mm frequency bands.
O. Fišer (2003 a,b) compared existing methods for meteorological rain type identification using Czech DSD data. A lot of work has to be done because existing criterions were devoted for tropical regions only and the mentioned study has an introductory meaning.
O. Fišer (2004) derived the radar reflectivity-rain intensity (Z-R) analytical approximation using the Czech DSD data.
O. Fišer (2006) has published a preliminary study showing DSD variability and its frequency dependence.
O. Fišer (2007) analysed the DSD moments for the estimation of the bilateral relationships between DSD products (outputs).
Jameson, A. R., A. B. Kostinski (2002) have found power law relation between rain rate and radar reflectivity factor. It is found that apparently realistic but spurious nonlinear power-law relations still appear among rainfall parameters even though the rain is not only statistically homogeneous but purely random as well.
Řezáčová D., Kašpar M., Novák P., Setvák M., (2007) have published an overview of published and measured DSDs of of rain (Best,Marhall-Palmer and other distributions).

2. Drop size distribution

The rain drop size distribution (DSD, quantity symbol N) represents the probability density of equivolumetric drop diameter D being in the unity volume. The product N(D) dD gives the number of rain of the diameter between D and D+dD in the unity volume. Only rain drops of diameters below 7 mm can exist for physical reasons. Example of the DSD measured in the Czech Republic is shown in Figure 1

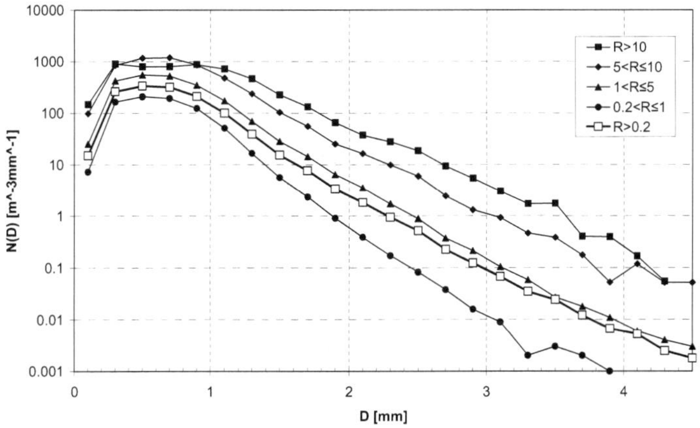

Fig. 1. DSD measured in Czech Republic (one year measurement, rain rate is the parameter of particular curves)

3. DSD measurement

Generally speaking, the measurement of the DSD is relatively rare. The greatest problem is the cost of distrometer. The most developed and user friendly device for the DSD measurement is a videodistrometer (videodistrometers developed at the Graz Technical University, Austria under ESA contract, are well known). Other electromechanical distrometer of the Joss-Waldvogel type, is used at the ETH Zurich and in Bavaria (in the DLR-Oberpfaffenhofen and in the DWD-Hohenpeissenberg). Optical distrometers are or were used in Oberpfaffenofen and in Dubna (Russia). The Rutherford Appleton Laboratory (UK), has performed a DSD measurement in Chilbolton (UK) and in Papua New Guinea. On the other hand, there is a lot of DSD data which is not processed or which is processed only partially.
The DSD measurement is not very simple and it is not performed so often in comparison with the rain rate measurement. Several DSD measurements were performed in the history (e.g. Marshall and Palmer, 1948, Joss and Waldvogel, 1968, Lakomá, 1971, Federer and Waldvogel, 1975, Li and Zhang, 1980) or more recently (e.g. Doelling et al., 1996, Hubert et al., 1999, Tokay et al., 1999, Tokay et al., 2002, Schönhuber et al., 2000, Bringi et al.,2003) using different types of measurement technique (filter or glass catchment, electromechanical distrometer, optical distrometer, videodistrometer and others). Many DSD measurements have been made by the electromechanical distrometer the concept of which was developed by Joss and Waldvogel (e.g. Joss and Waldvogel, 1967).
The results of the Czech DSD measurement (performed in 1998-1999) was published by Fišer et al.,2002, Fišer, 2002b and Fišer, 2004.

4. Analytical approximations of DSD

The exponential and Gamma distribution are the most frequently used analytical approximations of the DSD because of their satisfactory correspondence with the typical drop size distribution shape in the majority of experimental samples. There are also many other DSD models in the literature - for instance the log-normal model (Ajayi, Kozu, 1999). It should be remaindered that also many various factors like the rain type, time of integration and others influence the analytical DSD modelling.

The equation (1) expresses the Gamma model of distribution function (DSD)

$$N(D) = N_0 D^{\mu} \exp(-\lambda D) \quad (1)$$

where

D [mm] is the rain drop diameter
N(D) [m^{-3} $mm^{-1-\mu}$] is the number of drops per unit volume per drop diameter interval (dD)
N_0 [$m^{-3}mm^{-1-\mu}$] is the intercept parameter of DSD
λ [mm^{-1}] is the slope parameter.
μ [-] is shape of the DSD, to avoid a mismatch it is preferred to call it as "μ parameter"
Examples of numerical values of parameters are shown in Table 1 (Iguchi T., 1999) and plotted in Figure 2.

Gamma	N_o	λ	μ
Rain type	$mm^{-3-\mu}m^{-3}$	mm^{-1}	-
Convective	$6.29E5*R^{-0.416}$	$8.35R^{-0.185}$	3
Stratiform	$2.57E4*R^{0.012}$	$5.5R^{-0.129}$	3

Table 1. Examples of numerical values of the Gamma DSD model parameters (tropical region) where R is the rain rate in [mm/h]

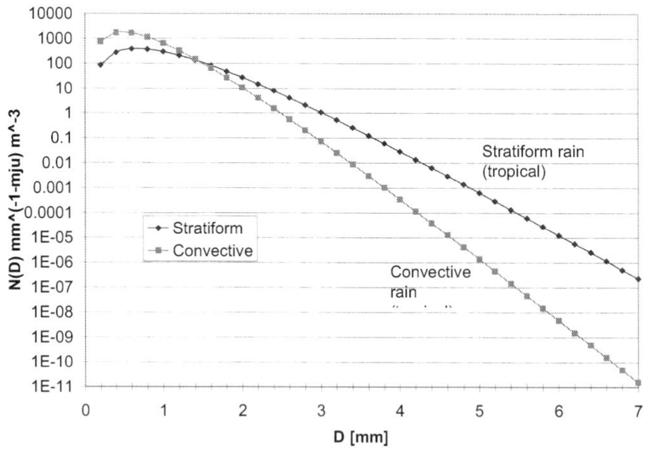

Fig. 2. Gamma DSD model using parameters from Table 1 for certain rain rate value

The simpler Exponential distribution is used in the form given by the following expression

$$N(D) = N_0 \exp(-\lambda D) \tag{2}$$

As it is obvious, the exponential DSD model can be considered as a special case of the gamma DSD where the parameter μ equals to zero.

The parameter λ in this model can be also expressed in the dependence on rain rate R:

$$\lambda \approx aR^b \tag{3}$$

where b=-0,21.
Typical exponential DSD for various rain types (drizzle, thunderstorm and average rain in mild climate) is shown in Figure 3, which is plotted after Table 2. (source: Joss J. and Waldvogel A., 1968).

Rain type	N_o	λ
	$mm^{-1}\,m^{-3}$	mm^{-1}
Thunderstorm or shower	1 400	$3 * R^{-0.21}$
Continuos rain	7 000	$4.1*R^{-0.21}$
Drizzle	30 000	$5.7*R^{-0.21}$
Average rain	8 000	$4.1*R^{-0.21}$

Table 2. Typical parameters of Exponential DSD model (Europe) where R is the rain rate in [mm/h]

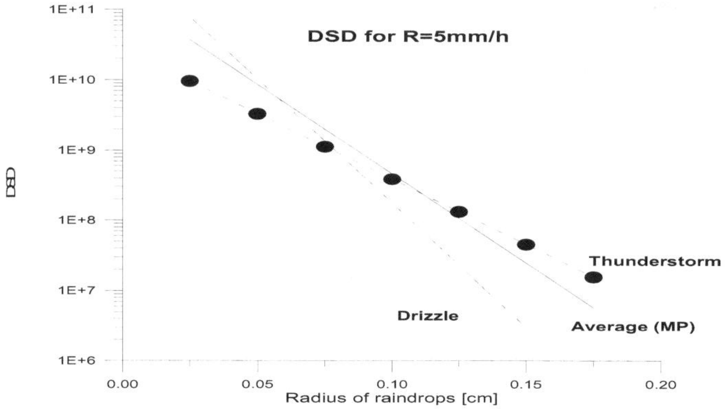

Fig. 3. Exponential DSD for various rain types (drizzle, thunderstorm and average rain in mild climate)

5. Determination of DSD parameters from measured values

In this part we describe two techniques determining the values of the model parameters N_0, λ, and μ. After logarithmic linearization of the N(D) approximation (1) the linear regression can be applied (least square method, Bartsch, 1996). The moment method uses the definition of the DSD moments, which, for the n-th moment, M_n, gives:

$$M_n = \int_0^\infty D^n N(D)dD \tag{4}$$

Using the method of moments following formulas for the parameters N_o and λ in the exponential DSD model (2) were derived:

$$N_o = 98.65 * M_3 * \left(\frac{M_3}{M_6}\right)^{\frac{4}{3}} \tag{5}$$

$$\lambda = 4.93 * \left(\frac{M_3}{M_6}\right)^{\frac{1}{3}} \tag{6}$$

where M_3 , and M_6 are the 3rd, and 6th DSD moments, respectively. Note that the same formulas were derived by Prof. Waldvogel (Waldvogel, 1974) where the liquid water content W was preferred to the 3rd moment. The third and the sixth moments were chosen to approximate the DSD in order to determine the rain intensity (being approximately proportional to M_3) and the radar reflectivity factor (proportional to M_6) from the DSD model as accurate as possible.

To determine the parameters of the Gamma approximation (1) the formulas of Tokay and Short (Tokay, Short, 1996) can be used. The expressions using M_3, M_4, and M_6 are as follows:

$$\mu = \frac{11G - 8 + [(G(G+8)]^{1/2}}{2(1-G)} \tag{7}$$

$$\lambda = \frac{(\mu + 4)M_3}{M_4} \tag{8}$$

$$N_0 = \frac{\lambda^{\mu+4} M_3}{\Gamma(\mu + 4)} \tag{9}$$

where the auxiliary G factor is given by

$$G = \frac{{M_4}^3}{{M_3}^2 * M_6} \tag{10}$$

6. Outputs (products) of DSD

The rain drop size distribution (some times it is called "rain drop spectrum") determines uniquely its outputs (outputs can be called also "products") as rain rate, radar reflectivity factor, rain attenuation and others (see next parts).

6.1 DSD products of no frequency dependence

a, Rain intensity (rain rate)

$$R_g = \frac{3.6}{10^3} \pi \int_0^\infty D^3 v(D)\ N(D)dD \tag{11}$$

where R_g [mm/h] is the rain rate (corresponding to the rain rate derived from rain gauge measurement)
v is the terminal falling velocity
N(D) is the drop size distribution
D is the equivolumetric drop radius.

b, Radar reflectivity factor z [mm^6m^{-3}]

The radar reflectivity factor z (small letter "z") is the 6th DSD moment, it can be computed from following expression:

$$z = \int_0^\infty D^6 N(D)dD \tag{12}$$

while for its logarithmic unit Z [dBZ] (capital letter "Z") it is used:

$$Z = 10 \log_{10} z = 10 \log_{10} \{ \int_0^\infty D^6 N(D)dD \} \tag{13}$$

If we study electromagnetic energy coming back from rain volume to the radar, we must be aware that the reflected energy is not dependent on frequency only in the "Rayleigh region" case (drop diameter D is much smaller in comparison with the wave length λ , i.e. $D \ll \lambda$), more precisely the Rayleigh region is defined

$\pi D/\lambda \ll 1$ for n =1 (n is the refractive index.) or $\pi nD/\lambda \ll 1$ for n > 1

If we suppose rain with rain drops having diameter up to 4 mm (typical for mild climate), the Rayleigh region holds for frequencies below 2,5 GHz. But if we consider the existence of maximum rain drop diameter (D=7 mm), the Rayleigh region is met at frequencies lower than 1,36 GHz. In practice, it is not so strict and the Rayleigh region is applicable for frequencies below 5 GHz.

6.2 DSD products dependent on frequency

The specific rain attenuation A is strongly dependent on frequency, but rain attenuation is negligible for frequencies below 4 GHz.
For the specific rain attenuation A in [dB/km] next equation is used:

$$A = 4{,}343 \cdot 10^{3} \cdot \lambda \cdot \operatorname{Im} \int \mathrm{f}\ (D) \cdot N(D)dD \tag{14}$$

where **f** is the complex forward scattering function (for exact definition see in part 7)
λ is the wave length of the used transmission
Im represents the imaginary part of complex number

The formula (14) is derived in part 9. As one can see, this expression strongly depends on the wave length λ (i.e. frequency) in contrast to the no frequency dependence of the radar reflectivity factor.

Through the numerical simulations it is possible to search for relationships between mentioned quantities (Z, R and A) called "DSD products." The term frequency means the radio frequency of pertinent technical application in this paragraph.

7. Scattering functions

In literature different definitions of scattering functions are used, see, for instance [Uzunoglu et al., 1977]. One of the most used definition of the scattering function is the following one:

----------->E i 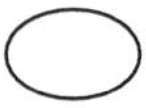 ---- r -------> E s

$$\boldsymbol{E}^{s} = \boldsymbol{E}^{i} \cdot f(\mathbf{K}_1, \mathbf{K}_1) \cdot r^{-1} \cdot e^{(jk_0 r)} \tag{15}$$

where:
$\boldsymbol{E}^{s}$ ... electric field of a scattered wave [V/m]
$\boldsymbol{E}^{i}$ ... electric field of a wave impressing on the raindrop [V/m]
k_0 ... free space propagation constant [m^{-1}]
r .... observation distance from the scattering drop [m]
$f(\mathrm{K}_1, \mathrm{K}_2)$ ... scattering function [m], $\mathbf{K}_1$ is the direction of the incident field, $\mathbf{K}_2$ is the direction of the scattered field
especially:
if $\mathbf{K}_1 = \mathbf{K}_2$, the forward scattering function is considered, or
if $\mathbf{K}_1 = -\mathbf{K}_2$, the backward scattering function is considered

If the propagating radio wave direction is not parallel with horizontal drop axis, formula (16) can be used (see geometry in Figure 4).

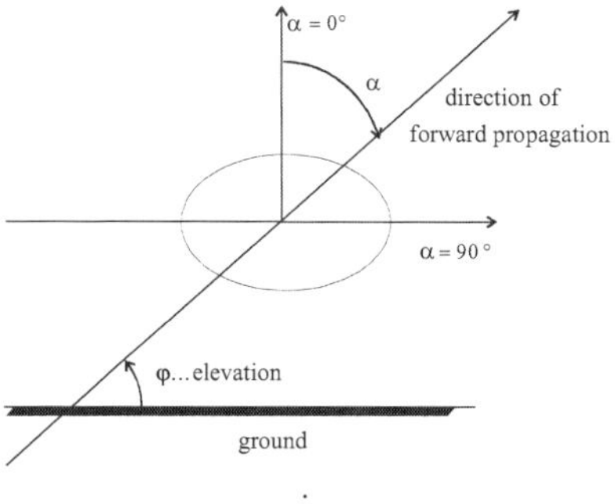

Fig. 4. Geometry of forward propagation direction in the general incidence angle α.

$$f_{h,v}(\alpha) \approx f_0(0^\circ)\cos^2(\alpha) + f_{h,v}(90^\circ)\sin^2(\alpha) \tag{16}$$

where

α is angle between direction of propagation and zenith axis [°]

$f_{h,v}(90^\circ)$ is the scattering function of horizontally respective vertically polarised wave for direction of propagation parallel with horizontal drop axis ($\alpha = 90^\circ$)

$f_0(0^\circ)$ is scattering function for the "zenith" propagation ($\alpha = 0^\circ$), it is not dependent on polarisation because of rain drop symmetry

8. Some methods computing scattering functions

8.1 Rayleigh scattering

The Rayleigh scattering theory was developed and published by Strutt (1871 a,b) - later Lord Rayleigh. The theory gives an approximation for the scattering of electro-magnetic radiation from spheres. It is valid for sphere diameters significantly smaller than the vacuum wavelength of the radiation., i.e. $\pi D/\lambda \ll 1$, see part 6.1. Lord Rayleigh derived a scattering function approximation for such case using elementary dipole theory.

The computation of the Rayleigh scattering is very simple (a handy calculator is sufficient) but we must be aware of strict limitations. It was proved that the frequency above about 5 GHz is owing to the usual rain drop diameter out of Rayleigh region.

The Rayleigh scattering helps us to study the frequency and temperature properties of attenuation on lower frequencies. It does not enable to compute depolarisation and angular dependencies.

8.2 Mie scattering

The Mie scattering theory was developed and published by Mie (1908). The scattering function **f** (subscript "f" for forward, "b" for backward scattering) for spherical dielectric particles is given by the next formula:

$$f_f = \frac{-j\lambda^3}{\pi^3 D^2}\left[\sum_{n=1}^{\infty} (2n+1)(a_n + b_n)\right]^*$$
$$f_b = \frac{-j\lambda^3}{\pi^3 D^2}\left[\sum_{n-1}^{\infty} (-1)^{(n+1)}(2n+1)(a_n - b_n)\right]^* \qquad (17)$$

where λ denotes the vacuum wavelength of the electro-magnetic radiation, j is the imaginary unit and D the diameter of the spherical drops, * is symbol for conjugate imaginary numbers. The coefficients a_n and b_n according to Mie depend on the complex relative refractivity $\varepsilon_r = \varepsilon / \varepsilon_0$ of the material (rain water in our case) and on the diameter D of the scattering sphere.
a_n, b_n are the Mie's coefficient , its evaluation need not to be very complicated.
The Mie scattering calculation is also possible at this web page:
http://omlc.ogi.edu/software/mie/
For the Mie scattering computation a simple programmable computer is needed under the condition that it can work with complex variables. The Mie algorithm can be quite simple if the Bessel and Legandre polynomials were replaced by simple complex goniometric functions. The infinite series (the above printed formula) can be limited to the n being about 10 (or even less) of a perfect accuracy, see (Fiser, O., 1993).

Mie scattering helps us to study the frequency and temperature properties of rain attenuation if we accept that rain drop shape is spherical (for larger rain drops it is not true). Mie scattering does not enable to compute depolarisation and angular dependencies. On the other hand, there is no frequency limitation like in the Rayleigh scattering computation case.

8.3 Other methods computing scattering functions

For full utilisation (angular dependence, polarisation properties), numerical methods computing the scattering functions, are required.

The point matching method is much more complicated and general and it enables to study not only the properties of scattering functions but also the incident angle dependence, the bi-static scattering and depolarization phenomena as well. See, for instance Oguchi, T., 1973. Some studies to derive the scattering function using numerical methods are used. For instance the MultipleMultiPole (MMP) method was used by Hajny, Mazanek and Fiser at the Czech Technical University Prague , cf. Hajny, M. et al, 1998.

9. Derivation of formula for specific rain attenuation in rain volume

To derive a formula for specific rain attenuation we prepared next collection of formulas related to the Figure 5 (E^s is scattered field, E^i is incident field, E^0 is "free space" field, z1->0 (infinitesimal length of the rain volume between two parallel slabs), resulting electrical field is summed at the point $P(z_o)$ - contributions of the original free space field E^0 and scattered field from rain drops in the thin rain volume. Incident wave is of the planar type in our

model (a very good approximation of spherical wave being very far from the transmitter). Here are the formulas:
Resulting electrical field $E(z_o)$ [V/m] is a sum of free space field and scattered field:

$$E(z_o) = E^s(z_o) + E^o(z_o) \tag{18}$$

From the definition of scattering functions we have:

$$E^s(z_o) = E^i(z)\ \mathbf{f}(D)\ \frac{e^{jkr}}{r} \tag{19}$$

For planar wave it is valid:

$$E^i(z) = E^i(0)\ e^{jkz} \tag{20}$$

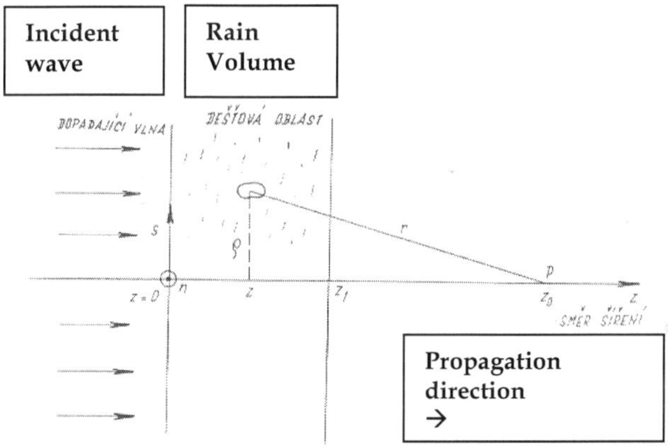

Fig. 5. On specific rain attenuation

After some numerical simplifications

$$E^s(z_o) \sim E^i(0)\ \mathrm{f}(D)\ \frac{e^{jkr}}{r} \tag{21}$$

$$E^s(z_o) \sim E^o(z_o)\ \mathrm{f}(D)\ \frac{e^{jk(r-zo)}}{z_o} \tag{22}$$

$$E^o(z_o) = E^i(0)\, e^{jkz_o} \tag{23}$$

$$r = (\rho^2 + z_o^{\;2})^{0.5} \sim z_o\,(1 + \frac{1}{2}\frac{\rho^2}{z_o^{\;2}}) \tag{24}$$

$$r - z_o \sim \frac{\rho^2}{2\,z_o} \tag{25}$$

we obtain:

$$E(z_o) = E^o(z_o) + E^o(z_o) \int_0^{\infty}\{\int_0^{z_1} \mathrm{f}(D)N(D)\mathrm{I}\,dz\}\,dD \tag{26}$$

where

$$\mathrm{I} = \int_0^{\pi}\{\int_0^{F} \rho\, e^{jp\rho^2}\, d\rho\}\, d\varphi \tag{27}$$

and where F is the radius of the first Fresnel zone. This integral expresses the contributions of all rain drops in the infinite plane perpendicular to the direction of radiowave propagation. In fact, the contribution is considered from ashlar of the infinitesimal length in the direction of propagation (z axis in our picture):

$$E(z_o) = E^o(z_o)\,[1 + j\frac{2\pi}{k_o} z_1\,\mathbf{C}\,] \tag{28}$$

where

$$\mathbf{C} = \int_0^{\infty} \mathbf{f}(D)N(D)\; dD \tag{29}$$

Finally

$$\tau = \{\lim_{\Delta z \to 0}(1 + j\frac{2\pi}{k_o}\Delta z\mathbf{C})^{\frac{1}{\Delta z}}\}^L = e^{j\frac{2\pi}{k_o}\mathbf{C}L} \tag{30}$$

where

τ is the ratio of the attenuated field strength to the non attenuated one (transmission)

f is the complex scattering function

Transmission τ was computed as a product of a number of (L/Δz) ashlars of infinitesimal lengths arranging in the series of the certain length L. The formula for specific rain attenuation α in the [dB/km] unit is then given as 10 $\log_{10}(\tau)$:

$$\alpha = 4.3434 \quad \lambda \ 10^{3} \int_{0}^{\infty} \operatorname{Im} \mathbf{f}(D) N(D) dD \quad [dB / km] \tag{31}$$

The next parametrical approximation for the drop size distribution - DSD (symbol N) after Marshall-Palmer is used very frequently for so called "average rain:"

$$N(D, R) = 8000 \cdot e^{\frac{-4.1 \cdot D}{R^{0.21}}} \quad [m m^{-1} m^{-3}] \tag{32}$$

where R is rain rate (or rain intensity) in [mm/h] units and D is equivolumetric rain drop diameter, see Marshall and Palmer, 1948. The same formula (31) was derived through similar way by Van de Hulst (1957).
The mostly used and very simple approximation for specific rain attenuation is this one:

$$\alpha \sim a \cdot R^{b} \quad [dB / km] \tag{33}$$

where a [k alternatively] and b [α alternatively] are constants depending on frequency, polarisation and temperature. An example is shown in next table (ITU-R report):

f [GHz]	10	**12**	15	20	25	30	**To interpolate:**
a	0,0094	**0,0177**	0,0350	0,0722	0,1191	0,1789	**Logarithmi-cally**
b	1,273	**1,211**	1,143	1,083	1,044	1,007	**linearly**

10. Variability of DSD

The majority of existing models estimating the radar reflectivity or microwave attenuation from rain intensity are rough ones (equations 33 or 34) because they are neglecting the DSD variability. By other words: two various rain events (shower and continuous rain, for instance) of the same rain intensity, say we 5 mm/h, can cause namely different numerical value of rain attenuation through the DSD variability (3 dB/km in shower and 2 dB/km in continuous rain in our example).

The radar reflectivity factor as well as the specific rain attenuation (of the radar signal, or of the microwave and mm wave link) depend on the rain rate only roughly. They both depend on the drop size distribution (DSD) primarily; this fact is frequently neglected (see equations 12, 13 and 14).

The DSD variability is obvious from Figure 6. This figure shows the probability density of the radar reflectivity factors computed through the equations 12 and 13 from measured DSDs corresponding to rain rates between 4.5 and 5.5 mm/h. After the usually used Marshal Palmer relation

$$Z=10\log(300\ R^{1.5})\quad [\mathrm{dBZ}] \tag{34}$$

the radar reflectivity factor would be 35,3 dBZ, but one can see, that Z varies from 27 to 42 dBZ (radar would announce rain rate between 1 and 6 mm/h in this case!)

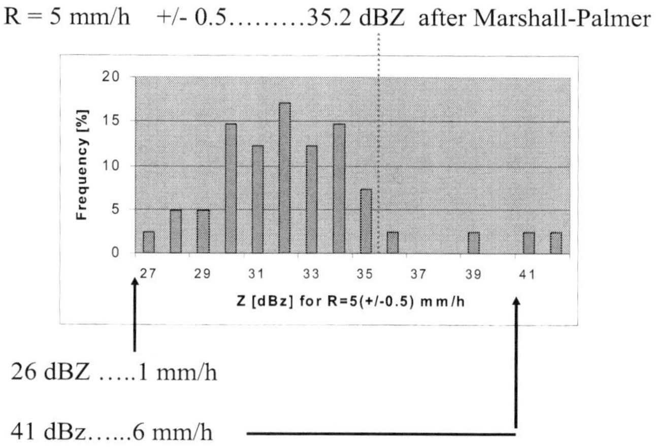

Fig. 6. Radar reflectivity factor histogram corresponding to the rain rate between 4.5 and 5.5 mm/h.

A big dispersion of rain rate values R corresponding to the observed values of the radar reflectivity factor Z is also obvious from scatterplots (Figure 7). Again, it is due to the DSD variability.

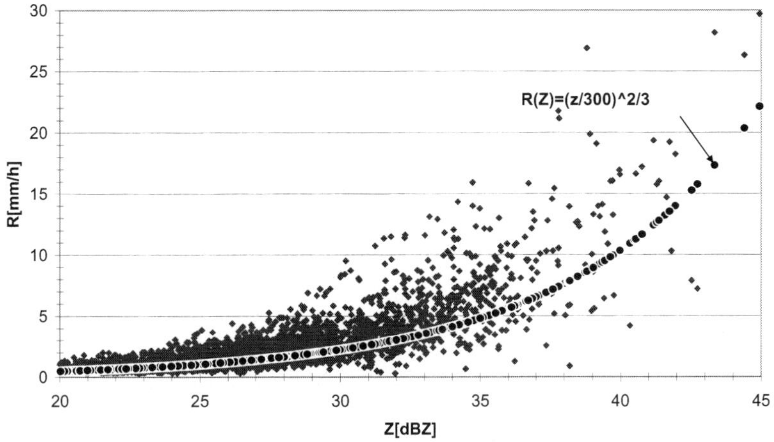

Fig. 7. Scatterplot of the R-Z relation (DSD data and equations 11 and 13 were used while integraion time being 1 minute)

Similar scatterplots "attenuation versus rain rate" were done considering frequencies in the 10 - 100 GHz region. A big dispersion is observed, too, but the dispersion depends on the frequency of radio communication link. It was found that the rain attenuation at frequencies close to 40 GHz depends on the rain rate quite uniquely.

In Figure 8 there are shown imaginary parts of forward scattering functions being responsible for the specific rain attenuation (cf. equation 3). It is obvious, that the slope is varying with the frequency.

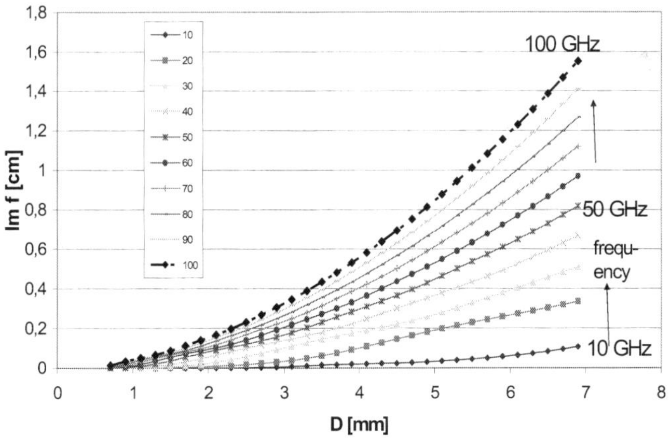

Fig. 8. Imaginary parts of forward scattering functions being proportional to the rain attenuation

Through the mathematical regression we have found the exponent "n" in the approximate relationship

$$\text{Im}\, f \sim D^n \tag{34}$$

By other words it means that the specific rain attenuation A is approximately proportional to the n-th DSD moment M_n (see equation 4) while "n" varies with the frequency f (unfortunately, there is used the same symbol "f" for both, frequency as well as for scattering function in the technical literature). The found exponent "n" for frequencies between 10 and 100 GHz is shown in Figure 9. Its value is decreasing with the frequency of the radio transmission.

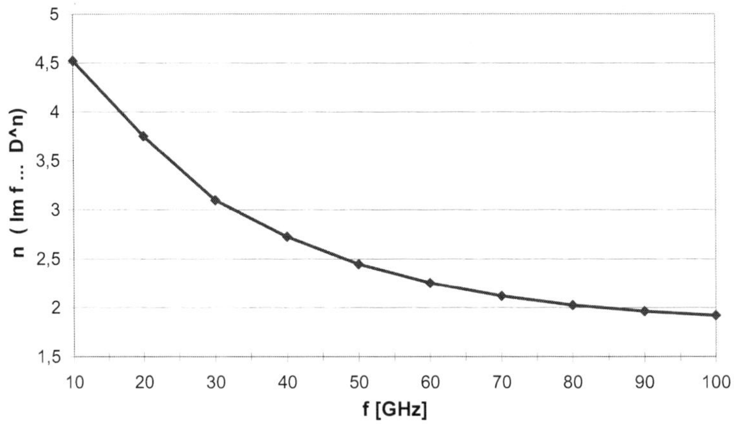

Fig. 9. Exponent "n" in the Im f ~ D^n dependence, where D is rain drop diameter

After some speculations it could be concluded that the DSD variability courses the measure of the correspondence (uncertainty) between one DSD product (Z,R and A) to another one, for instance the Z -R relation.

It is known that the rain rate is given by the 3.67th DSD moment. As the exponent "n" in equation 34 approximates the 3.67 value in the case of frequencies within the 18 - 42 GHz interval, the A - R relationship is quite unique for this frequency range. On the other hand, the 10 GHz specific rain attenuation approximates the 4.5th DSD moment, which is not very far from the 6th moment, i.e. from the radar reflectivity factor Z (equation 13). That's why the DSD variation does not very influence the A - Z relation in the 10 GHz case and this relationship is not so ambiguous.

The particular contribution of rain drops of certain diameters to the rain attenuation (after equation 14) is varying considering varying frequency. More concretely: the role of small rain drops is increasing with the frequency. The prevailing contribution is caused by drops of the equivolumetric diameter close to 0.7–1.5 mm (see Figure 10)

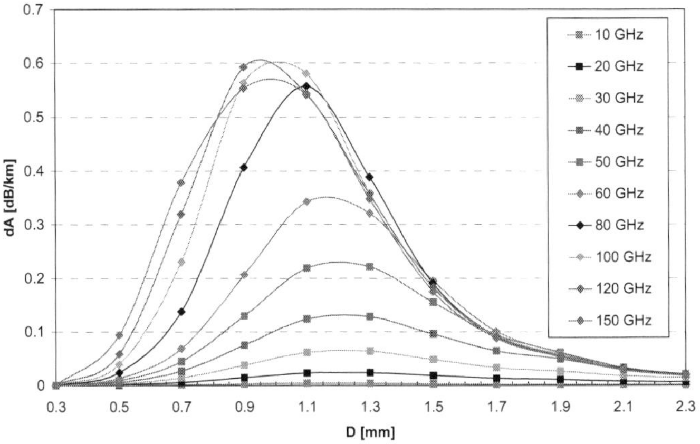

Fig. 10. Contribution of rain drops of diameter D to the specific rain attenuation; transmission frequency is a parameter

All results in this contribution were derived from the actual drop size distributions measurement performed by the videodistrometer of ESA, which was lent to the Institute of Atmospheric Physics Prague through its manufacturer Joaneum Research Graz (Austria) in the period July 1998–July 1999 [Fiser et al., 2002]. The time of integration was chosen to be 1 minute.

11. Conclusion

The applicability, importance and properties of the rain drop size distribution was demonstrated in this chapter. It was also shown that DSD determines rain rate, radar reflectivity and rain attenuation of microwave signal. A part of this chapter describes scattering functions describing radiowave reflection from rain drop.

One of the important results of this study is the following one: the radar reflectivity factor derived from the rain rate through the Z-R relation could be incorrect through the DSD variability. It is due to the fact, that the radar reflectivity is given by the 6th DSD moment while the rain rate depends on 3.67^{th} DSD moment. A bigger rain drop contributes thus much more to the radar reflectivity than to the rain rate.

The rain attenuation at 10 GHz depends on the radar reflectivity factor quite uniquely and similarly the dependence of 42 GHz rain attenuation on the rain rate is unambiguous. For other relationships between DSD products we recommend the utilisation of the dependence of one DSD product on two other DSD products (or DSD moments), for instance to estimate the specific rain attenuation from both rain rate and radar reflectivity factor.

Acknowledgement

This contribution was thankfully supported by the GACR grant 102/08/0851.

12. References

Ajayi G.O. and Kozu T., 1999: Rain drop size distribution during convective rainfall in Japan and Nigeria. Proc. of Abstracts, GA URSI, FP37.

Bartsch H-J (1996) " Mathematische Formeln," FINIDR Press.

Bringi, V. N., V. Chandrasekar, J. Hubbert, E. Gorgucci, W. L. Randeu, W. L. and M. Schönhuber,2003: "Raindrop Size Distribution in Different Climatic Regimes from Disdrometer and Dual-Polarized Radar Analysis" Journal of Atmospheric Sciences, vol. 60, 2003, Issue 2, pp. 354-365.

Bringi, V.N., M. Thurai, K. Nakagawa, G.J. Huang, T. Kobayashi, A. Adachi, H. Hanado, and S.Sekizawa, 2005:"Rainfall Estimation from C-Band Polarimetric Radar in Okinawa, Japan: Comparison with 2-D Video Disdrometer and 400 MHz Wind Profiler" 32nd Conference on Radar Meteorology, 22-29 October 2005, Albuquerque, Mexico, pp. 1-8.

Doelling I., Joss J. and Riedl J., 1996: Systematic variations of raindrop size distributions measured in northern Germany during 7 Years. Proc. 12th International Conference on Clouds and precipitation,Vol.2, Zurich, 1310-1313.

Federer B. and Waldvogel A., 1975: Hail and raidrop size distribution from a Swiss multicell storm. J.Appl.Meteorol., Vol.14, 91-97.

Fiser, O., 1993: A simple generator of forward scattering functions on spherical dielectrics. Radioengineering, 2(1), 21-22

Fišer O. and Hagen M., 1998: Analysis of distrometer data. COST 75 Final Seminar "Advanced Weather Radar Systems,". Proc.: "14 papers on Precipitation Estimates by Radar and on Analyses for Weather-Forecasting." Swiss Meteorological Institute, Locarno, 21-33.

Fišer O., Řezáčová D., Sokol Z., Školoud O. and Pešice P., 1998: An attempt to classify the basic rain types from the rain rate records. COST 255 Workshop "First International Workshop on Radiowave Propagation Modelling for SatCom Services at Ku-band and above", Noordwijk, the Netherlands, 155-157.

Fišer O., Schoenhuber M, Pešice P.,2002: First results of DSD measurement by videodistrometer in the Czech Republic in 1998-1999. Studia Geophysica et Geodetica,46, 3, 2002, 485-506.

Fišer O.,2002b: "The role of particular rain drop size classes on specific rain attenuation at various frequencies with Czech data example," ERAD02 series, Delft(NL), 17.-22.November 2002, pp 1-4.

Fišer, O.,2003a" :On Rain (DSD) Types Aimed at Radar Reflectivity and Attenuation Calculation. Proceedings of International Workshop on Precipitation in Urban Areas /6/. - Pontresina ,2003., p. 5.

Fišer O.,2003b: On methods distinguishing drop size distribution type for improved estimation of radiowave attenuation due to rain.Proc.Conf.COMITE 2003, Pardubice, Czech Rep.,September 23-24,2003,pp.21-24.

Fišer, O., 2004: "Z-R (Radar Reflectivity-Rain Rate) Relationships Derived from Czech Distrometer Data." In Proc. of the ERAD Conference, Visby, Sweden, 2004, pp. 233-236.

Fišer O.,2006: Improved Rain Attenuation Estimation Based on DSD and Radar Data. In Proc. of CNES Workshop on Earth - Space Propagation 2006, [CD-ROM], Toulouse,

Fišer, O.:, (2007): Selected DSD properties for meteo radar applications and microwave link attenuation in rain. In Wave Propagation in Communication, Microwave Systems and Navigation (WFMN07)- A conference of ITG commission 7.5 "Wave Propagation," pp. 1-4 (CD).

Hajny, M.; Mazanek, M.; Fiser, O.: Ku-Band Rain Scattering Parameters Calculated by MMP Method, Proc. In First International Workshop on Radiowave Propagation Modelling for SatCom Services at Ku band and above, 28-29 October 1998, ESTEC,The Netherlands

Hubbert J.C., V. N. Bringi, and M. Schönhuber,1999""2D-Video Distrometer Measurements: Implications for Rainrate and Attenuation Estimators."Preprints, 29th Conference on Radar Meteorology, Jul 12 - 16, 1999, Montreal, Quebec, Canada, American Meteorological Society, pp. 666-669.

Van De Hulst: Light Scattering by Small Particles, New York, J.Wiley pub., 1957

Iguchi T.: Personal communication, August 1999

Jameson, A. R., A. B. Kostinski, 2002: "Spurious power-law relations among rainfall and radar parameters" Q. J. R. Meteorol. Soc. (2002), 128, pp. 2045-2058.

Joss J. and Waldvogel A., 1967: Ein Spektrograph fuer Niederschlagstropfen mit automatischer Auswertung, Rev. Pure and Applied Geophysics,68, 240-246.

Joss J. and Waldvogel A., 1968: The variation of raindrop size distributions at Locarno. Proc. Int. Conf. Cloud. Phys., 369.

Joss J. and Gori E.G., 1978: Shapes of raindrop size distributions. J. Appl. Meteorol, 17, 1054-1061.

Lakomá - Řezáčová D., 1971: Ein Beitrag zur Genauigkeit der Bestimmung der Regenintanistaet aus einem Tropfenspektrum, Meteorologische Rundschau, 24, 161-171.

Li H.J. and Zhang Z.W., 1980: A study of raindrop size distributions in central and northwest China and their effects on some propagation parameters for wavelengths from 0.86 cm to 10 cm. Ann.Telec, 35, 405-410.

Marshall and Palmer, 1948: The distribution of raindrops with size. J. Meteorol., 5, 165.

Mie, G., „Beiträge zur Optik trüber Medien, speziell kolloidaler Metallösungen", Annalen der Physik, Vierte Folge, 25(3), pp. 377-445, 1908.

Oguchi, T., "Scattering Properties of Oblate Raindrops And Cross Polarization of Radio Waves Due To Rain: Caculations at 19.3 and 34.8 GHz", J.Radio Research Labs, 20(102), pp. 79-119, 1973.

Richter C.(1995): On the parametrisation of Drop Size Distributions - Case Studies with Distrometer , internal report of the Rutherford Appleton Laboratory (UK), 14th December, 1995

Řezáčová D., Kašpar M., Novák P., Setvák M., 2007: Fyzika oblaků a srážek (Physics of Clouds and Precipitation), Academia.

Schönhuber, M. H., Urban J. P. V., Poiares Baptista W. L., Randeu and Riedler W., 1994: Measurements of precipitation characteristics by a new distrometer. Proceedings of "Atmospheric Physics and Dynamics in the Analysis and Prognosis of Precipitation Fields", Rome, Italy, pp. 1-4

Schönhuber M., Urban H. E., Randeu W. L. and Poiares Baptista J. P. V., 2000: Distrometer results obtained in various climates and their application to weather radar data inversion. ESA SP-444 Proceedings, "Millennium Conference on Antennas & Propagation", Davos, Switzerland.

Strutt, J. W., 1871: On the Light from Sky, its Polarization and Colour (I). Philosophical Magazine, Series 4, 41(241), 107-120.

Strutt, J. W., 1871: On the Light from Sky, its Polarization and Colour (II). Philosophical Magazine, Series 4, 41(243), 274-279.

Tokay A. and Short D., 1996: Evidence from tropical raindrop spectra of the origin of rain from stratiform versus convective Clouds. J. Appl. Meteorol., 35, 355-371.

Tokay A., Thiele O. W., Kruger A. and Krajewski W. F., 1999: New measurements of drop size distribution and its impact on radar rainfall retrievals. Preprints, 29th Conference on Radar Meteorology (American Meteorological Society), Montreal, Canada, 659-662.

Tokay, A., A. Kruger, W.F. Krajewski, P.A. Kucera, and A.J. Pereira Filho,2002: "Measurements of drop size distribution in Southwestern Amazon basin" Journal of Geophysical Research-Atmospheres, 107(D20), LBA 19-1 to LBA 19-15, 2002.

Uzunoglu,Evans,Holt, 1977:Scattering of electromagnetic radiation by precipitation particles and propagation characteristics of terrestrial and space communication systems. Proc IEE,124,417

Waldvogel A., 1974: The N_o jump of raindrop spectra. J.Atmos.Sci., 31, 1067-1078.

24

Registration of radar and optical satellite images using multiscale filter technique and information measure

Qi Li[1], Bihong Fu[2] and Yanfang Dong[3]
[1]*National Institute of Advanced Industrial Science and Technology (AIST) Japan*
[2]*Institute of Geology and Geophysics, Chinese Academy of Sciences China*
[3]*Institute of Earthquake Science, China Earthquake Administration China*

1. Introduction

Since a spectacular series of missions in the context of the Earth Observing System (EOS) by NASA beginning from the late 1990's, the significance of the satellite remote sensing has been recognized all over the world (Kafatos & Qu, 2007; Kaufman et al., 1998). In particular, the applications on hazard mitigation and resource exploration have been widely regarded as one of basic approaches over the past years (e.g., Barrett et al., 1991; Chuvieco, 2008; Fu et al., 2004; Ninomiya et al., 2005; 2006; Realmuto, 2000; Sato et al., 2006; Teeuw, 2007; Urai et al., 2007). In general, remote sensing, from different points of view, includes many branches, or exactly speaking many application fields, such as environmental and ecological remote sensing, geological remote sensing, and military remote sensing. In this chapter, we focus our research on geological applications. However, the proposed algorithms and approaches might be applicable to every fields associated with image registration processing. Although remotely sensed optical images from satellite sensors can meet most needs in the practical applications, considerable weather-dependence limits its functional deployment under some circumstances. For instance, during the period of the devastating Ms 8.0 Wenchuan earthquake in the summer of 2008 (Fu et al., 2009), the most optical images from ASTER (Advanced Spaceborne Thermal Emission and Reflection Radiometer) (Yamaguchi et al., 1998) sensors on the NASA's Terra satellite can hardly be used to do some refined applications just because of heavy clouds contaminated. However, SAR (Synthetic Aperture RADAR) images are not influenced by climate and time. In practical applications, the optical satellite images, in particular with high resolutions from sensors such as SPOT (Chevrel et al., 1981) and IKONOS (Tanaka & Sugimura, 2001), provides excellent legibility, but they may be affected by the clouds and weather conditions. On the other hand, SAR images are not influenced by climate and they can be obtained day-and-night, but they suffer from a serious intrinsic speckle noise (Franceschetti & Lanari, 1999; Lampropoulos & Boulter, 1997).

Therefore, the joint application of these two different kinds of data information will be great interest for many geological problems associated with the remote sensing (e.g., Chen et al., 2003). The first major processing for such a combination is to finish an accurate registration. That is why registration of images coming from different sources is of increasing importance (Le Moigne et al., 2002; Li, 2006; Li et al., 2006c; Li et al., 2008; Schowengerdt, 2006).

Up to now, many registration methods have been proposed to register SAR and optical satellite images manually or (semi-)automatically (e.g., Ali & Clausi, 2002; Cheng et al., 2004; Curlander & Kober, 1990; Dare & Dowman, 1996; 2001; Galland et al., 2005; Hong & Schowengerdt, 2003; 2005; Inglada & Vadon, 2005; Lampropoulos et al., 2003; Lampropoulos et al., 2002; Li et al., 1993; 1995; Li et al., 2007c; Mao et al., 2007; Raucoules & Carnec, 1999; Shu & Tan, 2007; Shu et al., 2005; Thepaut et al., 1998; Vornberger & Bindschadler, 1992; Wang & Chen, 2003; Wegner & Soergel, 2008a; 2008b; Wu & Maitre, 1990; Yang et al., 2005; Zamora et al., 1998; Zhang et al., 2007; Zhang et al., 2004; Zhao & Chen, 2003). In general, these methods can be classified into two major categories: feature based method and intensity based method (Li et al., 2006c). The feature based registration method has been widely implemented into the commercial software, such as ENVI and ERDAS Imagine, initially for the mono-modality image registration, but it encounters significant difficulties when to-be-matched features are not easily extracted from different source images. For example, it is very difficult to extract the features (e.g., roofs and crossroads) from ASTER image for very rural area (e.g., Horonobe, Hokaido, Japan) rather than for metropolitan area (e.g., Tokyo). Furthermore, for the radar image (e.g., RADARSAT-2), this difficulty becomes even heavier than the optical images. Therefore, the corner reflector is often needed to install on the ground so as to enhance the accurate registration process (Li et al., 2009). On the other hand, the intensity based registration method may be easily implemented as a semi- or fully automatic manner, but it is seriously dependent on the choice of similarity measures or metrics, and usually needs large amount of computation (Inglada & Giros, 2004).

Due to the distinct and intrinsic differences of imaging properties between radar and optical images, it is not easy to extract corresponding features from them. Moreover, serious speckle noise existing in radar images further aggravates this problem. Thus, it is a very challenging problem to precisely and automatically register them with a satisfactory accuracy. Wu & Maitre (1990) proposed multiresolution approach to register the SPOT-XS and SEASAT SAR images by matching the contour lines in different scale space representations, but the problems of no distinct features in SAR images still remain. Vornberger & Bindschadler (1992) conducted multispectral research of ice sheets over an area of Greenland by registering Landsat TM and SAR imagery. They found that significant corrections to the SAR data were required to account for range-darkening, non-square pixel dimensions, speckle, and relief distortion. The exposed rock was available to be used as corresponding control points in one area, while it was absent in another area and lakes and streams were used. Li et al. (1993) used an elastic active contour to register the optical and SAR images. Using the contours from the optical image as the initial condition, accurate contour locations in the SAR image are obtained by applying an active contour model. They found that this snake method outperformed manual registration in terms of root mean square error at the control points. Dare & Dowman (1996) tried to develop an automatic system for registering SAR data to optical data by feature matching. In order to enhance the extraction of features from SAR images, they tested various speckle reduction filters and segmentation procedures to aid this procedure. Thepaut et al. (1998) proposed an automatic registration method of multidate and multisensory, i.e. ERS SAR and SPOT, images based on a multichannel consensual segmentation scheme. Zamora et al. (1998) also used the segmentation technique to register multi-sensor images by adaptive clustering prior to performing preprocessing and cepstrum operation to determine the translational displacement. Raucoules & Carnec (1999) proposed the use of local correlation to deal with the co-registration of ERS SAR and SPOT orthoimage by taking the interferometric phases as additional information. Dare & Dowman (2001) introduced an improved model, which is based on their previous work (1996), for automatic feature-based image registration, and presents a robust system for automatically registering SAR and SPOT imagery by incorporating multiple feature extraction and feature matching algorithms. Their experiments on both small and large images showed that in each case a large number of accurate tie points could be identified fully automatically across the images. Ali & Clausi (2002) conducted an investigation on the development of a fully automatic registration system for SAR and optical remote sensing images. The registration method is based on the extraction and matching of common features that are visible in both images. The algorithm involves the five steps, i.e. noise removal, edge extraction, edge linking pattern extraction and pattern matching. The experiments showed that accurate ground control points (GCPs) could be identified automatically. Hong & Schowengerdt (2003; 2005) proposed an automated precise registration approach of ERS2 SAR and Landsat5 TM images using visually-located control points based pre-registration and large-scale edge gradient contours based precise-registrtion. Lampropoulos et al. (2003) presents a proof of concept demonstration to perform image registration from dissimilar sources based on multiple transformations of two quite dissimilar images into new domains, where local or global similarities are extracted. Wang & Chen (2003) proposed an automatic SAR and optical image registration approach based on fuzzy linear feature extraction and neural network controlled resampling and image transformation. Cheng et al. (2004) proposed a SVM (support vector machine) supported edge feature extraction technique to register SAR and optical images. They used a modified Hausdorff distance as a similarity measure and genetic algorithm as the search strategy. Galland et al. (2005) first presented a SAR-to-optical image registration method by using a perfect knowledge of the sensor parameters. They suggested that the feature extraction can be used to perform the refinement of parameters, and the registration can be approximated by simple polynomial transformations. Inglada & Vadon (2005) proposed a fully automatic approach for the fine registration of SPOT5 and ENVISAT/ASAR IMS data. Shu et al. (2005) implemented a mutual information based registration method to match SAR and SPOT images. They used Gabor filters to extract orientation at four directions to enhance the registration process. The experimental results showed that the approach is much better than coarse manual registration. Yang et al. (2005) presents a SAR-to-optical image registration method by using extracted multiple features to improve the control points. Mao et al. (2007) used the approach being similar with Cheng et al. (2004) to register SAR and optical images based on linear features using Hausdorff distance and genetic algorithm. Shu and Tan (2007) proposed a mutual-information based method to register SAR and SPOT images. The novelty is to calculate local contrast of 5x5 windows centred at each point in both images, and to obtain two contrast images to be used to accurately register by assigning the contrast values to each pixel. Zhang et al. (2007) proposed a region-feature based automatic registration workflow for ENVISAT/ASAR and Landsat TM images. Wegner and Soergel (2008) proposed a semi-automatic registration approach based on line features for the improvement of the bridge scene of interest. They used a road extraction approach based on measuring spectral angels in the optical image, and thresholding and morphological operators for bridge extraction in the SAR image. The distance maps are calculated to transform discrete line segments to continuous two-dimensional information, and such distance maps are finally registered using a global transformation and a cross-correlation metric.

Summarizing the past research, we can find most studies used the feature based registration method (e.g., Wang & Chen, 2003; Wu & Maitre, 1990). The key problem of such kind of methods is how to enhance the extraction process of the corresponding features. The registration workflow is usually complicated, and it depends on many assistant means to lubricate the whole process. However, few researchers have tried to automate the SAR-to-optical image registration by using the intensity based approach (e.g., Mao et al., 2007; Shu et al., 2005). This kind of registration methods does not need many assistant means, and fully utilize the intensity information contained in images. The whole registration workflow is driven by an optimization process for similarity measure based object function. This search procedure for the transformation parameters usually needs large amount of computation. In this chapter, we put forth a novel intensity based method to register PALSAR (Kimura & Ito, 2000) and ASTER images based on our previous work (Li et al., 2007c). The intensity based mutual information is used as the similarity measure to automate the registration process. The multiscale steerable Simoncelli filter (Simoncelli & Adelson, 1990; Simoncelli & Farid, 1996; Simoncelli & Freeman, 1995; Simoncelli et al., 1992) is implemented to lubricate the registration process. A hybrid search technique is used to enhance the optimization process of transformation parameters. The experimental results showed that the proposed registration scheme is competent and feasible for PALSAR and ASTER images.

The whole chapter consists of five sections. The first section shortly reviews the research background of radar and optical image registration, and the latest registration techniques, and our novel registration methodology. The second section addresses a generalized mathematical model with mutual-information based similarity measure and multiscale steerable Simoncelli filter technique. The third section gives the description of PALSAR and ASTER satellite imagery. The fourth section discusses the experiment design, process, and results. The fifth section lays out the concluding remarks.

2. Algorithm

2.1 Image registration

The task of image registration is fundamental in image processing. Therefore, it is common problems in nearly all the scientific fields associated with image applications (Brown, 1992; Goshtasby, 2005; Maintz & Viergever, 1998; Pluim et al., 2003; Zitova & Flusser, 2003). In the literature, image registration is also termed image alignment or image matching (e.g., Chen et al., 2007; Chen et al., 2003; Szeliaki, 2005). From a mathematical point of view, image registration can be defined as a process to search a transformation which determines a mapping that is the best match of two or more images of the same object field, acquired by different sensors, or taken by the same sensor at different times (Li, 2006).

For an image pair (I_F, I_R) to be registered, the definition of registering the float image (I_F) to the reference image (I_R) can be expressed mathematically as (1):

$$I_R(x,y) = \zeta(I_F(T_\alpha(x,y))) \tag{1}$$

where T_α is a transformation function, which maps two spatial coordinates x and y, to the new spatial coordinates x' and y' by the set of parameters α as (2):

$$(x',y') = T_\alpha(x,y) \tag{2}$$

ζ is a one dimensional intensity or radiometric interpolation function.
The intensity based image registration can be analyzed as a non-convex optimization problem (Li, 2006; Modersitzki, 2004). This can be expressed mathematically as (3):

$$\alpha^* = \underset{T_\alpha}{Arg\,optima}(S_{T_\alpha}(I_F, I_R)) \tag{3}$$

where S is the similarity measure, and α^* is the optima estimated by the search algorithm. The generalized registration process using eqns. (1-3) for the intensity based method can always be depicted in Figure 1 (Li et al., 2006c; Yoo, 2004). It can be found that the intensity based image registration is an iterated process to search for the optimized transformation parameters.

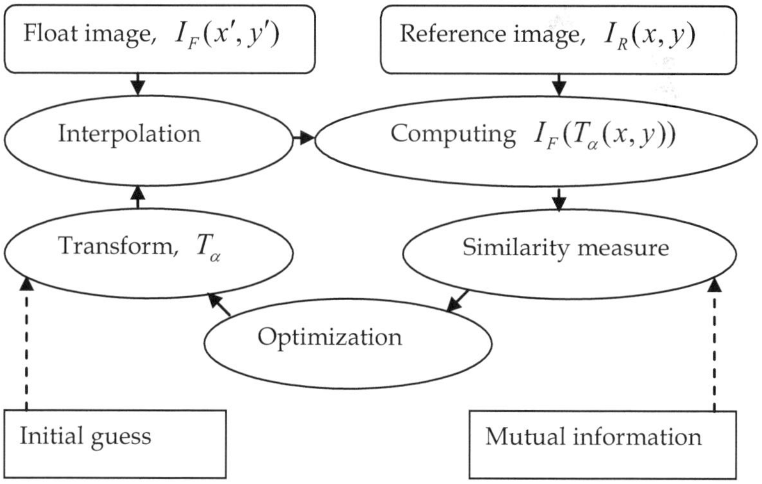

Fig. 1. A generalized framework for intensity based image registration

2.2 Mutual information

As pointed in Figure 1, similarity measure is an element of the image registration. It is used to construct an object function associated with image intensities for the optimization step. It should be noted that the term of similarity measure is different from the term of similarity metric (Cover & Thomas, 2006). In this chapter, we used the former. Starting in 1995, with the successful implementation of mutual information as a novel similarity measure to the multimodality medical image registration (Maes et al., 1997; Viola & Wells, 1995), the achievement of the intensity based automated image registration becomes possible. Based on the information theory (Cover & Thomas, 2006), the standard definition of mutual information, $MI(I_F, I_R)$, of two images I_F and I_R can be written as (4):

$$MI(I_F, I_R) = H(I_F) + H(I_R) - H(I_F, I_R) \tag{4}$$

where $H(I_F)$ and $H(I_R)$ are the marginal entropies of I_F and I_R, and $H(I_F, I_R)$ is their joint entropy. Considering the definition in (4), the mutual information is maximal when the two images are totally geometrically aligned by a certain transformation matrix. In practice, the normalized version (NMI) of the standard mutual information is popular. It may be defined as (5):

$$NMI(I_F, I_R) = \frac{H(I_F, I_R)}{H(I_F) + H(I_R)} \tag{5}$$

In order to compute of mutual information, the marginal entropies and joint entropy of the image pair should be calculated. These entropies can be calculated from the probability density functions. Furthermore, these probability density functions can be estimated from the histogram of images or other tricks such as Parzen windows (Glavinovic, 1996; Parzen, 1962). The detailed computation is left out in this chapter and it can be found in (Maes et al., 1997; Viola & Wells, 1995).

2.3 Simoncelli filter

The multiresolution decomposition techniques using wavelet-like filters are usually adopted to enhance the image registration (Le Moigne et al., 2002). Because the steerable Simoncelli filters are more robust to translation, rotation and noise than the standard Daubechies wavelet filters (Cole-Rhodes et al., 2003), it enables us to use it for registration of SAR and optical satellite images (Li et al., 2007c).

According to the definition of steerable Simoncelli filters (Simoncelli & Freeman, 1995), the steerable pyramid is a multiscale representation that is translation-invariant, but that also includes representation of orientation. Furthermore, the representation of orientation is designed to be rotation-invariant. The basis and projection functions are oriented (i.e., steerable) filters, localized in space and frequency. It is overcomplete to avoid aliasing. It is also "tight frame", i.e. the projection functions and basis functions are identical, though it is not an orthogonal representation.

The diagram for steerable Simoncelli pyramid may be depicted in Figure 2. The filters {Fhi0, Flo0} are used to initially split the image into a highpass residual band and a lowpass subband. This lowpass band Flo0 is then split into some lowerpass bands {Flo1, Flo2, ...}. {FB0, FB1, ...} represent the oriented subbands which ensure that the representation is rotation-invariant. In order to ensure some translation-invariance, the outputs of the high-pass filter and of the band-pass filters are not subsampled. The resulting transform is overcomplete by a factor of 4k/3, where k is the number of oriented band-pass filters. The scale tuning of the filters is constrained by the recursive system diagram. The orientation tuning is constrained by requiring the property of steerability (Cole-Rhodes et al., 2003; Li et al., 2007c; Simoncelli & Freeman, 1995).

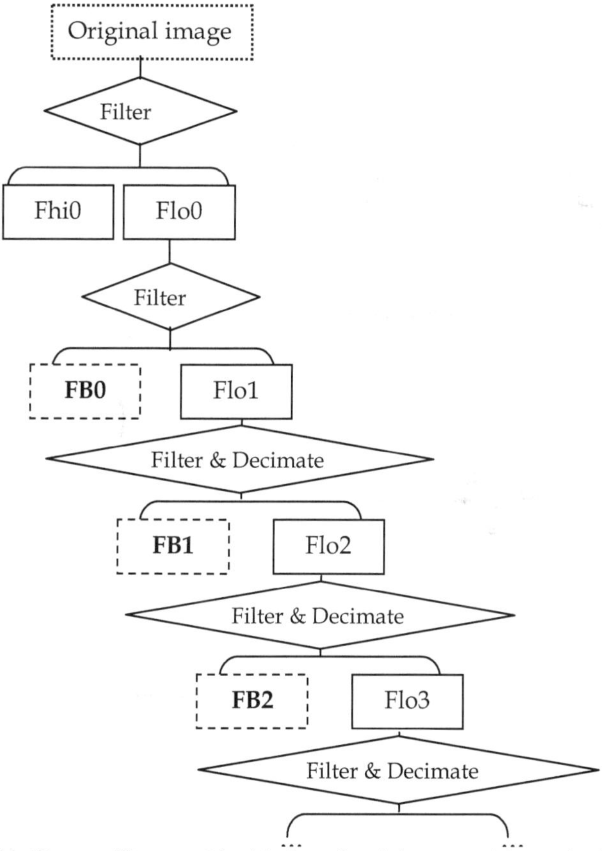

Fig. 2. A steerable Simoncelli pyramid with three-level decompositions of original image

2.4 Workflow

As shown in Figure 1, the computational process of intensity based image registrations can be partitioned into four major modules, i.e., intensity interpolation, mapping transformation, similarity measure, and optimization strategy for the parameter space. Till now, our developed registration system has implemented 7 interpolation algorithms, 25 similarity measures, 11 optimization algorithms, and it supported 7 transformations from rigid to polynomial mapping (Li et al., 2007a). The whole workflow of image registration in this chapter can be depicted in Figure 3.

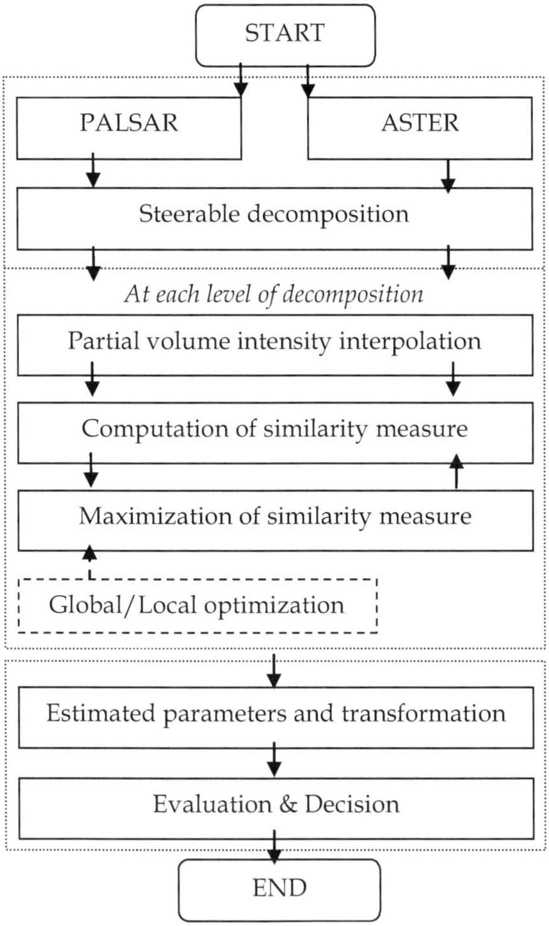

Fig. 3. Workflow of SAR-to-optical image registration

3. Data

3.1 ASTER

The Advanced Spaceborne Thermal Emission and Reflection Radiometer (ASTER) sensor instrument was launched in December 18, 1999, onboard the first NASA's EOS series of satellites, Terra. ASTER covers a wide spectral region with 14 bands from visible to thermal infrared with high spatial, spectral and radiometric resolution. Three visible and near infrared (VNIR) bands, six shortwave infrared (SWIR) bands, and five thermal infrared (TIR) bands have a spatial resolution of 15 m, 30 m, and 90 m, respectively. In addition, ASTER has a stereoscopic capability for the bands (bands 3N and 3B) in near infrared region (Iwasaki & Fujisada, 2005; Yamaguchi et al., 1998), and so far, it can generate Digital Elevation Model (DEM) data with high accuracy (Fujisada et al., 2005).
ASTER-TIR is the first satellite-borne multispectral TIR remote sensing system with spectral, spatial and radiometric resolutions adequate for geological applications (e.g., Ninomiya et al., 2005). Compared with two bands of Landsat TM or ETM in SWIR region (between 1.6 to 2.5 microns), ASTER sensor has 6 bands in this region and provides an opportunity to identify mineral component of surface rocks in the semi-arid to arid region (e.g., Rowan & Mars, 2003). Therefore, ASTER imaging system can provide some important capabilities to identify lithologic and mineralogic features on earth surface (e.g., Fu et al., 2007; Ninomiya et al., 2005). Especially, ASTER multispectral TIR sensor can provide an important tool for monitoring heat flow related to volcanic activities (e.g., Urai et al., 2007). Therefore, ASTER can provide a potential tool for mapping the wide multiple-aim geologic products from regional to global scales because high-resolution multispectral dada obtained by ASTER can cover almost throughout earth surface.

3.2 PALSAR

PALSAR is the abbreviation of Phased Array type L-band Synthetic Aperture Radar. It is an active microwave sensor using L-band frequency, operated at all-weather conditions regardless of day and night, launched with ALOS satellite on January 24, 2006 in Japan. It is improved based on SAR onboard the first earth observation satellite (JERS-1) with multi-mode observation functions of multi-polarization, variable off-nadir angle, and switching spatial resolution and swath width observation (Igarashi, 2000; Kimura & Ito, 2000). It provides higher performance than the JERS-1's SAR with a totally new advantageous observation mode (i.e., ScanSAR). PALSAR has incorporated many highly advanced observation technologies, and is expected to contribute greatly in areas such as resource exploration, environmental monitoring on earth and monitoring of natural disasters (e.g., Rosenqvist et al., 2007; Takada et al., 2009). The signals are recorded in complex notation on PALSAR sensor from which their amplitude and phases could be computed. The specifications of PALSAR sensor can be summarized in Table 1. It should be noted that PALSAR sensor can not observe the areas beyond 87.8 Degrees north latitude and 75.9 Degrees south latitude, when the off-nadir angle is 41.5 Degrees (http://www.eorc.jaxa.jp/ALOS/en/about/palsar.htm).

3.3 Image set

In this chapter, all image pairs were extracted from each of the full scene ASTER and PALSAR data. The ASTER L1B Band 1 data and PALSAR fine mode data were used in the experiments. The extracted sub-images for experiments have 128x128, 256x256, 512x512, 1024x1024 pixels, respectively. The research area covers the part of the city of Tokyo, Japan. One sub-image pair has been shown in Figure 4. The very clear difference in terms of visual appearance of features can be observed. The same spatial features can not be easily found from both images. PALSAR image is inevitably contaminated by the speckle noise and strongly scattered signals from any corners on the earth surface. In principle, physical properties and viewing geometries between ASTER and PALSAR images are intrinsic different. However, the effect resulted from the viewing geometry will be alleviated in flat regions.

Mode	*Fine*		*ScanSAR*	*Polarimetric**
Center frequency (MHz)	1270 (L-band)			
Chirp bandwidth (MHz)	28	14	14, 28	14
Polarization	HH VV	HH+HV VV+VH	HH VV	HH+HV+VH+VV
Incident angle (Degree)	8-60	8-60	18-43	8-30
Range resolution (m)	7-44	14-88	100 (Multi look)	24-89
Observation swath (km)	40-70	40-70	250-350	20-65
Bit length (bits)	5	5	5	3 or 5
Data rate (Mbps)	240	240	120, 240	240
NE sigma zero ** (dB)	<-23 (Swath width 70km) <-25 (Swath width 60km)		<-25	<-29
S/A *** (dB)	>16 (Swath width 70km) >21 (Swath width 60km)		>21	>19
Radiometric accuracy	Scene: 1dB / Orbit: 1.5dB			

* Due to power consumption, the operation time will be limited.

** Valid for off-nadir angle with 34.3 Degrees (Fine mode), 34.1 Degrees (ScanSAR mode), 21.5 Degrees (Polarimetric mode).

*** S/A level may deteriorate due to engineering changes in PALSAR

Table 1. Characteristics of PALSAR

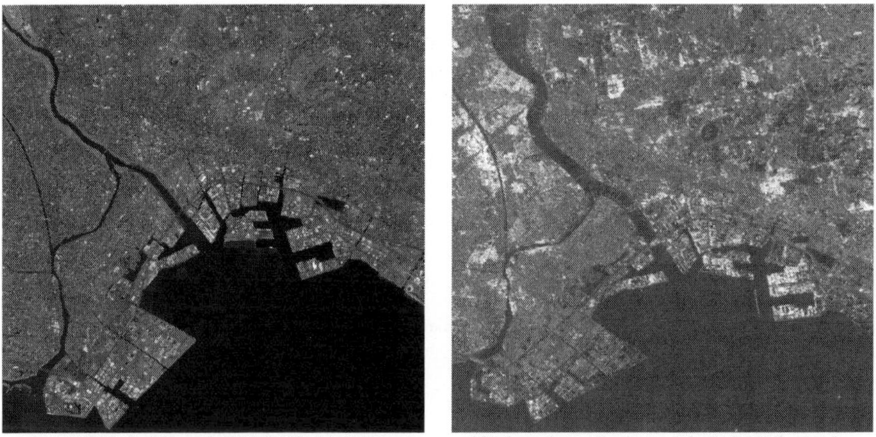

Fig. 4. ASTER (left) and PALSAR (right) images of Tokyo bay, Japan with 512x512 pixels

4. Experiments

The registration scheme of our proposed method (Figure 3) includes three major stages similar with our previous work (Li et al., 2007c). Both SAR and optical image are firstly decomposed into a steerable pyramid. Subbands FB*i* (Figure 2) are utilized to extract features in the provided image set. The partial volume intensity interpolation (Li et al., 2006a) is adopted for the estimation of probability density functions. The particle swarm optimization (Li & Sato, 2007) is used to globally search the parameter space of the registration function at the coarsest level. The local stochastic gradient search, which is conducted by a simultaneous perturbation stochastic approximation technique (Li et al., 2006b; 2007b), is implemented to optimize the registration function at other levels. The result of the global optimization is used as the initial guess of the local stochastic gradient search.

Figure 5 shows the visualization check of one registration result. By visual comparison, we can note that the registration of the proposed scheme is much better than the manual method conducted in some commercial software. To do a numerical comparison, we manually locate 5-10 pairs of check points with a good distribution and evaluate the registration accuracies according to RMSE (root mean squared error). The RMSE values are 5-7 pixels for manual registration and 1-3 pixels for our registration. This indicates that the registration accuracy is greatly improved by our proposed scheme. In particular, the results are better than our previous work on the registration of JERS-1 SAR and ASTER images (Li et al., 2007c).

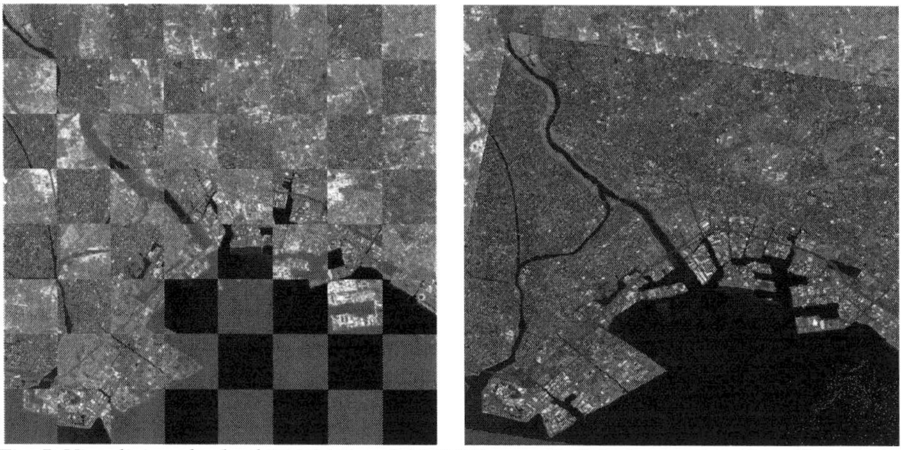

Fig. 5. Visualizing check of PALSAR and ASTER image (Left: before registration; Right: after affine registration)

5. Conclusions

The experimental results showed that proposed scheme can be capable to register radar and optical satellite images such as PALSAR and ASTER images. Compared to the traditional manual method, the scheme using multiscale filter technique and information measure can greatly enhance the registration process. The hybrid search/optimization approach is relatively less sensitive to initial guess, and make the registration process robust. Meanwhile, our method maintains comparable accuracy comparing with the traditional manual method. Except for the computing time, the scheme might encounter difficulties when the two images with very time differences.

Future work should include two parts. The first one is to conduct many more experiments on different regions. The other one is to incorporate some feature based technique to speed up the search process.

Acknowledgements

This research has been mainly funded under the research contract with the Ministry of Economy, Trade and Industry (METI), Japan as the part of the R&D of Remote Sensing Technologies for Non-renewable Resources (SEKITOKU). QL would like to thank I. Sato, Y. Murakami, M. Urai, Y. Ninomiya, F. Sakuma, and S. Okuyama (AIST) for all of their help. ASTER and PALSAR data were provided by the Earth Remote Sensing Data Analysis Centre (ERSDAC), Japan.

6. References

Ali, M.A. & Clausi, D.A. (2002). Automatic registration of SAR and visible band remote sensing images, Proceedings of the 22nd IEEE International Geoscience and Remote Sensing Symposium (IGARSS 2002)/24th Canadian Symposium on Remote Sensing - Remote Sensing: Integrating Our View of The Planet, pp. 1331-1333, 0-7803-7536-X, Toronto, Canada, June 24-28, 2002, IEEE, New York

Barrett, E.C.; Brown, K.A. & Micallef, A. (Eds.) (1991). Remote sensing for hazard monitoring and disaster assessment: Marine and coastal applications in the mediterranean region, Gordon and Breach Science Publishers, 2-8812-4809-8, Philadelphia, PA, USA

Brown, L.G. (1992). A survey of image registration techniques. ACM Computing Surveys, Vol.24, No.4, (December 1992) page numbers (325-376), 0360-0300

Chen, F.; Wang, C. & Zhang, H. (2007). Automatic matching of high-resolution SAR images. International Journal of Remote Sensing, Vol.28, No.16, (January 2007) page numbers (3665-3678), 0143-1161

Chen, G.H.; Tang, L.L.; Dai, C.D. & Jiang, X.G. (2003). Selection of fusing methods with SAR and TM data to monitor flood quickly, Proceedings of Conference on Image Processing and Pattern Recognition in Remote Sensing, pp. 162-170, 0-8194-4684-X, Hangzhou, China, October 25-27, 2002, SPIE, Bellingham, WA, USA

Chen, H.; Varshney, P.K. & Arora, M.K. (2003). Mutual information based image registration for remote sensing data. International Journal of Remote Sensing, Vol.24, No.18, (September 2003) page numbers (3701-3706), 0143-1161

Cheng, H.; Zheng, S.; Yu, Q.Z.; Tian, J.W. & Liu, J. (2004). Matching of SAR images and optical images based on edge feature extracted via SVM, Proceedings of the 7th International Conference on Signal Processing, pp. 930-933, 7-121-00215-9, Beijing, China, August 31-September 04, 2004, Publishing House Electronics Industry, Beijing

Chevrel, M.; Courtois, M. & Weill, G. (1981). The SPOT satellite remote-sensing mission. Photogrammetric Engineering and Remote Sensing, Vol.47, No.8, (August 1981) page numbers (1163-1171), 0099-1112

Chuvieco, E. (Ed.) (2008). Earth observation of global change: The role of satellite remote sensing in monitoring the global environment, Springer, 1-4020-6357-1, London, UK

Cole-Rhodes, A.A.; Johnson, K.L.; LeMoigne, J. & Zavorin, I. (2003). Multiresolution registration of remote sensing imagery by optimization of mutual information using a stochastic gradient. IEEE Transactions on Image Processing, Vol.12, No.12, (December 2003) page numbers (1495-1511), 1057-7149

Cover, T.M. & Thomas, J.A. (2006). Elements of information theory, Wiley-Interscience, 0-471-24195-4, New York

Curlander, J. & Kober, W. (1990). Innovative algorithms for SAR to optical-image registration, Proceedings of the 10th Annual International Geoscience and Remote Sensing Symposium: Remote Sensing Science for the Nineties, pp. 329-331, College Park, MD, USA, May 20-24, 1990, IEEE, New York

Dare, P.M. & Dowman, I.J. (1996). Automated procedure for registering SAR and optical imagery based on feature matching, Proceedings of Conference on Microwave Sensing and Synthetic Aperture Radar, pp. 140-151, 0-8194-2362-9, Taormina, Italy, September 23-26, 1996, SPIE, Bellingham, WA, USA

Dare, P.M. & Dowman, I.J. (2001). An improved model for automatic feature-based registration of SAR and SPOT images. ISPRS Journal of Photogrammetry and Remote Sensing, Vol.56, No.1, (June 2001) page numbers (13-28), 0924-2716

Franceschetti, G. & Lanari, R. (1999). Synthetic aperture radar processing, CRC, 0-8493-7899-0, Boca Raton, FL, USA

Fu, B.H.; Ninomiya, Y.; Lei, X.L.; Toda, S. & Awata, Y. (2004). Mapping active fault associated with the 2003 Mw 6.6 Bam (SE Iran) earthquake with ASTER 3D images. Remote Sensing of Environment, Vol.92, No.2, (August 2004) page numbers (153-157), 0034-4257

Fu, B.H.; Shi, P.L.; Wang, P.; Li, Q.; Kong, P. & Zheng, G.D. (2009). Geometry and kinematics of the 2008 Wenchuan earthquake surface ruptures around the Qushan town of Beichuan county, Sichuan: Implications for mitigation of seismic and geologic disasters. Chinese Journal of Geophysics-Chinese Edition, Vol.52, No.2, (February 2009) page numbers (485-495), 0001-5733

Fu, B.H.; Zheng, G.D.; Ninomiya, Y.; Wang, C.Y. & Sun, G.Q. (2007). Mapping hydrocarbon-induced mineralogical alteration in the northern Tian Shan using ASTER multispectral data. Terra Nova, Vol.19, No.4, (August 2007) page numbers (225-231), 0954-4879

Fujisada, H.; Bailey, G.B.; Kelly, G.G.; Hara, S. & Abrams, M.J. (2005). ASTER DEM performance. IEEE Transactions on Geoscience and Remote Sensing, Vol.43, No.12, (December 2005) page numbers (2707-2714), 0196-2892

Galland, F.; Tupin, F.; Nicolas, J.-M. & Roux, M. (2005). Registering of synthetic aperture radar and optical data, Proceedings of the 25th IEEE International Geoscience and Remote Sensing Symposium (IGARSS 2005), pp. 3513-3516, 0-7803-9050-4, Seoul, South Korea, July 25-29, 2005, IEEE, New York

Glavinovic, M.I. (1996). Comparison of parzen density and frequency histogram as estimators of probability density functions. Pflugers Archiv-European Journal of Physiology, Vol.433, No.1-2, (November-December 1996) page numbers (174-179), 0031-6768

Goshtasby, A.A. (2005). 2-D and 3-D image registration: For medical, remote sensing, and industrial applications, Wiley-Interscience, 0-4716-4954-6, New York

Hong, T.D. & Schowengerdt, R.A. (2003). Automated precise registration of radar and optical satellite images, Proceedings of Conference on Applications of Digital Image Processing XXVI, pp. 88-96, 0-8194-5076-6, San Diego, CA, USA, August 05-08, 2003, SPIE, Bellingham, WA, USA

Hong, T.D. & Schowengerdt, R.A. (2005). A robust technique for precise registration of radar and optical satellite images. Photogrammetric Engineering and Remote Sensing, Vol.71, No.5, (May 2005) page numbers (585-594), 0099-1112

Igarashi, T. (2000). ALOS mission requirement and sensor specifications. Advances in Space Research, Vol.28, No.1, (2001) page numbers (127-131), 0273-1177

Inglada, J. & Giros, A. (2004). On the possibility of automatic multisensor image registration. IEEE Transactions on Geoscience and Remote Sensing, Vol.42, No.10, (October 2004) page numbers (2104-2120), 0196-2892

Inglada, J. & Vadon, H. (2005). Fine registration of SPOT5 and Envisat/ASAR images and ortho-image production: A fully automatic approach, Proceedings of the 25th IEEE International Geoscience and Remote Sensing Symposium (IGARSS 2005), pp. 3510-3512, 0-7803-9050-4, Seoul, South Korea, July 25-29, 2005, IEEE, New York

Iwasaki, A. & Fujisada, H. (2005). ASTER geometric performance. IEEE Transactions on Geoscience and Remote Sensing, Vol.43, No.12, (December 2005) page numbers (2700-2706), 0196-2892

Kafatos, M. & Qu, J.J. (2007). Introduction to science and instruments, In: Earth science satellite remote sensing: Science and instruments, Qu, J.J.; Gao, W.; Kafatos, M.; Murphy, R.E. & Salomonson, V.V., (Eds.), page numbers (1-11), Tsinghua University Press and Springer, 7-302-12844-8, Beijing and Berlin

Kaufman, Y.J.; Herring, D.D.; Ranson, K.J. & Collatz, G.J. (1998). Earth observing system AM1 mission to earth. IEEE Transactions on Geoscience and Remote Sensing, Vol.36, No.4, (July 1998) page numbers (1045-1055), 0196-2892

Kimura, H. & Ito, N. (2000). ALOS/PALSAR: The Japanese second-generation spaceborne SAR and its applications, Proceedings of Conference on Microwave Remote Sensing of the Atmosphere and Environment II, pp. 110-119, 0-8194-3805-7, Sendai, Japan, October 09-12, 2000, SPIE, Bellingham, WA, USA

Lampropoulos, G.A. & Boulter, J.F. (1997). A new nonlinear speckle and noise reduction algorithm, Proceedings of 5th Conference on Infrared Spaceborne Remote Sensing, pp. 260-270, 0-8194-2544-3, San Diego, CA, USA, July 30-August 01, 1997, SPIE, Bellingham, WA, USA

Lampropoulos, G.A.; Chan, J.; Secker, J.; Li, Y. & Jouan, A. (2003). Automatic registration of electro-optical and SAR images, Proceedings of IEEE Workshop on Advances in Techniques for Analysis of Remotely Sensed Data held in Honor of David A Landgrebe, pp. 219-226, 0-7803-8350-8, Greenbelt, MD, USA, October 27-28, 2003, IEEE, New York

Lampropoulos, G.A.; Li, Y.F.; Secker, J.; Sevigny, L. & Beaudoin, A. (2002). Electro-optical and SAR image fusion for improvements on target feature estimation, Proceedings of the 5th International Conference on Applications of Photonic Technology (ICAPT 2002), pp. 214-225, 0-8194-4612-2, Quebec City, Canada, June 01-06, 2002, SPIE, Bellingham, WA, USA

Le Moigne, J.; Cole-Rhodes, A.; Eastman, R.; El-Ghazawi, T.; Johnson, K.; Kaewpijit, S.; Laporte, N.; Morisette, J.; Netanyahu, N.S.; Stone, H.S. & Zavorin, I. (2002). Multiple sensor image registration, image fusion and dimension reduction of earth science imagery, Proceedings of the 5th International Conference on Information Fusion (FUSION 2002), pp. 999-1006, 0-9721844-2-2, Annapolis, MD, USA, July 08-11, 2002, International Society of Information Fusion (ISIF), Sunnyvale, CA, USA

Li, H.; Manjunath, B.S. & Mitra, S.K. (1993). Optical-to-SAR image registration using the active contour model, Proceedings of the 27th Asilomar Conference on Signals, Systems and Computers, pp. 568-572, 0-8186-4120-7, Pacific Grove, CA, USA, November 01-03, 1993, IEEE, Computer Society Press, Los Alamitos, CA, USA

Li, H.; Manjunath, B.S. & Mitra, S.K. (1995). A contour-based approach to multisensor image registration. IEEE Transactions on Image Processing, Vol.4, No.3, (March 1995) page numbers (320-334), 1057-7149

Li, Q. (2006). Challenging registration of geologic image data - initialization, estimation, and decision. Seminar Note of Mathematical Sciences, Vol.10, (September 8-10, 2006) page numbers (78-87), ISSN-N/A

Li, Q.; Ito, K.; Tomishima, Y. & Seki, Y. (2009). Bridging satellite monitoring and characterization of subsurface flow: With a case of Horonobe Underground Research Laboratory, In: Prediction and simulation methods for geohazard mitigation, Oka, F.; Murakami, A. & Kimoto, S., (Eds.), page numbers (519-524), CRC Press, 978-0-415-80482-0, New York

Li, Q. & Sato, I. (2007). Multimodality image registration by particle swarm optimization of mutual information. Lecture Notes in Artificial Intelligence, Vol.4682, (August 2007) page numbers (1120-1130), 0302-9743

Li, Q.; Sato, I. & Murakami, Y. (2006a). Interpolation effects on accuracy of mutual information based image registration, Proceedings of 2006 IEEE International Geoscience and Remote Sensing Symposium / 27th Canadian Symposium on Remote Sensing - Remote Sensing: A Natural Global Partnership (IGARSS 2006), pp. 180-183, 0-7803-9510-7, Denver, Colorado, USA, July 31 - August 4, 2006, IEEE, New York

Li, Q.; Sato, I. & Murakami, Y. (2006b). Simultaneous perturbation stochastic approximation algorithm for automated image registration optimization, Proceedings of 2006 IEEE International Geoscience and Remote Sensing Symposium / 27th Canadian Symposium on Remote Sensing - Remote Sensing: A Natural Global Partnership (IGARSS 2006), pp. 184-187, 0-7803-9510-7, Denver, CO, USA, 31 July - 4 August, 2006, IEEE, New York

Li, Q.; Sato, I. & Murakami, Y. (2006c). Toward the establishment of robust automatic system for multimodal image registration, Proceedings of The 40th (2006 Spring) Annual Meeting of the Remote Sensing Society of Japan, pp. 101-102, Keyaki Kaikan, Chiba University, Chiba, Japan, May 18-19, 2006, The Remote Sensing Society of Japan (RSSJ), Tokyo

Li, Q.; Sato, I. & Murakami, Y. (2007a). Current status of automatic image registration system GSJ/AIR for geologic image integration analysis, Proceedings of The 42nd (2007 Spring) Annual Meeting of the Remote Sensing Society of Japan, pp. 221-222, College of Humanities and Sciences, Nihon University, May 10-11, 2007, The Remote Sensing Society of Japan (RSSJ), Tokyo

Li, Q.; Sato, I. & Murakami, Y. (2007b). Efficient stochastic gradient search for automatic image registration. International Journal of Simulation Modelling, Vol.6, No.2, page numbers (114-123), 1726-4529

Li, Q.; Sato, I. & Murakami, Y. (2007c). Steerable filter based multiscale registration method for JERS-1 SAR and ASTER images, Proceedings of 2007 IEEE International Geoscience and Remote Sensing Symposium (IGARSS 2007). Sensing and Understanding Our Planet, pp. 381-384, 978-1-4244-1212-9, Barcelona, Spain, 23-28 July 2007, IEEE, New York

Li, Q.; Sato, I. & Sakuma, F. (2008). A novel strategy for precise geometric registration of GIS and satellite images, Proceedings of 2008 IEEE International Geoscience and Remote Sensing Symposium - Geoscience and Remote Sensing: The Next Generation (IGARSS 2008), pp. II-1092-II-1095, 978-1-4244-2807-6, Boston, MA, USA, July 6-11, 2008, IEEE, New York

Maes, F.; Collignon, A.; Vandermeulen, D.; Marchal, G. & Suetens, P. (1997). Multimodality image registration by maximization of mutual information. IEEE Transactions on Medical Imaging, Vol.16, No.2, (April 1997) page numbers (187-198), 0278-0062

Maintz, J.B.A. & Viergever, M.A. (1998). A survey of medical image registration. Medical Image Analysis, Vol.2, No.1, (March 1998) page numbers (1-36), 1361-8415

Mao, H.C.; Yu, Q.Z. & Zhang, T.X. (2007). Matching SAR image to optical image using modified hausdorff distance and genetic algorithms, Proceedings of Conference on Pattern Recognition and Computer Vision, art. No. 678820, 978-0-8194-6952-6, Wuhan, China, November 15-17, 2007, SPIE, Bellingham, WA, USA

Modersitzki, J. (2004). Numerical methods for image registration, Oxford University Press, 0-1985-2841-8, New York

Ninomiya, Y.; Fu, B.H. & Cudahy, T.J. (2005). Detecting lithology with advanced spaceborne thermal emission and reflection radiometer (ASTER) multispectral thermal infrared "Radiance-at-sensor" Data. Remote Sensing of Environment, Vol.99, No.1-2, (November 2005) page numbers (127-139), 0034-4257

Ninomiya, Y.; Fu, B.H. & Cudahy, T.J. (2006). Detecting lithology with advanced spaceborne thermal emission and reflection radiometer (ASTER) multispectral thermal infrared "Radiance-at-sensor" Data" (vol 99, pg 127, 2005). Remote Sensing of Environment, Vol.101, No.4, (April 2006) page number (567), 0034-4257

Parzen, E. (1962). On estimation of a probability density function and mode. Annals of Mathematical Statistics, Vol.33, No.3, page numbers (1065-1076), 0003-4851

Pluim, J.P.W.; Maintz, J.B.A. & Viergever, M.A. (2003). Mutual-information-based registration of medical images: A survey. IEEE Transactions on Medical Imaging, Vol.22, No.8, (August 2003) page numbers (986-1004), 0278-0062

Raucoules, D. & Carnec, C. (2000). Use of interferometric phase for co-registration of ERS SAR and SPOT images, Proceedings of CEOS SAR Workshop, pp. 113-117, 92-9092-641-4, Toulouse, France, October 26-29, 1999, European Space Agency, Paris, France

Realmuto, V.J. (2000). The potential use of earth observing system data to monitor the passive emission of sulfur dioxide from volcanoes, In: Remote sensing of active volcanism, Mouginis-Mark, P.J.; Crisp, J.A. & Fink, J.H., (Eds.), page numbers (101-115), AGU, 0-87590-099-2, Washington DC, USA

Rosenqvist, A.; Shimada, M.; Ito, N. & Watanabe, M. (2007). ALOS PALSAR: A pathfinder mission for global-scale monitoring of the environment. IEEE Transactions on Geoscience and Remote Sensing, Vol.45, No.11, (November 2007) page numbers (3307-3316), 0196-2892

Rowan, L.C. & Mars, J.C. (2003). Lithologic mapping in the mountain pass, california area using advanced spaceborne thermal emission and reflection radiometer (ASTER) data. Remote Sensing of Environment, Vol.84, No.3, (March 2003) page numbers (350-366), 0034-4257

Sato, I.; Li, Q.; Nomura, L.; Bandibus, J.; Nishida, K.; Masuda, T. & Kita, Y. (2006). Research and development of resource fusion technology (part of advanced research of utilization technology of satellite images), In: Report on R&D of remote sensing technologies for non-renewable resources 1/3, page numbers (1.1.A.1-12), National Institute of Advanced Industrial Science and Technology (AIST), Tsukuba, Japan

Schowengerdt, R.A. (2006). Remote sensing: Models and methods for image processing, Academic Press, 0-1236-9407-8, New York

Shu, L.X. & Tan, T.N. (2007). SAR and SPOT image registration based on mutual information with contrast measure, Proceedings of 2007 IEEE International Conference on Image Processing (ICIP 2007), pp. 2681-2684, 978-1-4244-1436-9, San Antonio, TX, USA, September 16-19, 2007, IEEE, New York

Shu, L.X.; Tan, T.N.; Tang, M. & Pan, C.H. (2005). A novel registration method for SAR and SPOT images, Proceedings of 2005 IEEE International Conference on Image Processing (ICIP 2005), pp. 213-216, 0-7803-9134-9, Genova, Italy, September 11-14, 2005, IEEE, New York

Simoncelli, E.P. & Adelson, E.H. (1990). Nonseparable extensions of quadrature mirror filters to multiple dimensions. Proceedings of the IEEE, Vol.78, No.4, (April 1990) page numbers (652-664), 0018-9219

Simoncelli, E.P. & Farid, H. (1996). Steerable wedge filters for local orientation analysis. IEEE Transactions on Image Processing, Vol.5, No.9, (September 1996) page numbers (1377-1382), 1057-7149

Simoncelli, E.P. & Freeman, W.T. (1995). The steerable pyramid: A flexible architecture for multi-scale derivative computation, Proceedings of International Conference on Image Processing (ICIP 1995), pp. 444-447, 0-7803-3122-2, Washington, DC, USA, October 23-26, 1995, IEEE, New York

Simoncelli, E.P.; Freeman, W.T.; Adelson, E.H. & Heeger, D.J. (1992). Shiftable multiscale transforms. IEEE Transactions on Information Theory, Vol.38, No.2, (March 1992) page numbers (587-607), 0018-9448

Szeliaki, R. (2005). Image alignment and stitching, In: Handbook of mathematical models in computer vision, Paragios, N.; Chen, Y. & Faugeras, O., (Eds.), page numbers (273-292), Springer, 0-3872-6371-3, New York

Takada, M.; Mishima, Y. & Natsume, S. (2009). Estimation of surface soil properties in peatland using ALOS/PALSAR. Landscape and Ecological Engineering, Vol.5, No.1, (February 2009) page numbers (45-58), 1860-1871

Tanaka, S. & Sugimura, T. (2001). A new frontier of remote sensing from IKONOS images. International Journal of Remote Sensing, Vol.22, No.1, (January 2001) page numbers (1-5), 0143-1161

Teeuw, R.M. (Ed.) (2007). Mapping hazardous terrain using remote sensing, Geological Society, 1-8623-9229-3, London, UK

Thepaut, O.; Kpalma, K. & Ronsin, J. (1998). ERS SAR and SPOT images automatic registration in a multichannel consensual segmentation scheme, Proceedings of 1998 International Geoscience and Remote Sensing Symposium (IGARSS '98) on Sensing and Managing the Environment, pp. 1040-1042, 0-7803-4403-0, Seattle, WA, USA, July 06-10, 1998, IEEE, New York

Urai, M.; Geshi, N. & Staudacher, T. (2007). Size and volume evaluation of the caldera collapse on Piton de la Fournaise volcano during the April 2007 eruption using ASTER stereo imagery. Geophysical Research Letters, Vol.34, No.22, (November 2007) page numbers (L22318: 1-7), 0094-8276

Viola, P.A. & Wells, W. (1995). Alignment by maximization of mutual information, Proceedings of the 5th International Conference on Computer Vision, pp. 6-23, 0-8186-7042-8, Cambridge, MA, USA, June 20-23, 1995, IEEE, New York

Vornberger, P.L. & Bindschadler, R.A. (1992). Multispectral analysis of ice sheets using co-registered SAR and TM imagery. International Journal of Remote Sensing, Vol.13, No.4, (March 1992) page numbers (637-645), 0143-1161

Wang, Y.L. & Chen, Z. (2003). Automatic registration of SAR and optical images based on linear features and neural network, Proceedings of the 3rd International Symposium on Multispectral Image Processing and Pattern Recognition, pp. 553-557, 0-8194-5181-9, Beijing, China, October 20-22, 2003, SPIE, Bellingham, WA, USA

Wegner, J.D. & Soergel, U. (2008a). Bridge height estimation from combined high-resolution optical and SAR imagery. XXIth ISPRS Congress: Silk Road for Information from Imagery, Vol.XXXVII, No.B7, (July 2008) page numbers (1071-1076), 1682-1750

Wegner, J.D. & Soergel, U. (2008b). Registration of SAR and optical images containing bridges over land, Proceedings of EARSeL Joint Workshop "Remote Sensing - New Challenges of High Resolution", pp. 194-201, 978-3-925143-79-3, Bochum, Germany, March 5-7, 2008, EARSeL, Bochum, Germany

Wu, Y.F. & Maitre, H. (1990). A multiresolution approach for registration of a SPOT image and a SAR image, Proceedings of the 10th Annual International Geoscience and Remote Sensing Symposium: Remote Sensing Science for the Nineties, pp. 635-638, College Park, MD, USA, May 20-24, 1990, IEEE, New York

Yamaguchi, Y.; Kahle, A.B.; Tsu, H.; Kawakami, T. & Pniel, M. (1998). Overview of advanced spaceborne thermal emission and reflectionradiometer (ASTER). IEEE Transactions on Geoscience and Remote Sensing, Vol.36, No.4, (July 1998) page numbers (1062-1071), 0196-2892

Yang, W.; Han, C.Z.; Sun, H. & Cao, Y.F. (2005). Registration of high resolution SAR and optical images based on multiple features, Proceedings of the 25th IEEE International Geoscience and Remote Sensing Symposium (IGARSS 2005), pp. 3542-3544, 0-7803-9050-4, Seoul, South Korea, July 25-29, 2005, IEEE, New York

Yoo, T.S. (Ed.) (2004). Insight into images: Principles and practice for segmentation, registration, and image analysis, A K Peters, Ltd., 1-5688-1217-5, Wellesley, Massachusetts, USA

Zamora, G.; Dickens, M. & Mitra, S. (1998). A robust registration technique for multi-sensor images, Proceedings of 1998 IEEE Southwest Symposium on Image Analysis and Interpretation, pp. 87-90, 0-7803-4876-1, Tucson, AZ, USA, April 05-07, 1998, IEEE, New York

Zhang, D.R.; Yu, L. & Cai, Z.G. (2007). Automatic registration for ASAR and TM images based on region features, Proceedings of the 15th International Conference on Geoinformatics, art. No. 675240, 978-0-8194-6912-0, Nanjing, China, May 25-27, 2007, SPIE, Bellingham, WA, USA

Zhang, Z.H.; Pan, C.H. & Ma, S.D. (2004). An automatic method of coarse registration between multi-source satellite images, Proceedings of Intelligent Sensors, Sensor Networks and Information Processing Conference, pp. 205-209, 0-7803-8893-3, Melbourne, Australia, December 14-17, 2004, IEEE, New York

Zhao, Y. & Chen, Y.Q. (2003). A robust contour model for matching synthetic aperture radar (SAR) images with maps, Proceedings of Conference on Image Processing and Pattern Recognition in Remote Sensing, pp. 150-161, 0-8194-4684-X, Hangzhou, China, October 25-27, 2002, SPIE, Bellingham, WA, USA

Zitova, B. & Flusser, J. (2003). Image registration methods: A survey. Image and Vision Computing, Vol.21, No.11, (October 2003) page numbers (977-1000), 0262-8856

25

Magnitude and Extent of Six Years of Land Subsidence in Shanghai Revealed by JERS-1 SAR Data

[1]Peter Damoah-Afari, [2]Xiao-li Ding, [3]Zhong Lu, [4]Zhiwei Li and [5]Makoto Omura
[1,2]The Hong Kong Polytechnic University, Hong Kong
[3]U.S Geological Survey, USA
[4]Central South University, China
[5] Kochi Women's University, Japan

1. Introduction

Many provinces in China including Shanghai, Tianjin, Jiansu, Hubei, Hangzhou, Suzhou, Changzhou, Cangzhou, and Jianxing (Deng and Ju, 1994; Ding et al., 2005; Hu et al., 2004; Xue et al., 2005; Yang et al., 2005) are subsiding owing to rapid development and urbanization. The cause identified with these subsidence phenomena has been mainly due to excessive withdrawal of groundwater (Chai et al., 2004; Shi and Bao, 1984; Xue et al., 2005; Yang et al., 2005). The subsidence phenomenon in China occurs mostly in medium-seized and large cities with common characteristics of being slow, accumulative and irreversible (Xue et al., 2005). At least, 45 cities and municipalities (China Daily, 2003b; Deng and Ju, 1994; Hu et al., 2004) are experiencing this geological problem owing to massive exploitation of underground water to meet the needs of the rapid evolving industries and urbanisation. In the case of Shanghai, the massive construction of high-rise buildings in the downtown area has also contributed to the subsidence problem (China Daily, 2003b; Zhang et al., 2002). About 11 cities, including Shanghai, have recorded accumulative subsidence values exceeding 1 m at their subsidence centres (Hu et al., 2004; Yang et al., 2005; Zhang et al., 2002). The most affected cities, since the early 1900s, are Shanghai, Tianjin and Taiyuan, with each of them having accumulative subsidence value greater than 2m (China Daily, 2003b). The Chinese government has spent billions of Yuan (RMB$), as direct and indirect economic costs, in addressing land subsidence problems in China. Hu et al. (2004) mentioned that the estimated total economic loss due to land subsidence in Tianjin is RMB$189.6 billion. In the three areas of Suzhou, Wuxi and Changzhou, the total economic loss is estimated at RMB$46.87 billion (Yang et al., 2005). For Shanghai, the China Daily (2003b) and Yang et al. (2005) quoted the total economic losses of about RMB$290 billion, in the last 40 years, due to land subsidence and its related problems.

The city of Shanghai has been experiencing land subsidence mainly due to anthropogenic induced factors: (1) intensive exploitation of ground water (Chai et al., 2004; China Daily, 2003a; 2003b; Hu et al., 2004; Shi and Bao, 1984; Xue et al. 2005; Zhang et al., 2002), and (2) massive construction of high-rise buildings in the downtown area (China Daily, 2003a; 2003b; Gong et al., 2005; Xue et al. 2005; Zhang and Wei, 2005; Zhang et al., 2002). It is estimated that the high-rise buildings contribute about 30 to 40 per cent of the subsidence problem. According to Gong et al. (2005), there exist about 7000 high-rise buildings with different altitudes (storeys) in Shanghai; additional 2000 high-rise buildings are yet to be built. They provided the height distribution of high-rise buildings in Shanghai as follows: 53% have 9-17 storeys, 31% have 18-24 storeys, 11% have 25-30 storeys, and 5% have over 30 storeys. Results of dynamic monitoring indicates that, the city centre subsided by 36 mm between 1986 and 1990, 53.8 mm between 1991 and 1995, and 98.8 mm between 1996 and 2000 (Zhang and Wei, 2005). To support the assertion that the concentration of high-rise buildings have contributed to the subsidence problem in Shanghai, the China Daily, 11 December 2003 Edition, added that the Pudong New Area, which houses 1.4 million people in a 520 square-kilometre area, suffers most from severe subsidence.

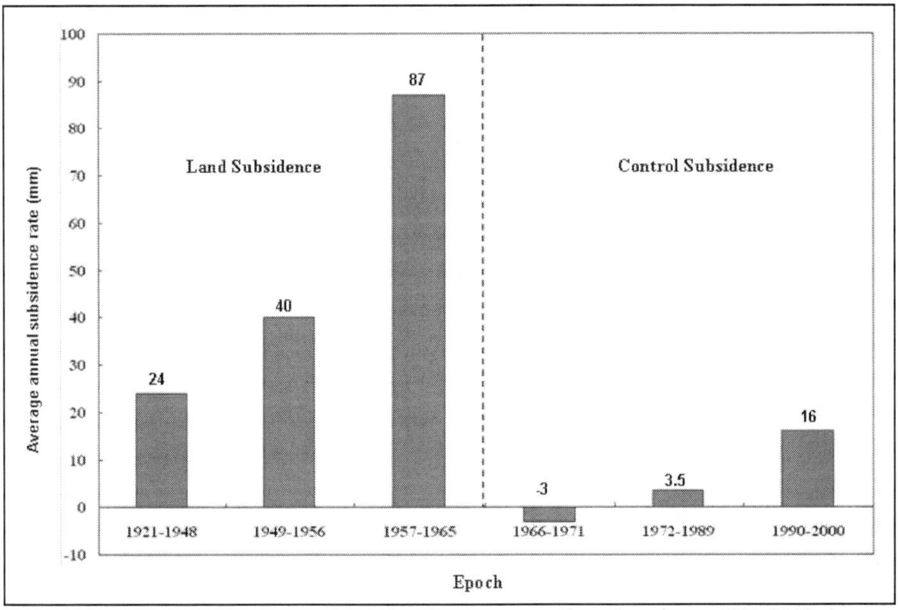

Fig. 1. Historical development of land subsidence in Shanghai from 1921-2000 (source: Zhang et al., 2002)

The problem of land subsidence in Shanghai was first reported in 1921 (Shi and Bao, 1984), and the problem continued till 1963 when its impact became more severe. A cumulative subsidence value of 2.63 m was observed with greatest subsidence occurring between 1956 and 1959, at an annual rate of 98 mm. The historical development of land subsidence in Shanghai at 6 epochs during the period 1921-2000 is provided in Figure 1. In 1963, some measures such as restriction and rational usage of ground water, artificial recharge of ground water, and adjustment of exploited aquifers (Deng and Ju, 1994; Shi and Bao, 1984), were put in place to check the subsidence problem. The enforcement of those measures resulted in considerable rebound of water level between 1963 and 1965. The Shanghai Municipal Government in 1995 came out with another policy that limited the usage of underground water of the whole city to less than 10 million cubic meters per year, and also demanded that all deep wells in the city are to operate with official permits (China Daily, 2003a; Xue et al., 2005). In spite of these measures, Shanghai is still experiencing some degree of subsidence (China Daily, 2003a; Xue et al., 2005). The average land subsidence rates, gathered from Shanghai Geological Survey Institute are 12.12 mm in 2000, 10.94 mm in 2001, and 10.22 mm in 2002 (China Daily, 2003a), and the targeted land subsidence rates as perceived by the Shanghai Municipal Land and Resources Administration Bureau are at most 10 mm by 2005 and 5 mm by 2010 (Zhang and Wei, 2005).

Subsidence phenomenon in Shanghai has been previously measured or monitored using the surveyor's method of precise levelling, and more recently the global positioning system (GPS). Although these methods provide precise measurements they are labour and time intensive, costly and are restricted to specific points in the terrain. They cannot therefore be used to provide information on detailed ground motions if the area of ground subsidence is large. Interferometric synthetic aperture radar (InSAR), a space geodetic method, has demonstrated the capability of mapping extensive areas, on pixel-by-pixel basis, and in a more convenient manner than the aforementioned geodetic methods. InSAR is capable of detecting ground surface elevation changes with very high precisions (Bürgmann et al., 2000; Gabriel et al., 1989; Zebker et al., 1997). Chen et al. (2000) argued that InSAR is presently the only technology capable of monitoring the deformation of the Earth's surface in a large area with dense points, quickly and cost-effective, in day and night under all weather conditions.

In this paper, we present results of using the L-Band Japanese Earth Resource Satellite 1 (JERS-1) SAR data to map the extent and magnitue of land subsidence in Shanghai with the differential interferometric synthetic aperture radar (DInSAR) technique. The choice of L-Band JERS-1 SAR data over the C-Band Earth Resources Satellite 1 and 2 (ERS-1/2) SAR data was based on analysis made using appropriate SAR parameters, and results obtained from coherence analysis performed using both the C-band and L-band SAR data over a section of Shanghai. The JERS-1 SAR has a longer wavelength, and hence a larger critical baseline, than the ERS-1/2 SAR. In addition, L-band SAR can penetrate vegetation to retrieve ground-surface, making it superior to C-band ERS SAR for mapping vegetated areas or low coherence areas such as Shanghai.

2. Coherence Analysis

Coherence which is a measure to check the quality of an interferogram is degraded from three different sources, namely spatial, thermal, and temporal decorrelations (Zebker and Villasenor, 1992; Zebker et al., 1992). Decorrelation significantly affects the accuracy of InSAR measurements (Zebker and Villasenor, 1992; Zebker et al., 1992), and it is a common problem associated with the repeat-pass InSAR technique as a result of discrepancies that arise between two related SAR signals over same location at different epochs. It is therefore necessary to assess the temporal coherence situation of scatterers prior to any serious studies involving the use of InSAR techniques. Temporal decorrelation occurs when scatterers change their behaviors over time. It is one of the major problems that affect interferograms covering Shanghai metropolis. Studies conducted by Ding et al. (2005), Damoah-Afari et al. (2007) and Damoah-Afari et al. (2008) show that coherence is not maintained in Shanghai metropolis after a period of about 10 months with the C-band ERS-1/2 SAR data. Therefore, it is problematic to use ERS-1/2 SAR data with long temporal baselines to monitor land subsidence characterized by very slow movement, as in the case of Shanghai.

Coherence analysis was performed in a section of Shanghai covering the downtown area where the problem of land subsidence is severe in order to assess the behaviour of scatterers in Shanghai over time. The region of interest selected for coherence analysis spans approximately 9 km in azimuth direction and 8 km in range direction. The coherence analysis was also performed to assess which of the SAR data, either the C-band or the L-band, would be suitable for mapping the subsidence phenomena in Shanghai. Figure 2 shows both the region of interest selected for the entire study and the coherence analysis using ERS-2 SAR amplitude image of Shanghai acquired on 31 March 1998 as a base.

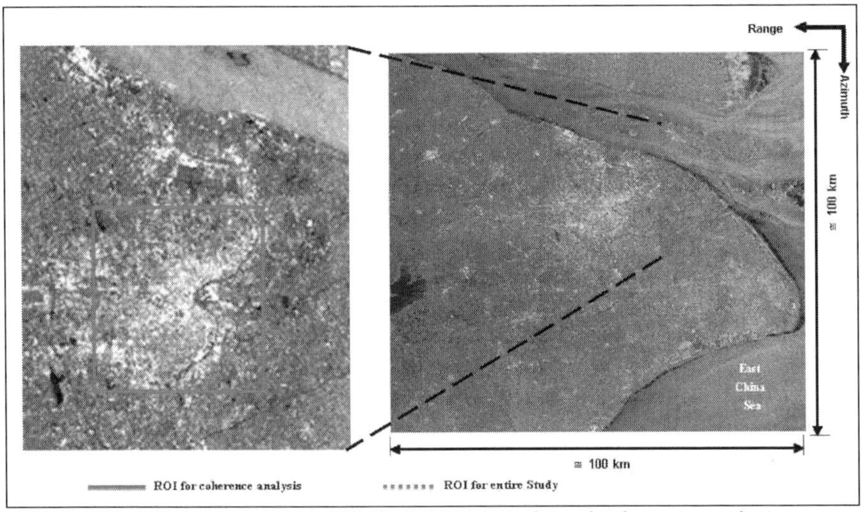

Fig. 2. Regions of interest selected for both the entire study and coherence analysis

2.1 Data selection, processing and results

Data from both the C-band ERS-1/2 and L-band JERS-1 SAR systems were used for the analysis. In order to minimize the effect of baseline decorrelation on the final results, only InSAR pairs having smaller normal baselines were selected. In the case of ERS-1/2 SAR data, interferograms having a normal baseline of at most 200 m were selected, whiles a cut-off baseline of 1350 m was used for the JERS-1 SAR data. Twenty-seven JERS-1 and 33 ERS-1/2 interferograms were generated for the coherence analysis. The distribution of normal baselines ($B_\perp$) and temporal baselines (B_t) from the two datasets used for the coherence analysis is presented in Table 1. The total coherence for each InSAR pair was estimated from InSAR processing using the EarthView InSAR v3.1 software. Temporal coherence for each InSAR pair was then estimated by calculating and removing the influence of both the thermal and spatial decorrelations from the total coherence using equation (1) (Zebker and Villasenor, 1992) and individual sensor parameters.

$$\gamma_{temporal} = \gamma_{total} \Big/ \left(1 - \frac{|B_\perp|}{B_{\perp c}}\right)\left(\frac{1}{1 + SNR^{-1}}\right); \quad B_\perp < B_{\perp c} \tag{1}$$

where $\gamma_{temporal}$ is temporal decorrelation; γ_{total} is overall coherence; $B_\perp$ is the normal baseline; $B_{\perp c}$ is the critical baseline, and SNR is the signal-to-noise ratio of the SAR sensor (SNR for ERS-1/2 SAR data = 12 dB and SNR for JERS-1 SAR data = 5 dB (Cuddy et al., 1993)).

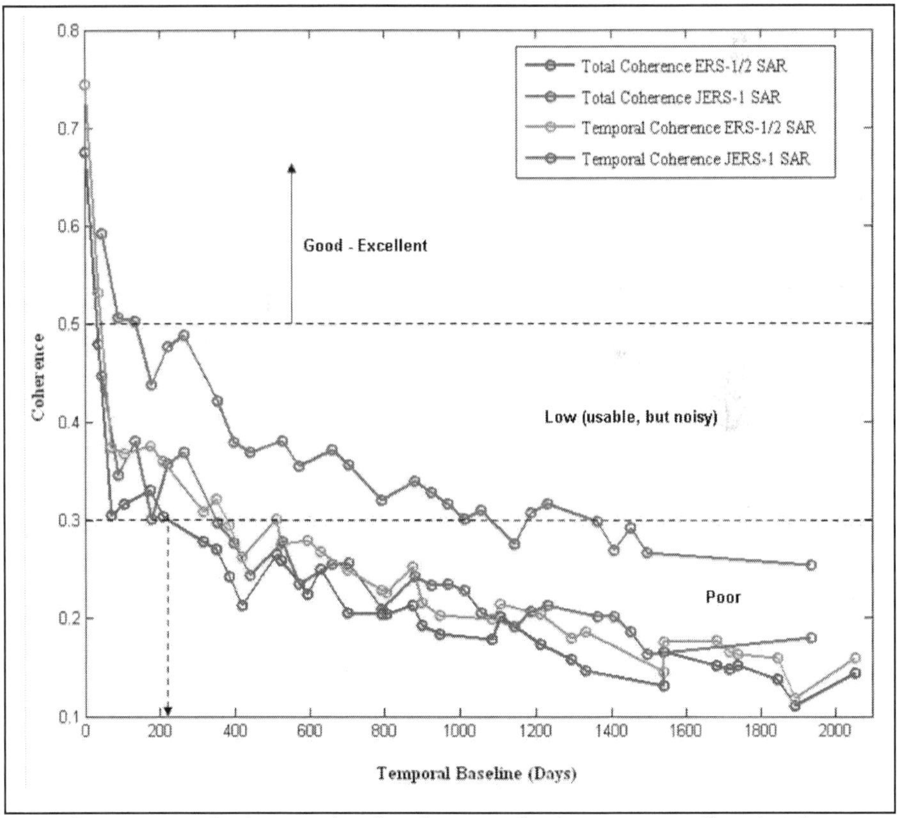

Fig. 3. Comparison of coherence (total and temporal) of scatterers in Shanghai downtown area between the C-Band ERS-1/2 and the L-Band JERS-1 SAR data

SAR Sensor	Normal Baselines (m)			
	Min.	Max.	Mean	Std Dev.
ERS-1/2	1	189	76	53
JERS-1	1	1313	517	363
	Temporal Baselines (days)			
ERS-1/2	1	2054	891	614
JERS-1	44	1936	794	507

Table 1. Distribution of normal and temporal baselines of SAR datasets used for the coherence analysis

The result of coherence analysis is presented in Figure 3. It can be seen from the figure that temporal coherence for both sensors are higher than their respective total coherences. It is also evident from the figure that JERS-1 SAR interferograms have higher coherence (both total and temporal) than the ERS-1/2 SAR over the study area, despite the larger normal and longer temporal baselines used in the case of JERS-1 SAR data. It can be seen that ERS-1/2 SAR data do not maintain good coherence levels after a temporal baseline of 7 months. It is evident from Figure 3 that majority of the interferometric pairs used in this study have coherence levels that fall below the optimum, 0.3.

3. InSAR qualitative mapping of land subsidence in Shanghai

The power of DInSAR is not only limited to its ability to do quantitative measurements, but can also consistently and accurately map out the extent or limits of ground surface deformation phenomena. This advantage makes DInSAR a powerful cartographic tool for qualitative mapping. DInSAR provides very dense and accurate observations with a very high level of precision (Gabriel et al., 1989). The results of DInSAR mapping can serve as reconnaissance or a base map for engineers in making decisions regarding the location and placement of benchmarks for GPS surveys or spirit leveling, and extensometers (Bawden et al., 2003). In this section, we present the results of using L-band JERS-1 SAR differential interferograms acquired during the period October 1992 to August 1998 to delimit and to track the stages of the land subsidence phenomenon of Shanghai. From the previous section, it was shown that most of the interferograms produced for coherence analysis had their coherence levels falling below the optimum, 0.3. However, they were very useful for qualitative analysis of the land subsidence phenomenon in Shanghai.

3.1 Data selection, processing and results

When using the DInSAR technique to monitor land subsidence that is characterized by a very slow velocity, such as the case of land subsidence phenomena in Shanghai, it is important to use InSAR pairs with longer temporal baselines. InSAR pairs with longer temporal baselines, however, have problem due to temporal decorrelation. It is therefore important to set criteria for selecting the kind of SAR data and SAR image pairs that can provide satisfactory interferograms for such a study. We used the critical baseline together with the result of coherence analysis (Figure 3) to support our choice of using the L-band

JERS-1 SAR data other than the C-band ERS-1/2 SAR data for this study. The critical baseline, B_c is defined as (Zebker and Villasenor, 1992)

$$B_c = \frac{\lambda r}{2R_y \cos^2 \theta} \tag{2}$$

where λ is radar wavelength, R_y is range resolution, r is the distance between the sensor and the center of the resolution pixel, and θ is the incidence angle.

Substituting individual sensor parameters (i.e., For ERS-1/2: θ = 23°, R_y = 25 m and r = 853 km; for JERS-1: θ = 39°, R_y = 18 m and r = 693 km) into equation (2) gives B_c of approximately 1150 m for ERS-1/2 SAR data and B_c of approximately 6750 m for the JERS-1 SAR data. Based on the above analyses, we interpret that JERS-1 SAR enables us to extend the normal baseline and the temporal baseline further than the ERS-1/2 SAR to achieve satisfactory interferograms. This in fact, is the advantage of longer wavelength SAR data over shorter ones for interferometric applications, especially in areas where there are known problems of loss of coherence.

A total of 19 L-band differential interferograms and 2 DEMs were generated to track the developmental stages of land subsidence in Shanghai. The data has minimum relative temporal baseline of 44 days, maximum relative temporal baseline of 1056 days, minimum relative normal baseline of 2 m, and maximum relative normal baseline of 971 m. One of the DEMs was used to check the effectiveness of using the other to remove topographic phase from raw interferograms generated. Table 2 presents JERS-1 DInSAR pairs used for the qualitative mapping of land subsidence phenomenon in Shanghai. A major problem associated with using JERS-1 SAR data for interferometric applications is with precise orbit data to facilitate the removal of flat-earth phase from raw interferograms. Unlike the ERS-1/2 satellites which have very good orbit information to aid in interferogram flattening process, the JERS-1 SAR system had inaccurate orbit data. It was therefore necessary to refine the baseline during data processing. It is worthy to mention here that ground control points could have been used to aid in the baseline refinement. Unfortunately there were no available ground control points at the time of undertaking this study. The interferogram flattening was therefore achieved by using a baseline refinement method embedded in the EV-InSAR software, where both the normal baseline and the slave yaw angle were adjusted intuitively and interactively. Figure 4 shows the results of JERS-1 interferogram flattening process using the baseline refinement tool. The figure presents JERS-1 differential interferogram pair of 19971024-19980715, with normal baseline of 207 m and temporal baseline of 264 days. Experiment indicated that a change in normal baseline of -48.5 m and a change in slave yaw angle of 0.0004 rad were good to provide suitable flattened interferogram.

Results of the JERS-1 differential interferograms are shown in Figure 5 and 6. A closed examination of the JERS-1 differential interferograms in Figure 5 and 6 revealed three stages of land subsidence in Shanghai during the period 1992 to 1998. The first stage stretches from October 1992 up to April 1995. The second stage emerged by the end of 1995 and continued till the ending of 1997, where the third stage began to progress. Figure 7 presents the developmental stages of land subsidence phenomenon in Shanghai for the period October 1992 to August 1998 revealed by L-band JERS-1 SAR data. The accumulative land subsidence map of Shanghai produced from land survey methods is shown in Figure 7 (**Bottom right**). It is evident from Figure 7 that the accumulative subsidence map of Stage 3, obtained from InSAR, has close similarities to that produced by other survey methods. However, the extents of the subsidence phenomena have been accurately mapped by InSAR technique. Some areas marked as stable with land survey methods have been mapped otherwise as deforming areas, and vice versa, by InSAR technique.

No.	Master Image	Slave Image	$B_\perp$ (m)	B_t (days)
DEM				
1	19960810	19960923	-893.835	44
2	19960923	19961106	-546.643	44
InISAR				
1	19921002	19940906	344.525	704
2	19921002	19950301	-418.261	880
3	19921002	19950414	442.313	924
4	19921115	19930327	814.734	132
5	19940906	19950414	97.118	220
6	19940906	19961106	971.135	792
7	19950301	19971024	167.897	968
8	19950301	19971207	-91.867	1012
9	19950301	19980120	861.803	1056
10	19950414	19961106	871.718	572
11	19950414	19971024	-691.515	924
12	19950414	19980120	1.882	1012
13	19961106	19980120	-870.420	440
14	19961220	19980601	248.301	528
15	19970501	19980601	-255.113	396
16	19970910	19980828	481.129	352
17	19971024	19980715	39.576	264
18	19971207	19980715	298.076	220
19	19980120	19980715	-655.105	176

Table 2. JERS-1 SAR DInSAR pairs used for mapping the extent and stages of land subsidence in Shanghai

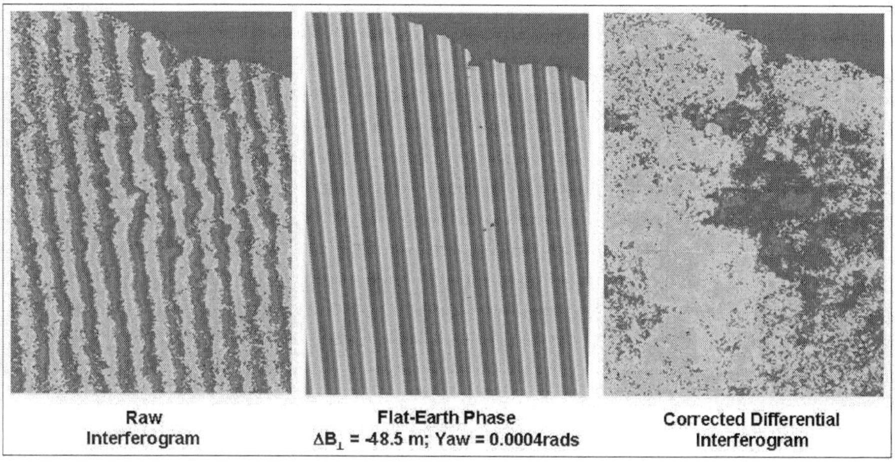

Fig. 4. Interferogram flattening process using JERS-1 differential interferogram pair 19971024-19980715; $B_\perp$ = 207 m, B_t = 264 days, and a DEM pair 19960810-19960923, $B_\perp$ = -894 m, B_t = 44 days

4. InSAR quantitative mapping of land subsidence in Shanghai

This section presents quantitative mapping of the land subsidence phenomenon in Shanghai using the DInSAR technique and datasets acquired by the L-band JERS-1 SAR interferometer. Our choice for using JERS-1 SAR data over ERS-1/2 SAR data is based on analyses and discussions presented in the previous sections.

4.1 Data selection, processing and results

JERS-1 SAR data acquired over Shanghai in descending orbit, for the period October 1992 to July 1998, were employed in the DInSAR quantitative mapping. Eight JERS-1 SAR images were selected based on the temporal and the normal baseline information obtained from Japanese Aerospace Exploration Agency (JAXA) website. The data has a minimum relative temporal baseline of 44 days and a maximum of 924 days, and a minimum relative normal baseline of 96 m and a maximum relative normal baseline of 893 m. They were selected such that differential interferograms generated could form a time series right from the first acquisition to the last.

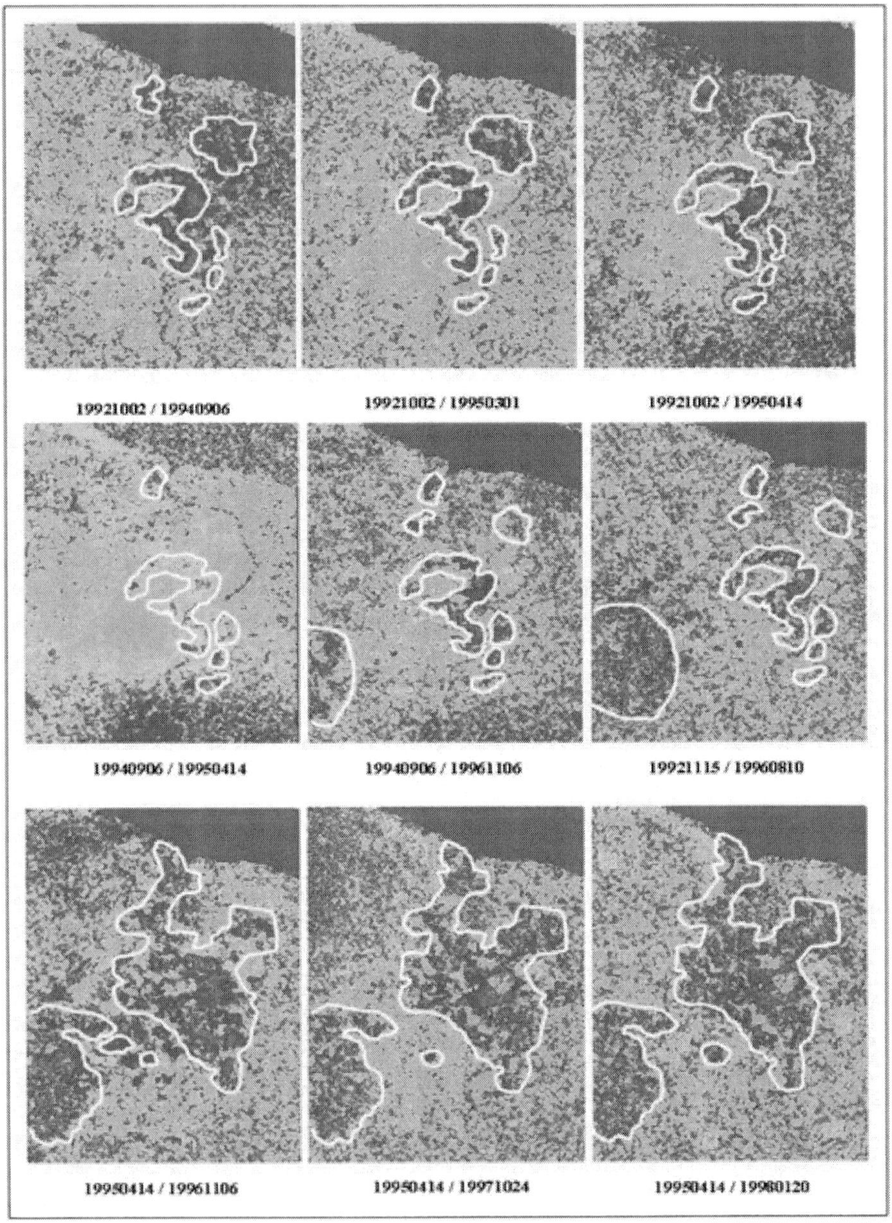

Fig. 5. Developmental stages of land subsidence in Shanghai mapped by L-band JERS-1 SAR data from October 1992 to January 1998

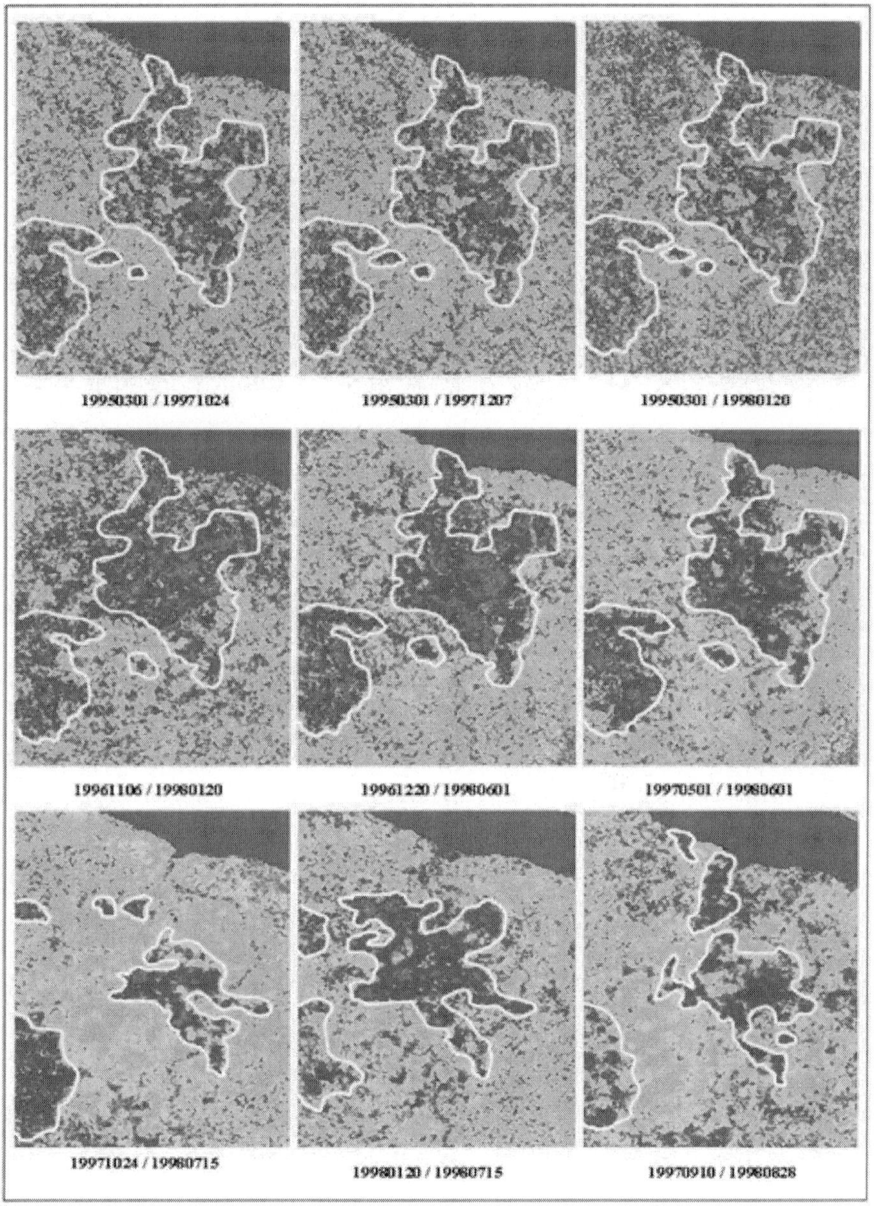

Fig. 6. Developmental stages of land subsidence in Shanghai mapped by L-band JERS-1 SAR data from March 1995 to August 1998

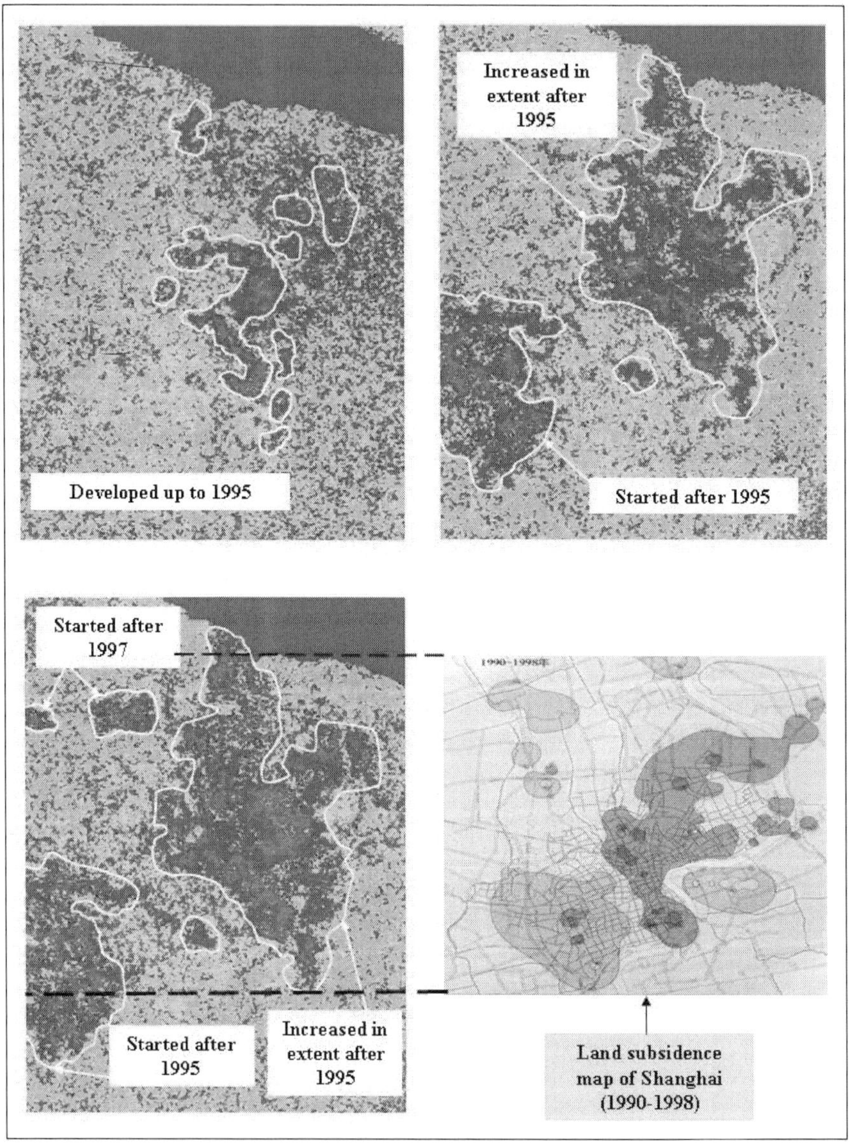

Fig. 7. Developmental stages of land subsidence in Shanghai from 1992-1998 mapped by JERS-1 SAR data. **(Top left)** Stage 1: accumulative subsidence up to April 1995; **(Top right)** Stage 2: accumulative subsidence up to October 1997; **(Bottom left)** Stage 3: accumulative subsidence map up to August 1998; and **(Bottom right)** accumulative subsidence map from 1990-1998 obtained from other survey methods (Source: Zhang et al., 2002)

Three SAR scenes (acquired on the following dates: 10 August 1996, 23 September 1996, and 6 November 1996), forming 2 possible interferometric pairs with temporal baselines of 44 days were selected to generate digital elevation models (DEMs) for the removal of topographic fringes from all the interferograms. One of the DEMs was used to check the effectiveness of the other to remove topographic fringes from the raw interferograms. Six out of the 8 images were selected to generate 6 possible differential interferograms using the two-pass DInSAR method with the DEMs generated above. Table 3 is a summary of the JERS-1 SAR interferometric pairs produced for the quantitative study. The EarthView InSAR (EV-InSAR) v.3.1 Software, a product of VEXCEL Corporation, Canada, was used for all interferometric processing. Unfortunately there were no available ground control points for baseline refinement process at the time of undertaking this study. We therefore removed flat-earth fringes from raw interferograms using the approach discussed in Section 3.1.

No.	Master Image	Slave Image	$B_{\perp}$ (m)	B_t (days)
DEM				
1	19960810	19960923	-893.835	44
2	19960923	19961106	-546.643	44
InISAR				
1	19921002	19940906	346.613	704
2	19921002	19950414	442.934	924
3	19940906	19950414	95.672	220
4	19950414	19961106	870.043	572
5	19961106	19980120	-869.364	440
6	19980120	19980715	-653.750	176

Table 3. JERS-1 interferometric pairs used for quantitative mapping of land subsidence in Shanghai

Subsidence maps from the 6 DInSAR pairs were further processed with MATLAB to produce the final individual subsidence maps. A region of 1200 pixels in azimuth direction by 1300 pixels in range direction was selected for final presentation. Two separate accumulative subsidence maps of land subsidence covering the downtown area of Shanghai, over the period October 1992 to July 1998, were generated from two different time series obtained from the stack of DInSAR pairs. Table 4 shows the two time series and the DInSAR pairs forming them.

Time Series	DInSAR Pairs or Subsidence maps
I	19921002-19940906; 19940906-19950414; 19950414-19961106; 19961106-19980120 and 19980120-19980715.
II	19921002-19950414;19950414-19961106; 19961106-19980120 and 19980120-19980715.

Table 4. Two time series produced from the stack of JERS-1 DInSAR pairs.

Figure 7 presents the subsidence maps obtained for the 1st, 4th and the 6th DInSAR pairs of data in Table 3. The result of the two accumulative subsidence maps of Shanghai for the period 1992-1998, obtained from the stack of above time series are presented in Figure 8.

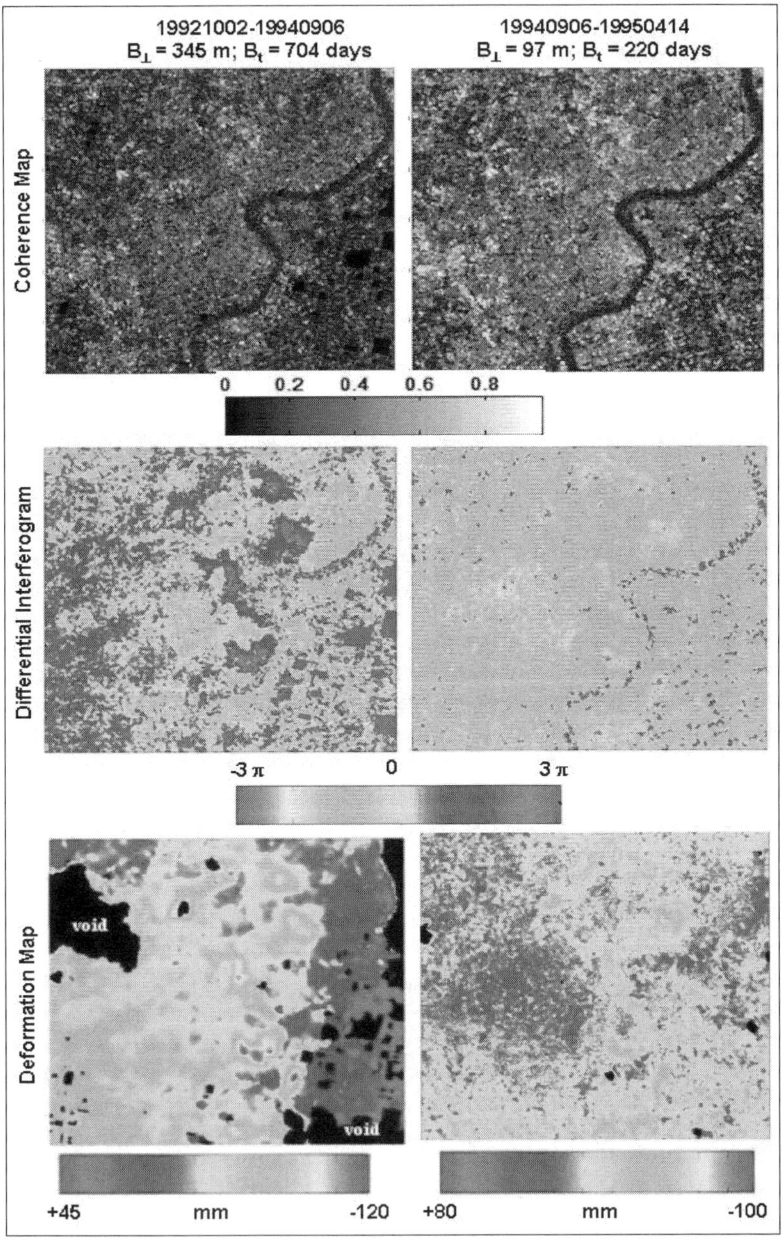

Fig. 8. JERS-1 SAR coherence maps, differential interferograms and deformation maps of DInSAR pairs 19921002-19940906 and 19940906-19950414

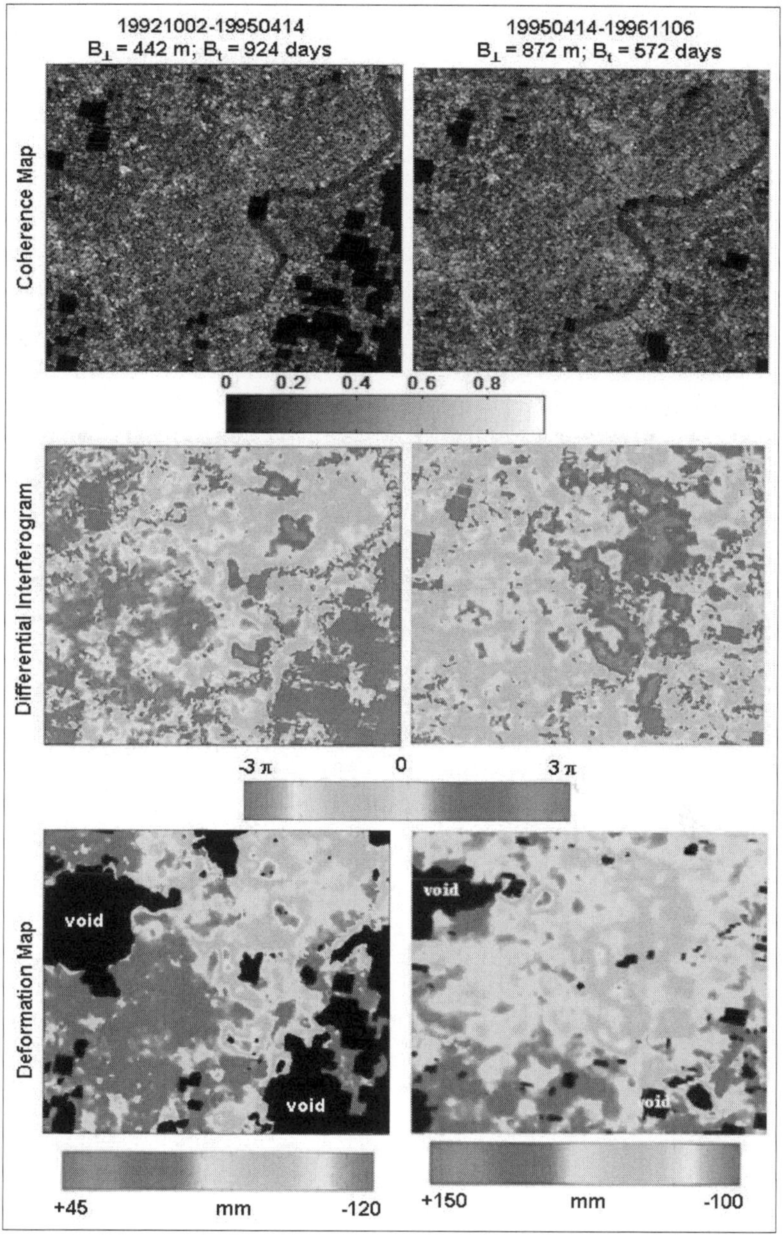

Fig. 9. JERS-1 SAR coherence maps, differential interferograms and deformation maps of DInSAR pairs 19921002-19940414 and 19940414-19961106

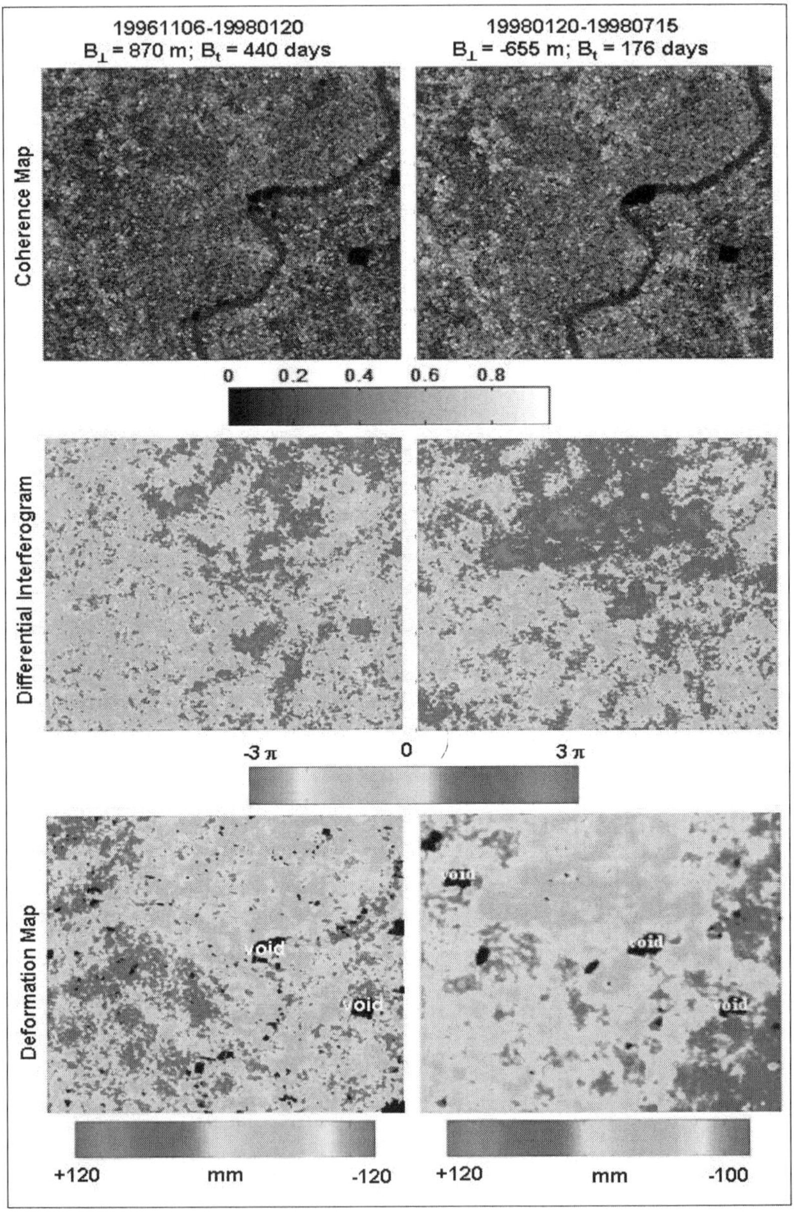

Fig. 10. JERS-1 SAR coherence maps, differential interferograms and deformation maps of DInSAR pairs 19961106-19980120 and 19980120-19980715

Figure 8, 9 and 10 present the coherence maps, differential interferograms and deformation or subsidence maps obtained for DInSAR pairs presented in Table 3. The result of the two accumulative subsidence maps of Shanghai for the period 1992-1998, obtained from the stack of above time series are presented in Figure 11. Figure 12 compares pictorially the accumulative subsidence map obtained from Time Series I and the land subsidence map of Shanghai for the period 1990-1998 produced from other survey methods. It is evident that, the three maps show common characteristics, and the subsidence values in general are in agreement. It is also evident that some places presented in the map obtained by conventional survey methods and described as stable are actually subsiding. The map of deformation rates obtained for the period 1992-1998 is presented in Figure 13.

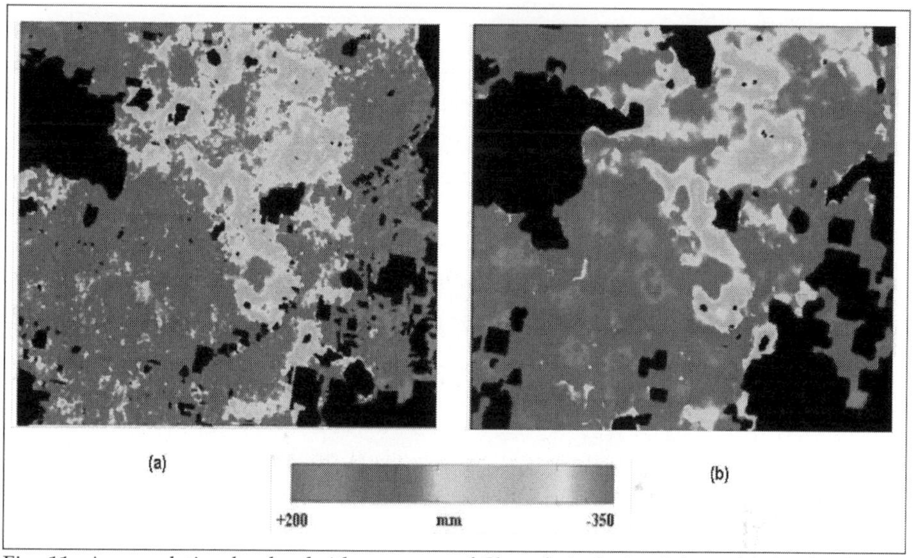

Fig. 11. Accumulative land subsidence map of Shanghai obtained from **(a)** Stack of Time Series I for the period 1992-1998; **(b)** Stack of Time Series II for the period 1992-1998

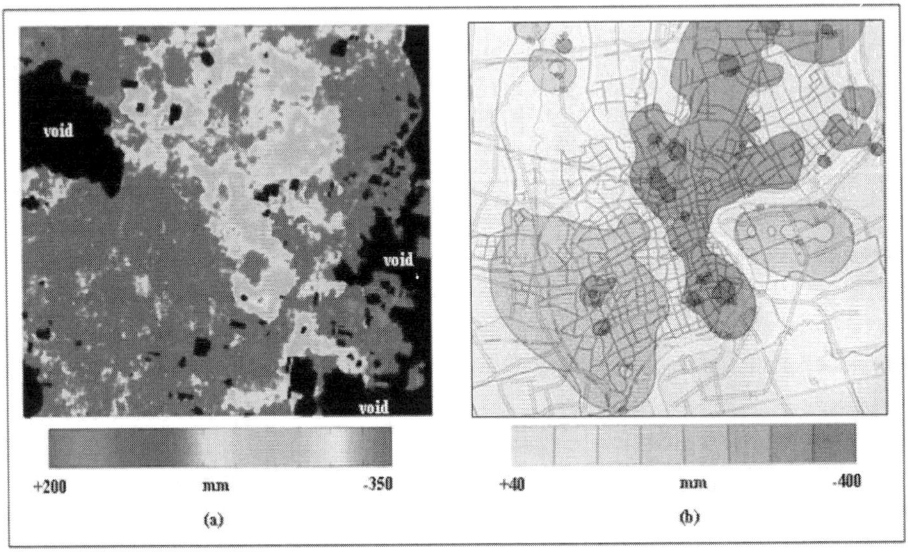

Fig. 12. Accumulative land subsidence map of Shanghai obtained from **(a)** Stacking Time Series I for the period 1992-1998; **(b)** Land surveys for the period 1990-1998 (Source: Zhang et al., 2002).

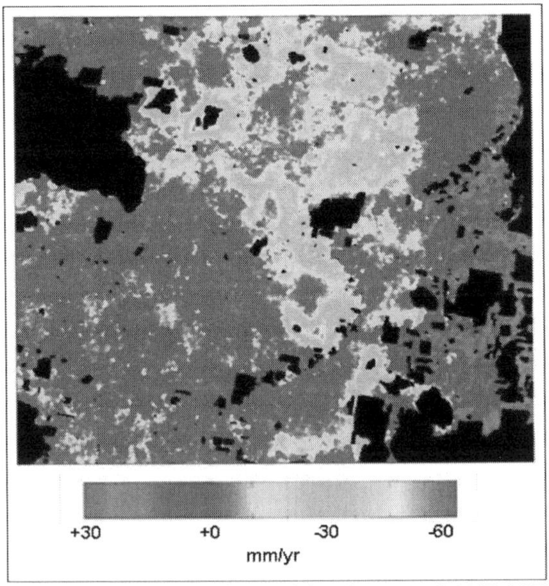

Fig. 13. The map of deformation rates obtained for the period 1992-1998 by Stacking Time Series I (see Table 3 and Table 4)

5. Conclusion

In this paper, we have shown the capability of using InSAR techniques to map land subsidence phenomenon in both qualitative and quantitative terms. DInSAR techniques make it possible to study the magnitue, extent and pattern of land subsidence phenomenon more efficiently than any other method available today. The problem of loss of coherence in Shanghai has also been assessed in this study with both the C-band ERS-1/2 and L-band JERS-1 SAR data. The strength of using JERS-1 SAR data for DInSAR application in places where there are known problems of temporal decorrelation has been demonstrated. Results obtained from DInSAR quantitative measurements using the L-band JERS-1 SAR data were in agreement with the land subsidence map of Shanghai obtained from other survey methods. DInSAR qualitative analysis using JERS-1 differential interferograms revealed three stages of land subsidence in Shanghai from 1992 to 1998. The results indicated that the extent of land subsidence increased from 1992 through to 1998. It follows therefore that, activities taking place in the city might, in one way or the other, influence the progress of land subsidence in Shanghai as asserted by some researchers.

Acknowledgement

ERS-1 and ERS-2, and JERS-1 SAR images are copyright © 1992-1998 of European Space Agency (ESA), and Japan Aerospace Exploration Agency (JAXA), respectively, and were provided by the ESA and JAXA. The work was supported by the Research Grants Council of the Hong Kong Special Administrative Region Government (Project No.: PolyU 5157/05E). The first author would like to express his sincere thanks to the Hong Kong Polytechnic University for providing him a studentship to enable him to pursue PhD studies.

6. References

Badwen, G.W., Sneed, M., Stork, S.V. and Galloway, D.L. (2003). Measuring human-induced land subsidence from space. *U.S. Geological Survey Fact Sheet 069-03*, December 2003.

Bürgmann, R., Rosen, P.A. and Fielding, E.J. (2000). Synthetic aperture radar interferometry to measure Earth's surface topography and its deformation. *Annual Reviews of Earth and Planetary Sciences*, 28, pp. 169-209.

Chai, J.C., Shen, S.L., Zhu, H.H. and Zhang, X.L. (2004). Land subsidence due to groundwater drawdown in Shanghai. *Géotechnique*, 54(2), pp. 143-147.

Chen, Y.Q., Zhang, G.B., Ding, X.L. and Li, Z.L. (2000). Monitoring Earth surface deformations with InSAR technology: principle and some critical issues. *Journal of Geospatial Engineering*, 2(1), pp. 3-21.

China Dialy (2003a). Shanghai puts up a fight to stop sinking. *China Daily*, 16 July 2003 Edition.
http://www2.chinadaily.com.cn/en/doc/2003-07/16/content_245660.htm.

China Daily (2003b). Cites Sinking due to Excessive Pumping of Groundwater. *China Daily, Hong Kong Edition*, 11 December 2003.
http://www.chinadaily.com.cn/en/doc/2003-12/11/content_289290.htm

Cuddy, D., Chen, M.J. and Bicknell, T. (1993). Alaska SAR processor implementation for JERS-1. *IEEE Proceedings of the International Geoscience and Remote Sensing Symposium*, 1993, 4, 2130-2132.

Damoah-Afari, P., Ding, X.L., Lu, Z. and Li, Z.W. (2008). Detecting ground settlement of Shanghai using interferometric synthetic aperture radar (InSAR) techniques. *Proceedings of Commission VII, International Society of Photogrammetry and Remote Sensing (ISPRS2008) Congres*, Beijing, 2008, vol. XXXVII, part B7, pp. 117-124.

Damoah-Afari, P., Ding, X.L., Li, Z.W., Lu, Z. and Omura, M. (2007). Six years of land subsidence in Shanghai revealed by JERS-1 SAR data. *IEEE Proceedings on International Geoscience and Remote Sensing Symposium*, 2007, pp. 2093-2097.

Deng, A.S. and Ju, J.H. (1994). Land subsidence, sinkhole collapse and Earth fissure occurrence and control in China. *Journal of Hydrological Sciences*, 39(3), pp. 245-256.

Ding, X.L., Li, Z.W., Damoah-Afari, P., Liu, G.X. and Huang, C. (2005). Study of ground subsidence in Shanghai with satellite radar remote sensing. *Proceedings of the First International Symposim of Cloud-prone and Rainy Areas Remote Sensing (1st CARRS)* Hong Kong, 2005, pp. 298-316.

Gabriel, A.K., Goldstein, R.M. and Zebker, H.A. (1989). Mapping small elevation changes over large areas: differential radar interferometry. *Journal of Geophysical Research*, 94(B7), pp. 9183-9191.

Gong, S.L., Wu, J.Z. and Yan, X.X. (2005). Analysis of land subsidence due to construction engineering in soft soil region of Shanghai. In: *Proceedings of the 7th International Symposium on Land Subsidence*, Shanghai, China, Vol. 1, pp. 82-87.

Hu, R.L., Yue. Z.Q., Wang, L.C. and Wang, S.J. (2004). Review on current status and challenging issues of land subsidence in China. *Engineering Geology*, 76, pp. 65-77.

Shi, L.X. and Bao, M.F. (1994). Case History No. 9.2 - Shanghai, China. In Poland, J.F. ed. *Guidebook to studies of land subsidence due to groundwater withdrawal*, UNESCO, Paris. http://wwwrcamn.wr.usgs.gov/rgws/Unesco/PDF-Chapters/Chapeter9-2.pdf.

Xue, Y.Q., Zhang, Y., Ye, S.J., Wu, J.C. and Li Q.F. (2005). Land subsidence in China. *Environmental Geology*, 2005, 48, 713-720.

Yang, G.F., Wei, Z.X. and Huang, C.S. (2005). Advances and challenge in research on land subsidence in China. In: *Proceedings of the 7th International Symposium on Land Subsidence*, Shanghai, China, Vol. 1, pp. 3-9.

Zebker, H.A., Rosen, P.A. and Hensley, S. (1997). Atmospheric effects in inteferformetric synthetic aperture radar surface deformation and topographic maps. *Journal of Geophysical Research*, 1997, 102(B4), 7547-7563.

Zebker, H.A. and Villasenor, J. (1992). Decorrelation in interferometric radar echoes. *IEEE Transactions on Geoscience and Remote Sensing*, 30(5), pp. 950-959.

Zebker, H.A., Villasenor, J. and Madsen, S.N. (1992). Topographic mapping from ERS-1 and SEASAT radar interferometry. *Proceedings of the IEEE International Geoscience and Remote Sensing Symposium*, 1992, 387-388.

Zhang, A.G., Luo, D.Y., Shen X.G., Lü, L.A., Chen, H.W., Wei, Z.X., Yan, X.X., Fang, Z., and Shi, Y.Q. (2002). *Shanghai Geological Environmental Atlas: Editorial Board* (SGEAEB). Geological Publishing House, Beijing, pp. 130-133.

Zhang, A.G. and Wei, Z.X. (2005). Prevention and cure with Shanghai land subsidence and city sustaining development. In: *Proceedings of the 7th International Symposium on Land Subsidence*, Shanghai, China, Vol. 1, pp. 10-17.

Impact of Daily Melt and Freeze Patterns on Sea Ice Large Scale Roughness Features Extraction

Eric Hudier and Jean-Sébastien Gosselin
University of Quebec
Canada

1. Introduction

Improvement of large scale models developed to study the impacts of global changes in polar regions requires a better knowledge of the variability of the different parameters that control fluid dynamics and thermodynamics budgets at the interfaces. In the part of a fluid influenced by the presence of a solid interface, surface roughness is one of the key parameters that control momentum as much as heat (or salts at the ice-water interface) vertical exchanges through the boundary layer. This, in turn, makes roughness a data of primary importance when coupling ocean and atmosphere.

First year sea ice is characterized by large scale roughness features resulting from the accumulation of ice blocks that are piled above and below the surface under the compression and shear stresses induced by winds and currents. They develop linear sinuous features several hundred meters in length which on satellite images draw a network of bright structures. When illuminated by a SAR beam, the orientation variability of ice block faces in ridges increases the probability to receive an enhanced signal resulting from specular reflection on point sources. It follows that conditions that enhance specular reflection over diffuse scattering translate into a better resolution of the ridge network. This is mostly achieved at spring, when, with the snow wetness content increasing, the penetration depth into snow and ice becomes negligible. It has been observed that overnight wetness changes may cause strong resolution contrast on the ridge network. Late in the afternoon, increased wetness and the presence of a liquid film on ice and snow surfaces results in an increased forward scattering on most surfaces and an improved resolution of the pressure ridge network. On the contrary, on morning the reduced wetness allows some volume scattering to occur in the snow surface layer causing an increased diffused scattering that tends to reduce the backscattered signal from ridges and increase the signal from flat ice surfaces away from ridges.

In addition, with coherent imaging systems such as SAR, multiplicative random noise can combine to generate a signal strong enough to partly mask large roughness features signature. The speckle issue is for ridges of utmost importance. Extending only over several meters

---

Based on "Diurnal SAR variability due to ice and snow air interface wetness overnight changes.", by Hudier E., J-S. Gosselin & D. Febres, which appeared in the *Proceedings of the IGARSS, Barcelona 23-27 july 2007*.

across, ridges are best resolved at the highest available resolution. However, the reduction of speckle is first achieved through simultaneous processing of multiple looks at the expense of the spatial resolution. In fact, any attempt at getting information on ridges is hindered by the difficulties at extracting the very location of individual ridges out of a speckled image.
As a consequence, improving roughness extraction techniques requires to thoroughly understand and differentiate the physics behind signal coherence. In this chapter we will review the geophysical parameters that enhance the detection of the specular coherent component backscattered from ridges. In the conclusion we will review speckle properties emphasizing the physics differences between coherent summation and specular coherence as a mean to filter speckle from polarimetric SAR and improve pressure ridge resolution.

2. Background

The snow cover is an important component of the ice-atmosphere interface. It is known to accumulate in the wake of ridges creating mesoscale surface roughness features (Déry & Yau, 2001; Gallée et al, 2001). Whereas, during the winter season, the snow cover is mostly transparent to microwaves (Garity et al, 1990), at spring, the solar radiation and temperature changes cause the metamorphosis of snow crystals which in turn reduce the penetration depth and translate into volume scattering (Barber et al, 1998). Later on, as the snow wetness content increases, the dielectric properties changes of the wet snow medium reduce furthermore the penetration depth of radar electromagnetic waves and eventually turn the snow surface into an opaque medium (Koskinen, 2001; Barber et al, 1998).
At the beginning of the melt season, pressure ridges are mostly enveloped or coated with snow. As volume scattering still dominates, up to 4 % of volumetric wetness, (Koskinen, 2001; Guneriussen, 1997), they get drowned by the signal component back-scattered by is the snow layer and, as a consequence, are not well delineated on satellite images.

Between 5 and 8% humidity, surface scattering becomes the main component of the received signal and adds surface slope as a major factor in the analysis of microwave imagery (Hallikainen & Winebrenner, 1992). More exactly, the parameter to consider is the actual angle that the radar incident beam makes with the normal to snow or ice surfaces. After Gohin et al. (1994), and Lewis et al. (1994), the back-scattered intensity σ_0 (in decibels) is well described by a linear function of θ, the angle made between the normal to the iceor snow interface and the radar incident beam:

$$\sigma_0(\theta) = \sigma_0(0°) + C_i\,\theta \qquad (1)$$

Kim et al. (1984) showed that σ_0 (0°) and the slope C_i are characteristic of the surface roughness. In the case of electromagnetically smooth surfaces, such as an ice block in a pressure ridge, the radar return is large at normal incidence and σ_0 decreases rapidly with θ as most of the radar energy scatters forward. In contrast, rough surfaces show a reduced dependence with θ.
In addition, at spring, the orientation of snow and ice surfaces relative to sun rays can, depending on solar irradiation intensity, translate into a variability in surface snow wetness, eventually the development of a film of liquid water on ice surfaces and, as irradiance decreases over the day, the re-crystallization and build-up of frost flower structures where liquid films formed on ice blocks. Thermodynamically, the snow ice medium is during this

period of the year at the fresh water freezing point (Gow & Tucker III, 1990; Wadhams, 2002; Petrenko & Whitworth, 1999). While within the snow layer the heat budget is only balanced by melt or freezing of ice crystals or interstitial liquid water, at the air snow interface, evaporation comes into play to remove or add heat. As a consequence, while the water content within the snow layer changes slowly, at the air interface evaporation can quickly shift the surface temperature below freezing turning a thin layer of snow into a dry medium then allowing microwaves to penetrate and generate volume scattering.

Beside the incident angle, the most important parameter when analyzing SAR images is the surface roughness of the snow-air interface (Ulaby et al, 1990; Fung, 1994). Fundamentally, the electromagnetic wave interaction with a snow surface is determined by the scattering efficiency of surface roughness *rms* and correlation height. The transition from electromagnetically smooth to rough surface depends on an arbitrary criterion that is proportional to $[\lambda/\cos\theta]$, where λ is the radar wavelength and θ the radar incident angle. On terrains that offer surface slope variability such as ice block faces or snow cover surfaces, θ must be seen as the local incident angle relative to a given area. The proportionality factors vary from 1/8 (the Rayleigh criterion) to 1/32 (the Fraunhofer criterion). It results that at normal incidence ($\theta=0°$) a surface roughness *rms* between 0.15 cm and 0.6 cm causes a transition from smooth to rough surface. On ice surfaces a thin film of water, such as observed at spring during sunny afternoons, turn ridge blocks into specular reflectors. This allows for an increased contrast between ridges and the snow covered background due to the variability in surface orientations and occurrence of multiple reflection on several faces in ridges (Hudier, 2006; Carlstrom & Ulander, 1995). On the other hand, it can also cause an enhanced forward scattering on snow surfaces, generating dark stripes or low backscattering areas where surface slopes cause the SAR incident beam to reach the snow interface at a large angle.

3. Field site and image processing.

Images analyzed in this paper were recorded as part of a field work conducted on the east coast of the Hudson Bay offshore of Kuujjuarapik, Quebec, Canada (Figure 1). This area is covered with a first year landfast sea ice about 1m30 to 1m60 thick. By mid-April the first signs of melt can be observed with an increase of snow wetness. By mid-May, most of the snow is melted away with some left where sheltered from sun rays on the north side of ridges. Figure 2 shows an aerial photograph of the study area.

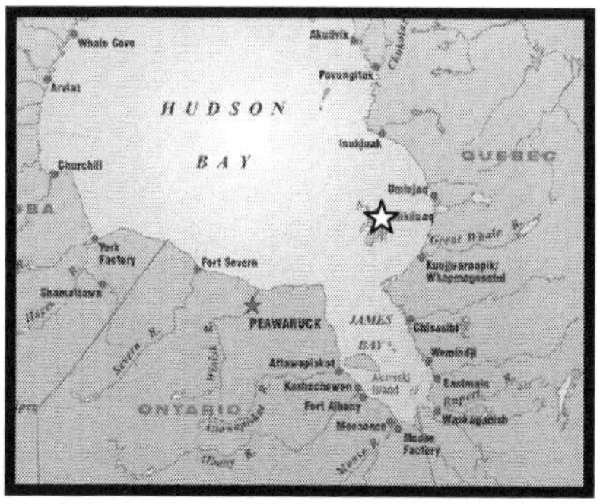

Fig. 1. Geographic location of the site where field works were done (white star).

Fig. 2. Early May picture of the ice sheet offshore Kuujjuarapik, Quebec, Canada.

Overall thirteen Radarsat-1 SAR fine beam mode scenes were processed "table I". In its fine beam mode, the Radarsat antenna looks to the right of the spacecraft at an angle between 35° and 49° with a resolution 8-9 m and a swath area of 57x50 km. The average local overpass times were 5:30 am (descending) and 4:57 pm (ascending). Processing of the data were done using ENVI-6.2 software. Fields works were carried out offshore of Kuujjuarapik on the East coast of the Hudson Bay. Detailed information about data collection and image processing can be found in Hudier (2006).

From the data set listed below, 7 images (in bold) were recorded after wet conditions were observed. All images recorded early in the morning were recorded after a night below freezing. Qualitative analysis shows that morning images offer a poor resolution of the ridge network. In all instances afternoon images offer a better resolution. Isolating areas that could be identified as ridges and areas that didn't show any sign of a ridge, we computed the average change in the backscattered signal between morning and afternoon images. Our results show a 4 dB decrease in ridged areas with a 5.5 dB std. In areas away from ridges an increase of 1.9 dB with a 2.5 dB std was computed.

Recording date	Mode	Incident angle	Orbit
April 23, 1997	F2	40.3-42.5	Descending
May 10, 1997	**F1**	**36.8-39.9**	**Descending**
April 04, 1998	F1	36.8-39.9	Ascending
April 11, 1998	F1	36.8-39.9	Descending
April 28, 1998	**F1**	**36.8-39.9**	**Ascending**
May 05, 1998	**F1**	**36.8-39.9**	**Descending**
March 26, 2001	F1	36.8-39.9	Descending
April 26, 2001	**F2**	**40.3-42.5**	**Descending**
April 29, 2001	**F3**	**41.1-43.1**	**Ascending**
April 15, 2005	F1	36.8-39.9	Ascending
April 22, 2005	**F1**	**36.8-39.9**	**Descending**
April 25, 2005	F4	42.3-44.5	Ascending
May 02, 2005	**F2**	**40.3-42.5**	**Ascending**

Table 1. SAR images acquisition

4. Low back-scattering areas paralleling pressure ridges.

4.1 Methods

Data used in this research work were collected over four field experiments: 1997, 1998, 2001 and 2005. Field work was done from mid-April to mid-May which covers the melt initiation period for this region of the bay. A detailed survey of the snow layer and ridge features was performed systematically along a total of eight transects which start and end points were geo-referenced. 87 measurement sites were selected along these transects to sample a range of ridge orientations relative to SAR incident beams. The mapping of the snow layer morphology included: structure orientation, slope, size, lateral extension of the structures relatively to the ridge, and normal vs. satellite incidence angle. Ridge characteristics measurements included: height, width, block size and distribution. In order to study the relation between surface orientation and the backscattering reduction from these regions, we measured the angle between the radar incident beam's orientation and the perpendicular to the slope. At each sampling location, data were collected along 4 transects parallel to ridge axes at distances of 5, 10, 15 and 20 meters. Measurements were made at regular intervals along the different transects. Also, ridge blocks surface characteristics were documented as a function of face orientations. In addition, snow temperature and wetness were measured using a thermocouple sensor and a snow fork (dielectric measurement device). These last data were collected for each SAR scene on the day and time (+/- 30 minutes) the images were recorded.

Selected sites were first extracted from the available set of images and sampled areas located. Backscattering data were then computed over a set of 3 pixels wide strips oriented parallel to ridge axes, a first strip localized over the ridge itself, and a second next to it in the area of reduced backscattering. This operation was performed manually by moving a test strip across the ridge in order to localize the areas of maximum (the ridge) and minimum backscattering. Backscattering statistics and the distance between the areas of reduced backscattering and the adjacent ridge were then computed. It must be underlined that for a site to be classified as having a "dark band" there had to be continuity between the band of minimum backscattering and the middle of the adjacent floe.

4.2 Low backscattering regions associated with ridges.

As anticipated, areas of reduced backscattering were observed paralleling ridges (Figure 3). This result confirms previous observations. We also confirmed that these dark structures get more contrasted as the spring melt goes on.

Overall, they were observed on 61% , 53 out of 87, of our sample locations. However, in most of the 53 sites they showed up only on images recorded later at spring (table 2). In the case of 3 images recorded before mid-April, no such structures were spotted. This period of the season is more typical of the winter conditions. The maximum snow wetness amounted then to 0.8 % in which conditions volume scattering and other boundary surface scattering become important. Snow wetness was observed to be highly variable depending on the sampling point location. Surfaces perpendicular to sun rays display the highest surface wetness content, which was measured in the first 5 mm of the snow layer. However, when sampling a thicker layer (10cm) the snow drift geometry becomes a factor. We observed a lower wetness content on crests and maximums in troughs. Table 3 summarizes the average backscattering value computed from ridges, dark band areas and on floes away from sampled ridges.

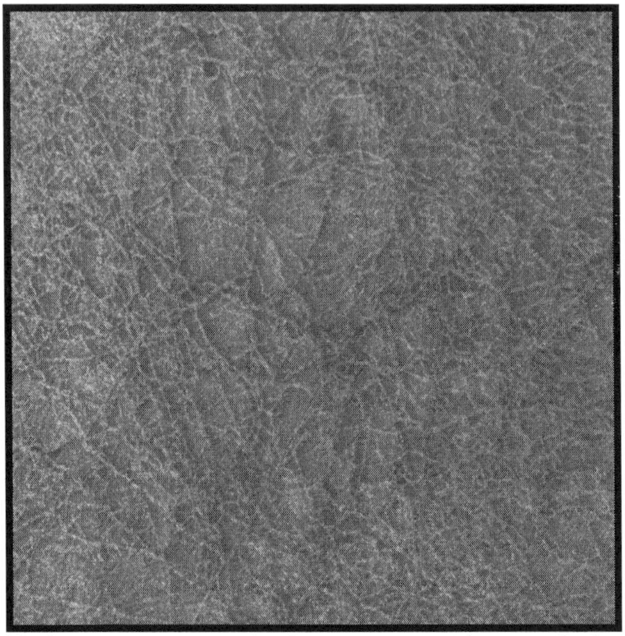

Fig. 3. Reduced back-scattering in areas bordering pressure ridges on a RadarSAT 1 image of a first year sea ice.

Recording date	Nb of test sites	Sites with dark bands
April 23, 1997	12	4
May 10, 1997		7
April 04, 1998	23	0
April 11, 1998		0
April 28, 1998		14
May 05, 1998		14
March 26, 2001	20	0
April 26, 2001		12
April 29, 2001		14
April 15, 2005	32	15
April 22, 2005		17
April 25, 2005		17
May 02, 2005		16

Table 2. Percentage of sampled locations where dark bands were observed.

σ0 (dB)	Average	Average distance to ridge
Ice floes away from ridges	-16.2	
Bright ridges	-13.5	
Dark strip	- 17.7	22 (m)

Table 3. First year sea ice signatures

In the lee of pressure ridges snow accumulates behind obstacles to form linear features oriented after the wind that developed them. It creates a wave pattern with crests heights tapering away from ridges. These structures offer a wide range of surface slopes that differentiates this region from the rest of the ice floe. As underlined by data summarized in table 4, in cases where physical obstacles may generate slope patterns, pressure ridges can not only draw a network of bright linear features but also a network of dark bands parallel to these bright features. This also underlines how, when spring melt turns the snow cover into an opaque medium, snow slopes and orientation relative to the sun and the satellite beam can be as an important parameter as surface roughness.

σ0 (dB)	Slope average	Std
Ice floes away from ridges	38°	15°
Ridge vicinity where σ0 is reduced	72°	28°
Ridge vicinity where σ0 is not reduced	44°	35°

Table 4. Snow surface slope characteristics in three regions: (1) away from ridge influence, (2) in the vicinity of the ridge where reduced backscattering was measured and, (3) in the vicinity of the ridge where no-reduction of the backscattering intensity was measured.

5. Resolution variability due to overnight wetness changes.

5.1 Overnight resolution loss

Figure 4 and 5 give an example of resolution variability on two ridge networks. In the 2 situations the day prior to the image recording was sunny and warm, liquid films were observed on snow and ice surfaces during mid-afternoons with water contents over 8% in the first centimeter of the snow layer. Figure 4 was recorded at 4 pm with temperatures close to the air temperature peak for the day (Figure 6). In the situation pictured on figure 5, the image was recorded at 4 am on the morning, after the night temperature went down below freezing (Figure 7). At that time, surface wetness was measured at 3.1% and frost flower observed where liquid films were spotted the previous afternoon. Similar diurnal "patterns of daily melt and nightly freeze" were also reported by Ashcraft and Long (2005) to cause a strong increase of the absorption coefficient of the snow due to the introduction of liquid water into the snowpack.

Fig. 4. SAR image recorded at 4 pm after a warm and sunny day, East coast of the Hudson bay, Canada.

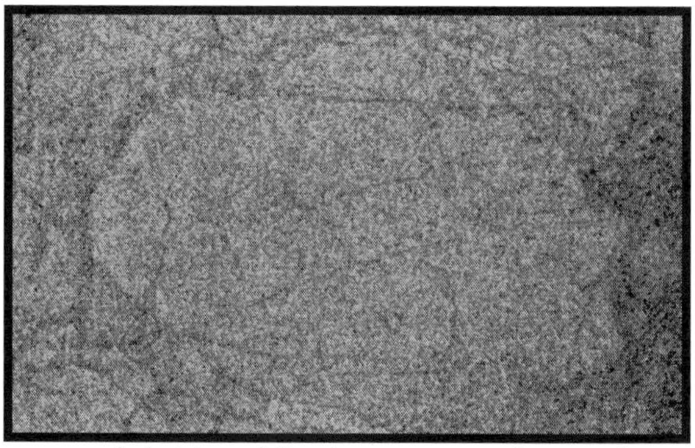

Fig. 5. SAR image recorded at 4 am following a warm and sunny day, and after a night with temperatures below freezing, East coast of the Hudson bay, Canada.

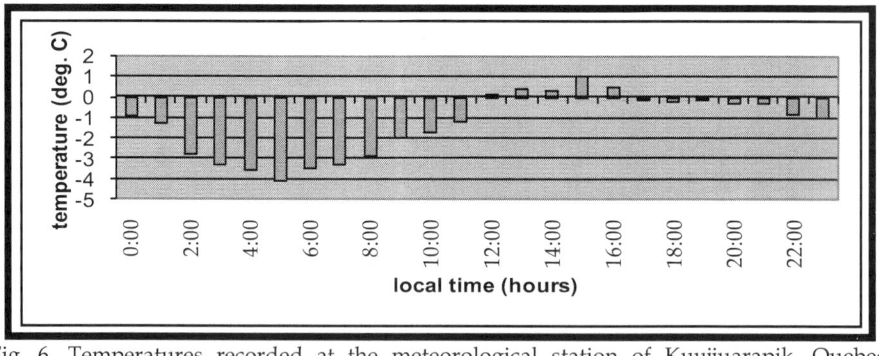

Fig. 6. Temperatures recorded at the meteorological station of Kuujjuarapik, Quebec, Canada, on the day the image displayed on figure 2 was recorded.

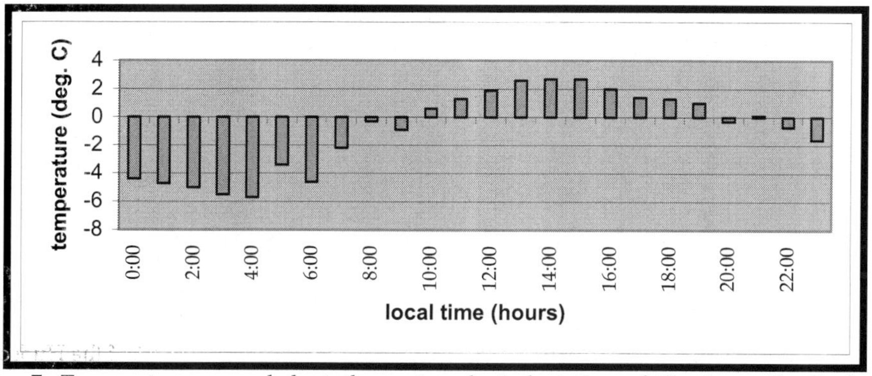

Fig. 7. Temperatures recorded at the meteorological station of Kuujjuarapik, Quebec, Canada, on the day the image displayed on figure 3 was recorded.

5.2 Field data and discussion

The wetness content was measured in the first centimetre of the snow cover and at 5 centimetres below the surface. At the end of the afternoon we observed a light dependence between surface wetness and the orientation of the surface slope relative to the zenith. As expected, surfaces subject to a higher irradiation showed a higher wetness content. In the first centimetre the wetness averaged 7.5 with a std of 2.1 at +/-30° from normal sun incidence and 6.3 with a std of 3 for larger incidences. At 5 cm from the surface, surface slope orientation does not show any significant influence on the water content. We measured an average 3.9 with a std of 2.1. Data collected at 4 am give wetness contents of 3.1 in the first centimetre with a std of 2.9 and no clear slope dependence. It is to be noted that a liquid film was observed on both on most ice and snow surfaces in sampling completed in the afternoon.

The reduced resolution on morning images is the result of two things. First a reduced backscattering in ridged areas and secondly an increased backscattering from areas away from ridges. This second phenomenon is well explained by an increased penetration depth and the resulting volume scattering due to a reduced wetness in the top layer and the metamorphosis of the snow cover that is associated the succession of thaw and freezing that occurs overnight (Barber et al., 1998).

In ridged areas, where some bare wet ice faces were observed mid-afternoon, all ice blocks faces were dry on morning. Besides, no "pure ice faces" were observed. Selected ice blocks sampled daily showed that where bare wet ice can be found at some point during the day, a thin layer of flower frost develops at night with measured *rms* values in and above the smooth to rough transition range for SAR Radarsat 1.

At spring, when the solar irradiation cycle causes diurnal temperatures to rise above freezing during the day and sink below freezing at night, our observations show a clear resolution difference between images recorded early in the morning and at the end of warm and sunny days. Wetness changes overnight are in the value range that causes the snow cover to become opaque to microwave lengths. During the day, the increase of the absorption coefficient cause surface scattering to become dominant over volume scattering (Ulaby et al., 1990) and, therefore surface slope and roughness to control the microwave energy back-scattered toward the antenna. With the development of a liquid film on surfaces, microwaves do not penetrate anymore into snow or ice mediums, reflectivity increases and forward scattering occurs. As a consequence, all but the surfaces which lie almost at a right angle with the satellite beam return a weaker signal. As the probability that an ice surface may lie perpendicular to the incident SAR beam is much higher in ridges, this increased forward scattering on most surfaces results in an improved resolution of the pressure ridge network. As shown on figure 8, most ice blocks are enveloped in a thin layer composed of metamorphosed snow crystals, ice and air bubbles. With no porous medium

Fig. 8. Two ridges that gave similar backscattering intensity on the satellite image illustrated on figure 4. The probability density that an ice surface lies perpendicular to the incident ice beam increases with smaller ice blocks.

beneath, liquid water resulting from surface melt creates a continuous film that reduces surface roughness as well as increases its dielectric constant. Microwaves penetration is then negligible and most of the SAR beam energy reflected according to Snell's law. The development of a liquid film on ice blocks' faces turns each of them into an opaque electro-magnetically smooth surface which increases furthermore their brightness on satellite images.

During the night, temperature below freezing pumps heat out of the top layer of the snow cover causing wetness to drop. On morning, at the time the image illustrated on figure 5 was recorded, the reduced wetness content allows some volume scattering to occur in the snow surface layer to completely erase contrast between ridges and flat ice. In ridges, freezing removes the liquid film from ice block's faces turning them into an highly diffusive dry ice - snow crystals and air bubbles medium. This allows part of the microwave energy to penetrate into it causing a drop in the surface scattering component. Besides, some growth of snow flower crystals at the surface increases roughness to cause back-scattering from surfaces oriented perpendicularly to the incident SAR beam to drop furthermore. It should be observed that despite a loss in ridge resolution, some ridges remain visible (Figure 5) mainly because of the areas of reduced back-scattering as discussed in section 3.

6. Conclusion and polarimetric potential for ridge extraction

Results presented in section 4 and 5 underline the critical impact of wetness changes during spring time. With snow temperatures at the freezing point, any change in the snow-atmosphere heat flux initiates melt or freezing in the top layer of the snowpack. While, within the snowpack, water content undergoes only slight variations over a 24 hours period, snow crystals within the top layer can turn from wet to dry overnight. More importantly, where an ice surface is exposed to the atmosphere, melt and freezing cause the cyclic development of a liquid film and a re-crystallized highly diffusive dry layer of snow, ice and air bubbles. As liquid water is removed, part of the microwave energy penetrates into this layer to be absorbed and scattered. The fact that the ridge network can completely vanish overnight emphasizes the importance of specular reflection on electro-magnetically smooth surfaces for the extraction of ridges.

Surface roughness changes not only increase (or decrease) drastically the back-scattering from surfaces oriented perpendicularly to the SAR beam but also cause the back-scattered electromagnetic wave to be the result of an incoherent (or noncoherent) or coherent reflection. To explore furthermore the implications we need to clarify the concept of incoherence. In principle we measure the combined reflections from all scatterers within the scene which in term of physics is coherent summation. However, incoherence, in the literature, refers to the nature of the scene rather than to the physics of electro-magnetic waves. If the scene is viewed twice through the exact same geometry, the same speckle pattern should be observed (Raney, 1998). Conversely, due to the random distribution of individual scatterers, a slight change in the position of the satellite would give a totally different speckle pattern despite the fact that the statistics over the scene would remain unchanged. This is the incoherence nature of speckle which turns a mostly uniform area into a salt an pepper image. On the other hand, coherent specular reflection describes a highly correlated phase structure within part of a scene, which implies that multiple measurements from this same scene would reveal the presence of a specular reflector. This requires that the satellite position remains the same in order for the relative geometry of the SAR beam angle and target to be about the same.

Polarimetric SAR are systems that acquire two simultaneous images (HH and VV) of a same scene. While the geometry of the relative positions of the satellite and scene are unchanged, the physics of coherent summation from multiple scatterers implies that the two images would show a different speckle pattern. Conversely, the nature of specular coherent reflection implies that a specular reflector would be pinpointed on both images. It follows that the cross-correlation of single look HH and VV channels could help filter bright pixels generated by speckled coherent summation while revealing the pressure ridge network.

7. References

Ashcraft I.S. & Long D.G. (2005). Differentiation Between Melt and Freeze Stages of the Melt Cycle Using SSM/I Channel Ratios. *Geoscience and remote sensing IEEE Transactions*, vol. 43, No.6, pp.1317-1323.

Barber D.G., Fung A.K., Grenfell T.C., Nghiem S.V., Onstott R.G., Lytle V.I., Perovich D.K. & Gow A.J. (1998). The role of snow on microwave emission and scattering over first-year sea ice, *Geoscience and Remote Sensing, IEEE Transactions*, vol. 36, I.5, pp.1750-1763.

Carlstrom A. & Ulander L.M.H. (1995). Validation of backscatter models for level and deformed sea-ice in ERS-1 SAR images, *Int. J. Remote Sens.*, V.16, No.17, pp.3245 - 3266.

Déry, S.J. and Yau M.K. (2001). Simulation of blowing snow in the Canadian Arctic using double moment model, *Boundary-Layer Meteorol.*, 99, 297 - 316.

Fung A. (1994). Microwave scattering and emission models and their applications, *Artech House*, 573 p.

Gallée, H., Guyomarc'h, G & Brun, E. (2001). Impact of snow drift on the Antarctic ice sheet surface mass balance: possible sensitivity to snow-surface properties, *Boundary-Layer Meteorol.*, 99,1-19.

Garrity C., Ramseier R.O. & Rubinstein I.G. (1990). Snow wetness and SSM/I brightness temperature for the Weddell Sea. *Proceedings of the IGARSS, 1990*, pp.1521-1524

Gohin F., A. Cavanié & R. Ezraty (1998). Evolution of the passive and active microwave signatures of a large sea ice feature during its 2½-year drift through the Arctic Ocean. *J. Geophys. Res.*, 103, 8177-8189.

Gow A.J. & Tucker III W.B. (1990). Sea Ice in the Polar Ocean, In: *Polar Oceanography, Part A, Physical Science,* Walker O. Smith Jr., (Ed.), 47-122, Academic Press, ISBN 0-12-653031-9, San Diego.

Guneriussen T. (1997). "Backscattering properties of a wet snow cover derived from DEM corrected ERS-1 SAR data", *Int. J. Remote Sensing*, Vol. 18, No. 2, pp. 375-392.

Hallikainen, M. & Winebrenner D. P. (1992), The Physical Basis for Sea Ice Remote Sensing. In: F.D. Carsey, Editor, *Geophysical Monograph* **vol. 68**, American Geophysical Union, Washington, pp. 29–46.

Hudier E. (2006). Low back-scattering bands paralleling pressure ridges on first year sea ice, *Proceedings of the IGARSS, 2006*, Denver, pp.731-734, july 2006.

Kim Y.S., Moore R.K. & Onstott R.G. (1984). Theoretical and experimental study of radar backscatter from sea ice. *Remote Sensing Laboratory, Univ. Kansas Rept.,* RSL TR 331-337.

Koskinen J. (2001). Snow monitoring using microwave radars, *Thesis Doctor of Technology*, Helsinki University, 31p.

Lewis E.O., Livingstone C.E., Garrity C. & Rossiter J.R. (1994). Chapter 2: Properties of snow an ice, in *Remote Sesing of Sea Ice and Icebergs*, S.Haykin, E.O. Lewis, R.K. Raney and J.R. Rossiter (Ed.), John Wiley & Sons, Inc., 21-96.

Petrenko V.F. & Whitworth R.W. (1999). Physics of Ice. Oxford University Press, 373 p., ISBN 0-19-851895-1, New-York.

Raney R.K. (1998). Chapter 2: Radar fundamentals: Technical perspectives, in Principles and applications of imaging Radar; Manual of remote sensing Volume II, F.M. Henderson & A.J. Lewis (ED.), John Wiley & Sons, Inc., 9-130, ISBN 0-471-29406-3.

Ulaby F.T., Moore R.K. & Funk A.K. (1990). Microwave remote sensing active and passive, Norwood, M.A, Artech House, Volume III.

Wadhams P. (2002). Ice in the ocean, Gordon and Breach Science Publishers, 351p., ISBN 90-5699-296-1